Estimating for Collision Repair

Michael Crandell
Carl Sandburg College

DELMAR
CENGAGE Learning™

Australia • Brazil • Japan • Korea • Mexico • Singapore • Spain • United Kingdom • United States

DELMAR
CENGAGE Learning™

Estimating for Collision Repair
Michael Crandell

Publisher: Alar Elken
Acquisitions Editor: Vernon Anthony
Editorial Assistant: Betsy Hough
Fast Cycle Production Editor: Dianne Jensis
Marketing Manager: Mona Caron

Library of Congress Control Number: 98–22159

ISBN-13: 978-0-7668-0891-1

ISBN-10: 0-7668-0891-2

Delmar
Executive Woods
5 Maxwell Drive
Clifton Park, NY 12065
USA

Cengage Learning is a leading provider of customized learning solutions with office locations around the globe, including Singapore, the United Kingdom, Australia, Mexico, Brazil, and Japan. Locate your local office at **international.cengage.com/region**

Cengage Learning products are represented in Canada by Nelson Education, Ltd.

For your lifelong learning solutions, visit **delmar.cengage.com**

Visit our corporate website at **www.cengage.com**

Printed in the United States of America
6 7 8 9 10 14 13 12 11 10

ED134

Contents

Preface ... v
Acknowledgments ... vi

Chapter 1 **Introduction to Damage Report Writing** ... **1**
1.1 Design Changes • **2** 1.2 New Equipment and Skills • **2** 1.3 Sound Business Management • **3** 1.4 The Importance of Damage Reports • **3** 1.5 Insurance and Claims Adjustments • **4** 1.6 Insurance Company Procedures • **6** 1.7 Building Professional Relationships • **8** 1.8 The Customer—A Mutual Concern • **9** 1.9 Ethics in Collision Repair • **10** 1.10 The Right Mental Approach • **13** 1.11 Estimating Guides • **13** 1.12 Tools and Equipment • **15** Review Questions • **16**

Chapter 2 **Modern Vehicle Construction** ... **17**
2.1 Vehicle Design—Past and Present • **17** 2.2 Modern Body-Over-Frame Construction • **20** 2.3 Semiunitized Construction • **23** 2.4 Modern Unitized Construction • **23** 2.5 FR Unibody Structure • **26** 2.6 FF Unibody Structure • **29** 2.7 Component Assemblies • **32** 2.8 Automotive Metals • **32** 2.9 Automotive Plastics • **37** 2.10 Automotive Sealers and Adhesives • **39** 2.11 Automotive Glass • **41** Review Questions • **43**

Chapter 3 **Collision Damage to Body and Frames** ... **45**
3.1 Impact Effects • **46** 3.2 Collision Conditions • **46** 3.3 Types of Collision Damage • **49** 3.4 Impact Effects on BOF Vehicles • **52** 3.5 Impact Effects on Unibody Vehicles • **54** 3.6 Inertia Damage • **57** 3.7 Basic Sheet Metal Damage • **57** Review Questions • **64**

Chapter 4 **Auto Collision Repair Methods** ... **67**
4.1 Realigning the Vehicle • **67** 4.2 Stress Relieving • **68** 4.3 Removing Structural Panels • **77** 4.4 Installing New Panels • **79** 4.5 Welding • **80** 4.6 Metalworking • **90** 4.7 Shrinking Metal • **95** 4.8 Applying Body Filler • **96** 4.9 Surface Preparation • **97** 4.10 Applying the Finish • **98** 4.11 Plastic Repairs • **102** Review Questions • **104**

Chapter 5 **Making a Damage Appraisal** ... **107**
5.1 The Estimator's Tools • **107** 5.2 A Sequence of Inspections • **108** 5.3 Zone 1 Primary Damage • **109** 5.4 Zone 2 Secondary Damage • **110** 5.5 Zone 3 Mechanical Damage • **121** 5.6 Zone 4 Passenger Compartment • **125** 5.7 Zone 5 Exterior Trim and Finish • **126** 5.8 Fire, Flood, and Vandalism Damage • **126** Review Questions • **126**

Chapter 6 **Estimating Labor and Parts Costs** ... **129**
6.1 Repair or Replace? • **129** 6.2 Labor Times • **132** 6.3 Labor Operations • **134** 6.4 Adjustments to Operation Times • **137** 6.5 Estimating Frame Straightening Labor • **139** 6.6 Estimating Panel Straightening Labor • **140** 6.7 Estimating Plastic Repair • **141** 6.8 Parts Prices • **141** 6.9 Betterment • **145** Review Questions • **147**

Chapter 7 **Assessing Damage to the Front Body** ... **149**
7.1 Bumpers • **150** 7.2 Grille and Lights • **155** 7.3 Radiator Support • **158** 7.4 Hood Panel • **159** 7.5 Fenders • **162** 7.6 Aprons and Wheelhouses • **164** 7.7 Frame Members • **166** Review Questions • **171**

Chapter 8 **Assessing Damage to the Body Shell** ... **173**
8.1 Cowl and Dash Assembly • **173** 8.2 Windshield • **176** 8.3 Body Side Panels • **178** 8.4 Roof and Back Glass • **181** 8.5 Quarter Panel • **184** 8.6 Front and Rear Doors • **190** 8.7 Rear Body and Lights • **193** 8.8 Luggage Lid • **196** 8.9 Tailgates and Liftgates • **197** 8.10 Body Section • **199** 8.11 Interior Trim and Upholstery • **199** 8.11 Restraint Systems • **201** Review Questions • **207**

Chapter 9 **Assessing Mechanical Component Damage** 209
9.1 Cooling System • **210** 9.2 Air Conditioning • **212** 9.3 Electrical System • **214** 9.4 A Front Wheel Drive Drivetrain • **218** 9.5 Transaxle • **219** 9.6 Rear Wheel Drive Drivetrain • **222** 9.7 Exhaust System • **225** 9.8 Front Suspension • **227** 9.9 Other Suspension Systems • **231** 9.10 Wheels • **232** 9.11 Steering Systems • **233** 9.12 Brakes • **236** Review Questions • **238**

Chapter 10 **Estimating Refinishing Costs** 239
10.1 The Refinishing Process • **239** 10.2 Refinishing Time • **243** 10.3 Costs of Refinishing Materials • **248** 10.4 Entire Car Painting • **249** Review Questions • **250**

Chapter 11 **Preparing Damage Reports/Appraisals** 253
11.1 Factors Influencing the Repair • **254** 11.2 Preparing a Damage Report/Appraisal • **259** 11.3 Writing Up the Report • **267** Summary and Review • **277** Review Questions • **278**

Chapter 12 **Computerized Damage Reports and Shop Management** 279
12.1 Damage Report Writing • **280** 12.2 Job Costing • **282** 12.3 Supplements and Job Control Functions • **282** 12.4 Accounting and Business Analysis • **283** 12.5 Computer Components • **284** Review Questions • **290**

Glossary **293**
Index **299**

Preface

"Nothing happens until someone sells something" is an old maxim that is very applicable to the business of repairing collision damaged vehicles. The job of selling the services of a collision repair facility belongs to the estimator. He or she must analyze the extent of damage incurred by a vehicle, determine the cost of restoring the vehicle to its preaccident condition, convince the vehicle owner to hire the shop to repair the vehicle, and justify the cost of repairs to an insurance company's claims adjuster. To accomplish these tasks, the shop estimator must be knowledgeable of vehicle construction and design and familiar with methods of repairing and replacing damaged panels, and must understand how to determine labor allowances and parts costs from published collision estimating guides. This expertise must then be focused on writing a thorough, accurate damage report. A professionally prepared damage report is the estimator's best selling tool; it provides the customer with an objective basis on which to choose a repair shop.

Although the vehicle owner is the potential customer, the owner's insurance company usually pays the bill. That is where the job of the insurance company's claims adjuster becomes an important factor in the estimating process. A claims adjuster must also be able to write a comprehensive damage report, or appraisal. It is his or her job to make sure that all repairs necessary for restoring the customer's vehicle to its preaccident condition are included on the damage report. However, the adjuster must also ensure that no unnecessary or unrelated work is performed at the insurance company's expense. He or she is responsible for ensuring that the vehicle is repaired in the most economical manner possible without compromising the preaccident value or safety of the vehicle.

Estimating For Collision Repair will teach both the collision repair facility estimator and the claims adjuster how to write accurate, comprehensive damage reports and appraisals. Chapter 1 begins by discussing the changes that have occurred in the collision repair industry. It explains the importance of accurate report writing and the use of the damage report as a selling and negotiating tool. The first chapter also discusses at length some of the ethical considerations facing the collision repair industry.

The estimator and adjuster must be familiar with how today's vehicles are constructed. Chapter 2 explains the construction of body-over-frame and unibody vehicles. The "stressed hull structure" concept is explored and body parts nomenclature for various unibody structures is given. The use of sheet metals, plastic, and glass in vehicle construction is also discussed.

Changes in auto body design have altered the way a vehicle reacts to collision forces. Chapter 3 discusses the types of damage both full-frame and unibody vehicles incur in front-, side-, and rear-end collisions. An explanation of how metal panels react to and are deformed by collision stresses is also given in Chapter 3.

For the student with little or no experience in collision repair, Chapter 4 explains the basic procedure for pulling damaged frames and repairing buckled or dented sheet metal. Methods of repairing plastics and installing glass are also discussed.

Equipped with a knowledge of vehicle construction and collision repair techniques, the student is ready to analyze a vehicle to determine the extent of damages incurred in an accident. Chapter 5 explains how to analyze a vehicle zone by zone, how to spot signs of secondary damage, and how to measure for frame misalignment.

After the damage analysis is complete, an estimate of labor, parts, and materials must be made. Chapter 6 explains how to determine replacement parts costs and labor allowances for many operations when using published collision estimating guides. Guidelines are also given for estimating frame straightening and metal-working labor hours.

In order to write an accurate, thorough damage report or appraisal, the estimator and adjuster must understand which operations are included in the labor allowances quoted in the collision estimating guides; operations that are not included must be listed separately on the damage report. Chapters 7, 8, and 9 explain included/not included operations for most body panels, mechanical systems, and interior trim.

Chapter 10 deals with determining the labor and materials needed to refinish repaired or replaced panels. Included/not included operations and reductions to paint labor times are also explained.

Preparing a written damage report or appraisal is the subject of Chapter 11. This chapter explains how to determine a vehicle's cash value, when to "total" a vehicle, what factors to consider when deciding whether to repair or replace a panel, and how to prepare a damage report or appraisal.

Most insurance companies and an increasing number of collision repair facilities use a computer system to prepare appraisals and damage reports. Chapter 12 gives an overview of the several types of computer estimating systems in use and explains how they can be a valuable tool in preparing professional looking, accurate damage reports.

Acknowledgments

The author would like to thank all of the employees of the following companies for their assistance in making this book possible:

- Dave's Autobody — Galesburg, IL
- Uftring Chevrolet Body Shop — Washington, IL
- Pekin Auto Storage Pool — Pekin, IL
- Robertson's Used Auto Parts — Mapleton, IL

Also, I would like to thank Bill Harrison for his review of the project and my wife, Janine, who typed the manuscript and did the initial editing.

The author is grateful for the contribution of the following companies:

ADP
American Honda Motor Company
Atlantic Pneumatic
Beeline Company
Blackhawk Division, Applied Power, Inc.
Chief Automotive Systems
Chrysler Corporation
Country Companies Insurance
DuPont Company
Garmat Corporation
Insurance Institute for Highway Safety
Kansas Jack, Inc.
Maaco Enterprises, Inc.
Mitchell International
Moog Automotive
Motor Publications, Inc.
National Automobile Dealers Used Car Guide Co.
Nicator, Inc.
Nissan Motor Corp.
Rotary Lift
Tech Cor
3M
Toyota Motor Corp.
Wedge Clamp International, Inc.

Chapter 1

Introduction to Damage Report Writing

Objectives

After reading this chapter, you should be able to:

- **Define the job tasks of collision repair facility estimators and insurance company adjusters and appraisers.**
- **Describe how changes in modern vehicle design and materials have altered the way damage is appraised and repaired.**
- **Explain the legal rights and obligations of collision repair facilities, insurance companies, and customers in the repair process.**
- **Recognize certain ethical dilemmas that can occur in the present system of collision repair.**
- **Recognize the need for mutual cooperation, communication, and training within the collision repair industry.**
- **List the qualities of a professional collision damage estimator and adjuster.**
- **Understand the value of damage reports in selling jobs, negotiating settlements, and controlling costs.**
- **List the tools needed to perform a complete damage inspection.**
- **Explain the important role estimating guides play in the collision repair industry.**

Key Terms

Body-over-frame
Unibody
MIG
Hazardous waste
OEM
Salvage
Aftermarket
Estimating guides
Flat rate
Estimate
Appraisal
Damage report
Estimator
Appraiser
Technician
Insurance adjuster
Supplement

In the time it takes you to read this sentence, three vehicle accidents will occur somewhere in the United States. Each day, close to thirty thousand minor and tragically major accidents take place on our streets and highways, in crowded parking lots or garages, or in other nontraffic-related situations. Vehicles may be only slightly damaged and not be repaired. Other unrepaired vehicles may be so heavily damaged that they are declared a total loss by the insurance company and sold as salvage. The vehicles that are repaired are the lifeblood of collision repair facilities (independent or franchised) and collision repair departments of new car dealerships.

In addition to the 12.8 million collision repairs performed annually, collision repair facilities also perform noncollision-related work such as new paint jobs, installation of sun roofs and body accessories, car customizing and detail work, and other customer initiated repairs. In all, collision repair constitutes a $16.9 billion a year industry.

But high work volume, impressive dollar amounts, and even projected growth do not guarantee financial success. This is particularly true of independent shops that must rely on their own marketing and managerial skills. There are approximately thirty-six thousand collision repair facilities in operation today. That number increases annually at a rate of 1.7 percent. For facilities to prosper, owner and employees must meet an ever-changing list of challenges and demands.

1.1 DESIGN CHANGES

Faced with the need to build downsized, lightweight, fuel efficient vehicles to compete with ever increasing foreign competition, American automakers abruptly abandoned traditional body-over-frame (BOF) construction in favor of unitized vehicle construction in the late 1970s and early 1980s. This rapid retooling to produce unibody vehicles created a series of problems in the collision repair industry, many of which have not been fully resolved.

However, body-over-frame and full-frame vehicles, such as pickup trucks and many sport utility vehicles (SUVs) have become increasingly popular. Approximately 44 percent of the 1998 model vehicles are full-frame. As these vehicles become more common, collision repair technicians, trained to repair unibodies will need to learn how to repair full-frame vehicles.

UNDERSTANDING UNIBODY DAMAGE

Full-frame vehicles are made using a heavy steel frame. Separate sheet metal body components, such as cabs, beds and fenders, are bolted to this rigid frame. In a unibody vehicle, the structural shell or body is made by welding together frame and body panels into a single, rigid, hull-like structure. This radical departure from full-frame assembly methods requires extra care in locating and estimating collision damage. Added importance must be placed on accurately assessing secondary damage and alignment problems.

Unibody construction uses lightweight, high-strength steels requiring highly specific welding techniques using MIG welding equipment. The new vehicle designs and steels have resulted in tighter guidelines concerning where vehicles can be repaired using sectioning or clipping methods. The weakening of high-strength steels at high temperatures has also reduced the amount of heat straightening that can be performed during repairs. Unitized vehicles require exact alignment for proper operation of steering, suspension, and other mechanical components.

INCREASE IN PLASTICS

The use of thermoplastics (weldable) and thermosetting plastics (nonweldable) has dramatically increased. Plastics are now used to form exterior body panels, bumper covers, gravel deflectors, filler panels, mouldings, radiator shrouds, dash panels, grilles, and many other vehicle components traditionally made of sheet metal.

In the 1990 model year, GM introduced a line of all-purpose vans having a steel skeleton covered with a plastic skin. All exterior body panels are made of plastic. They are bonded to the steel skeleton using newly developed adhesives that are so strong they allow working loads to be shared by both skin and frame. The principle is the same as the frame members and body panels working together to create strength in an all-steel unibody car. But instead of welding, special adhesives are used. Other composite vehicles—made of a combination of steel and plastic—include the Chevrolet Camaro and Saturn models.

Escalating use of plastics and adhesives presents an entirely new set of repair challenges in the next decade.

1.2 NEW EQUIPMENT AND SKILLS

These changes have had a profound impact on the collision repair industry. To survive, collision repair facilities have had to invest in new equipment such as MIG welding units, straightening equipment capable of handling unibody vehicles and more efficient paint booths and spraying equipment (Figure 1–1). The capital investment needed to equip a modern body repair shop is substantial and has forced many smaller shops out of business. The cost of properly handling and disposing of hazardous wastes, such as paints and thinners, now has a definite impact on the shop's balance sheets. Shops using computerized estimating systems and cost management programs are no longer the exception to the rule. Progressive facilities have seen the value of investing in computerized estimating and cost management systems to improve operating efficiency.

FIGURE 1–1 Pulling and straightening equipment is just one example of the technological advances made in the collision repair industry. (Courtesy of Car-O-Liner Company.®)

But modern equipment and large investments do not make repairs. Skilled technicians do. Finding, training, and keeping skilled workers is a major concern. Repairing today's vehicles is a science and an art, not a predictable assembly line operation. The need for knowledgeable workers in the collision repair and related insurance industry is vital. Without proper training no one can effectively do their job, whether it is damage report writing, frame or unibody straightening, metalwork, surface preparation, or painting. In a modern collision repair facility that relies on a fair profit for each labor hour, "small" mistakes in estimating and repair work can quickly result in lost money on the job.

And it is not merely the new employee who requires training. Established workers need training in the new equipment and repair techniques. The "unlearning" of old habits and procedures that no longer work is often the case with experienced technicians. For example, a painter may have worked successfully for years, but new paint systems and formulas create the need for additional training. The same holds true for measuring and straightening techniques. And management techniques have advanced a long way from handwritten entries on an estimating sheet or business ledger to today's computerized estimating and job costing systems.

1.3 SOUND BUSINESS MANAGEMENT

Poor management will bankrupt even the busiest shops. The average profit margin for most collision repair facilities is approximately 10 percent. To maintain this profit, the facility's owner must meet a target profit on each repair job. In order to accomplish this, the owner or manager must consider how efficient the repair technicians are, how much materials actually cost, and how to schedule work. They must also be concerned about fixed costs, such as materials (cost varies per job), insurance, license fees, equipment payments, and building overhead. Today's consumers and insurance companies are well educated and rightly demand high-quality repairs. The modern collision repair facility must operate with profit and customer satisfaction as its top priorities.

1.4 THE IMPORTANCE OF DAMAGE REPORTS

Every job a collision repair facility performs should produce a profit. Think about this statement and two points become clear.

1. A collision repair facility cannot depend upon rough estimates, approximations, or educated guesses when determining the cost of doing a repair. There is no excuse for inaccurate bids that cost the facility money.
2. Not all vehicles brought to the facility for repair will make a profit for the shop. Accurate damage report writing allows the facility to look at each job in terms of potential profits and to turn down repairs that may cost the facility money or be break-even jobs at best.

A surprising number of collision repair facilities ignore these business principles. They believe they are in business to write as many estimates as possible in the hopes of being the lower bidder on enough jobs to keep the facility busy. They are also frightened of turning away work for any reason. They see refusing a job as refusing money, not as avoiding a costly headache. The only way to assure that the facility makes money on every job it accepts is to write accurate damage reports and maintain a system of job costing.

A damage report is a detailed listing of every part, operation, labor hour, and miscellaneous material expense it will take to complete the repair. It is a complete blueprint for the repair of the vehicle (Figure 1–2).

Many times a vehicle has hidden damage. The extent of damage cannot be determined until the vehicle is disassembled for repair. In these cases, the facility begins repair and, if additional damage is noted, the insurance company or owner is notified. The responsible party will reinspect and authorize additional repairs. These are known as supplements. Supplements are a way of life at a collision repair facility. The insurance company understands that an estimate or damage report lists visible and obvious damage. Total accuracy can only be obtained by removing damaged parts and checking for hidden damage.

DAMAGE REPORTS AS SELLING TOOLS

A detailed, accurate damage report is a powerful selling tool. Collision repair facilities often forget that the customer is responsible for selecting a repair shop, even with repairs involving insurance claims. Sell the customer on the facility's skills, thoroughness, and attention to detail and you often sell the job (Figure 1–3). Price is important, but it is only one of many factors.

A professional damage report builds customer confidence in the facility. It explains the job from beginning to end. It tells the customer exactly what he or she is receiving for each repair dollar. Given the choice between a low-priced, vague estimate and a higher-priced, step-by-step report on how their vehicle is going to be returned to its preaccident condition, most people will choose the second option. The best way to close a repair sale is to generate a complete damage report and then walk the customer through the repair, pointing out each item to be repaired or replaced and explaining the equipment and work methods that will be used. The most recent surveys indicate that the average collision repair facility prepares twenty estimates per week and converts slightly over half of these into jobs. With a consistent system of damage report writing and selling, the percentage of repair inquiries that turned into actual jobs would be much higher.

DAMAGE REPORTS AS NEGOTIATING TOOLS

When the repair involves an insurance settlement, the collision repair facility must reach agreement with the insurance company over the cost of the repair. The facility's damage report writer must be able to justify to an adjuster each dollar charged for labor, parts, and materials. The negotiating process is easier when each repair operation is listed separately on the damage report, along with the corresponding costs of parts and materials.

1.5 INSURANCE AND CLAIMS ADJUSTMENTS

The collision repair industry has always been closely involved with the insurance industry. Over 85 percent of all collision repair bills are paid through insurance companies, either the customer's company or the company of another party involved in the accident. The three main types of auto insurance are liability insurance, collision insurance, and comprehensive insurance. Most states require that motorists carry liability insurance. Banks and lending institutions require that all three types of insurance be carried before they will approve auto loans.

Liability Insurance. Liability insurance normally protects the insured against claims resulting from an accident caused by the operation of the insured's vehicle. It protects the insured's personal possessions from being lost through a lawsuit filed by the owner of the vehicle or property with which the insured's vehicle has collided. Liability insurance also pays for the repair of the vehicle belonging to the other motorist who is not at fault in the collision. This type of insurance will not cover the cost of having the insured's vehicle repaired.

Collision Insurance. Collision insurance is used to pay for the cost of repair to the insured's vehicle if the insured person causes the damage. Also, the insured vehicle is covered for the cost of repairs if it is damaged by an uninsured motorist.

Most collision insurance policies contain a "deductible" clause. This means that the insured is required to pay a predetermined amount of the total cost of the repair. Deductible amounts usually range between $100 and $500 (although $1,000 is common now), with higher deductibles decreasing the size of the premium payments on the policy. Any repair costs over and above the deductible are paid by the insurance company.

Comprehensive Insurance. Comprehensive insurance covers the insured for damage to the vehicle

Description: 1996 Chevrolet Caprice Impala SS
Body Style: 4D Sed — Drive Train: 5.7L Inj 8 Cyl AO
VIN: 1G1BL52P9TR167053

Line Item	Entry Number	Labor Type	Operation	Line Item Description	Part Type/ Part Number	Dollar Amount	Labor Units
1	600930	BDY	REMOVE/INSTALL	FRT BUMPER ASSY			1.1 #
2	900500	BDY*	REPAIR	BUFF FRONT COVER	Existing		0.5*
3	600101	BDY	REMOVE/INSTALL	GRILLE ASSY			0.3 #
4	601470	BDY	REPAIR	GRILLE	Existing		1.0*#
5		REF	REFINISH/REPAIR	GRILLE			1.0*
6	600001	BDY	REMOVE/INSTALL	GRILLE EMBLEM	Existing		0.2*#
7	602140	BDY	REMOVE/INSTALL	R MARKER LAMP ASSEMBLY	Existing		0.2*
8	602150	BDY	REMOVE/REPLACE	L MARKER LAMP ASSEMBLY	5976555 GM PART	50.75	0.2
9	602410	BDY	REMOVE/INSTALL	HOOD ASSY			0.5
10	602432	BDY	REPAIR	HOOD PANEL	Existing		4.0*#
11	AUTO	REF	REFINISH	HOOD OUTSIDE			C 3.1
12				ROCK CHIPS STRIP HOOD			
13	604590	BDY	REPAIR	R FENDER PANEL	Existing		1.0*#
14	AUTO	REF	REFINISH	R FENDER OUTSIDE			C 2.1
15	604600	BDY	REPAIR	L FENDER PANEL	Existing		1.0*#
16	AUTO	REF	REFINISH	L FENDER OUTSIDE			C 2.1
17	604750	BDY	REMOVE/INSTALL	R FENDER ANTENNA MAST	Existing		0.5*
18	605150	BDY	REMOVE/INSTALL	R FENDER W/O MLDG			0.2
19	605160	BDY	REMOVE/INSTALL	L FENDER W/O MLDG			0.2
20	600321	GLS	REMOVE/REPLACE	W/SHIELD GLASS	Sublet	445.00*	0.0*#
21				DAVE'S GLASS 555-1212			
22	612900	BDY	REMOVE/REPLACE	UPR W/SHIELD REVEAL MOULDING	10248871 GM PART	34.25	1.8
23	900500	BDY*	ADD'L LABOR OP	BUFF TO MATCH TEXTURE	Existing		2.5*
24	AUTO	REF	ADD'L OPR	CLEAR COAT			2.1
25	933003	REF	ADD'L OPR	TINT COLOR			0.5*
26	AUTO		ADD'L COST	PAINT/MATERIALS		228.90*	
27	AUTO		ADD'L COST	HAZARDOUS WASTE DISPOSAL		5.00*	

* Judgment item
\# Labor note applies
C Included in clear coat calc

I. Labor Subtotals	Units	Rate	Add'l Labor Amount	Sublet Amount	Totals
Body	15.2	36.00	0.00	0.00	547.20
Refinish	10.9	36.00	0.00	0.00	392.40
	Nontaxable Labor				939.60
Labor Summary	26.1				939.60

III. Additional Costs			Amount
Taxable Costs			233.90
Sales Tax	@	6.500%	15.20
Total Additional Costs			249.10

II. Part Replacement Summary			Amount
Taxable Parts			85.00
Sales Tax	@	6.500%	5.53
Nontaxable Parts			445.00
Total Replacement Parts Amount			535.53

IV. Adjustments	Amount
Customer Responsibility	0.00

I. Total Labor:	939.60
II. Total Replacement Parts:	535.53
III. Total Additional Costs:	249.10
Gross Total:	1,724.23
IV. Total Adjustments:	0.00
Net Total:	1,724.23

This is a preliminary estimate.
Additional changes to the estimate may be required for the actual repair.

ESTIMATE RECALL NUMBER: 6/15/98 08:31:13 290

Mitchell Data Version: JUN_98_A

UltraMate is a Trademark of Mitchell International

FIGURE 1–2 Damage reports and appraisals must list all parts and labor needed to make the repair. Computerized systems have resulted in clear, accurate printouts. (Courtesy of Mitchell International.)

FIGURE 1–3 The shop representative should always review every line of the damage report with the customer.

not covered by collision or liability insurance. Examples would be vehicle damage caused by hail, wind, fire, and other acts of nature. Damage and loss through theft and vandalism are also covered by comprehensive insurance.

No-Fault Insurance

Often one of the most difficult and time-consuming aspects of settling an insurance claim is determining which motorist is at fault in the accident. To help eliminate these hassles, some states allow their motorists to carry 100 percent no-fault insurance. With no-fault policies, no effort is made to determine which motorist is at fault if both parties carry this type of insurance. Each party's insurance company pays his or her vehicle and/or personal injury costs, regardless of who causes the accident. The application of no-fault provisions to vehicle and property damage liability varies from state to state.

1.6 INSURANCE COMPANY PROCEDURES

Insurance companies have their own representatives write damage reports on most claims. These technicians are usually called insurance adjusters or appraisers. These individuals may be employees of the insurance company or independents who work or several insurance companies on a per-vehicle basis. An insurance appraiser evaluates damaged vehicles and

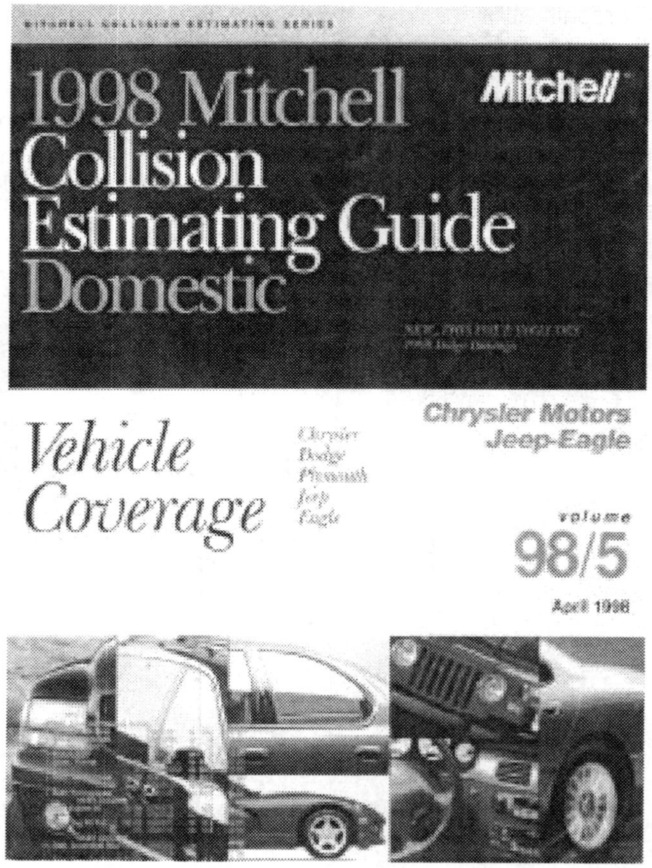

FIGURE 1–4 Collision estimating guide used to prepare a damage report.

writes estimates or appraisals. An insurance adjuster also evaluates damaged vehicles and writes appraisals and also negotiates with customers and collision repair facilities and authorizes payments. To help avoid confusion between the damage reports written by collision repair facility personnel and the damage reports written by insurance representatives this text will refer to insurance damage reports as *appraisals.*

A typical adjuster will have a set territory and will be very familiar with the collision repair facilities in that territory. When an adjuster writes an appraisal, he or she is doing the same job as the facility's damage report writer—determining a game plan or restoring the vehicle to its preaccident condition. The adjuster determines the straightening, repair, replacement, and paint labor he or she believes is required to repair the vehicle. The adjuster also determines the types of parts (Original Equipment Manufacturer (OEM), salvage, or aftermarket) he or she feels can be used to make a quality repair.

Labor times and new OEM parts prices are determined using data found in crash estimating guides. Several major publishers compile collision estimating guides that are continually updated to list the latest parts, prices, and labor times (Figure 1–4). Both colli-

TAURUS/SABLE 1996–97

FRONT FENDER

Use Procedure Explanations 7 and 28 with the following text.

Refinish Fender Outside	R/L	2.0
Add to Edge Fender	R/L	.5
R&I Fender	R/L	#2.9

#Includes R&I Front Cover Assy & Hood Assy & R&I/R&R Front Rocker Panel Moulding & Headlamp Assemblies

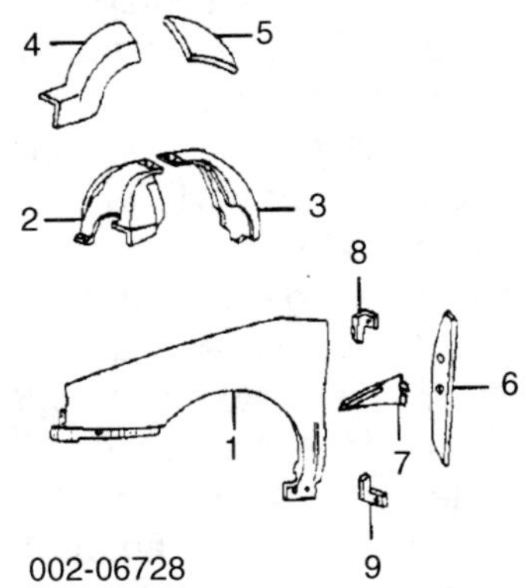

1 Fender	R	F6DZ	16005 A	**#3.2**	119.60
	L	F6DZ	16006 A	**#3.2**	119.60
#Includes R&I Front Cover & Assy & Hood Assy & R&I/R&R Front Rocker Panel Moulding & Headlamp Assemblies					
2 Liner, Fender Front					
Sable					
GS Model	R/L	F6DZ	16102-3 AA	**#.5**	35.07
LS Model	R	F6DZ	16102 BB	**#.5**	53.12
	L	F6DZ	16103 AA	**#.5**	35.07
Taurus					
G, GL Model	R/L	F6DZ	16102-3 AA	**#.5**	35.07
LX, SHO Model	R	F6DZ	16102 BB	**#.5**	53.12
	L	F6DZ	16103 AA	**#.5**	35.07
3 Liner, Fender Rear	**R/L**	**F6DZ**	**16102-3 A**	**#.5**	**57.53**
4 Shield, Liner Front	**R/L**	**F6DZ**	**16071-2 AB**	**#.5**	**9.48**
5 Shield, Liner Rear	R/L	F6DZ	16071-2 AA	**#.5**	13.53
#R&R One Side Complete, Included in R&R Fender					
6 Shield, Fender					
Left Only		F6DZ	16103 BB	**.1**	4.85
7 Bracket, Fender Rear	R	F6DZ	16134 AA	**.3**	6.85
	L	F6DZ	16135 AA	**.3**	8.33
8 Brkt, Fender Rear Upr	R/L	F6DZ	5402732-3 A	**.3**	11.90
9 Brkt, Fender Rear Lwr	R/L	F6DZ	5402722-3 A	**.3**	11.75

FIGURE 1–5 Collision estimating guides include exploded views of components, part numbers and prices, labor times, and special notes. (Courtesy of Mitchell International.)

sion repair facility damage report writers and insurance adjusters rely heavily on these guides. Information in the guides also has been formatted into computerized estimating programs that have greatly increased the speed and accuracy of damage report writing.

If the adjuster calls for the use of salvage parts, the salvage yard source and price should be included on the damage report. Sources and prices of aftermarket parts must also be given if they are used in the repair.

The labor times listed in the collision estimating guides are determined from OEM information and data collected from labor time studies performed in the collision repair facilities (Figure 1–5). The labor rate in dollars per labor hour is determined by the geographical area's prevailing labor rate for automobile collision repair.

After the adjuster has finished the written appraisal, it is reviewed with the collision repair facility owner or manager, who usually has the facility's written damage report. It is rare that the insurance appraisal and the collision repair facility damage report will agree. No two people approach and view a repair in exactly the same manner. But if the insurance adjuster and collision repair facility representative are well-trained professionals, the difference between the appraisal and damage report should not be that great. Most differences arise over labor allowances. And if they are both professionals who understand modern repair work and respect each other's point of view, the adjuster and collision repair facility rep should be able to work out these slight differences and agree upon a fixed repair cost or bid.

As stated earlier, the negotiating phase of the repair process is an area where thorough damage reports can help collision repair facilities to maintain steady profits. Like a knowledgeable customer, a skilled adjuster is trained to be wary of vaguely worded "estimates." Insurance companies deal in facts and figures, and it is much easier for a collision repair facility to defend and justify detailed labor times and operations than to defend a lump-sum labor figure that is not itemized. If nothing else, a good damage report or insurance appraisal states that party's point of view and gives a clear-cut starting point for working out a settlement.

Once the collision repair facility and insurance company are in agreement over the method and cost of returning the vehicle to its precollision condition, the insurance claim can be processed and the repair work begun. The adjuster or appraiser is often only responsible for writing up the insurance appraisal and negotiating an agreement with the collision repair facility.

The final bid is then given to an insurance company claims manager, who oversees the handling of the claim. The claims manager investigates the accident, contacts law enforcement officials if "fault" must be established, contacts and deals with the second party insurance company if needed, authorizes and routes the settlement check to the proper party, and closes out the file on the accident.

If additional or hidden damage not listed on the approved bid is found as the repair is being made, the collision repair facility must file a supplement to the original bid. The insurance adjuster must be notified and must personally confirm the accuracy and need for the supplement.

Some insurance companies give their field adjusters the power to authorize settlement on claims below a set amount. In a few states, insurance companies allow their agents to settle minor claims, but this is the exception, not the rule. Generally, claims are handled by the insurance company's claims department, not agents.

Drive-in claims centers are also popular. If the vehicle is safe to drive, the owner can take it to a drive-in claims center to have the damage appraised and the claim settled on the spot, including a check written for the damage amount. These centers allow insurance companies to process a greater number of smaller claims in a given day. They also free adjusters to visit shops doing major repairs on nondrivable vehicles.

From the collision repair facilities' point of view the drive-in claims centers have a major drawback. They eliminate or make it very difficult for the collision repair facility to negotiate differences of opinion with the insurance company. The customer arrives at the collision repair facility with the insurance company appraisal and settlement check and fully expects the shop to do the repair for that amount. If the shop believes the appraisal is incorrect or incomplete, correcting it or filing supplements can take more time than the job is worth. Obviously not all drive-in claim settlements are inaccurate, but this is one area where business-wise repair facilities are carefully weighing their options before accepting the job.

1.7 BUILDING PROFESSIONAL RELATIONSHIPS

Unfortunately, collision repair facilities and insurance companies do not have a long history of mutual cooperation and respect. Viewing each other as adversaries rather than partners in the repair process has led to questionable practices and policies on both sides. The keys to relieving this tension and fostering cooperation are increased training and communication.

The insurance industry faces the same ongoing need for training as the collision repair facilities. Unibody construction, modern steels, plastics, and paint systems, and new repair methods affect the bottom line on appraisals as well as damage reports.

Many problems exist because of poorly trained workers in both industries. Ask a group of truly professional collision repair facility owners what their most aggravating problem is, and the answer might very well be "dealing with incompetent insurance adjusters and unbending company policies" that

1. Do not understand my business, its overhead expenses, and how it differs from other collision facilities in quality and cost.
2. Do not know how to write a complete, accurate appraisal.
3. Are only interested in doing the job "cheap" to reduce claims payments and increase company profits.
4. Use the collision guides as a "bible" even when they know errors or omissions in the guides shortchange collision repair facilities.

Pose the same question to a group of accomplished adjusters and their answer might be "dealing with incompetent facilities" that

1. Do not understand the adjuster's job and the important role insurance companies play in assuring that customers can finance collision repairs that otherwise might never be done.
2. Do not know how to write a complete, accurate damage report.
3. Are only interested in padding profits at the expense of insurance company policyholders.
4. Do no understand the proper method of using the collision guides, even when it would be to their advantage.

A number of states now require that insurance adjusters/appraisers be licensed and pass competency testing. In some states, collision repair shops must be registered or licensed through state agencies. Licensing of adjusters and repair facilities can help assure a certain level of skill and professionalism in the collision repair industry, but it is not the only answer to the problems of poorly trained workers.

The growth and influence of various trade associations within the industry has led to increased communication and the sharing of knowledge and ideas from all sides. The Inter-Industry Conference on Auto Collision Repair (I-CAR), Society of Collision Repair Specialists (SCRS), the Automotive Service Association (ASA), and the National Institute for Automotive Service Excellence (ASE) are several examples of organizations that promote professional and consumer education and awareness (Figure 1–6).

Because both the insurance industry and collision repair facilities are interested in profit margins that rely on the same customer dollars, their relationship is somewhat adversarial by nature and will probably always remains so. But this does not mean the relationship must be completely negative.

One of the most important factors in establishing professional relationships is to downplay the personality confrontations that occur during negotiations between adjusters and collision repair facilities. Fortunately, recent trends are for more professional communications on both sides. It is important to take personalities out of any business situation and approach it with an unbiased attitude.

FIGURE 1–6 Various trade associations work to improve training and communication between all those involved in the collision repair industry.

Adjusters must avoid the "take-it-or-leave-it" approach to negotiation. If the collision repair facility presents a strong case in defense of its labor estimates, listen and be flexible. The facility has the right to be compensated for all work needed to return the vehicle to its preaccident condition.

On the other hand, collision repair facilities must realize that insurance companies are not a roadblock to profitability. The collision repair business has grown to be cash intensive, and profits are produced by operating efficiently. Accurate costing is a must. Poorly managed facilities are not entitled to assured profits. Mutual respect and cooperation is gained by displaying competence and fairness on the job.

DRP

Direct Repair Programs (DRP) are defined relationships between an insurance company and a collision repair facility. The insurance company and the approved collision repair facility agree to the repair criteria. The criteria include labor rates, material costs, aftermarket parts usage, repair times, quality of repair, and warranty of repair. In a DRP the customer takes the damaged vehicle to an approved facility. The collision repair facility writes the estimate, photographs the damage using a digital camera, and forwards the information to the insurance company. Payment from the insurance company is received after the repair facility has made the needed repairs. There are significant advantages for each of the participants:

- Customer—The customer does not have to spend the extra time getting two or more estimates. The customer simply takes the damaged vehicle to a DRP facility of their own choosing.
- Insurance Company—The insurance company is not required to look at each damaged vehicle. The collision repair facility decides what repairs are needed. The insurance company periodically makes spot checks of the facility to ensure that the facility is following specified criteria.
- DRP Facility—The collision repair/DRP facility can start the repairs immediately and depend upon a steady flow of work.

At the present time, there are several insurance companies that have established DRP relationships.

1.8 THE CUSTOMER—A MUTUAL CONCERN

Collision repair facilities and insurance companies share a common goal. They both want a satisfied customer whose vehicle has been repaired to preaccident condition. Use the following criteria to evaluate a preaccident condition:

1. Safety
2. Durability
3. Fit
4. Finish

Safety, the most important repair goal, means that the vehicle will drive straight, air bags, seat belts, lights, and horn all function properly, and the parts do not fall off. If the vehicle is involved in another crash, it will crush and protect the passengers as the manufacturer intended. Welded repairs are subject to rapid corrosion, it not properly treated. Durability means that the repaired vehicle will maintain its structural integrity as long as an undamaged vehicle. Doors, hoods, and deck lids that open and shut easily and panel gaps that are uniform are the essence of proper fit. Finish means that the repair is invisible. The repainted area is free of sand scratches, lumps, dirt, mismatches, tape edges, and overspray.

THE CUSTOMER IS THE CUSTOMER

The average collision repair facility performs hundreds of insurance claim repairs per year and most likely deals with a relatively small group of insurance companies on a regular basis. This constant contact, plus the fact that the insurance companies are involved in settling the price of repair, might lead some facility owners to believe that the insurance company is the true customer.

It is an easy trap to fall into, especially when the vehicle owner does not become involved in the key decisions of the repair process. Most customers are unfamiliar with the steps involved in repair, including their rights and responsibilities.

This lack of knowledge often results in the customer "letting the insurance company take care of it." Many people feel more comfortable having a "knowledgeable" insurance representative deal with the shop. The vehicle owner becomes a silent customer, unless something goes wrong or the repair does not meet his or her expectations.

An uninvolved vehicle owner makes the insurance company a very large, strong "customer" often with strict policies and procedures not always in the shop's or the customer's best interests.

Successful collision repair facilities always remember that the vehicle owner, not the insurance company, is their customer. The collision repair facility must take much of the responsibility for educating collision repair customers. By state law, insured motorists have many rights that can work to the collision repair facility's advantage.

SELECTING A REPAIR SHOP

The vehicle owner has the right to select the collision repair shop of his or her choice. Low bids are not the sole determining factor. Customers entering a collision repair facility for what they think is an "estimate" should be sold on the facility's ability to do the job and the quality of its work. Selling the shop and following it up with a professional damage report is the best way to close the repair sale.

In those cases where a drive-in claims center or a DRP facility is not available, the insurance company will ask the customer to get two or three estimates or bids from different repair facilities. The insurance company then has the option of paying the low bid.

As in any type of skilled work, the least expensive bid is not always the best bid. Modern shops with heavy investments in proper repair equipment and trained personnel are forced to compete with inadequately equipped and poorly trained shops on a dollar-for-dollar basis.

This bid process also discourages accurate damage report writing. "Courtesy" bids and faked estimates are often worked up simply to meet insurance company needs. Collision repair can also find themselves swamped with customers shopping for second and third estimates. Often these customers have no intention of bringing the job to that shop.

Laws in most states *do not* require that the customer secure two or three estimates. Every collision repair facility should know the laws in their state. Inform prospective customers of their rights. Many customers getting two or three estimates will be surprised to find out that they are probably doing unneeded comparison shopping for their insurance companies.

An insurance company might request multiple estimates when it is not practical to involve an adjuster in the repair. For example, the claim might be a small one and there are no drive-in claims centers nearby. In this case, the insurance company might authorize the agent to settle the claim. In turn, the agent might ask for at least two estimates to give the company some flexibility in settling the claim. But again, customers should ask if they are legally bound to secure more than one estimate.

In several states, the insurance company might have several options:

1. It can pay for damage loss in money based on an appraisal.
2. It can replace the damaged vehicle or property with one of like kind or quality.
3. It can elect to repair the vehicle at a facility of its own choosing.

Even with these options, most companies select option number one. But if a company has serious doubts about the shop's ability to make a safe, quality repair, or otherwise cannot reach a final agreement with the shop, they might select option number three. When an insurance company elects to repair the vehicle, it assumes total responsibility for restoring the vehicle to its preaccident condition within a reasonable time period. Electing to oversee the repair of a vehicle increases insurance company involvement and can make it liable to customer complaints concerning quality and other possible problems. Normally, this is a position most companies like to avoid.

The ultimate responsibility or the quality and safety of the repair work lies with the collision repair facility. When writing up damage reports and appraisals, nothing should ever be done to compromise the safety and structural integrity of the vehicle.

1.9 ETHICS IN COLLISION REPAIR

No business is immune from unethical behavior and practices. The collision repair industry has many "gray" areas where abuses can and have occurred. Disputes between collision repair facilities and insurance companies can fuel a general bending of the rules on both sides. Obviously, if every collision repair facility were well equipped, well trained, and repaired vehicles using the most economically practical methods available, there would be no need for insurance adjusters and appraisers. And if all adjusters were equally as competent and understood all the factors affecting a facility's profit and its right to that profit there would be no need for collision repair facility damage reports. Such is not the case.

Unfortunately the collision repair industry has the dubious honor of being famous for unethical behavior. This reputation, often exaggerated, is not only due to a small segment of dishonest shops and insurance representatives, it is also the result of unethical customers eager to strike deals and arrangements in their favor. Above-board shops and insurance representatives avoid these unethical situations.

UNAUTHORIZED CHANGES FOR PROFIT

If a collision repair facility is being paid to *replace* damaged parts, but *repairs* them instead at a lower cost to the shop, it is deceiving both the customer and the insurance company. There is nothing wrong with repairing components when it does not affect the quality of repair or safety of the vehicle. The key is following the specified repair plan on the damage report. If the repair plan changes, all parties must be informed about changes in part and labor costs.

The same principle holds true if a collision repair facility installs lower cost salvage or aftermarket parts when new original equipment manufacturer (OEM) parts are specified on the damage report. Once again the facility is profiting by straying from the original repair bid agreement made with the customer and the insurance company.

SAVING DEDUCTIBLES

Saving the customer's deductible is the most common way customers pressure body shops into an unethical situation. For example, on a repair valued at $800, the customer wants the facility to bid $1,000 so that he or she will not have to pay a $200 deductible. Because an experienced adjuster would spot excess labor times, these "deals" can also involve changing replace/repair and OEM/salvage part specifications without informing the insurance company. This is illegal in some states.

IMPROPER CUSTOMER "PROFITS"

Customers might also be eager to use an insurance claim as an all-expense-paid opportunity to return their vehicle to its like-new condition. They might pressure the collision repair facility to repair damage not related to the accident, but include its cost on the insurance bid. Insurance companies are very careful about assessing damage not caused by the accident and keeping it off or deducting it from the repair bid. But a series of small abuses in this regard can add up to substantial losses for insurance companies.

A similar situation can occur when the vehicle has received major damage. In this case the customer might pressure the collision repair facility or appraiser to declare the vehicle a total loss, even though it could be repaired at a cost below the total figure. The customer wants to turn the repair situation into an opportunity to finance a new car purchase.

Another clear case of a customer improperly profiting occurs when a customer attempts to collect two insurance claims for the same damage. For example, a vehicle receives slight damage to a door panel. The owner settles the claim at a drive-in claim center, but never has the car repaired. Several months later, the car is involved in another more serious accident that damages the same general area of the vehicle. The car is appraised and repaired but part of the repair cost is for damage received in the first minor accident. The figures might be as follows:

Damage in Accident #1 Unrepaired
Settlement #1: $ 300.00
Settlement #2: $1,000.00
Total: $1,300.00

Damage in Accident #1 Repaired
Settlement #1: $ 300.00
Settlement #2: $ 800.00
Total: $1,100.00

As you can see, if the damage from accident #1 goes undetected by the insurance adjuster or collision repair facility, the customer stands to gain $200 at the insurance company's expense.

Regardless of what most customers think, insurance companies do not have unlimited amounts of money to pay claims. What is more, insurance companies tend to monitor their profit margins much more closely than the average collision repair facility. Increased costs due to falsified claims and unethical practices are not paid out of company profits, they are passed on to customers through increased premiums and to collision repair facilities through tighter procedures and policies.

The insurance industry must also share the responsibility for ethical behavior and policies. Hiding behind corporate policies that cannot be changed by adjusters or regional managers is no excuse for unfair treatment of collision repair facilities or claimants.

DIRECTING REPAIR WORK

Directing business to "preferred" collision repair facilities is a charge frequently leveled at insurance companies. Blacklisting shops over settlement disputes, personality clashes, failure to give suggested parts price discounts, or other factors not related to the shop's ability to do the job is unethical.

PROFITING FROM CUSTOMER IGNORANCE

Failure to inform claimants of their rights in a collision repair often works to the insurance companies' advantage. Assuming an attitude of "what the customer does not know will not hurt him (or her)" is improper. Having the customer shop for a low bid when one bid is all that is needed is one example.

Insisting that a customer take the vehicle to a drive-in claims center is another example of how customer rights can be ignored. The customer is only obligated to notify their insurance company and tell them where the vehicle can be examined by their adjuster. Customers are required to cooperate reasonably with insurance companies. Notice of vehicle location, such as the shop of the customer's choice, is reasonable cooperation.

PROMOTING CUTTHROAT COMPETITION

Placing collision repair facilities in a no-win position concerning going labor rates, included services, and expected discounts and labeling this healthy, free-market competition is not ethical.

As mentioned earlier, insurance companies usually pay labor at the area's prevailing rate. The insurance company may determine the area's prevailing rate by calculating the median rate most shops in the area are charging. The median rate is not necessarily the true average rate and can be manipulated through selective sampling.

For example, in a given area there are ten collision repair facilities; five that charge $22 per hour, three charging $24 per hour, and two charging $26 per hour. The median labor rate in this area is $22 per hour, well below what half the shops need to cover their expenses.

The true average labor rate in this area is found by adding the labor rates of each facility and dividing by the number of facilities. In this case, the average rate is $23.40, still below what some facilities require.

Surveys, median figures, and averages do not take into account important differences between individual facilities. Facilities that invest in the best people, training, and state-of-the-art equipment will obviously need to charge more per hour than poorly equipped "bump-and-grind" facilities incapable of handling all but the easiest of repairs.

Insurance companies might argue that modern equipment and training increases productivity to a point where modern facilities can compete at lower rates. But increased work output does not always offset the cost. A value must be placed on having the tools and people capable of doing the job.

Insurance companies might also cite "prevailing practices" in a given area as reason not to pay for certain operations, even though they know the work takes time and is required to return the vehicle to its preaccident condition. For example, vacuuming up broken glass from a vehicle can be an aggravating, time-consuming task. But if several facilities in an area do not charge for this work, the insurance company might deem this a prevailing practice and refuse to pay *any* facility in the area for this work. It does not matter why the facilities are not charging for this work. Insurance companies can say it is pride of workmanship to include such operations for free, when in reality it might be ignorance, forgetfulness, or fear of a confrontation on the part of the collision repair facility.

Insurance companies might also seek discounts on crash parts from collision repair facilities. These discounts can substantially cut into the facility's profit margin. For example, most car dealerships offer a 25 percent discount on the crash parts they sell to collision repair facilities. Insurance companies frequently ask facilities for a 10 percent discount. So if the final customer retail price is $100 for a part, the facility pays the dealership $75, and the insurance company pays the collision repair facility $90. The collision repair facility's profit on the part is reduced from $25 to $15, a 40 percent drop.

Once again, if several facilities in an area grant the insurance company's "request" for a discount, it can quickly become a prevailing practice. Many collision repair facility owners feel unethical insurance representatives promote cutthroat competition among area through misinformation and intimidation. They see insurance companies having too much control over how their business is operated and the prices they can charge.

One solution for the collision repair facilities is more open communications between owners and managers. Facilities must stop viewing each other as pure rivals and start working together to ensure they are all compensated for the work they do. This does not mean fixing a local labor rate, which is an illegal practice and prosecutable crime. It means making sure all facilities in the area know the work they are giving away for free is billable.

It must be stressed that not all collision repair facilities and insurance companies engage in unethical behavior. They are in the minority. But the ethical dilemmas confronting the industry cannot be ignored.

Resolving the problems that lead to these dilemmas is the solution. There are usually valid points to be made on each side of the arguments among collision repair facilities, insurance companies, and customers. As stressed earlier in this chapter, communication and cooperation are the solution.

1.10 THE RIGHT MENTAL APPROACH

Successful damage report writers and appraisers are knowledgeable, precise, thorough, and ethical. Attention to detail and the ability to resist quick decisions or assumptions are essential. Few medical patients would be comfortable if their doctor advised major surgery after a 30-second examination and consultation. The same holds true for a customer faced with a major vehicle repair. A professional damage report takes time and thought.

The most effective method of writing damage reports or appraisals is to establish a working pattern and then write all reports using this same technique. This approach creates a consistent style of writing reports. It results in more uniform, accurate, and understandable reports. Each damaged vehicle should be examined in a logical, organized way. The modern vehicle contains thousands of separate parts and components. A haphazard approach to identifying and pricing components will lead to mistakes and miscalculations.

To the untrained eye, the modern vehicle presents a baffling mass of metal and plastic parts assembled in mysterious ways with hidden fasteners. But the skilled damage report writer or appraiser sees any vehicle as a series of much smaller subassemblies. A subassembly with twenty separate parts is much easier to understand and analyze for damage than one with hundreds of separate items.

Estimating guides and computer estimating programs break down vehicles into these smaller assemblies. But damage report or appraisal writers must understand relationships between the subassemblies and how damage can move through assemblies during impact. They must be able to visualize the entire repair process, including operations not covered in the estimating guides.

1.11 ESTIMATING GUIDES

Collision estimating guides (also called collision manuals or crash guides) are the single most important tools used by damage report writers and appraisers. The Mitchell Collision Estimating Guide, published by Mitchell International Inc., and the Motor Crash Estimating Guide published by the Hearst Corporation are the two most popular estimating guides in use today. They contain vital crash and repair information on both domestic and import passenger cars, vans, and light trucks.

Publishers constantly update and reissue their guides to keep abreast of changing prices, revised labor times, and new and discontinued components. With up to sixteen updates per year for current models, it is important for all involved in the industry to use the most up-to-date manuals when compiling and comparing estimates. The guides list the OEM parts prices and information for the past seven or eight model years. Early model guides are also available for vehicles up to fourteen years old.

As mentioned earlier, publishers strive to organize their manuals in a logical, easy-to-use manner. Although slight differences exist between manuals, information is generally first divided by vehicle year and vehicle model. Within each of these divisions, information is then categorized under major assemblies (such as the bumper or hood) or major systems (such as the air conditioner, heater, or exhaust). An exploded-view illustration of the assembly or system shows the relative position of the parts and gives their industry-accepted names and manufacturer's reference numbers. The parts in the exploded views are keyed to the pricing and flat rate labor allowances that accompany each illustration (Figure 1–7).

It is important to remember that the flat rate labor times provided by the manuals give times for the removal and replacement of new and undamaged parts on undamaged vehicles. Actual labor times needed to remove heavily damaged parts or parts frozen by rust or corrosion may vary.

It is impossible to establish accurate labor times for all operations. Jobs such as transferring brackets, braces, reinforcements, or ornaments from old to new parts vary in difficulty depending on the particular job. Removing undercoating, restoring corrosion protection, and filling or plugging unneeded holes on new parts are other examples of labor allowances not included in the published labor times.

A full listing of times not included in published labor times is given in the collision estimating guide. Guides also do not list any labor times for the straightening or repair of frame, sheet metal components, or plastics. Labor allowances for these repairs are referred to as judgement labor time. Allowances for judgement labor time and other nonincluded operations are determined through experience and careful calculation of all factors involved. Reaching an agreement concerning nonincluded labor allowances is often the most difficult part of shop/adjuster negotiations.

Part prices listed in the manuals are for new OEM parts. Glass listings give both the manufacturer's and National Auto Glass Specifications (NAGS) part numbers and prices.

Individual crash guides also include most, if not all, of the following:

WINDSHIELD

Use Procedure Explanation 13 with the following text.

Refinish Upper Inside Garnish Mldg 3
R&I Windshield .... #2.5
#Includes Clean Up Old Adhesive as Necessary
R&I/R&R Reveal Mouldings Complete 5

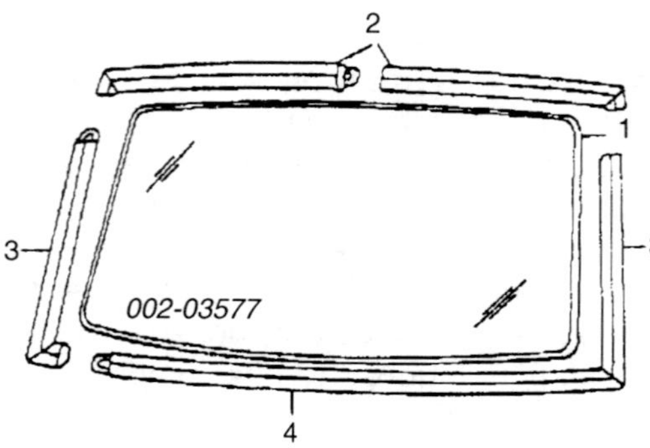

	Description		Part No.	Time	Price
1	**Glass, Windshield**				
	Ford-Tinted		FOTZ 1503100 A	**#2.2**	773.67
	NAGS				
	Clear		DW01099CLN	**#2.2**	773.70
	Green		DW01099GTN	**#2.2**	773.70
	Green/Blue Shade		DW01099GBN	**#2.2**	792.60
	#Includes Clean Up Old Adhesive as Necessary				
	Urethane		EOAZ 19562 A		d63.61
	Moulding, Reveal				
2	Upper				
	Bright	R/L	EOTZ 1003144-5 A	**#.2**	16.50
	Black	R	F1TZ 1503144 AJ	**#.2**	17.37
		L	F1TZ 1503145 AJ	**#.2**	16.08
3	Side				
	Bright	R	EOTZ 1003136 A	**#.2**	18.15
		L	EOTZ 1003137 A	**#.2**	19.70
	Black	R/L	F1TZ 1503136-7 AJ	**#.2**	14.55
4	Lower				
	Bright		EOTZ 1003148 A	**#.2**	40.48
	Black		F1TZ 1503148 AJ	**#.2**	42.63
	#R&R Complete .5				
	Moulding, Inside Garnish				
	Upper (P)	92-93	E8TZ 1503606 A	.3	26.20
		94-96	F4TZ 1503606 A	.3	17.73
	Side¶				
	1992-94				
	w/Headliner	92-93 R/L	F2TZ 1503598-9 E	**#.3**	12.73
		94 R/L	F4TZ 1503598-9 F	**#.3**	12.73
	w/o Headliner	92-93 R/L	F2TZ 1503598-9 A	**#.3**	12.73
		94 R/L	F4TZ 1503598-9 B	**#.3**	12.73
	1995-96	R/L	F4TZ 1503598-9 F	**#.3**	12.73
	(P) Paint to Match				
	¶Blue listed, Order by Application				
	#w/Upper Removed				
	Mirror, Rear View				
	Day/Nite		F2UZ 17700 A	.2	39.43
	Electrochrome		F4TZ 17700 A	.2	226.40

FIGURE 1–7 The exploded views illustrated in the guide are actually the components contained in an organized subassembly or related parts. Note the pricing for the windshield includes manufacturer and NAGS figures. (Courtesy of Mitchell International.)

FRONT FENDER

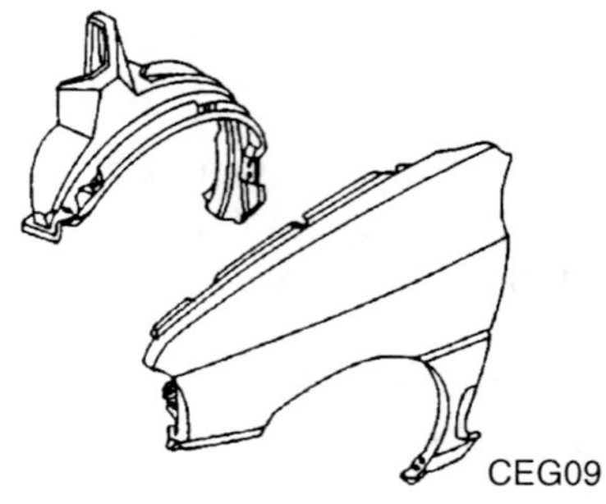

FRONT FENDER R&R

INCLUDED OPERATIONS

- Remove and install or replace:
  - Headlamp assembly if attached to fender
  - Cornering lamp if so equipped
  - Side marker lamp if so equipped
  - Turn indicator lamp if so equipped
  - Parts attached to fender except those listed in not included section
- Replace clip type moulding for base model vehicle

NOT INCLUDED OPERATIONS

- Refinish front fender
- Aim lamps
- Remove and install front bumper
- Remove and install or replace:
  - Front header panel
  - Skirt or liner
  - Wheel
  - Antenna
  - Rocker moulding
- Remove and install adhesive exterior trim; add to clean and retape
- Replace new adhesive exterior trim; deduct one-half of R&R time
- Install stripes, decals, transfers or overlays
- Drill holes for installing exterior trim
- Cut holes for installing lamps

FIGURE 1–8 Important instructions are included in the front of the estimating guide. (Courtesy of Mitchell International.)

- Detailed step-by-step instructions on how to use the guide effectively (Figure 1–8).
- Definition of the terms used in the guide.
- Guidelines for computing labor times for overlapping operations.
- Abbreviation explanations.
- Guidelines for vehicle identification, including how to interpret vehicle identification numbers (VINs).
- Cooling capacity information.
- Air conditioner refrigerant capacity listings.
- Underhood diagrams.
- Conversion tables for quickly determining labor prices.
- Sectioning information (when available from OEM).
- Listings of part numbers that can be interchanged between related models.

It is easy to see why damage report writers and appraisers must be totally familiar with the information in the most up-do-date guides. It is provided to make your job easier. Unfortunately, one of the most common complaints about incompetent adjusters and facility personnel is the inability to use the collision estimating guides.

The information in the guides also serves as the data base for most computerized estimating programs. Computerized estimating vastly reduces the time needed to prepare a damage report. It also reduced the likelihood of math errors, omitted operations, and inaccurate overlap computation. More will be said concerning these systems in chapter 12 of this text.

WARNING: Verbal, nonwritten estimates should never be made. They often lead to misunderstandings that generate customer disloyalty and possible legal action.

1.12 TOOLS AND EQUIPMENT

The following tools and equipment will make damage inspections easier and more accurate:

- Although it is not always practical, having an indoor bay or area set aside for inspections and write-ups is ideal. Any indoor area used for inspections should be well-lighted, preferably with lights in the walls to illuminate the lower sections of the vehicle.
- Because close inspection is often needed to spot slight damage, it is sometimes necessary to wash away dirt. Having a hose and floor drain setup is handy for quick cleanups.
- Placing the vehicle on a hydraulic lift is the best way to fully inspect for undercar damage (Figure 1–9). While not as convenient, a floor jack and jack stands can do the same job. A creeper is needed for easier access to undercar areas (Figure 1–10).

FIGURE 1–9 Checking the underbody for damage is much easier if the vehicle is raised on a lift. (Courtesy of Rotary Lift.)

WARNING: Always make certain the vehicle is fully and safely supported before making an undercar inspection. Be sure damage to the vehicle has not affected the lifting equipment's ability to hold the vehicle.

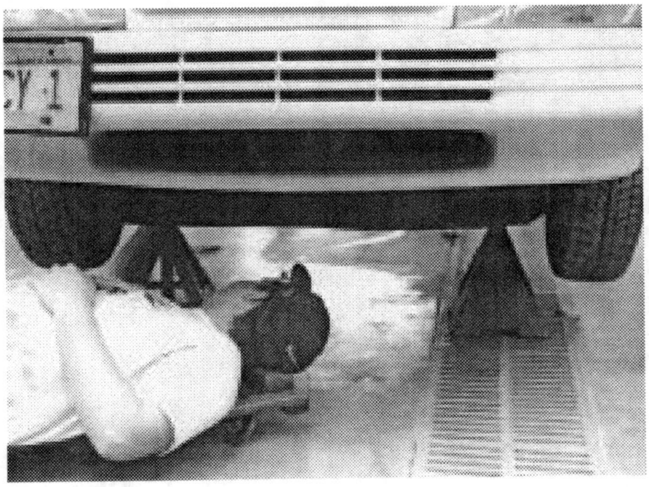

FIGURE 1–10 Always be on the lookout for hidden damage.

- A flashlight or trouble lamp is a must for inspecting enclosed or shadowed areas.
- A tape measure and tram gauge (Figure 1–11) should be used to make spot checks between alignment control points on the vehicle.
- The specification sheets with control point measurement information should be on hand.
- A small box of common hand tools is often useful. Include a magnet for identifying metal, plastic, and aluminum parts. A paint mil thickness gauge is also an excellent tool to have. If a vehicle has been painted to a thickness exceeding specifications, the problem can be spotted and documented on the damage report. This will make it easier to justify the strip time needed to repaint the vehicle to specifications.

FIGURE 1–11 *A tram gauge can be used to quickly spot under-hood misalignment.*

REVIEW QUESTIONS

1. By law, a vehicle owner is required to obtain ________ estimates before the insurance company can settle the claim.
   a. one
   b. two
   c. three
   d. four
2. Technician A says that the customer, not the insurance company, determines which facility will make the repairs. Technician B writes estimates to hide the customer's deductible. Who is right?
   a. technician A
   b. technician B
   c. both A and B
   d. neither A nor B
3. Insurance agents can settle claims in
   a. all states.
   b. some states.
   c. no states.
4. Technician A says that liability insurance will not pay for repairs to the insured vehicle. Technician B states that collision insurance usually has a deductible. Who is right?
   a. technician A
   b. technician B
   c. both A and B
   d. neither A nor B
5. The initials OEM mean ________________________________.
6. Technician A says that the most important aspect of collision repair is to restore the safety of the vehicle. Technician B says that the vehicle must be returned to preaccident conditions. Who is right?
   a. technician A
   b. technician B
   c. both A and B
   d. neither A nor B
7. Insurance companies decide what operations they will pay for, depending on the prevailing practices in a region.
   True False
8. Technician A was paid to replace a quarter panel; he elects to straighten it and does not inform the customer, nor does he reduce his bill. Technician B uses aftermarket parts when the customer has paid for OEM. Who is right?
   a. technician A
   b. technician B
   c. both A and B
   d. neither A nor B
9. The labor time in estimating guides is called ______________ labor time.
10. Technician A uses the collision estimating guide to find sectioning information. Technician B uses the collision estimating guide to find air conditioning system capacity. Who is right?
    a. technician A
    b. technician B
    c. both A and B
    d. neither A nor B

Chapter 2

Modern Vehicle Construction

Objectives

After reading this chapter, you should be able to:

- **Explain the major differences between body-over-frame and unibody vehicle design.**
- **Describe the general evolution of vehicle body design from early body-over-frame to present-day frame, semiunitized, and unitized construction.**
- **List the major design characteristics of modern body-over-frame and modern unibody construction and how they affect estimating and repair procedures.**
- **Identify the major structural components, sections, and assemblies of body-over-frame vehicles.**
- **Identify the major structural components, sections, and assemblies of unibody vehicles, including front engine/rear drive and front engine/front drive.**
- **Identify the types of low-carbon and high-strength steels used in modern vehicle design.**
- **List where these various steels are commonly used, and explain how their characteristics affect estimating and repair procedures.**
- **Identify the major types of plastics used in vehicle construction, their repair methods, and estimating criteria.**
- **Describe the major types of sealers used in modern vehicle construction and their function and importance.**
- **Describe the major types of glass used in modern vehicle construction.**

Key Terms

FF (front engine/front drive)
FR (front engine/rear drive)
Ladder frame
Perimeter frame
Space frame
HSLA (high-strength, low-alloy)
HSS (high–tensile strength)
Martensitic steel

The goal of collision repair is to restore the vehicle to its preaccident condition. To accurately estimate the parts, labor, and related materials needed to perform any repair, estimators and appraisers must fully understand how the vehicle is designed and constructed. They must be capable of accurately identifying all damaged components and the options available for their repair or replacement. And they must know the materials used in constructing the vehicle and how these materials may affect the repair process.

2.1 VEHICLE DESIGN—PAST AND PRESENT

To fully comprehend the challenges faced by estimators, appraisers, and repair technicians in today's collision repair market, consider the history of vehicle design.

With most passenger cars and minivan models, the conventional body-over-frame construction that served the industry for over sixty years has given way to unitized construction, a totally different concept in vehicle design that requires new assembly techniques, new materials, and a completely different approach to collision repairs. In unibody designs, heavy-gauge cold-rolled steels have been replaced with lighter, high-strength steel alloys requiring new handling, straightening, and welding techniques. The ever-increasing use of plastics, modular glass, and advanced paint systems is also changing the face of vehicle repair. On unibody vehicles, certain mechanical systems, such as suspension and steering, rely on the proper positioning of unibody components for alignment and smooth operation.

But it is not as simple as abandoning an old design concept for a new one. Body-over-frame construction is still being used on one-half and three-quarter ton pickup trucks, some sport utility vehicles (SUVs), and most full-size vans. While literally all passenger car imports built during the 1980s featured unitized construction, some larger domestic models maintained traditional body-over-frame setups. Following is a brief look at how the two major vehicle designs used today have evolved over the years.

A

B

C

D

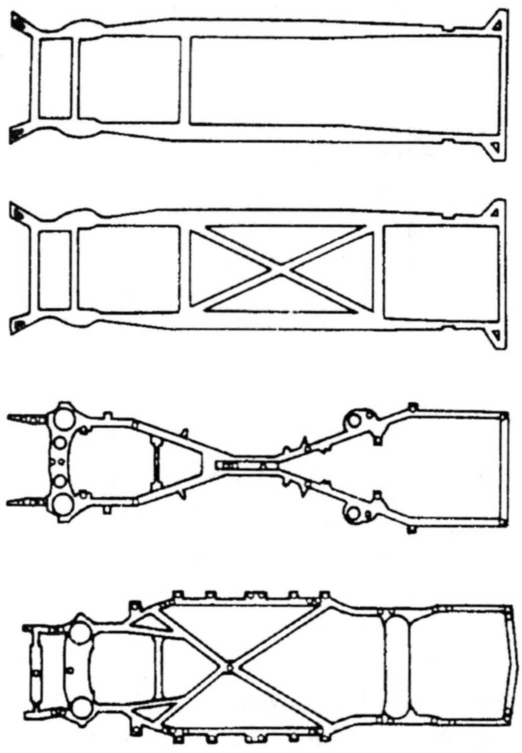

FIGURE 2–1 Types of early frame designs: (A) ladder; (B) X-shaped; (C) hourglass; (D) combination.

EARLY FRAME CONSTRUCTION

The earliest automobiles and trucks were constructed with heavy, rigid, ladder-shaped iron frames (Figure 2–1A). The massive legs and cross members provided great strength and served as a foundation to which body panels and components were mounted using bolts. The vehicle sat quite high off the ground, the ride was rough, and the high center of gravity meant the vehicle could not handle sharp turns at higher speeds.

During the 1930s, steps were taken to solve these problems. Frame members were lightened to provide the flex needed to accommodate increased speeds and handling demands. The center of gravity was lowered by mounting the power train between the side rails, not on top of them. This improved stability allowed styling changes, such as lower hood lines. The introduction of coil spring suspension, ball joints, and more sophisticated steering systems created a softer ride and improved handling and safety.

But reducing frame member weight decreased overall frame strength. To compensate for this loss, an X-shaped cross member was added to the basic ladder frame for increased rigidity (Figure 2–1B).

This reinforced X frame briefly gave way to an hourglass frame shape (Figure 2–1C). This frame design quickly proved ineffective and complicated to repair and was soon abandoned.

The next generation of frames relied on a combination design that further reduced frame weight and strength (Figure 2–1D). Automakers compensated for this loss by increasing the inherent strength of the body components. As we will see, providing vehicle strength through body panel design is the overriding principle in unibody construction. But with the combination design the frame and body remained separate units, bolted together. And frame and panel size and weight were still massive compared to modern standards.

EARLY UNIBODY DESIGNS

The first true domestic unibody car was produced by Nash in 1940. This design featured many of the concepts that still hold true in today's unitized vehicles. It also contained features that caused major repair headaches, some of which still plague modern collision repair facilities.

The concept of unibody design had first proved successful in the aircraft industry, where lightweight, high-strength fuselages were needed. Adapted to the auto industry, unitized construction eliminated the need for a separate, rigid frame. The strength and rigidity of the vehicle was integrated into the body shell, which was formed by welding sheet metal into a box- or egg-like configuration. Strength is achieved

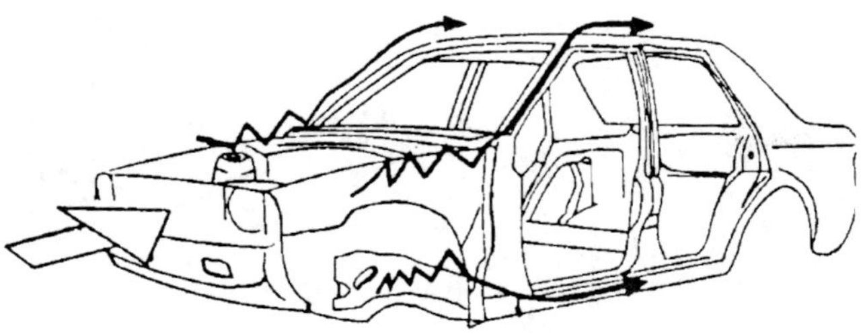

FIGURE 2–2 Unibody design is based on an egg shape or box shape, in which all components add to the structure's support.

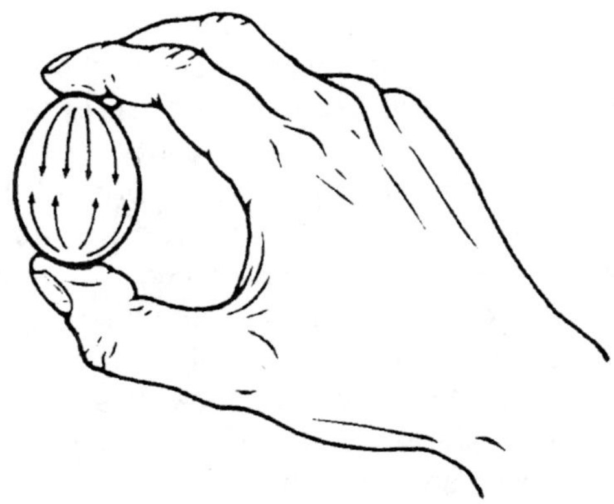

FIGURE 2–3 The shape of the egg spreads the force through the entire shell.

through shape and design instead of mass and weight. The entire inner structure acts as the support structure, with individual panels depending upon one another for structural integrity (Figure 2–2).

The eggshell concept is an ideal model for understanding the principle of unibody design. Even when pressing hard on an eggshell, it is comparatively difficult to destroy, since all the strength applied by the fingers is not concentrated in one place but is dispersed effectively through the entire shell (Figure 2–3). In engineering this concept is called a stressed hull structure.

A major advantage of stressed hull structuring is increased passenger compartment protection. Unlike heavier, body-over-frame designs that tend to dampen and localize collision damage, the stiffer, lightweight sections used in unibodies tend to transmit and distribute impact energy throughout more of the vehicle. The ability to absorb and dissipate impact energy helps keep passenger compartments intact during many types of collisions. Another advantage is the lighter weight of unibody vehicles. This means that smaller, more fuel-efficient engines are able to adequately power these vehicles.

But Nash's and other first-generation "unibody" vehicles presented several challenges and problems. The elimination of the separate frame created noise and vibration transfer from the road surface. Because suspension and steering components mounted directly to the body shell, setting and adjusting wheel alignment became a major problem, particularly during repairs. It was evident that a new generation of steering and suspension technology was needed to accommodate unibody design. Collision repair techniques were instantly outdated. There are no frame members to supply strong pulling points for straightening bent pieces. Individual body panels could not be unbolted and repaired or replaced. It was comparable to repairing a cracked egg with a pry bar.

To American automakers these disadvantages outweighed the pluses gained through unibody construction. Although frames still appeared the best way to attach running gear and support the power train, some of the concepts of unitized design were adopted and are evident in the modern perimeter frame and stub frame designs described later in this chapter. While domestic manufacturers resisted a full-scale changeover to unibody for over thirty years, European automakers realized that unibody design provided a means of producing small, safe, fuel-efficient cars for their gas-starved markets. That is why many mechanical systems used on modern unibody vehicles, such as rack-and-pinion steering, MacPherson strut suspensions, and constant velocity driven front wheel drive first gained acceptance through the import market.

THE CATALYST FOR CHANGE

Through the 1960s and early 1970s, virtually all American automobiles were manufactured with the following design characteristics:

- Body/frame construction
- Rear drive
- Independent front suspension
- Symmetrical design

In the mid-1970s, a variety of events took place that rocked the foundations of the automobile industry. First of all, the government placed very strict fuel economy and emission control laws and standards on auto manufacturers. The economy standard was known as CAFE (Corporate Average Fuel Economy), and the emission control standard was set by the Environmental Protection Agency (EPA). This meant that American automakers had to start designing more efficient methods of combustion and emission control systems for engines.

There was also increased pressure for better fuel economy from the public sector. The Arab oil embargo destroyed the American myth of a cheap, endless supply of gasoline for our big, heavy, fuel-hungry cars.

In addition to being found fuel inefficient, American vehicles were also being proclaimed unsafe in a series of media and consumer protection group reports.

Foreign automakers who had embraced unitized construction years earlier captured an increased share of the domestic new car market with their lines of smaller, lighter, more fuel-efficient vehicles.

The so-called disadvantages of unitized construction that had kept American carmakers entrenched in body-over-frame construction no longer outweighed the advantages, and the wholesale push to unibody was on.

The five construction areas where domestic automobiles have changed since the mid-1970s are

- Use of unibody construction
- Weight (average fleet)
- Metal composition
- Suspension/steering
- Engine location/drive

In 1977, most cars used a perimeter-type frame. They averaged around 4500 pounds, used 18 gauge mild steel, and were still conventional in design. In 1978, modern unitized construction was used for the first time on American-made cars. Body weight began to decrease, thinner-gauge metal was used, and the first American-made transverse engine, front wheel drive, strut suspension cars were introduced.

By 1981, unibodies were used in almost half the American-made cars. Fleet average weight decreased 600 pounds, 22 gauge high-strength steel was used in construction, the steering changed to rack and pinion, the suspension to MacPherson strut type, and the drive changed from rear to front wheel drive.

Today 56 percent of new vehicles are unibodies. All but a small handful of domestic cars produced are unibodies constructed of 24 gauge high-strength steel. Average vehicle weight is approximately 3,000 pounds. MacPherson struts, rack-and-pinion steering, and front wheel drive are standard features on many vehicles. The full-frame vehicles, 44 percent of new vehicles, are primarily pickup trucks, full-size vans, and some SUVs.

2.2 MODERN BODY-OVER-FRAME CONSTRUCTION

The following sections will take a closer look at the modern body-over-frame designs used on many of today's pickup trucks, full-size vans, some SUVs, and some full-size passenger cars.

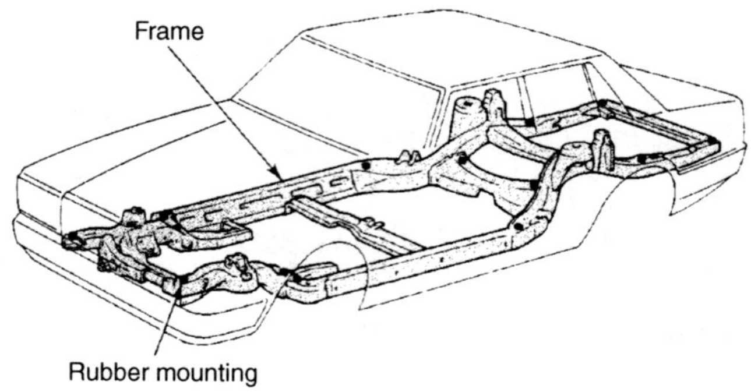

FIGURE 2–4 The location of rubber "biscuits" or mounts used on conventional body-over-frame vehicles. Pickup trucks and SUVs are similar.

In the conventional body-over-frame construction, the frame is the vehicle's foundation. The body and all major parts of a vehicle are attached to the frame. It must provide the support and strength needed by the assemblies and parts attached to it. The frame must also be strong enough to keep the other car parts in alignment should a collision occur.

The conventional frame is an independent, separate component, because it is not welded to any of the major units of the body shell. The body is generally bolted to the frame and large, specially designed rubber "biscuits," or mounts, are placed between the frame and body structure to reduce noise and vibration from entering the passenger compartment (Figure 2–4). Quite often, two layers of rubber are used in the mounting pads to provide a smoother ride. In higher-priced vehicles, a shock absorber is mounted between the body and the rear frame, which serves to minimize vibrations when the car is traveling at high speeds.

Today, the strong steel frame side members of the modern conventional design are normally made of U-shaped channel sections or box-shaped sections (Figure 2–5). Cross members of the same material reinforce the frame and provide support for the wheels, engine, and suspension systems. Various brackets, braces, and openings are provided to permit installation of the many parts that make up the automotive chassis. The various cross members, brackets, and braces are welded, riveted, or bolted to the frame side rails.

Most conventional frames are wide at the rear and narrow at the front. The narrow front construction enables the vehicle to make a shorter turn. A wide frame at the rear provides better support of the body. Other characteristics are as follows:

- Load-induced vibrations are transferred to the body via the frame, resulting in a smooth ride.

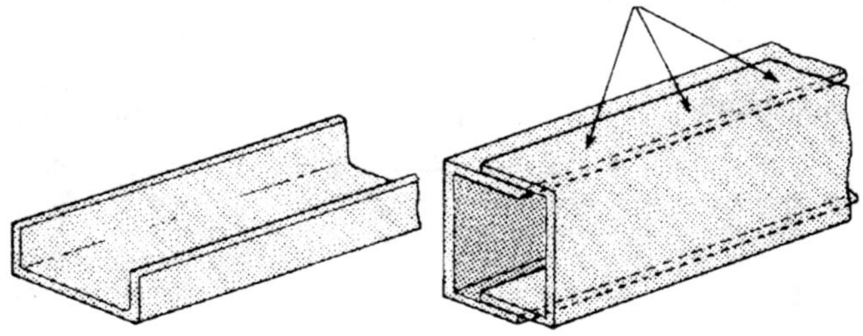

FIGURE 2–5 (Left) U-channel frame section and (right) box channel frame section used in body-over-frame construction.

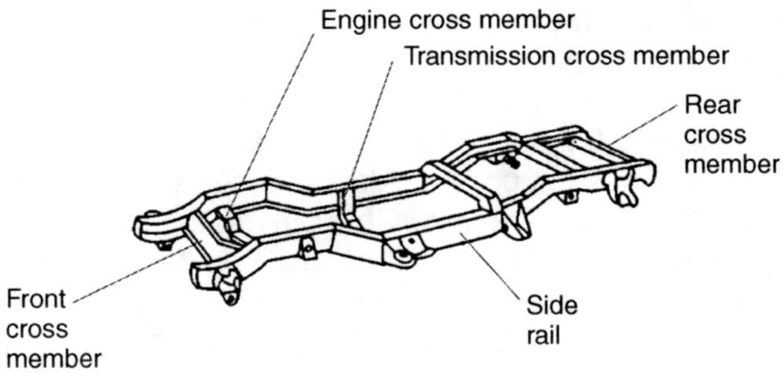

FIGURE 2–6 The modern ladder frame used on pickup trucks.

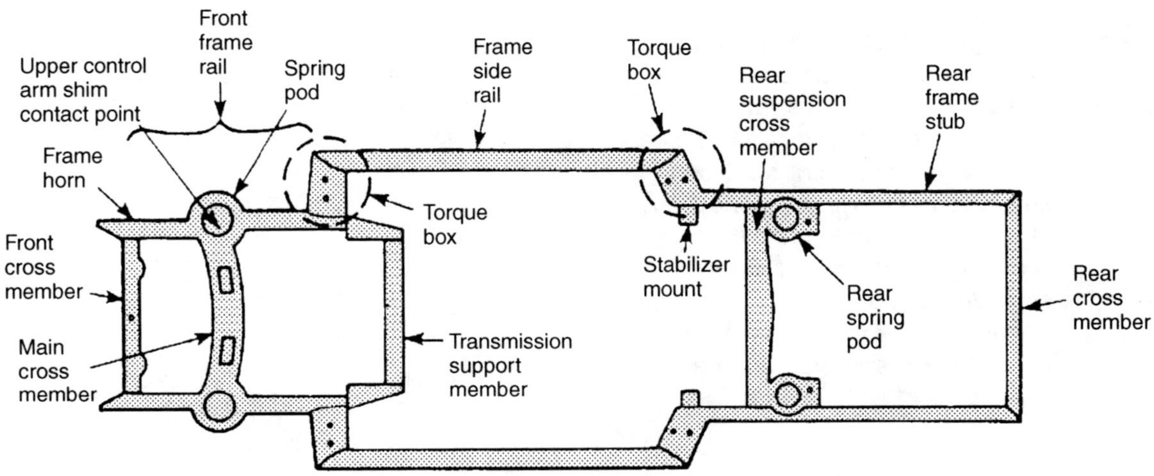

FIGURE 2–7 The major components of a modern perimeter frame.

- Rubber mountings between the body and frame insulate it from vibrations, providing a quiet interior.
- High amounts of energy are absorbed during a collision.
- Undersurfaces of the body are protected over rough roads.
- Suspension and power train parts can be quickly assembled on the basic frame.
- Heavy frame made of thick sheet metal is approximately 3/16 to 1/8 inch.

Modern Ladder Frames. While the forerunner of all frame designs is no longer used for passenger vehicles because of its hard ride, variations of the basic ladder frame are still used on some trucks (Figure 2–6). This strong design consists of two side rails, not necessarily parallel, connected to each other by a series of cross members.

Perimeter Frames. The perimeter frame (Figure 2–7) is similar in construction to the ladder frame, but there are significant differences. The full-length side rails support the body at its greatest width, which provides more protection to the passengers in the case of a side impact to the body. The areas behind the front wheels and in front of the rear wheels are stepped to form a torque box structure. In a head-on collision, the stepped areas absorb much of the energy, and in a side-impact collision, since the center frame rail is near the front floor side member, the passenger compartment is protected from collapse. In rear-end collisions, the rear cross members and kickup absorb the shock. As for twisting and bending, strategic areas are reinforced with cross members. The design changes help absorb and dissipate collision shock and energy.

The perimeter frame shown in Figure 2–8 features a center rail that passes very close to the inside of the front floor side member. Because of this, the passenger compartment floor can be made lower. As a result, the center of gravity is lower, the vehicle is enlarged, and the vehicle profile is lowered.

Most of the conventional frames used today are of the perimeter design.

BODY COMPONENTS

In a modern body-over-frame vehicle, the front body section is made up of the radiator support, front fender, and front fender apron. These components (Figure 2–9) are installed with bolts and form an easily disassembled structure. The radiator support is made up of the upper support, lower support, and left and right side supports welded together to form a single structure. The front fender of the separate frame–type vehicle differs from the front fender of the unibody. The support panels in the upper inside and rear ends of the fender are spot welded. This not only increases the fender's strength and rigidity, it also works with the fender apron to reduce vibration and noise transmitted to the passenger compartment and helps to prevent damage to the suspension and engine from side impacts.

The main body (Figure 2–10) is made up of the dash panel, underbody, roof, and so on, to form the passenger and luggage compartments, and is similar in structure to that of a true unibody vehicle. The dash panel is made up of the left and right front body pillars, cowl side panels, and cowl panel. The front of the underbody has a propeller shaft tunnel built into it, forming a channel cross section through the center of the floor pan, and cross members are welded to it where it joins the frame. Thus, the passenger compartment, roof side rail, door, and side body are protected from side impact collisions. In addition, the front, back, and left and right sides of the floor pan are made uneven in the stamping process, increasing the rigidity of the floor pan itself, which reduces vibration.

Pickup trucks, some SUVs, and full-size vans have body-on-frame construction. They do not need to meet the same safety standards as passenger cars, so most of these vehicles lack door crash beams and bumper isolators. The front end of a pickup truck is shown in Figure 2–11. Trucks either have the radiator support and aprons welded together as in a unibody or the radiator support bolted to the frame and fenders. If the radiator support is bolted, the fenders are reinforced and take the place of the apron. The hood, fenders, and bumper are all bolted in place. Metal or plastic fender liners prevent debris from accumulating inside the fenders.

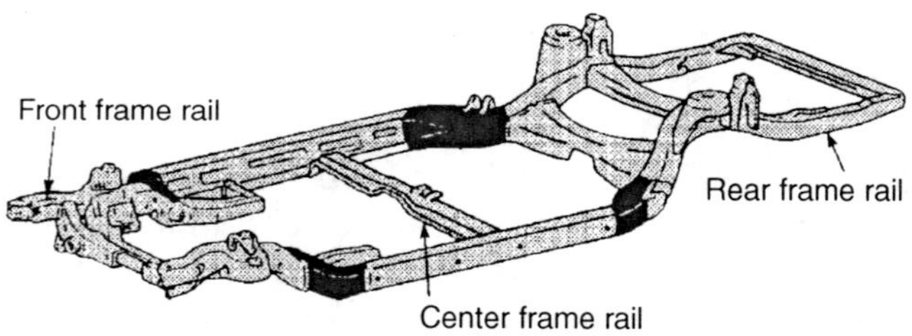

FIGURE 2–8 A perimeter frame design featuring a center rail; shaded areas are torque box sections.

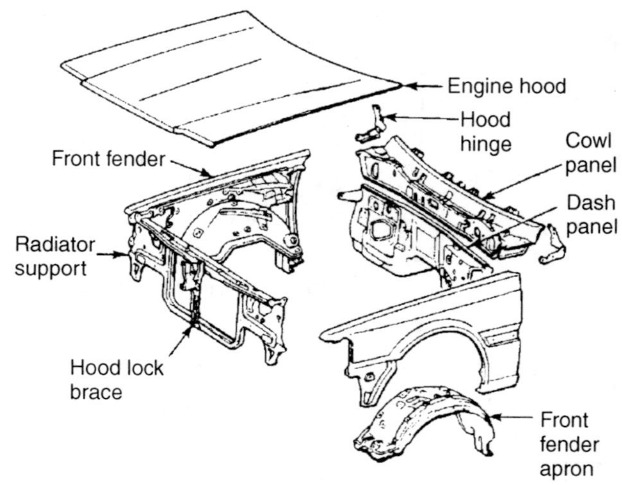

FIGURE 2–9 Typical front body components of a body-over-frame vehicle.

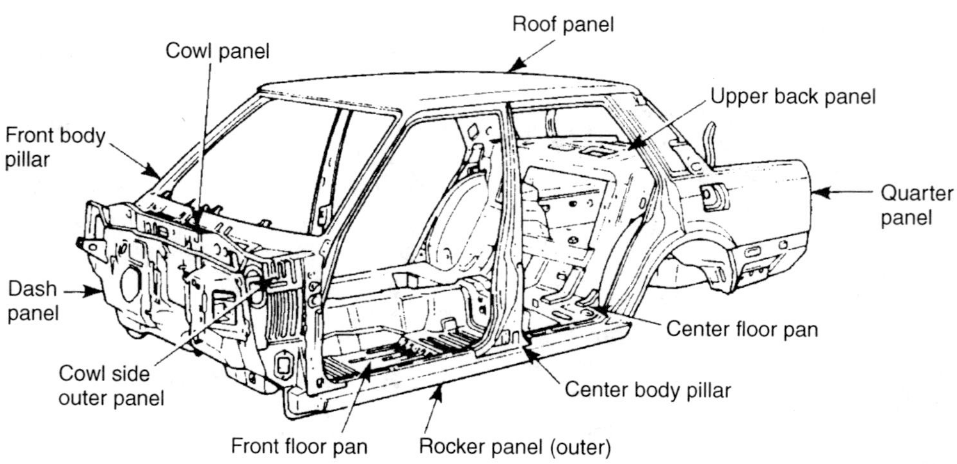

FIGURE 2–10 Typical main body components of a body-over-frame vehicle.

The cab of a pickup truck, SUV, or van is bolted to the frame (Figure 2–12). Frame vibration transfer to the cab is minimized by rubber mounting biscuits. The cab is made of several welded-together stamped steel parts. Many of these parts can be replaced individually if required. There is a tunnel in the floor for transmission clearance. Cab size can be standard—one set of seats—or extended—two sets of seats.

Extended cabs have either two, three, or four doors. The doors have either bolted or welded hinges. The entire cab can be removed by disconnecting all linkages and electrical wires and unbolting the cab from the frame.

The bed or box of a pickup truck is also removable as an assembly. Like the cab, the box is made of several stamped steel pieces welded together. Individual parts can be replaced if damaged. There are two types of beds. The most common type, styleside, has no external fenders. The other type, flare side, has external, removable fenders. Most often a flare side is found on a pickup equipped with dual rear wheels.

FIGURE 2–11 This Ranger pickup has a welded-together radiator support and aprons. Other trucks have a bolted radiator support and reinforced fenders.

FIGURE 2–12 This extended cab pickup truck features a bolted-to-the-frame box, cab, and radiator support.

Access to the bed is controlled by the tailgate. A topper, usually made from aluminum or fiberglass, is often installed to keep bed cargo dry and protected.

2.3 SEMIUNITIZED CONSTRUCTION

The semiunitized design provides a true frame only in areas where it is needed, such as in the front steering and suspension areas. Like the perimeter frame vehicle, cars and trucks with a semiunitized design are often considered "second generation unibodies."

This design uses heavy-gauge steel "stub" rails or a subframe that is welded or bolted to the front and rear of the body or platform structures (Figure 2–13). The suspension system, engine, transmission, and, of course, rear axle assembly are either attached to the stub rails or sections of the platform type construction. Between the front and rear stub rails is the underbody structure with no frame underneath to act as a support for the unitized body shell or platform. Many of the stub rails and their cross members are welded to both body shell and sheet metal components to form a single integral unit.

2.4 MODERN UNITIZED CONSTRUCTION

Modern or completely unitized design is sometimes called "third generation unitized construction," although it is the first true unitized design found in modern domestic vehicles.

As mentioned earlier in this chapter, the unitized vehicle has no separate frame (Figure 2–14) but is a stressed-hull structure in which each of the components supplies structural support and strength to the entire vehicle.

In a car body, there is no complete stressed-hull structure. In general, a body with a structure that integrates the frame and body to receive and hold

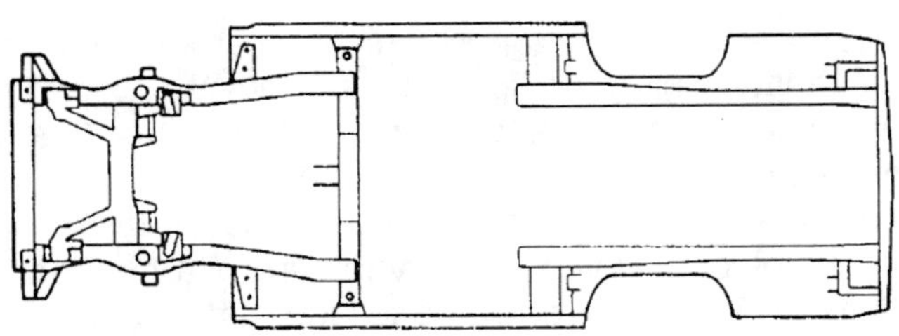

FIGURE 2–13 The bolt-on stub frame.

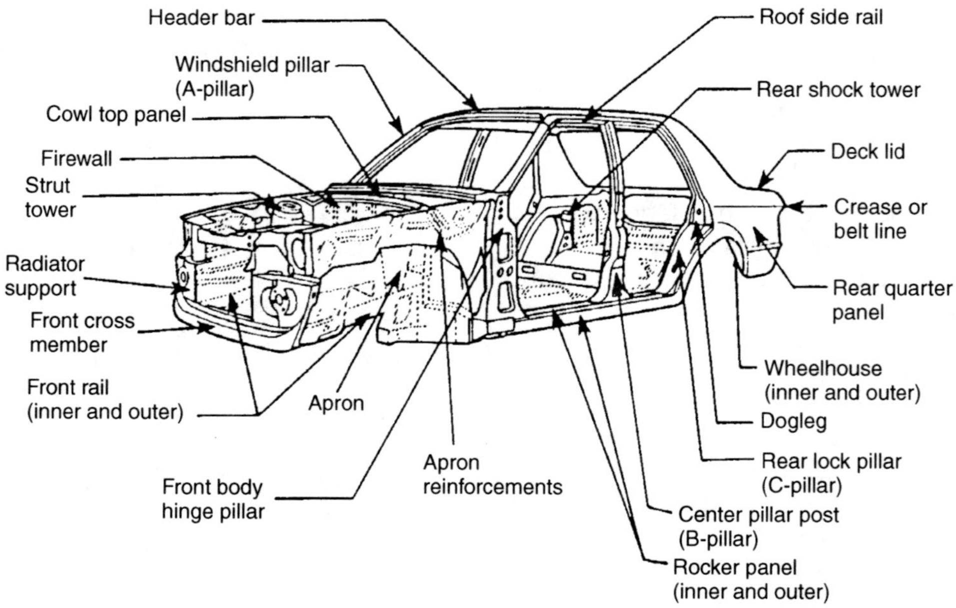

FIGURE 2–14 *The major components of a fully unitized body shell.*

outside forces is called a *unibody* and has the following characteristics:

- It is made by combining pieces of thin sheet metal pressed to form panels of various shapes and joined into an integrated structure by spot welding. This structure is lightweight and highly rigid against bending or twisting.
- The bulk taken by the frame can be used to make the car more compact.
- Vibration and noise from the drivetrain and suspension enter the floor pan and are amplified by the body, which acts as an acoustic chamber, making it necessary to add extra components to the body to suppress vibration and noise.
- Once deformed, a great deal of time may be needed to restore damaged components to their original shapes. Calculating accurate labor times for this work is essential to writing professional damage reports and appraisals.
- With the thin sheet metal body close to the road surface, adequate measures must be taken to prevent the deterioration in strength from corrosion. This is particularly important when dealing with reinforcing materials that make up the underbody.

The major advantage of unibody vehicles is that they tend to be more tightly constructed because the major parts are all welded together. This design characteristic that helps protect the occupants during a collision causes damage patterns that differ from those of body-over-frame vehicles. Rather than dampen and localize damage, the stiffer sections used in unibody design tend to transmit and distribute impact energy throughout more of the vehicle.

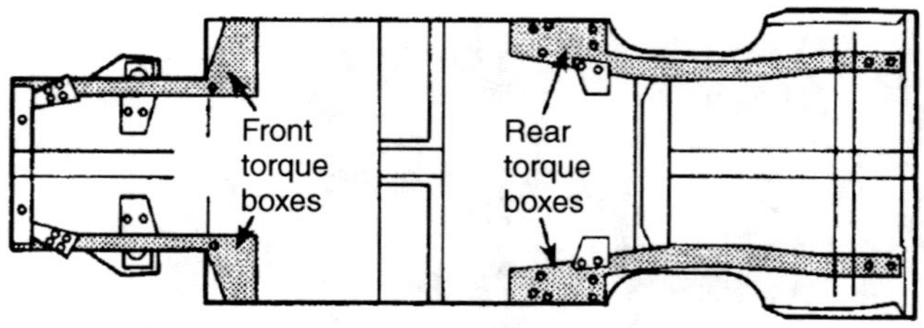

FIGURE 2–15 *Torque box locations on unibody vehicles.*

Because of this fact, estimators and appraisers must analyze collision damage to the unitized vehicle somewhat differently. Because all sections are welded together, the force of impact can cause misalignment in areas remote from the impact point. Even sections that are buckled or torn loose might have passed along heavy force before deforming. Unless the estimator is aware of this fact, it is extremely easy to overlook this secondary damage. For example, the force of a front impact may be sufficient to knock rear end components out of alignment, a problem that can be missed in casual inspections. Such a problem can be sufficient to cause handling or power train problems after the vehicle is repaired.

In some unibody designs, torque boxes similar to those on perimeter frames are used to help absorb collision impact forces (Figure 2–15). In unibody vehicles the complexity and stiffness of the structure is especially critical in the front end, which houses not only the front suspension and steering linkage, but also the entire drivetrain—engine, transaxle, drive shafts, and constant velocity U-joints (Figure 2–16). To keep all these in proper alignment requires support, including that supplied by the front end sheet metal. Accord-

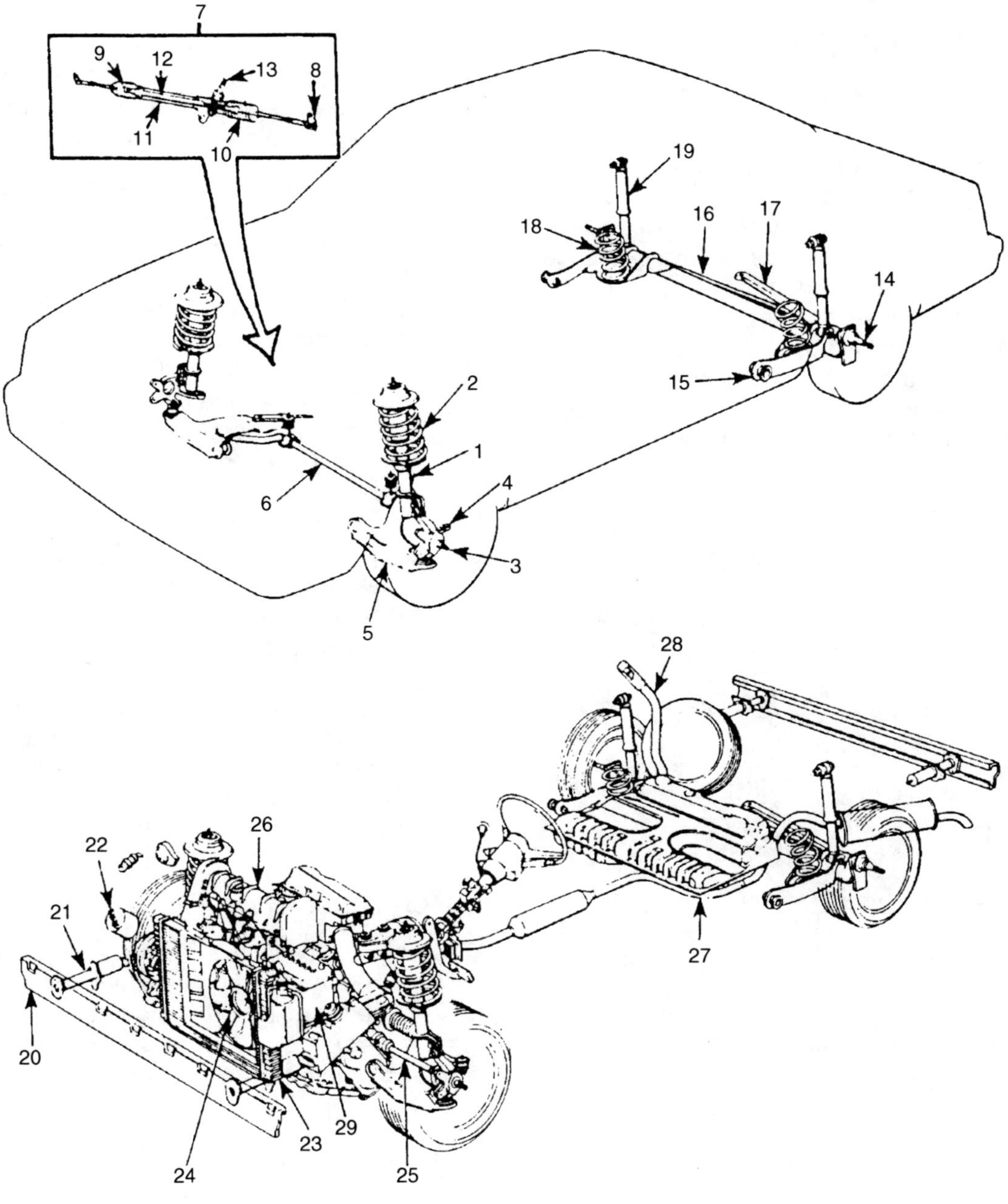

1. MacPherson strut
2. Front coil spring
3. Steering knuckle/front spindle
4. Steering arm
5. Lower control arm
6. Stabilizer bar
7. Rack and pinion assembly
8. Outer tie rod end
9. Inner tie rod
10. Boot
11. Housing
12. Rack
13. Pinion gear
14. Rear spindle
15. Rear control arm
16. Rear axle support beam
17. Track bar
18. Rear coil spring
19. Shock absorber
20. Bumper reinforcement
21. Energy absorber
22. Fuel vapor canister
23. Radiator
24. Fan and shroud
25. Drive axle
26. Engine and transmission assembly
27. Fuel tank
28. Fuel filler neck
29. Battery tray

FIGURE 2–16 Typical mechanical components and locations on a unibody vehicle.

ingly, a damaged unibody vehicle requires a more *thorough* damage analysis than a similar impact would require in a conventional frame/body car. Otherwise, it could mean that after a car is returned to its owner it might later show unsafe handling qualities, water leaks, or a new family of strange noises in the power train.

There are two basic unibody structures: front engine/rear drive (FR) vehicle and front engine/front drive (FF) vehicle.

2.5 FR UNIBODY STRUCTURE

The front engine/rear drive vehicle (Figure 2–17) has the engine mounted in the front and the driving wheels in the rear. The body of an FR vehicle is divided into four main sections: front body, passenger compartment (side body), underbody, and rear body. The engine, transmission, front suspension, and steering equipment are installed in the front body, and the differential and the rear suspension are installed in the rear body. Also, since all impacts from the road surface are transmitted to the entire body through the front and rear wheels, the body must have enough strength to fulfill its function as an automobile during normal driving. This strength is supplied by the side rails and cross members welded to the floor.

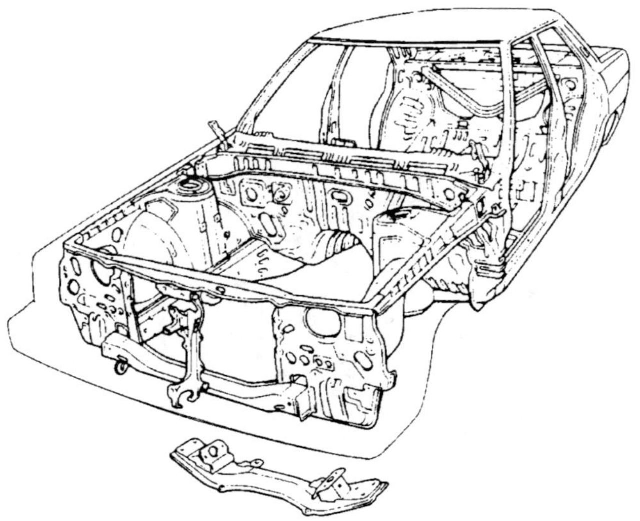

FIGURE 2–17 The main body or shell components of a unitized FR vehicle.

FR vehicles are characterized as follows:

- Since the engine, transmission, and differential are in separate positions, weight can be distributed uniformly between the front and rear wheels, lightening the steering force.
- In an FR vehicle, the engine is placed longitudinally with respect to the vehicle, and most vehicles have a single suspension cross member placed laterally between the front side members at about the middle of the front body, which supports the engine:
- Since it is possible to remove and install the engine, propeller shaft, differential, and suspension independently, body restoration and repair workability are good.
- Since rear wheel drive equipment is necessary, a tunnel in the floor is necessary, which decreases interior space.
- Since the engine's output is transmitted to the rear wheels by the differential mechanism through the propeller drive shaft, the vehicle, vibration, and noise sources are widely distributed over the front and rear.

FRONT BODY

The engine, suspension, and steering equipment are all mounted on the front fender apron and the front side member of the front body (Figure 2–18). This part is very important, because it influences front wheel alignment and the amount of vibration and

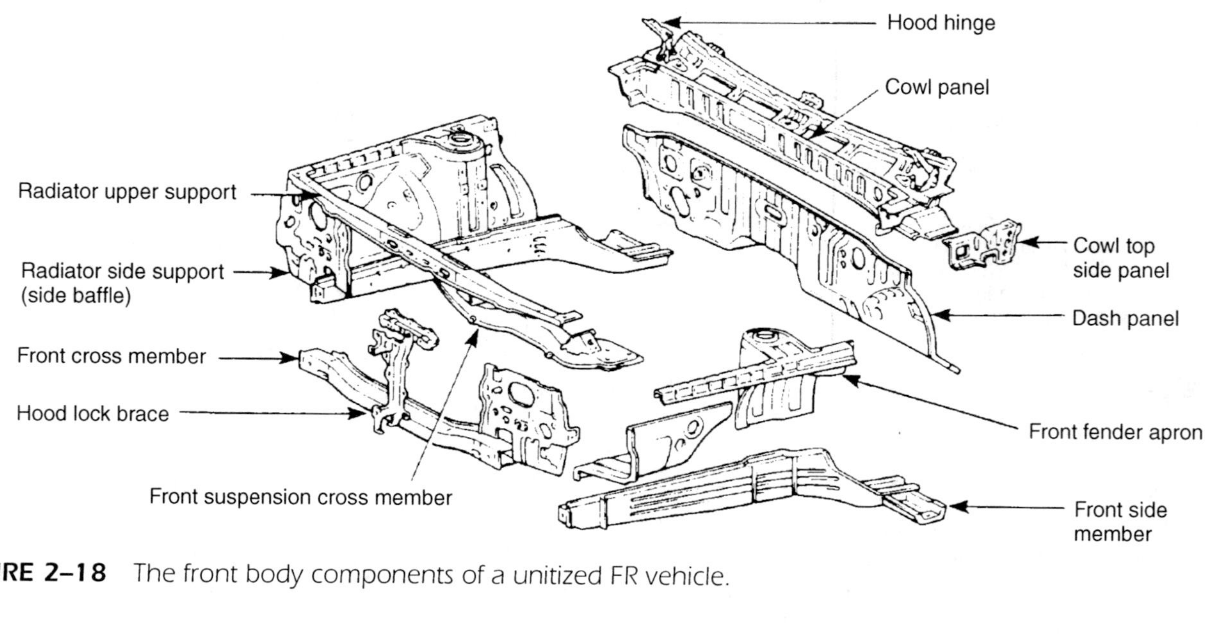

FIGURE 2–18 The front body components of a unitized FR vehicle.

noise that is transmitted into the passenger compartment. Therefore, it must be made with great accuracy and strength. With the exception of outer shell parts, such as the engine hood, front fenders, and front valance panel (installed with nuts and bolts), all other exterior parts are welded together, reducing body weight and increasing body strength.

SIDE BODY

The side body (Figure 2–19) is joined to the front body and roof panel to form the passenger compartment. During travel, these panels distribute the loads from the underbody to the upper part of the vehicle and prevent bending of the left and right sides. The side body members also serve as door supports and maintain the integrity of the passenger compartment if the vehicle should overturn. Since the sides are weakened by large door openings, they are reinforced by joining the inner and outer panels, which forms a very strong boxed-type structure.

UNDERBODY

The front side member, rear floor side member and floor pan, and cross member (Figure 2–20) are the most important structural components. The front and side members resemble the perimeter of a frame structure. The basic arrangement and shapes of these members and floor pan will vary slightly, depending on the size and shape of the suspension and the underbody structure.

The underbody of a unibody vehicle can be further broken down into front, center, and rear sections.

Front Section. Since the front side members and front cross members of the front underbody section directly affect front wheel alignment, they are formed into a boxed section made of high-strength steel (Figure 2–21). To prevent the collapse of the passenger compartment in a head-on collision, the front side members are made with a kickup so that all the members will bend and absorb shock loads.

Center Section. The center underbody section is mainly composed of the floor pan, cross member, and main floor side member. The center of the floor pan contains the propeller shaft tunnel (Figure 2–22), which prevents the floor from twisting. In addition, the main floor side member and cross members below the front seats and in front of the rear seats strengthen the left and right sides and prevent the floor from folding in the event of a side collision.

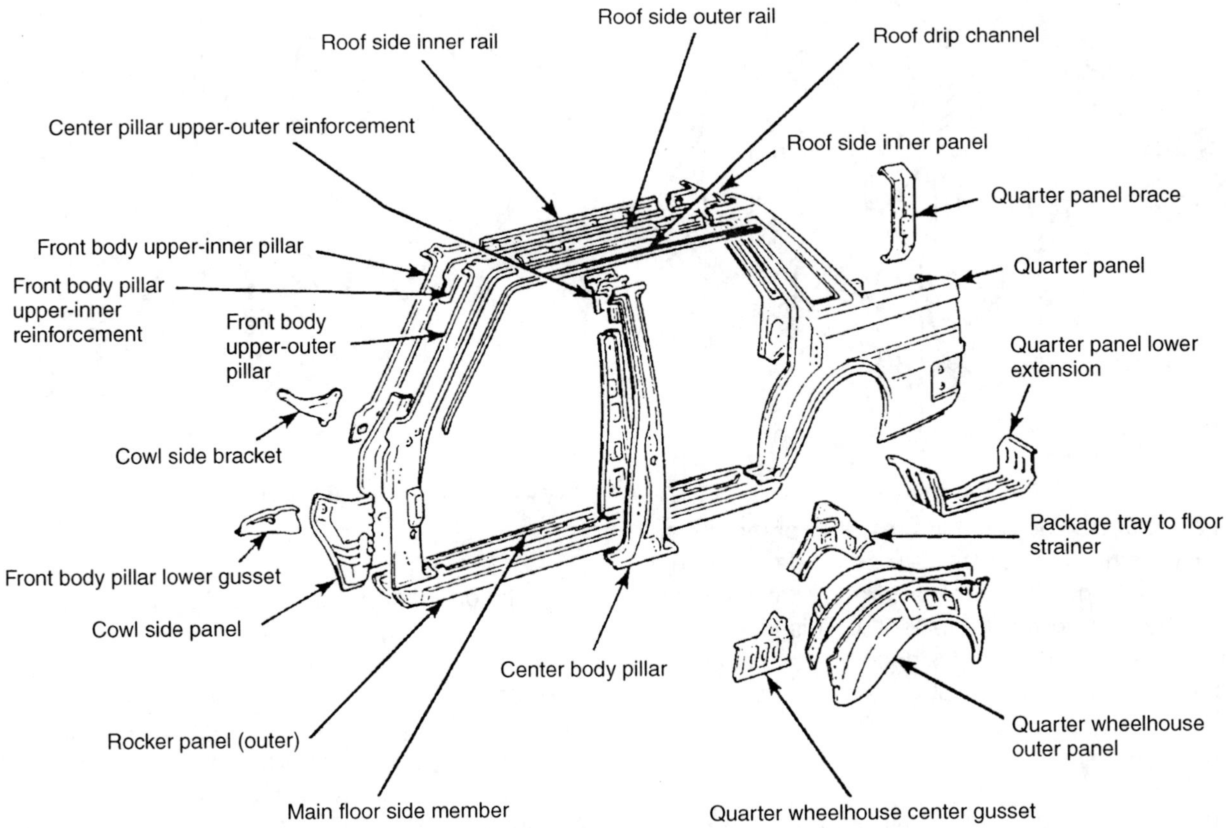

FIGURE 2–19 The side body components of a unitized FR vehicle.

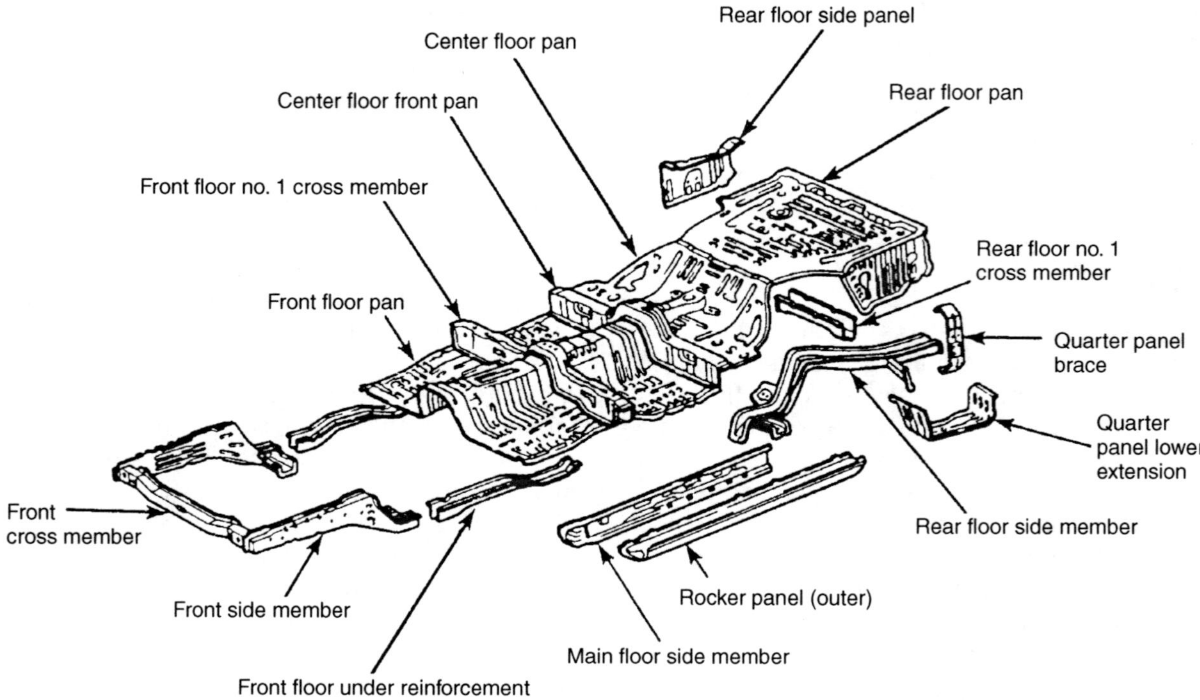

FIGURE 2–20 The underbody structural components of a unitized FR vehicle.

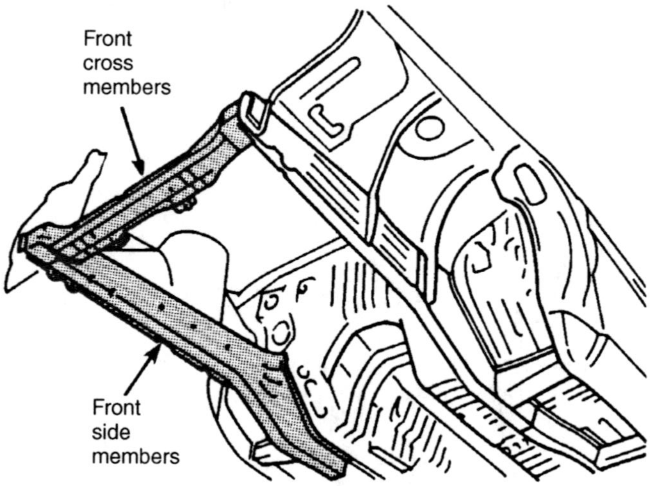

FIGURE 2–21 Close-up of underbody front section.

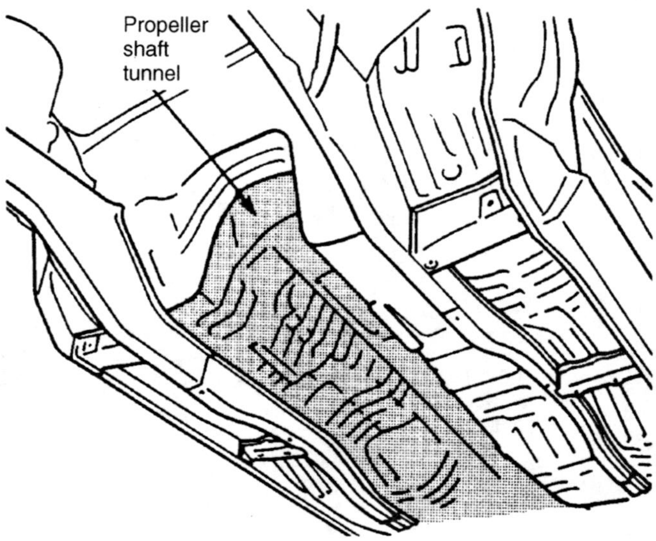

FIGURE 2–22 Close-up of underbody center section.

Rear Section. The rear side member of the underbody extends from under the rear seat to a point near the rear axle, where it forms a large kickup and extends to the rear floor (Figure 2–23). This kickup, like the front side members, is designed to absorb the energy of a rear-end collision.

REAR BODY

The rear body sections are divided into two categories: sedans (Figure 2–24); with the luggage compartment and the passenger compartment separated, and station wagons and liftbacks, with no separation between the luggage compartment and passenger compartment. The upper back panel and rear seat cushion support brace in sedans are joined at the side body and floor pan. The back panel prevents the body from twisting. In station wagons and liftbacks (Figure 2–25) that do not have a rear partition, body rigidity is enhance by adding enlarged roof side inner rear panels and a back window upper frame and by extending the roof side inner panels to the quarter panels.

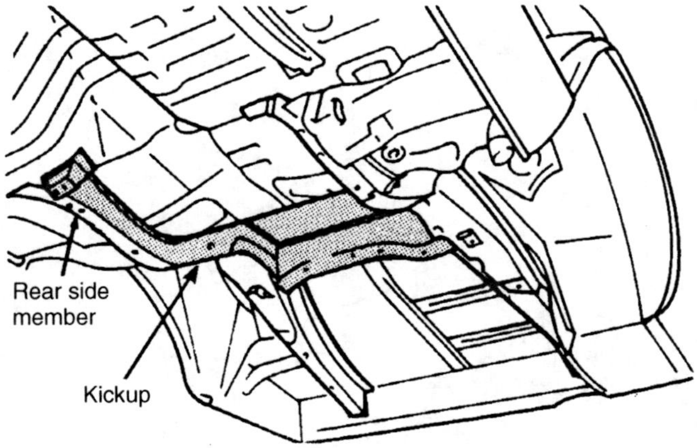

FIGURE 2–23 Close-up of underbody rear section.

2.6 FF UNIBODY STRUCTURE

In an FF passenger car (Figure 2–26), the engine is mounted in the front of the vehicle and the engine drives the front wheels. It is also called a *front wheel drive (FWD)* vehicle. In the space ordinarily taken up by the drive line, the passenger compartment can be enlarged and the rear suspension simplified, resulting in substantial weight reduction. Since the engine, transaxle, front suspension, and steering equipment are all located in the front body section, the methods of reinforcement are much different than those used in the FR vehicles.

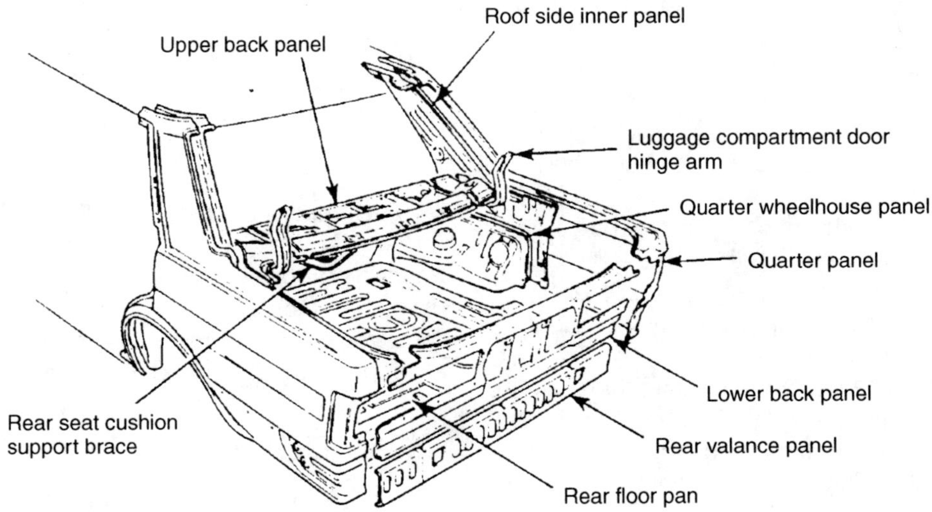

FIGURE 2–24 Rear body structural components of a unitized FR vehicle.

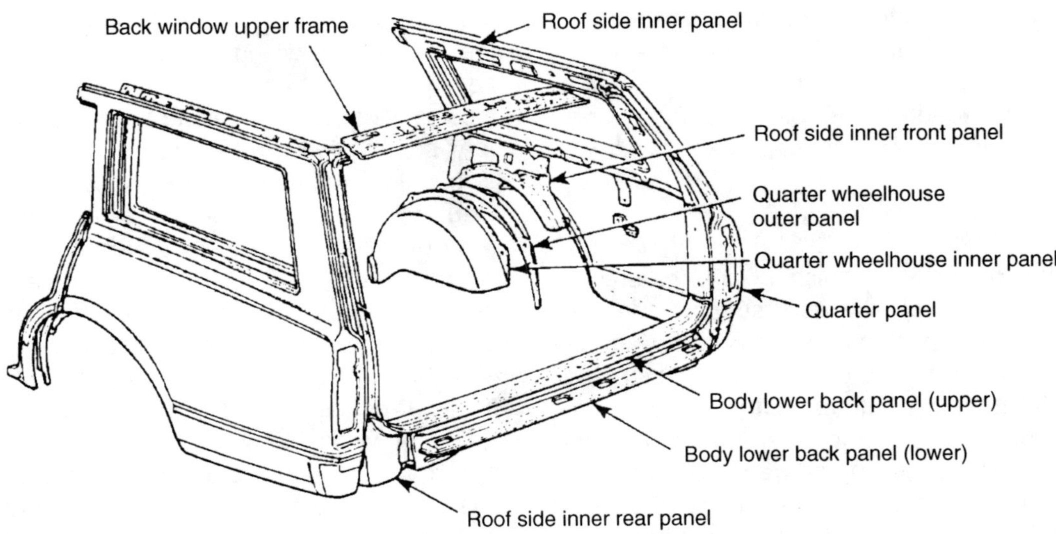

FIGURE 2–25 Rear body structural components of a unitized FR station wagon.

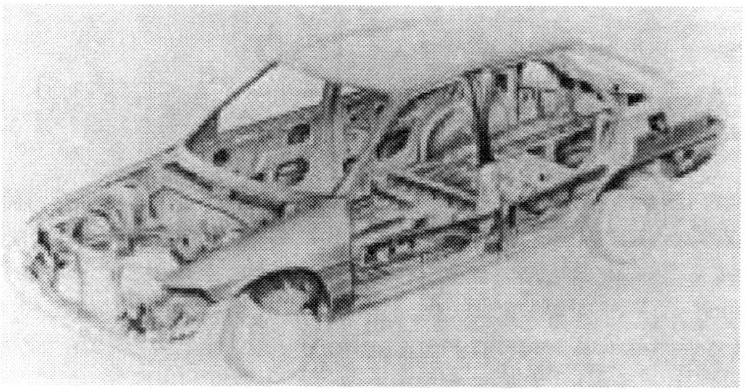

FIGURE 2–26 The main body or shell components of a unitized FF vehicle.

FF vehicles are characterized by the following:

- The transmission and differential are combined, and the propeller shaft is eliminated, providing a substantial weight reduction.
- Overall noise and vibration are reduced because they are confined to the front of the vehicle.
- Since the engine and transmission are located in the front, the load on the front suspension and tires is increased.
- The interior of the vehicle is larger because there is no need for a propeller shaft or rear drive axle.
- Since the fuel tank can be placed under the center of the vehicle, the luggage compartment can be large and flat.
- Because of the location of the engine, there is a greater forward inertial weight in a head-on collision. Therefore, engine mounting components are reinforced accordingly.

In front engine/front drive (FF) vehicles the engine is mounted transversely. The engine is supported at four points, the front and rear of the engine mounting center member, which is positioned longitudinally through the vehicle's center, and at the left and right front side members (Figure 2–27). In the same applications, the drive train is mounted in an engine cradle or subframe. The cradle resembles a conventional frame. It is bolted to the radiator support and cowl.

FRONT BODY

The front body components, which include the engine hood, front fenders, radiator upper support, radiator side support, front cross member, front side members, front fender apron, and dash panel are stamped from thin sheet metal. A high-strength structure consisting of reinforced side members and motor mounts capable of supporting the engine transaxle and suspension loads is used in the front section of the vehicle. Lightweight single structure plastic bumpers are used on FF and FR models.

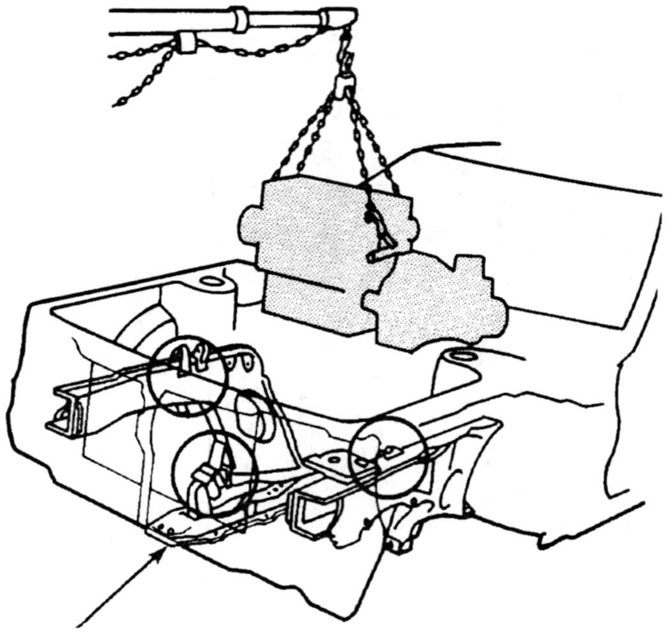

FIGURE 2–27 Position of FF transversely mounted engine supports.

The front suspension in FF and FR vehicles is almost identical. Both vehicles use an independent strut–type front suspension. The accuracy of the front body has a direct effect on front wheel alignment, and it is extremely important to check wheel alignment after performing front body repairs (Figure 2–28).

REAR BODY

The rear body section (Figure 2–29) consists of the back door panel, lower back panel, quarter panel, quarter wheelhouse outer panels, quarter wheelhouse inner panels, rear floor pan, and the rear floor side members. Since the entire power train is located in the front of FF vehicles, the fuel tank is located below the center floor, which allows the rear floor side member to be lower than in FR vehicles. The lower part of the rear floor side members is then connected to the rear suspension arm. An independent strut suspension is used to improve handling performance and driving stability; therefore, a rear-end collision has a greater influence on rear wheel alignment than it would in an FR vehicle. As a consequence, rear wheel alignment should be checked whenever repairs are performed on a rear body section.

In the case of an FR vehicle, the front of the rear floor pan is joined to the end of the center floor pan with spot welds. However, in an FF vehicle, the center and rear floor pan are joined and reinforced with an interlocking structure. The rear body of an FF is similar to that of an FR vehicle and no further description is necessary (Figure 2–30).

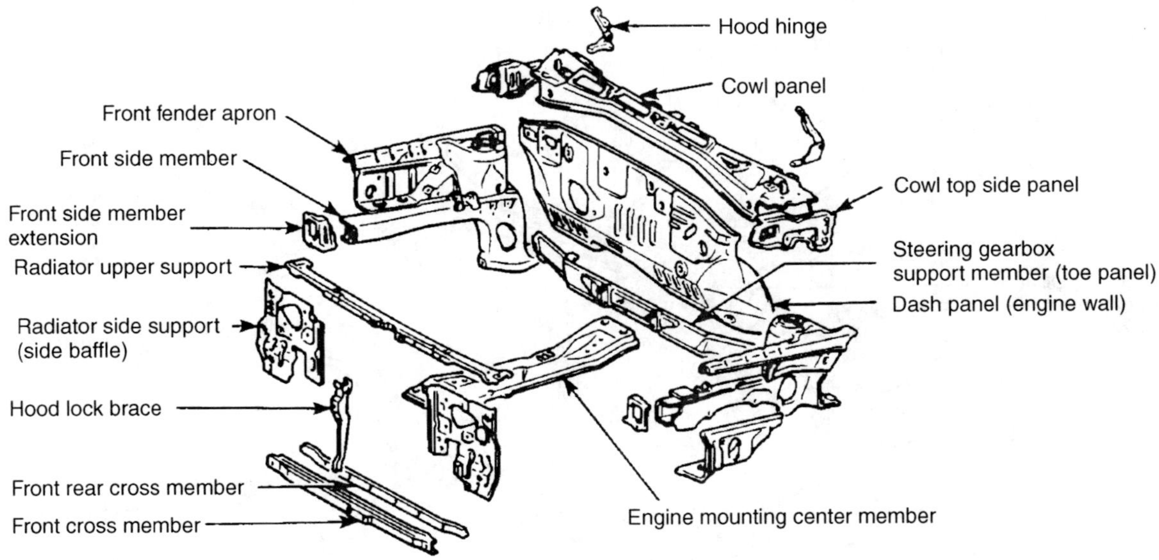

FIGURE 2–28 Front body structural components for a typical transversely mounted engine of an FF vehicle.

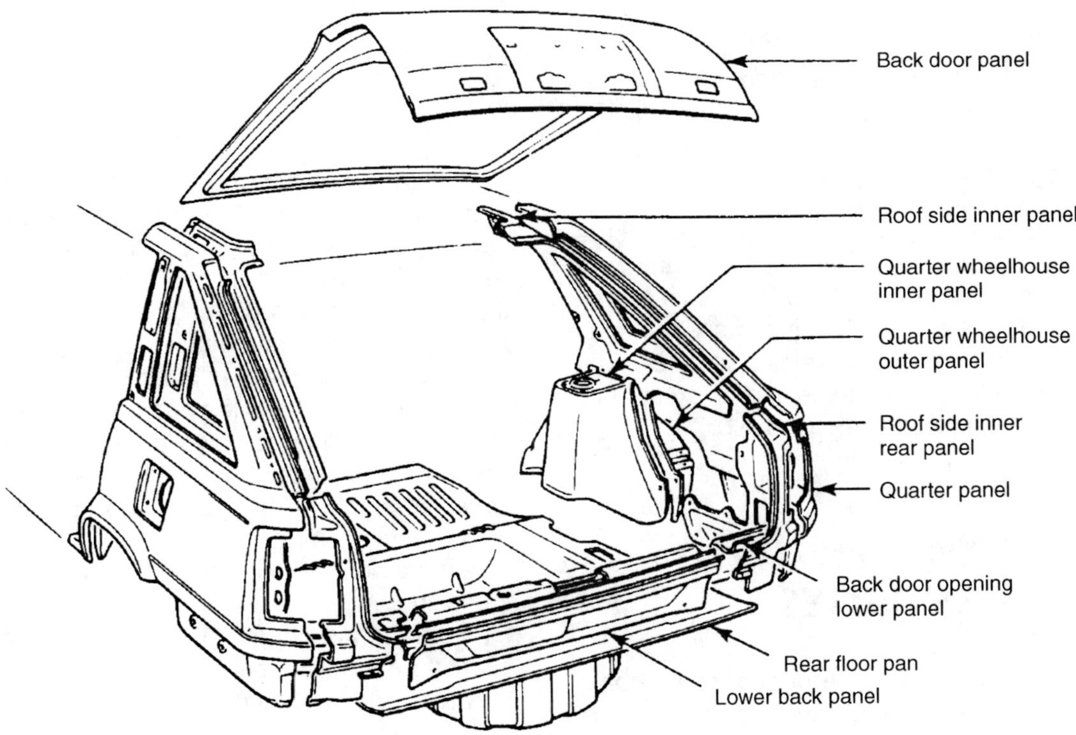

FIGURE 2–29 Rear body structural components of a typical FF vehicle.

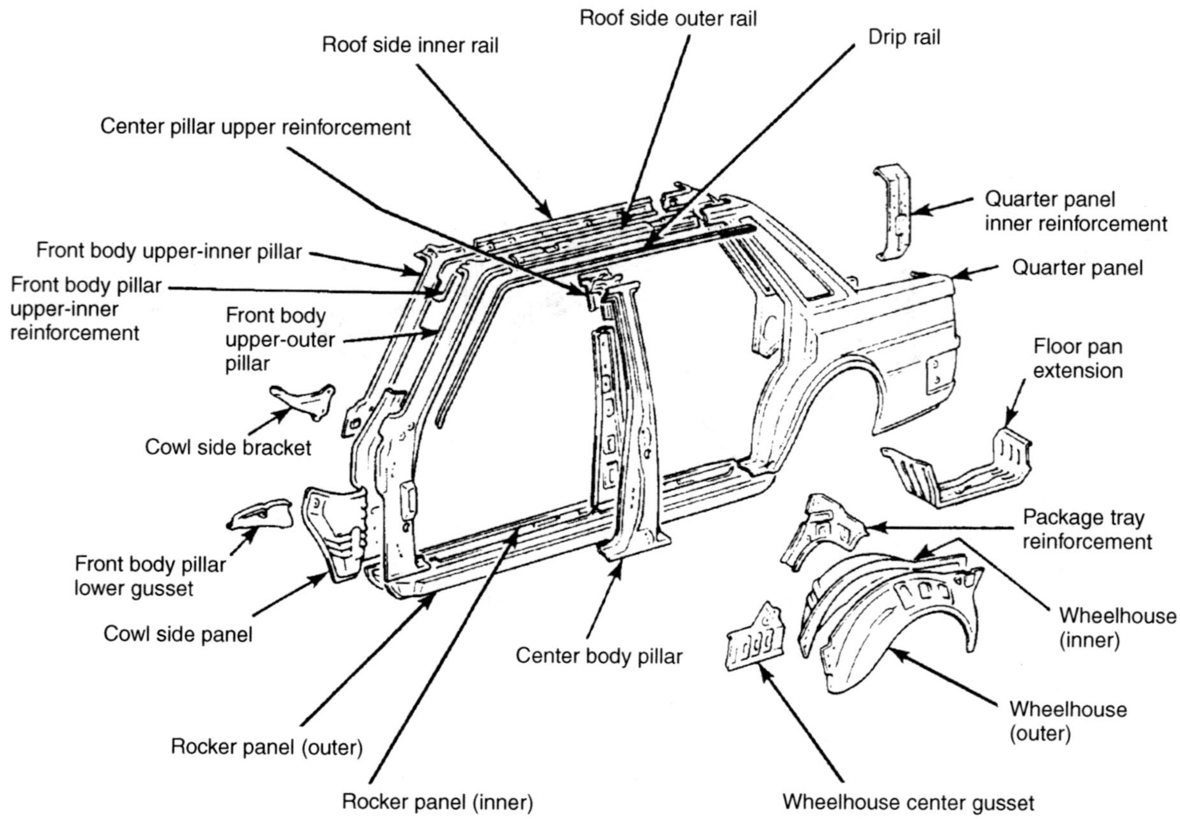

FIGURE 2–30 Side body structural components of a typical FF vehicle.

SPACE FRAME

Space frame vehicle are unibody in design (Figure 2–31). They consist of a lightweight, welded framework covered with a bonded or bolted skin. The skin can be made of plastic or metal. A Saturn has a space frame. The roof, hood, and deck lid are steel; all other outer panels are plastic. A Lumina van is similar except that all panels are sheet molded compound (SMC).

2.7 COMPONENT ASSEMBLIES

As shown in the illustrations throughout this book, many of the major body structures and sections explained on the previous pages are further divided or broken down into smaller units called assemblies. These assemblies are, in turn, divided into components or parts (Figures 2–32 and 2–33).

Both the shop manuals provided by auto manufacturers and the collision estimating guides used throughout the collision estimating and repair industry use this "component assembly" technique in their illustrations and parts list breakdowns.

It is a technique that should be used by every estimator and appraiser. The modern vehicle has thousands of component parts. Preparing an accurate damage report without a systematic approach to these vast numbers of individual parts is next to impossible. The estimator must break the vehicles down into more manageable "bites": the major sections must be broken down to component assemblies, and they must be broken down to individual parts. Consider damage report writing and analysis as writing a series of smaller damage reports on the subassemblies that make up the entire vehicle.

A more detailed breakdown of typical assemblies is found in chapters 7 through 9 of this book. The remainder of this chapter will cover the various construction materials used in today's vehicles and how they affect repair methods.

2.8 AUTOMOTIVE METALS

The sheet metal used in modern vehicles is made of either low-carbon (mild) steels or special alloy high-strength steels. All types of steels are manufactured

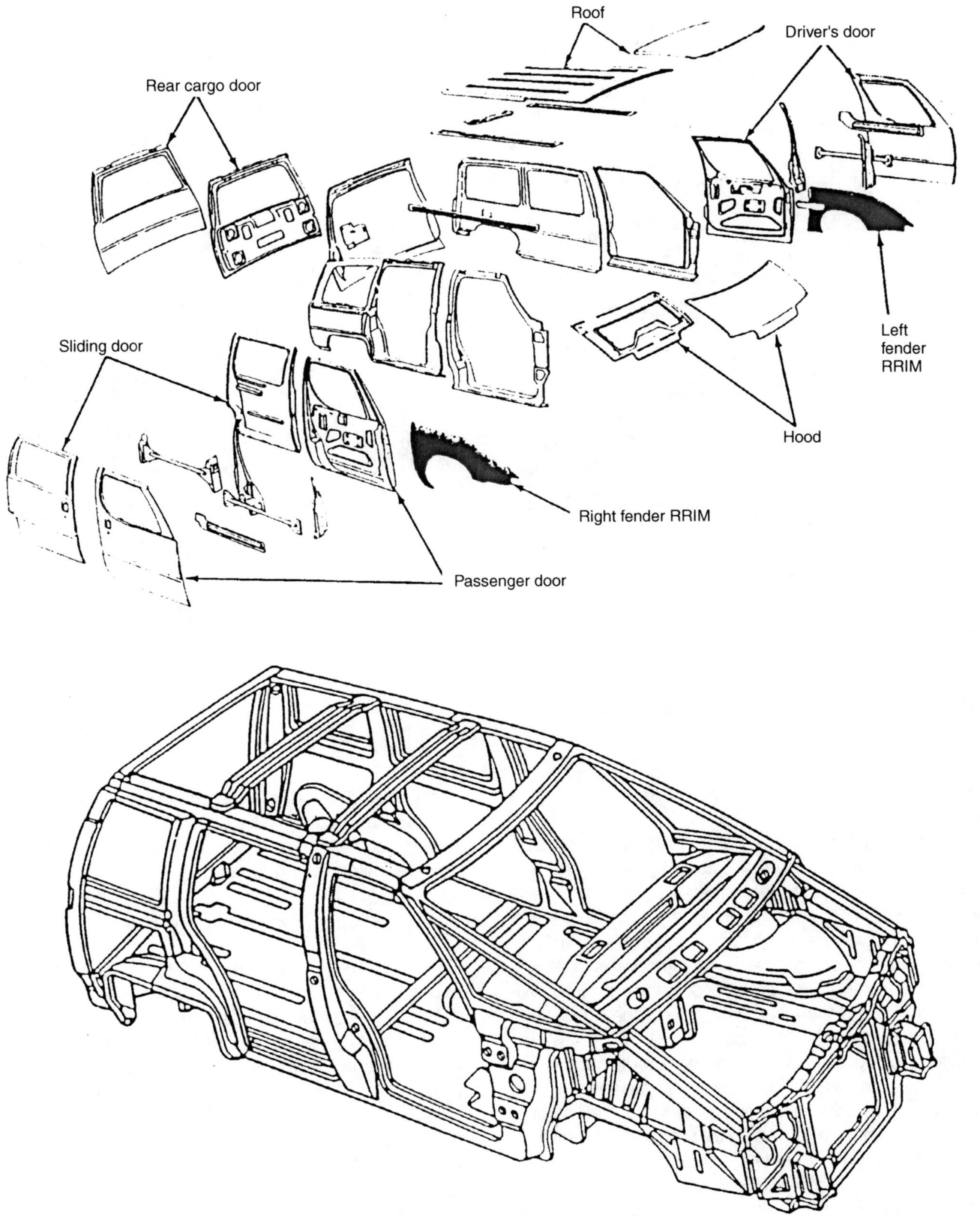

FIGURE 2–31 Space frame of a Lumina van.

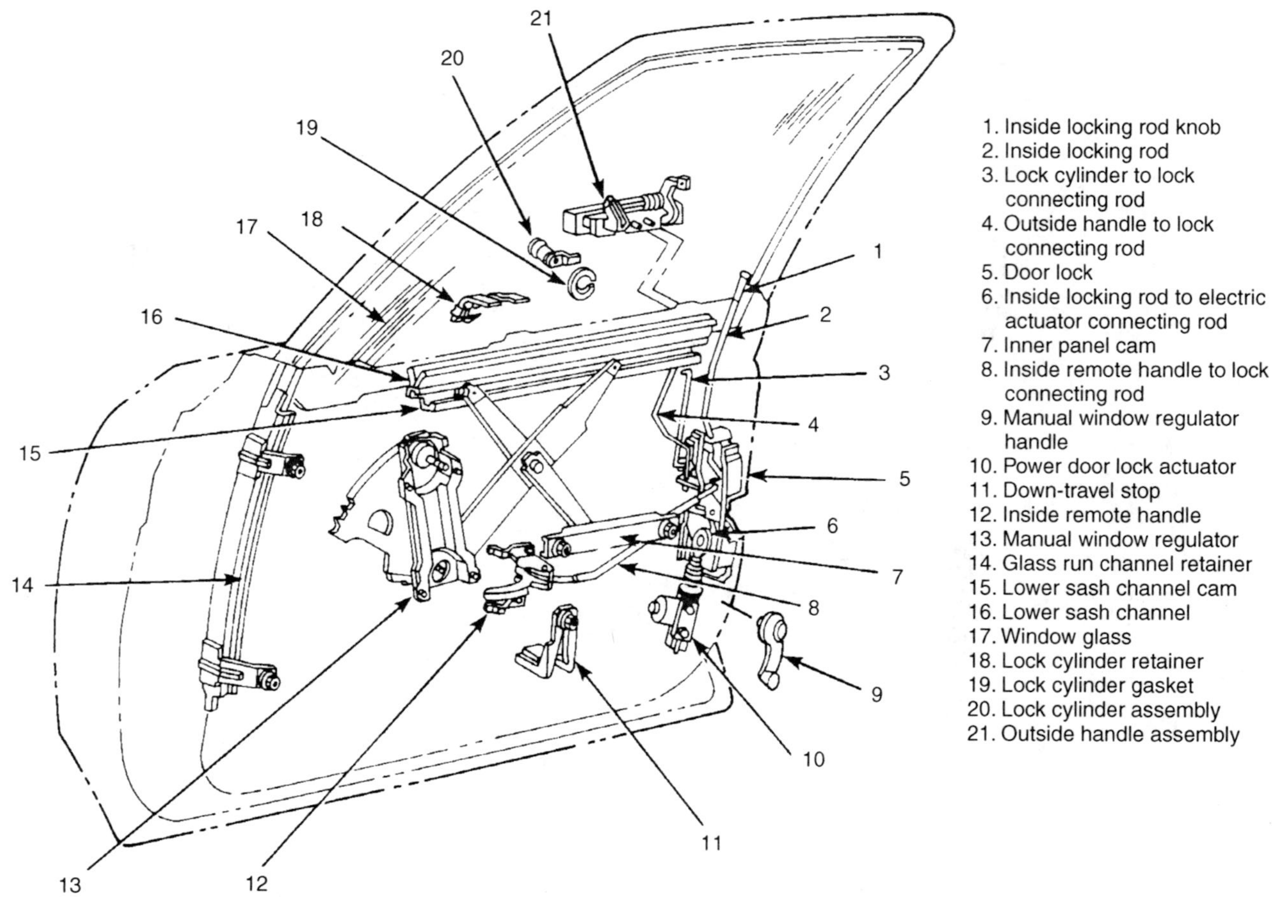

FIGURE 2–32 *Internal parts of a manual door.*

using one of two production methods: hot rolling or cold rolling.

Hot-Rolled Steels. Hot-rolled sheet metal is made by rolling at temperatures exceeding 1475 degrees Fahrenheit. It has a standard manufacturing thickness range of one-sixteenth to five-sixteenths inches, and it is used for comparatively thick components such as frame rails and cross members. Since it is rolled at high temperatures, the surface develops a black oxidized film. The workability of hot-rolled sheet metals is poorer than cold-rolled steels.

Cold-Rolled Steels. Cold-rolled sheet metal is actually hot-rolled steel that has been passed through an acid rinse, cold rolled to the desired dimensional thickness, and then annealed to improve workability.

Because the metal is worked by cold rolling, it has a dependable thickness accuracy, surface quality, and good press workability. Most unibodies are made from cold-rolled steels.

Corrosion Protection. In areas requiring corrosion resistance, such as the underbody sections, both hot- and cold-rolled sheet metals can be surface treated to resist rust. Galvanized sheet metal (plated with zinc) is used in large volume because it is economical and offers high corrosion resistance. Zinc-plated sheet metals give off small amounts of fumes during the welding process, and adequate safeguards must be taken during repairs.

LOW-CARBON (MILD) STEELS

For many years low-carbon or mild steel was the overwhelming choice for automotive sheet metal fabrication. Low-carbon steel is relatively soft and easy to work. It can be safely welded, heat shrunk, and cold worked without seriously affecting its strength.

However, low-carbon steels are relatively heavy and are also easily deformed. For these reasons, use of low-carbon steels has declined significantly, particularly in load-carrying vehicle components.

ALUMINUM

Vehicle weight reduction and resultant gas mileage increase is an important design goal. Lightweight aluminum hoods have been installed on the Ford F150, Buick Riviera, and Mazda RX7. The Acura NSX has an all-aluminum unibody. This means that the floor,

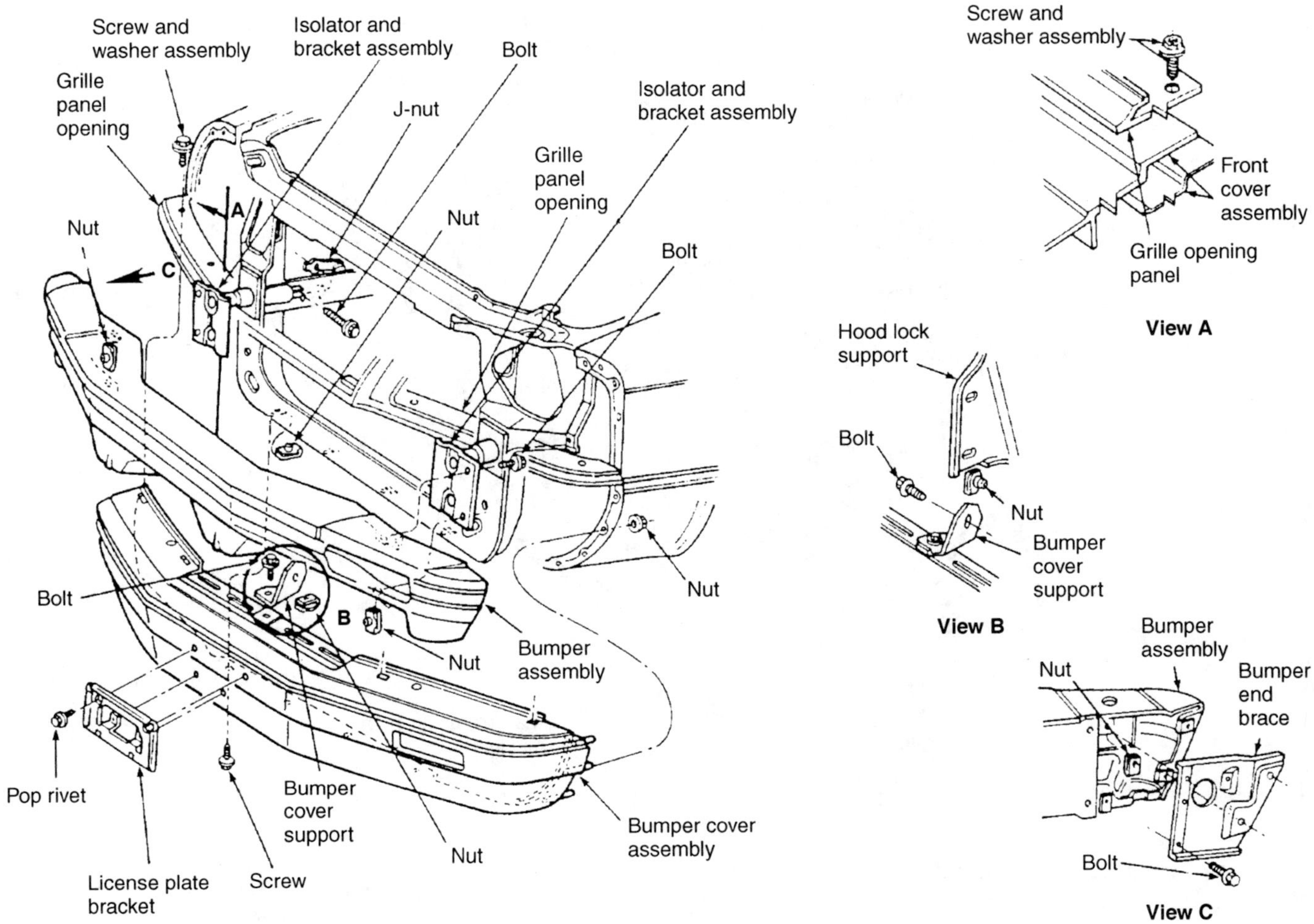

FIGURE 2–33 Front bumper and isolators.

rocker panels, cowl, roof, quarter panels, and frame rails are all made of aluminum. It is expected that the use of aluminum components, even entire unibodies, will increase in the future.

HIGH-STRENGTH STEELS

Beginning in the mid-1970s automakers faced the problem of reducing vehicle weight while retaining or actually increasing the strength of the vehicle. The resulting unitized vehicles produced were usually smaller than those they replaced. Front structures were designed to handle potentially far greater load-carrying and energy absorption requirements. Higher-strength steel was used to solve both problems.

Today's unitized construction means that a component such as an apron or rail assembly, which is no longer just a splash shield, has to not only be lightweight but also carry a portion of the suspension reaction load and support a transversely mounted engine, battery, ignition module, and bumper system. It is also designed with buckling sections to limit damage transmitted to the passenger compartment. In order to successfully achieve these objectives, many manufacturers made this new apron/rail assembly of stronger, lighter weight, high-strength steel. Figure 2–34 shows some typical body parts where high-strength steels are used.

However, these same properties that contribute strength and light weight offer some unique challenges to the repair industry. When high-strength steel is deformed on impact, it is more difficult to dimensional restore than mild steel. Normal repair procedures, which dictate using heat to either aid in relieving the stresses of pulling or welding in a new part or sectional clip, must be carefully performed to avoid structural damage to the high-strength steel components.

When calculating repair costs, the types of steel used in the construction of the damaged components must be positively identified. Characteristics of the steel affect the types of repair procedures that can be used, a fact that has a direct bearing on the number of labor hours involved in the repair. Although many types of steels are generally classified as high-strength steel, three major types are most often found

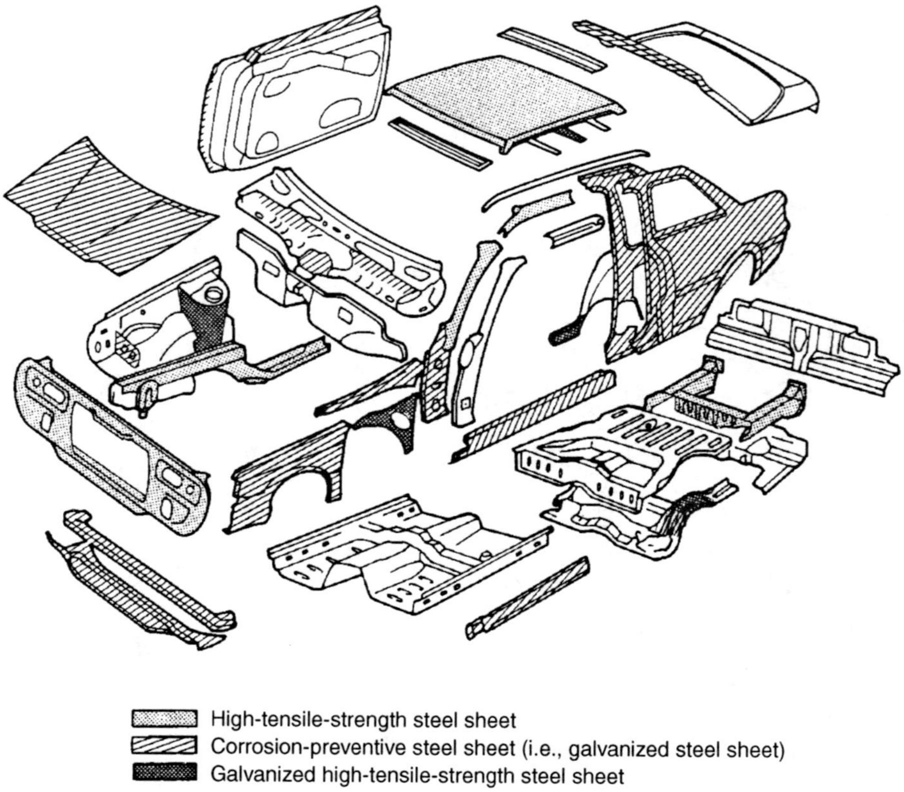

FIGURE 2–34 Typical areas where high-strength steel is used on unibody vehicles.

in automotive structures. But before explaining the differences between these three types, it is important to understand the definition of steel strength.

There are two types of steel strength, and they both relate to the steel's ability to resist permanent deformation.

- *Yield Strength.* Yield strength, or yield stress, is measured as the minimum force per unit of area that causes the steel to permanently change its shape.
- *Tensile Strength.* Tensile strength, or tensile stress, is measured as the maximum force, per unit of area, that causes a complete fracture or break in the material.

Traditionally, high-strength steels with high yield and tensile strengths offered poor workability and inferior welding strength when compared to low-carbon steels. But recently high-strength thin sheet metal with better formability and weldability has been developed. These sheet metals are usually divided and classified according to the strengthening process used during manufacturing.

High-Strength, Low-Alloy (HSLA) Steels. HSLA steels are used in many domestic cars for body structural components such as front and rear rails, rocker panels, bumper face bars, bumper reinforcements, door hinges and lock pillars. Their strength is mainly due to the addition of special chemical elements.

CAUTION: Use extreme care when using heat to relieve stress on HSLA panels. After exposure to a temperature of 1200 degrees Fahrenheit or higher, for several minutes, the special hardening elements are absorbed by larger, softer elements in the heated area, resulting in lower strength.

To avoid substantially reducing the ability of the structure to react to normal road loads or collision forces, never exceed the factory-recommended temperatures. A safe rule of thumb is to never heat over 700 to 900 degrees Fahrenheit and do not apply heat longer than three minutes. MIG welding, using AWS-E-70S-6 wire, is an acceptable practice. Automobile manufacturers do not recommend using arc welding or oxyacetylene to weld HSLA.

High–Tensile Strength Steels (HSS). High–tensile strength steel (HSS) is stronger than low-carbon or mild steel because of heat treatment. HSS has a yield strength of up to 35,000 psi and a tensile strength of more than 45,000 psi. Most Japanese import cars contain HSS in body structural components. Conventional heating methods will not adversely affect the strength level of HSS. During a collision, HSS will experience an increase in stress, exceeding the yield strength, as it is deformed during a collision. When heat is applied to an HSS component to assist in straightening, the stresses resulting from collision are decreased, thereby restoring the strength to a lower or normal level.

Normal stress-relieving practices, including oxyacetylene, can be used to aid in restoring HSS components. However, extreme caution must be exercised when using oxyacetylene. Temperature indicating methods should be applied around the area to be heated with an oxyacetylene torch to restrict these temperatures to 1200 degrees Fahrenheit. Door reinforcement guards and some bumper reinforcements cannot be straightened and should be replaced.

All new or used replacement parts should be MIG welded using AWS-E-70S-6 wire, which has the same strength level of HSS.

Martensitic Steels. Martensitic steel is alloy-free, with tensile strength almost ten times that of typical mild steel. All door reinforcement guards and some bumper reinforcements are martensitic, or ultra high-strength steels. In these steels, the unusually high-strength properties are a result of a special grain or crystalline composition imparted during forming and fabricating. Any reheating for repair destroys this unique composition and reduces strength to that typical of mild steel. In addition, these steels are so hard that they cannot be straightened cold with typical shop equipment. Therefore, a damaged martensitic, or ultra high-strength, steel member must not be repaired; it must be replaced. New parts should be installed by MIG plug welding using AWS-70S-6 welding wire.

2.9 AUTOMOTIVE PLASTICS

In recent years, more and more plastic has been used in various parts of car bodies, particularly in the front end: in bumper and fender extensions, in soft front fascia, fenders, hoods, grille panels, stone shields, instrument panels, trim panels, and elsewhere (Figure 2–35). Because these parts are much lighter in weight than sheet metal, they have become an important part of every American manufacturer's fuel saving, weight reduction program. And because of the high strength-to-weight ratio of plastic, the weight decrease does not mean a decrease in strength. Plastic body parts are here to stay, and new applications for plastic components are developed for each model year. General Motors is the leader among domestic manufacturers in finding new applications for plastics. The Corvette model began in 1953 with a fiberglass body. Starting in 1984, SMC was used instead of fiberglass. Saturns have plastic body panels bolted to a steel space frame. The Lumina vehicle (APV) (Figure 2–36) has an all-plastic exterior that is glued to a steel space frame.

This increased use of plastic has resulted in new approaches to collision damage repair and cost estimating. Many damaged plastic body parts can be repaired more economically than they can be replaced, especially if the part does not have to be removed from the vehicle. Cuts and cracks, gouges, tears, and punctures are all repairable; and, when necessary, some plastics can also be re-formed after distortion from their original shapes. Opting for repair over replacement is often quicker, since plastic replacement parts are not always readily available.

PLASTIC TYPES

There are two basic types of plastics used today in automotive production.

Thermoplastics. These plastics are capable of being repeatedly softened and hardened by heating and cooling. They soften or melt when you apply heat and, therefore, are weldable.

Thermosetting Plastics. These plastics are materials that undergo a chemical change by the action of heating, a catalyst, or ultraviolet light leading to an infusible state. Catalyst and resin are mixed to form a new product. Thermosets are not weldable.

Figure 2–37 explains the relationship of heat to plastics more fully. In general the repair techniques are chemical adhesive bonding for thermosetting plastics and welding for thermoplastics.

PLASTIC IDENTIFICATION

As with automotive sheet metals, the first step in determining accurate plastic repair costs is to identify the type of plastic damaged and the appropriate repair method. An improper repair based on improper identification can quickly delaminate, crack, or discolor—a mistake that can lead to increased labor, parts, and materials costs for the collision repair facility.

Many plastic components are stamped with standardized identification codes, which are becoming increasingly popular with domestic automakers. Unfortunately, a part often must be removed from the vehicle before the code can be found. Many manufacturers also list the types of plastics used in their cars in the shop manual or technical literature supplied to repair shops. Table 2–1 lists the standardized symbols along with the most common usages for the various types of plastics in use today.

REPAIR/REPLACE DECISIONS

Estimating plastic repair costs is based on repair/replace decisions. Repair to the existing component is usually the option when the problem is a minor crack, tear, gouge, or hole. Repair is often the answer when

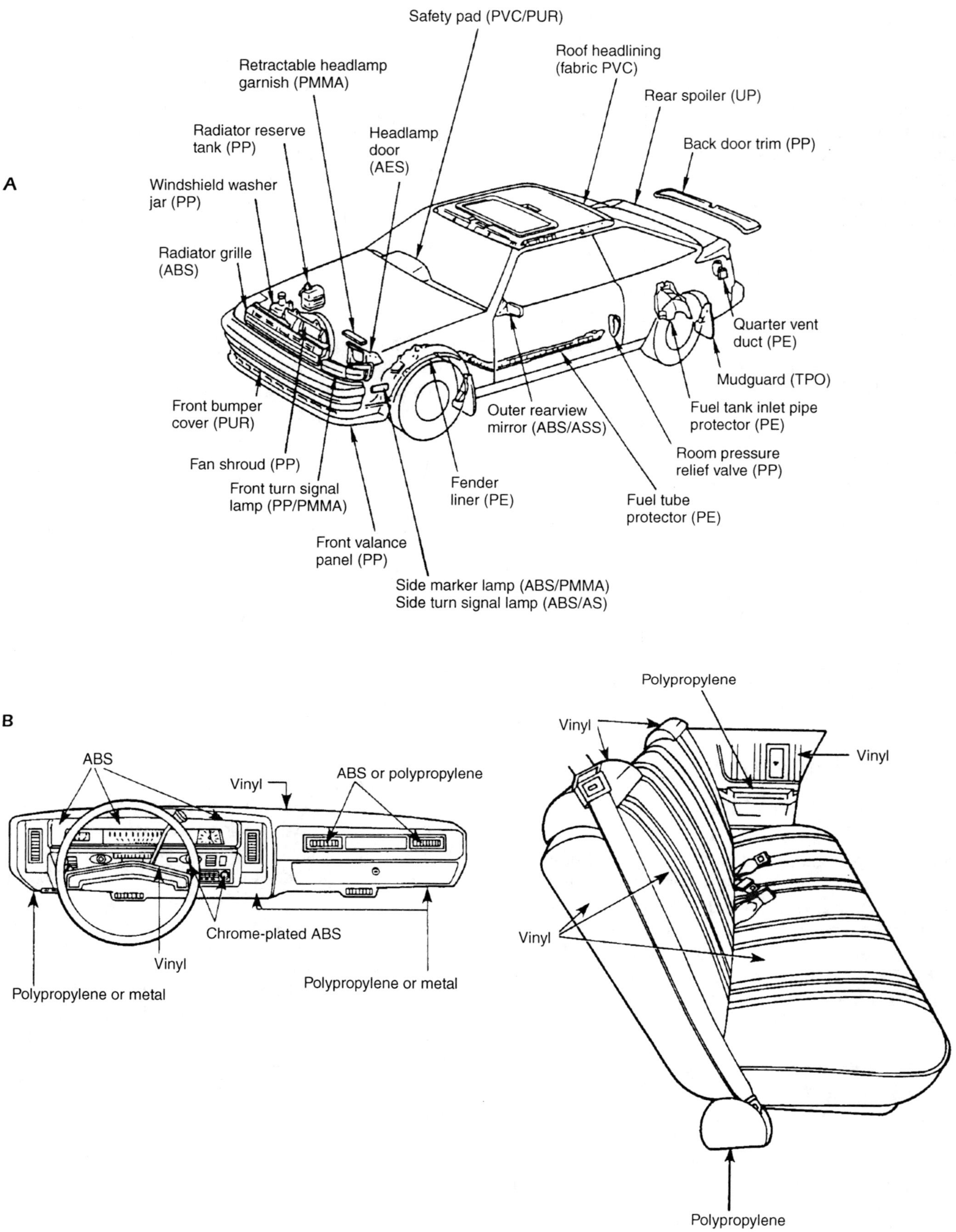

FIGURE 2–35 *Areas where plastics are used in vehicles: (A) exterior and (B) interior.*

FIGURE 2–36 The Lumina van has plastic exterior panels glued to a steel space frame.

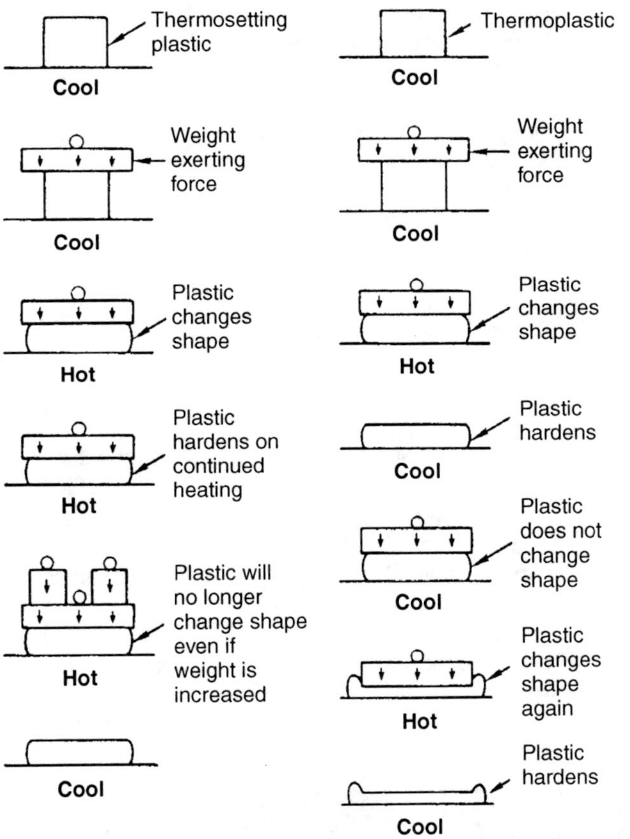

FIGURE 2–37 The effects of heat on automotive plastic.

the component is difficult to remove from the vehicle or is difficult to locate as a replacement part.

Repairs can also be made to components removed from the vehicle. This method can be less expensive than installing a new part, but labor for removal, replacement, and refinishing (if needed) must not be omitted.

Replacement with a new or like component is more likely to be the choice when the component is extensively damaged, is easy to remove and replace, or is inexpensive and readily available.

More information on estimating plastics repair using plastic welding and chemical adhesive techniques can be found in chapter 4.

2.10 AUTOMOTIVE SEALERS AND ADHESIVES

Sealers and adhesives are playing an increasingly important role in the construction and repair of modern vehicles (Figure 2–38). Body panel joints and seams require special sealing protection because they are areas highly vulnerable to corrosion. Welding at the joints increases the metals' susceptibility to corrosion. Water, snow, dirt, and road salts also have a tendency to collect at joint areas. As a general rule, a body sealant must be applied over all joints to eliminate gaps between the material and the panel surface and to increase corrosion protection.

Sealers are also used to help dampen vibration and noise transmission throughout the vehicle. Sealants must be paintable and have good adhesion to bare and primed metal. Sealers must also be flexible to withstand the motion associated with operational forces. This is particularly important in unibody construction.

There are four types of seam sealers that are commonly used for auto corrosion protection and dampening applications.

- *Thin-bodied sealers* are designed to fill seams under one-eighth inch wide. This sealer will shrink slightly to provide definition to the joint, while remaining flexible to resist vibration. Adhesion is good to both primed metal and bare metal surfaces. Since many of the seams are on a vertical surface, sag control is important so the sealer does not run out of the seam.
- *Heavy-bodied sealers* are used to fill seams from one-eighth to one-quarter inch wide. These sealers can be tooled to hide the seam or can be left in bead form. Shrinkage should be minimal, with good resistance to sagging, and high flexibility to resist cracking in service. Heavy-bodied sealants are used on both coach joints and overlap seams. They are typically dispensed from cartridges (Figure 2–39). Some products are available in squeeze tubes.
- *Brushable seam sealers* are used on interior body seams where appearance is not important. These seams are normally hidden and not seen by the customer, unless under close examination. Brushable sealers are designed to

TABLE 2–1: STANDARD SYMBOL, CHEMICAL NAME, TRADE NAME, AND DESIGN APPLICATIONS OF MOST COMMONLY USED PLASTICS

Symbol	Chemical Name	Common Name	Design Applications	Thermosetting or Thermoplastic
ABS	Acrylonitrile-butadiene-styrene	ABS, Cycolac, Abson, Kralastic, Lustran, Absafil, Dylel	Body panels, dash panels, grilles, headlamp doors	Thermoplastic
ABS/MAT	Hard ABS reinforced with fiberglass	—	Body panels	Thermosetting
ABS/PVC	ABS/Polyvinyl chloride	ABS Vinyl	—	Thermoplastic
EP	Epoxy	Epon, EPO, Epotuf, Araldite	Fiberglass body panels	Thermosetting
EPDM	Ethylene-propylene-diene-monomer	EPDM, Nordel	Bumper impact strips, body panels	Thermosetting
PA	Polyamide	Nylon, Capron, Zytel, Rilsan, Minlon, Vydyne	Exterior finish trim panels	Thermosetting
PC	Polycarbonate	Lexan, Merlon	Grilles, instrument panels, lenses	Thermoplastic
PE	Polyethylene	Dylan, Fortiflex, Marlex, Alathon, Hi-fax, Hosalen, Paxon	Inner fender panels, interior trim panels, valances, spoilers	Thermoplastic
PP	Polypropylene	Profax, Olefo, Marlex, Olemer, Aydel, Dypro	Interior mouldings, interior trim panels, inner fenders, radiator shrouds, dash panels, bumper covers	Thermoplastic
PPO	Polyphenylene oxide	Noryl, Olefo	Chromed plastic parts, grilles,	Thermosetting
PS	Polystyrene	Lustrex, Dylene, Styron, Fostacryl, Duraton	headlamp doors, bezels, ornaments	
PUR	Polyurethane	Castethane, Bayflex	Bumper covers, front and rear body panels, filler panels	Thermosetting
PVC	Polyvinyl chloride	Geon, Vinylete, Pliovic	Interior trim, soft filler panels	Thermoplastic
RIM	"Reaction injection molded" polyurethane	—	Bumper covers	Thermosetting
R RIM	Reinforced RIM-polyurethane	—	Exterior body panels	Thermosetting Thermosetting
SAN	Styrene-acrylonitrite	Lustran, Tyril, Fostacryl	Interior trim panels	Thermosetting
TPO	Thermoplastic olefin	TPO	Filler panels, mud guards	Thermoplastic
TPR	Thermoplastic rubber	—	Valance panels	Thermosetting
TPUR	Polyurethane	Pellethane, Estane, Roylar, Texin	Bumper covers, gravel deflectors, filler panels, soft bezels	Thermoplastic Thermosetting
UP	Polyester	SMC, Premi-glas, Selection Vibrin-mat	Fiberglass body panels	Thermosetting

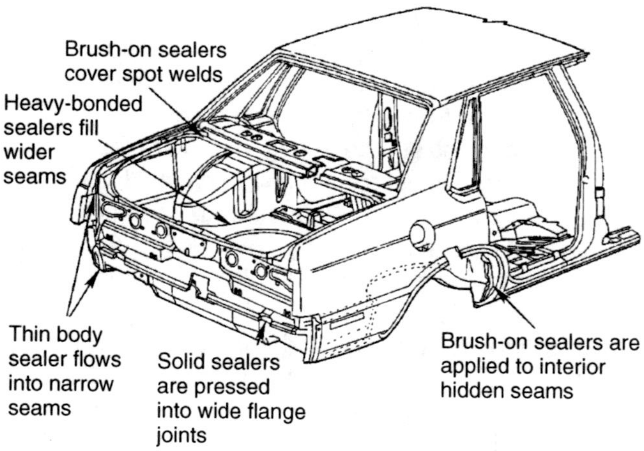

FIGURE 2–38 The various sealer types and a sample of typical application locations.

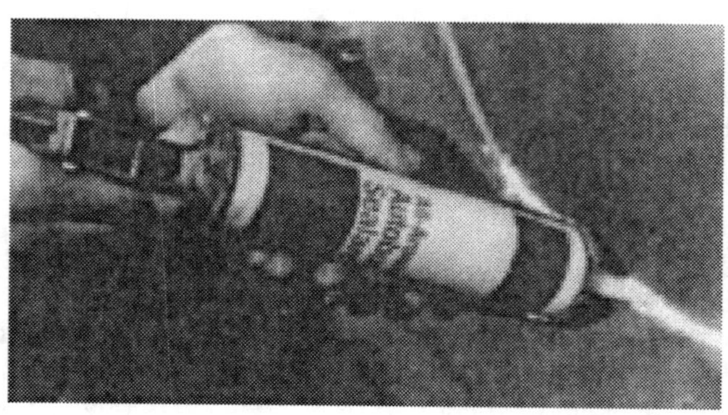

FIGURE 2–39 Applying a heavy-bodied sealer.

hold brush marks and to resist salt and automotive fluids such as gasoline, transmission fluid, and brake fluid. Any seams such as those under the hood and under the carriage that might be exposed to automotive fluids should have a brushable seam sealer. Applied with a brush, it normally has overlap seams (Figure 2–40).
- *Solid seam sealers* containing 100 percent solids are used to fill larger voids at panel joints or holes. This product comes in strip caulking form, designed to be pressed into place with the thumb.

It is important to remember that the flat-rate labor times quoted in estimating guides normally do not include labor time for applying new sealing materials or removing old sealers so repairs can be made.

FIGURE 2–40 Applying a brush-on sealer.

2.11 AUTOMOTIVE GLASS

Modern vehicles are built using more glass area than in the recent past. Use of glass affords greater visibility for driving and sightseeing. Auto glass also plays an important role in modern aerodynamic styling. The newest glass classification, modular or encapsulated glass, drastically reduces air drag by fitting the glass accurately to the sheet metal and eliminating metal trim. It also reduces wind noise and provides a sleek, uncluttered appearance.

Most of the glass built into the vehicle is centered in the front windshield, the doors, and the rear window (Figure 2–41). Obviously, glass damaged by collision, flying gravel, or vandalism must be replaced. But undamaged glass must also be removed from areas of major damage before pulling and straightening work can begin. Cleanup of broken glass from the vehicle's interior is not always a fast, simple task, and these labor times must be considered in an accurate damage report.

The glass used in modern vehicles is one of two types: laminated and tempered. Both are considered safety glass because of special construction characteristics.

- *Laminated glass* is used to make all windshields. It consists of two thin sheets of glass with a thin layer of clear plastic between them. When this

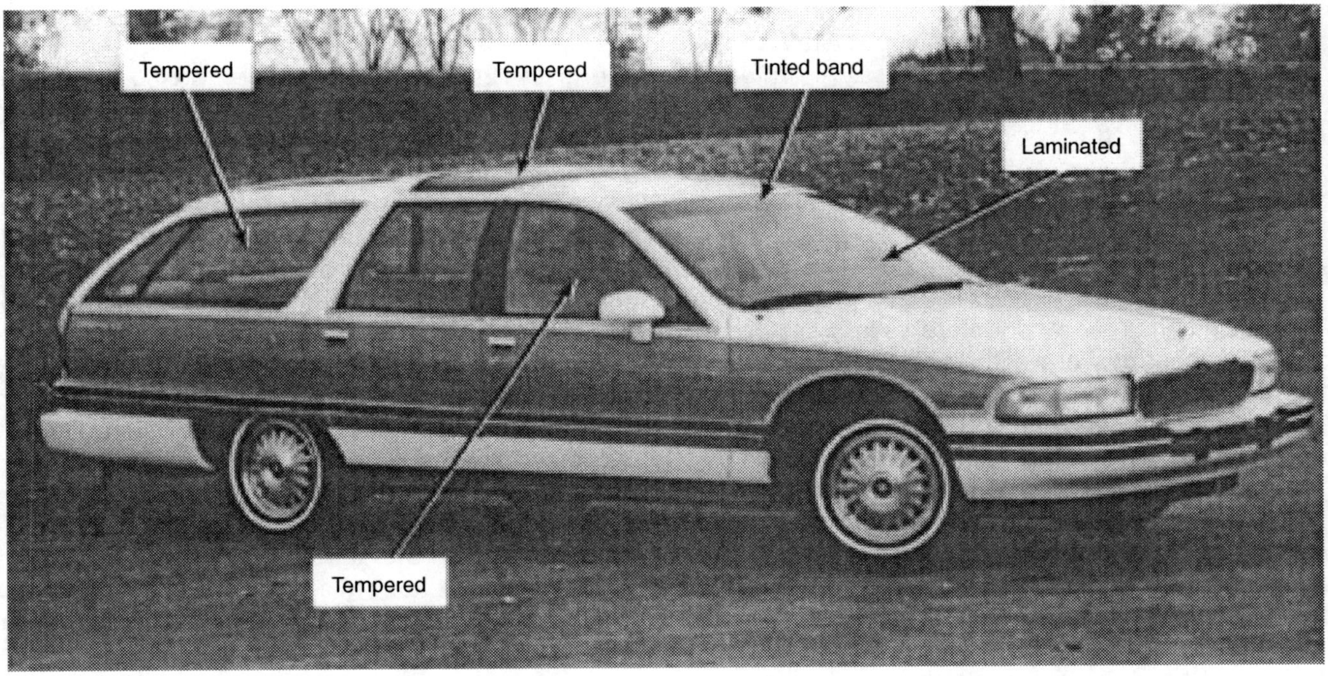

FIGURE 2–41 Types of automotive glass.

type of glass is broken, the plastic material will tend to hold the shattered glass in place and prevent it from causing injury.

Antilacerative glass is similar to conventional laminated glass, but it has one or more additional layers of plastic affixed to the passenger compartment side of the glass. This glass is used in the front windshield only and is added protection against shattering and cuts during impact.

- *Tempered glass* is used for side and rear window glass but rarely for windshields. It is a single piece of heat-treated glass and has more resistance to impact than regular glass of the same thickness. When tempered glass is broken, the pieces of glass are small and have a granular texture. The shattered glass has an interlocked structure to it that obstructs visibility. This is one reason why tempered glass is not used in windshields. Another reason it is not used is that this type of glass does not readily give and would, therefore, cause more severe head injuries in a collision.

Tinted glass is a special type of either laminated or tempered glass. Tinted laminated glass contains a light green sheet of vinyl material to filter out most of the sun's glare. Glass can also be tinted by adding minute quantities of metal powder to the other normal ingredients of glass to give it a particular color. The addition of cobalt gives the glass a blue tint. Iron gives the glass a reddish tint.

Glass can also be fitted with a defrost circuit or antenna. Defrost glass is common in rear windows. Before heat treatment, metal powder, which conducts electricity, is printed on the glass surface in the form of heating wires. The metal powder is baked on the surface during the tempering process. An antenna wire for radio reception may be placed between the layers of laminated glass (windshield) or printed on the surface of the glass (rear window). Some windows have antenna wires and heating wires side by side.

GLASS INSTALLATION

Windshields and rear windows are usually secured in place by rubber gasket weatherstripping or by an adhesive. Generally, mouldings are used on the interior and exterior of the body around the glass opening. Interior mouldings are called *garnish mouldings* and exterior mouldings are called *reveal mouldings.*

The gasket installation, more predominate in older vehicles, still finds use in present-day vehicles. In one design, the gasket is grooved to accept the glass, the sheet metal pinch weld flange, and sometimes the exterior reveal moulding (Figure 2–42A). On other older designs, a locking strip fits into the groove between

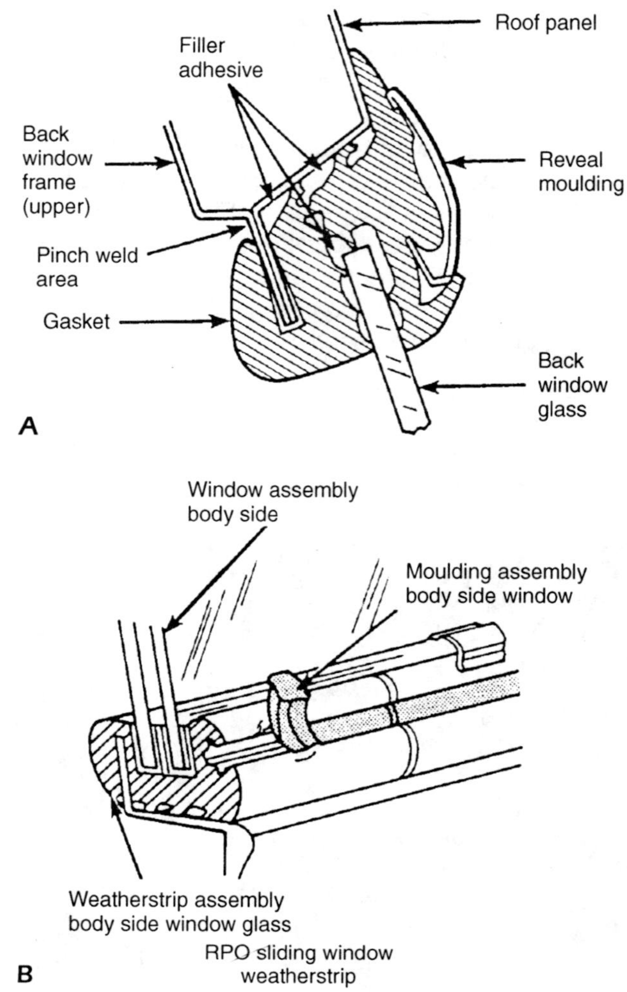

FIGURE 2–42 (A) Cross section of a typical rear window gasket installation and (B) an example of a locking strip gasketed installation.

the glass and pinch weld flange, forcing the gasket tightly against the glass and flange (Figure 2–42B).

The adhesive-type installation, as the name implies, uses an adhesive material to secure the glass in place in lieu of a gasket (Figure 2–43). The adhesive bond has virtually replaced the gasket method of attachment for the windshield. The use of adhesive permits the windshield to be mounted flush with the roof panel, decreasing wind drag and noise. Adhesive-bonded windshields also increase the overall rigidity of the vehicle, minimizing body twist and helping to keep the glass in place during a collision. Rubber stops and spaces separate the glass from the metal. Reveal mouldings are held in place by adhesive grooves in the body or by clips. This type is more advantageous than the gasket installation because the pinch welds do not have to be as exact, since no pressure is exerted on the glass outside of the reveal mouldings. The adhesive can be extruded from a cartridge or it can be applied in strip or tape form. Typical

FIGURE 2–45 Two-piece modular glass units have one moulding applied at the factory. The second is snapped in place after installation.

FIGURE 2–43 Cross section of an adhesive-bonded windshield installation.

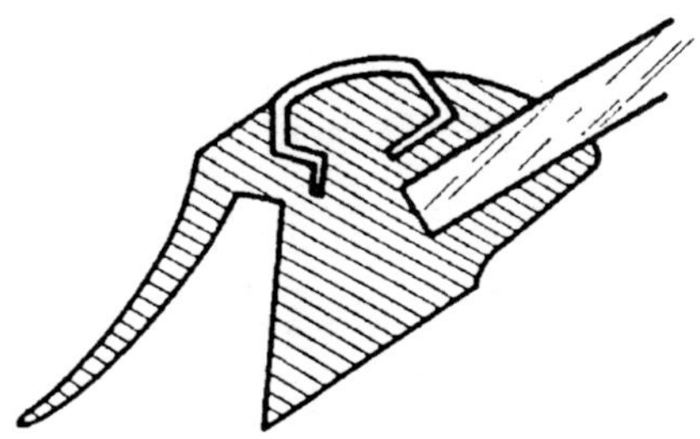

FIGURE 2–44 One-piece modular glass units come with a one-piece plastic moulding applied to the glass. No other moulding is applied after installation.

adhesives in use include polysulfide, urethane and butyl rubber. Note that glass originally installed with urethane adhesive must be reinstalled with urethane to maintain the structural integrity of the vehicle. This is especially important in vehicles equipped with air bags.

MODULAR GLASS

The new modular glass assemblies introduced on many late models fit the contours of the vehicle more closely. Modular glass also differs from ordinary auto glass in that it comes with a plastic trim moulding attached to the edge of the glass. There are some variations in moulding design. Some are one piece (Figure 2–44) while others are two piece (Figure 2–45). On two-piece designs, one moulding piece is glued to the glass at the time of manufacturing. The other moulding piece is snapped into place after the glass is installed, either at the assembly factory or in the repair shop. When the moulding is part of the glass assembly, it saves time and money on the factory assembly line but creates some problems for collision repair facility and glass installers.

Modular glass is not easy to work with. Its close fit makes removal of undamaged glass a tricky procedure requiring special tools for best results. In most cases glass shops do not guarantee removal without damage, although the success rate for removing undamaged side glass and windshields can be quite high if the proper tools and careful techniques are used. The possibility of breakage should be noted on the estimate. Agree beforehand on who will purchase new glass if the glass is broken during removal.

One type of special tool for this work is a vibrating power knife equipped with several blades to fit the various joint areas between the glass from the body. On several models, a string or wire is molded into the adhesive. When pulled, it cuts much of the adhesive, making removal easier.

REVIEW QUESTIONS

1. The two types of vehicle frames are full frame and __________.
2. Technician A says that many sport utility vehicles have full frames. Technician B says that cross members connect the side rails on a full-frame vehicle. Who is right?
   a. technician A
   b. technician B
   c. both A and B
   d. neither A nor B
3. Which of the following frame designs is no longer used in automobile manufacturing:
   a. perimeter
   b. stub
   c. hourglass
   d. ladder

4. Technician A says that unibody vehicles weigh more than full-frame vehicles. Technician B says that unibody vehicles get better gas mileage than full-frame vehicles. Who is right?
   a. technician A
   b. technician B
   c. both A and B
   d. neither A nor B
5. The two unibody designs are front engine/front drive and front engine/rear drive.
   True False
6. Technician A says that martensitic steel must be replaced if damaged. Technician B says that a full frame is made of one-fourth inch–thick steel. Who is right?
   a. technician A
   b. technician B
   c. both A and B
   d. neither A nor B
7. A unibody vehicle requires a more thorough damage analysis when compared to a similar impact on a full-frame vehicle.
   True False
8. Technician A says that if the yield strength is exceeded, permanent deformation results. Technician B always straightens damaged HSLA steel with tension and 1500 degree Fahrenheit heat. Who is right?
   a. technician A
   b. technician B
   c. both A and B
   d. neither A nor B
9. Aluminum is used in vehicles because it is
   a. stronger.
   b. lighter.
   c. cheaper.
   d. weaker.
10. Technician A welds PC plastic. Technician B welds RRIM plastic. Who is right?
    a. technician A
    b. technician B
    c. both A and B
    d. neither A nor B

Chapter 3

Collision Damage to Body and Frames

Objectives

After reading this chapter, you should be able to:

- **List the major damage conditions.**
- **List the six major types of body and frame collision damage in BOF and unibody vehicles.**
- **Explain the typical damage sequence in collisions involving BOF and unitized vehicles and the importance it plays in damage repair.**
- **Give two examples of possible inertia damage to vehicles.**
- **Explain the effects of elastic deformation, plastic deformation, and work hardening on sheet metal damage.**
- **State the difference between direct and indirect sheet metal damage.**

Key Terms

Sidesway
Sag
Mash
Diamond
Twist
Buckle
Elastic deformation
Plastic deformation
Work hardening
Direct damage
Indirect damage
Crown
Pressure area
Tension area

Chapter 2 covered the basics of modern body-over-frame (BOF) and unibody vehicle construction and reviewed the materials used to manufacture these vehicles. In Chapter 3, we will take a closer look at what happens to these frame and body structures during a collision. The differences in body-over-frame and unibody shell designs result in different types of collision damage, damage assessment techniques, and repair procedures.

Understanding the type(s) of damage a vehicle has sustained is vital to accurate cost estimating. This is particularly true when assessing moderate to major collision damage involving the repair of bent, crushed, or twisted body and/or frame members. These repairs usually involve significant amounts of straightening setup and pulling time that cannot be accurately estimated using collision manual information. Repair or replace decisions that can have a substantial impact on the final repair costs must be justified. Collision estimating guides list replacement time for structural parts at factory seams and, if allowed by the OEM, sectioning of structural parts. While flat-rated times can be used when estimating replacement of structural panels at factory seams, the same does not hold true for repairs involving sectioning or clips. Plus, the estimator or appraiser must know when sectioning can be safely and practically performed.

As stated many times in this text, the importance of complete, accurate damage diagnosis cannot be overstressed. A poor or incomplete damage appraisal will undoubtedly result in inaccurately low estimates that might win the collision repair facility jobs but lead to financial losses and/or headaches with cost revisions and add-ons. An even more serious concern is that poor damage evaluation could lead to improper repairs, compromising the vehicle's structural strength, handling, and overall safety.

3.1 IMPACT EFFECTS

The body of a modern vehicle is designed to withstand the shocks of normal driving and to provide safety for the occupants in the event of a collision. Special consideration is given to designing the body so that it will collapse and absorb the maximum amount of energy in a severe collision, while at the same time minimizing the effects on the occupants. For this purpose, the front body and rear body are to some extent made to deform easily, forming a structure that absorbs impact energy and, at the same time, is tough enough to preserve the passenger compartment area. During a head-on collision with a barrier at 30 miles per hour (mph) the engine compartment compacts by about 30 to 40 percent of its length, but the passenger compartment is compacted by only 1 to 2 percent of its length.

As described in Chapter 2, there are two basic types of automotive construction:

- Body-over-frame (BOF) cars
- Unibody vehicles

In the body-over-frame construction, the passenger (payload) area is enclosed with cosmetic panels of steel attached to a structural frame that also supports most of the drivetrain and mechanical accessories. In the unibody construction, the metal body panels are welded together making a structural unit as opposed to traditional construction where the cosmetic sheet steel body rests on top of a structural steel frame. With BOF vehicles, the collision damage can be cosmetic and/or structural. However, when a unibody car is hit, the collision usually results in *both cosmetic and structural damage.*

3.2 COLLISION CONDITIONS

Assessing the types and extent of damage is easier if the estimator has some idea of how the collision occurred. While obtaining this information is not always possible, it is helpful to know the following items:

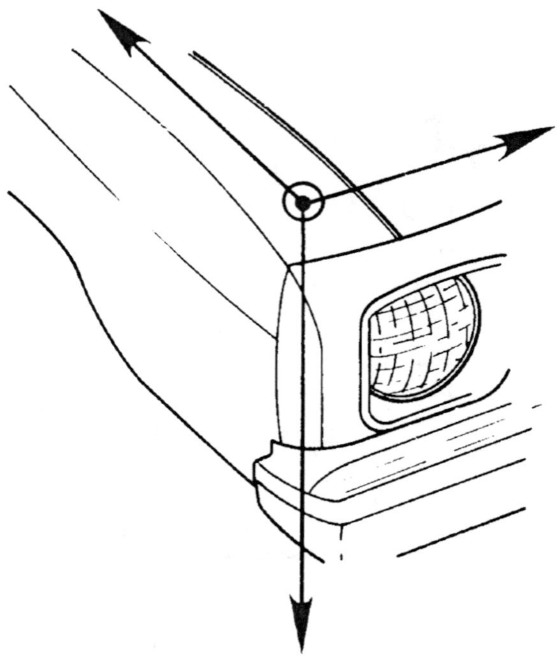

FIGURE 3–1 The direction and angle of impact have a direct bearing on how impact forces cause damage and dissipate through the vehicle.

- The size, shape, position, and speed of the vehicles involved in the collision
- Speed of the vehicle at the time of the collision
- Angle and direction of the vehicle at the time of the impact
- The number of passengers and their positions at the time of impact

IMPACT STRENGTH

During a collision, the direction of the impact force always occurs at a particular angle to the exact point of impact. Because of this, the total force of impact is broken down into several smaller forces, which are dissipated throughout the vehicle in different directions.

For example, if an impact occurs at the right front fender at a certain vertical and lateral angle, the total force breaks down into separate smaller vertical, horizontal, and lateral forces, which are absorbed by vehicle components (Figure 3–1). Vertical force pushes down on the fender. Horizontal force pushes the fender sideways toward the center of the hood. Longitudinal force pushes the fender back toward the rear of the car. The strength of these individual forces and the damage they cause depends on the angle of impact.

The amount of damage inflicted with a given impact force also depends on the direction of impact with regard to the vehicle's center of gravity. If the direction of impact is not along the vehicle's center of gravity, some of the impact force will be used up spinning the

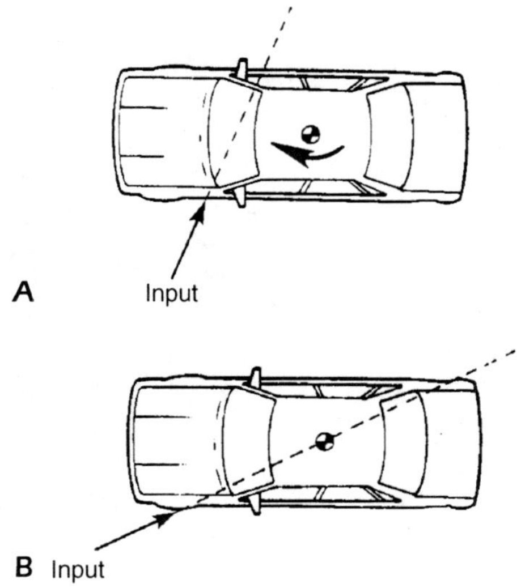

FIGURE 3–2 (A) Off-center hits use more energy to move the vehicle, not damage it. (B) Impacts along the vehicle's center of gravity normally cause greater damage to the vehicle.

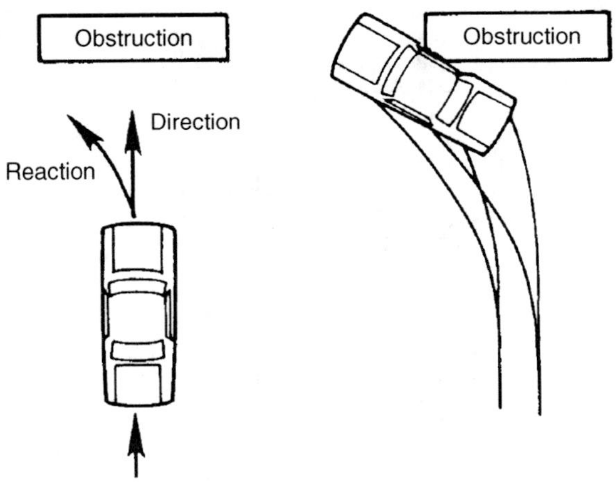

FIGURE 3–3 The driver's first reaction is to turn away from impact, resulting in sidesway damage.

vehicle around this center of gravity. The force is used to move the vehicle, and less force remains to damage components (Figure 3–2A).

On the other hand, if the impact force acts toward the vehicle's center of gravity, the vehicle will resist spin. More energy will be absorbed by body components, and damage will be greater (Figure 3–2B).

A driver's instinctive reactions often affect impact angles, particularly in front-end collisions. The first reaction of a driver, who realizes a collision is inevitable, is to turn the vehicle to one side to avoid a head-on collision (Figure 3–3). The damage that results from this action is called *sidesway,* the type of damage that usually occurs first in the sequence of collision damage types.

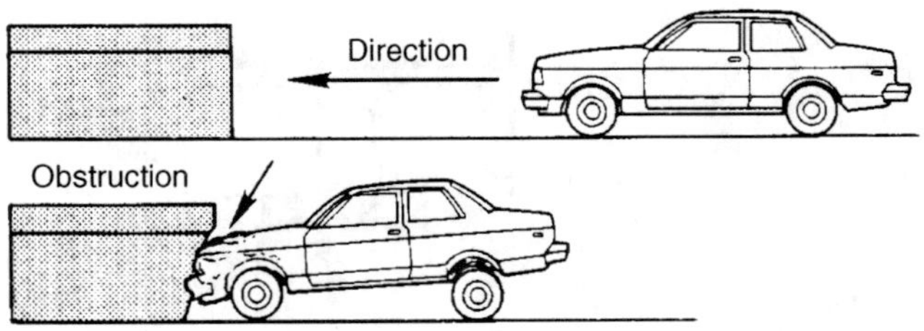

FIGURE 3–4 The driver's second reaction, skid lockup, forces the nose down, resulting in sag damage. Notice how the rear of the vehicle raises up. This causes inertia damage to the roof.

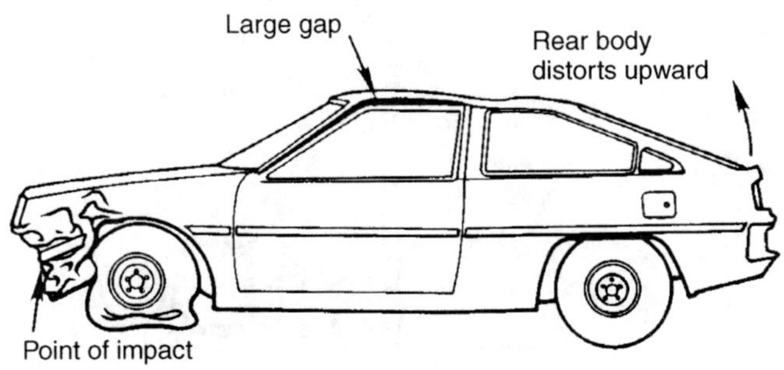

FIGURE 3–5 Typical results of a front impact.

The driver's second reaction is to attempt to panic stop. This places the vehicle into a skid, forcing its front end to dive down. The collision occurs with the front end lower than normal (Figure 3–4). The damage resulting from this driver reaction is called *sag,* and it usually occurs immediately after sidesway damage has begun taking place. Sag in a frontal collision can result in an impact point high on the vehicle's front end. This could cause the cowl and roof to move rearward and the rear of the vehicle to move downward. If the point of impact is low at the front, the inertia of the body mass could cause the rear of the vehicle to distort upward, forcing the roof forward. This would leave an excessively large opening between the front upper part of the door and the roofline (Figure 3–5).

DISSIPATION AREA

Given vehicles with similar weights traveling at about the same speed, vehicle damage will vary significantly depending on what is struck, for example, a telephone pole or a wall. If the impact is spread over a larger area, the damage will be minimal (Figure 3–6A).

Conversely, the smaller the area of impact, the greater the severity of the damage (Figure 3–6B). In this example, the bumper, hood, radiator, and so forth have been severely deformed. The engine has been

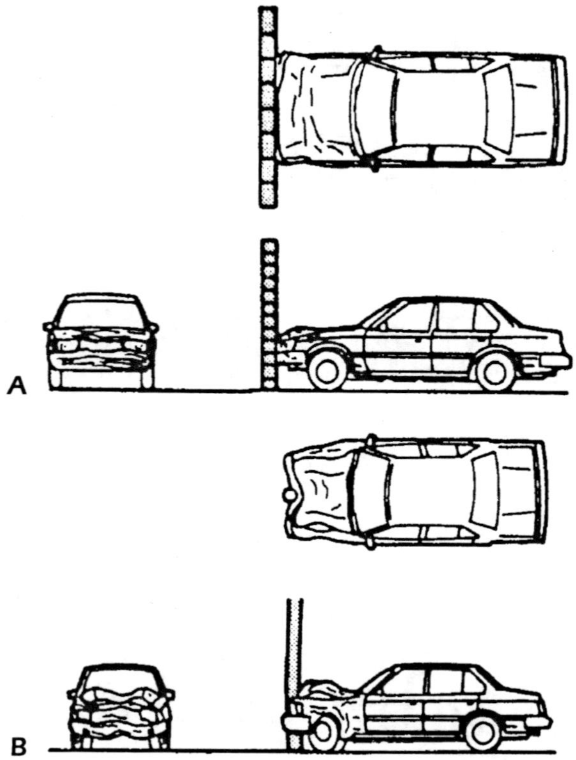

FIGURE 3–6 (A) Damage resulting from a large impact area. (B) Damage resulting from a small impact area.

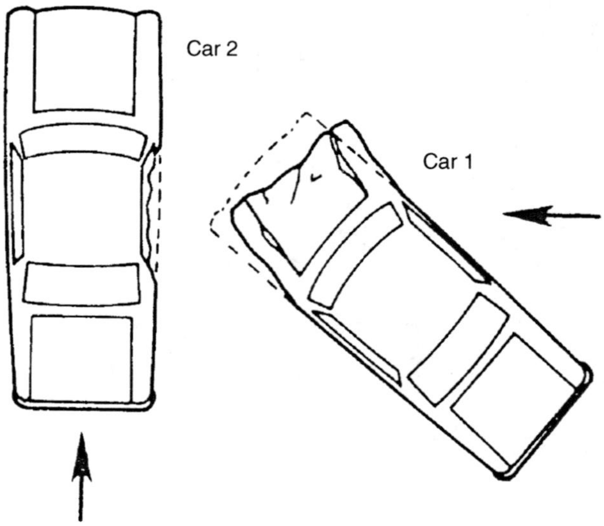

FIGURE 3–7 Damage resulting from a typical broadside collision. Arrows indicate which direction each car was moving before the crash.

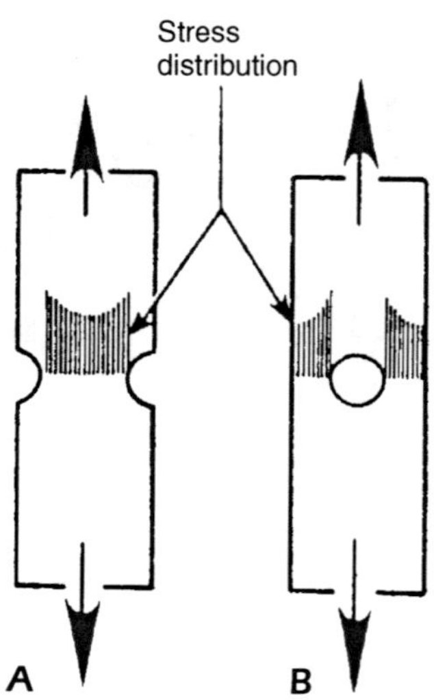

FIGURE 3–8 Differences in design and configuration change the concentration of stress.

pushed back, and the effect of the collision has extended as far as the rear suspension.

Another consideration is when one car hits another while moving. If car Number 1 in Figure 3–7 drives into the side of car Number 2 while Number 2 is moving, the motion of the first car will drive the front end of the car back, but the motion of car Number 2 will also "drag" that same front end to the side. There is only one collision, but the damage is in two directions. On the other hand, there might be two collisions in only one direction. This is a fairly common occurrence in freeway pileups. Or, a car that collides with another car and then leaves the road to hit a pole or guardrail ends up with two completely separate types of damage.

There are many other variables and possible combinations of damage types, and it is important to determine what actually happened before an accurate diagnosis can be made. Get as many facts as possible and combine them with physical measurements and centerline gauge readings to determine exactly what types of damage have occurred and which body panels and components have been distorted or broken.

CONCENTRATION OF STRESS

When any object is stretched or compressed by a force, the stress created will be evenly distributed throughout the object unless the structure of the object changes the stress distribution.

For example, Figure 3–8A shows a strip of metal with semicircular notches cut from both sides. When each end of the strip is pulled, the stress concentrates between the two notches. Figure 3–8B shows a strip with a circular hole cut in the center. Pulling on both ends of the strip creates a different stress pattern within the metal.

Unibody vehicles use this principle of structural variation to control stress damage in front side members and upper reinforcements of the front fender

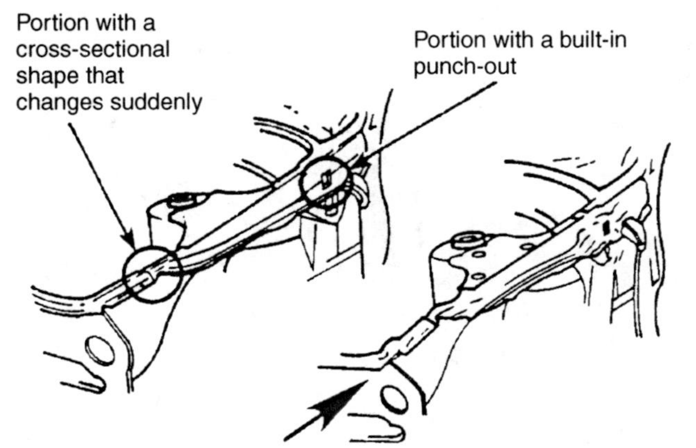

FIGURE 3–9 Example of convoluted design used in unitized construction to control stress concentration.

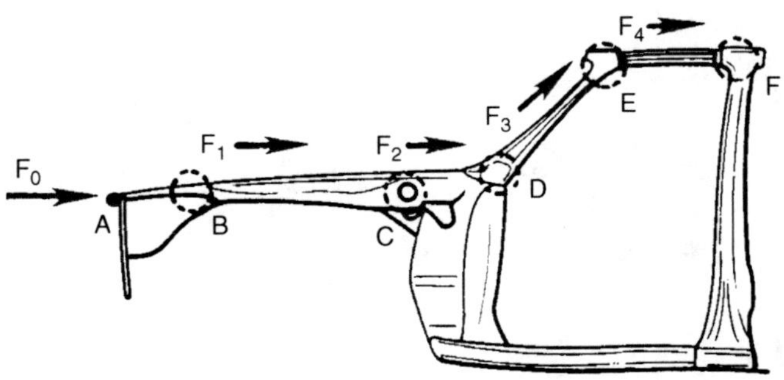

FIGURE 3–10 Example of how a given impact force divides and dissipates through a unitized structure.

aprons. These structural variations are known as convolutions or convoluted support members, and they more effectively control and absorb the impact forces in a collision (Figure 3–9).

TRANSMISSION OF IMPACT FORCES

The shells of modern vehicles contain many welded seams that act as rigid joints in the overall structure. These rigid joints help transmit impact forces to adjacent panels and components throughout the vehicle, thereby decreasing the total amount of structural deformation.

For example, as shown in Figure 3–10, when a given force F_0 is applied to the front corner of a vehicle, area B will deform, decreasing the impact energy of F_1. Next, the remaining impact force travels to point C, the metal deforms, reducing the energy to F_2. From C, the energy path changes directions and arrives at point D, further reducing its force when it gets to F_3. From that point, the energy path changes directions again and strikes the base of the front body pillar and roof, reducing its force to F_4 at point E as the sheet metal in the roof buckles slightly. By point F the energy has been completely dissipated, and no more structural deformation occurs. The energy of a collision is largely absorbed by the components along the path of the shock. Rigid joints and structural members and panels all offer resistance. Not only are the parts deformed that received the impact directly, the parts connected to that point and even parts on the side opposite from the point of impact could be deformed or knocked out of alignment.

This principle of tracing impact forces throughout the vehicle is essential in locating the total complete damage that occurs in today's vehicles, particularly unibody designs (Figure 3–11). Without this approach, slight damage that can cause major problems in vehicle handling and operation can occur.

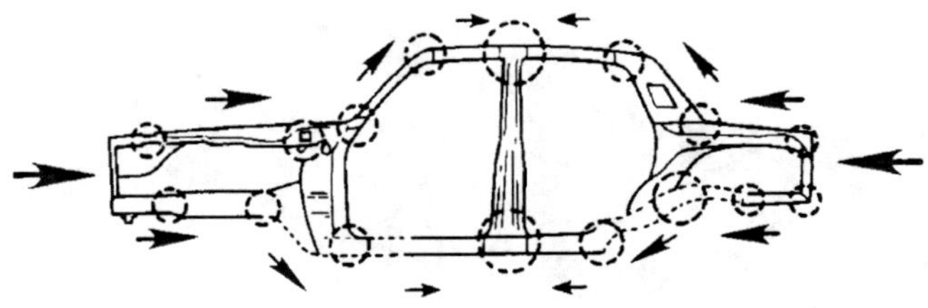

FIGURE 3–11 Impact dissipation paths and energy-absorbing areas for front and rear collisions in unibody vehicles.

3.3 TYPES OF COLLISION DAMAGE

Collision damage to frame and unitized shell structures can be classified in a number of distinct categories. Each of these damage types has its own set of features that makes them relatively easy to distinguish.

SIDESWAY

Sidesway is a condition in which the front, center, or rear portion of the vehicle has been pushed sideways by impact and is out of alignment. Figure 3–12A represents a vehicle with front sidesway. The impact force actually caused one side of the "vehicle" to lengthen and the other side to shorten.

Sidesway can also occur in the center or rear portions of the vehicle (Figure 3–13). Sidesway can be recognized by visual indications, such as a gap at the door on the long side and wrinkles on the short side. Inspect for impact damage from the side. Hood or deck lids that do not fit into the proper openings are a sign of sidesway damage.

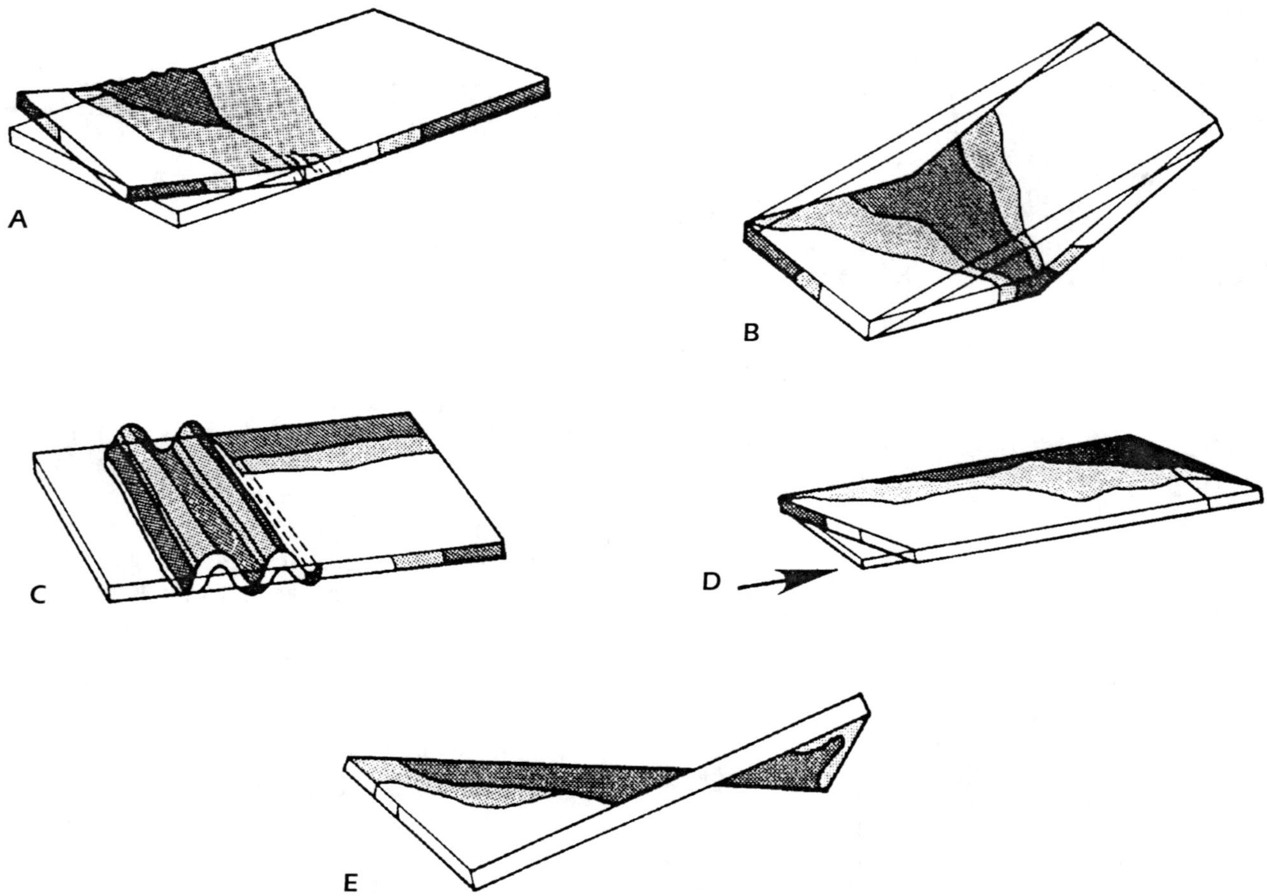

FIGURE 3–12 Various classifications of frame and body damage: (A) sidesway, (B) sag, (C) buckle or mash, (D) diamond, and (E) twist.

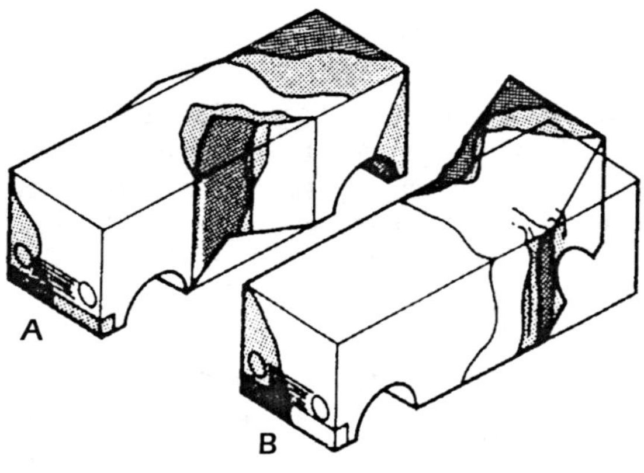

FIGURE 3–13 Examples of center and rear sidesway.

On body-over-frame vehicles, sidesway can be indicated by buckles on the inside of one frame rail and buckles on the outside of the opposite frame rail. A sidesway condition can also occur in unitized body shells.

SAG

Sag is a condition in which the cowl areas of the vehicle are lower than proper specifications. The damaged frame or shell structure has a swayback appearance. Sag damage is caused by a direct impact from the front or from the rear. It can occur on one side of the vehicle or on both sides (see Figure 3–12B). It can usually be detected visually by a gap between the fender and the door being narrow at the top and wide at the bottom. Also look for the door appearing to hang too low at the striker. Sag is the most common type of damage and occurs in most vehicles that are involved in an accident. Enough sag can be present in the frame to prevent body panel alignment, even though wrinkles or kinks are not visible in the frame itself.

BUCKLES AND MASH

A *buckle* is a slight or minor bend in the frame (body-over-frame) or a side member (unibody). Closely note the location of the buckle in relation to the frame rail

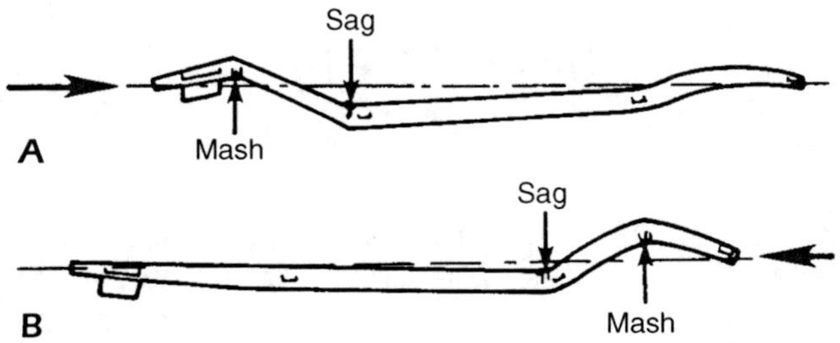

FIGURE 3–14 Frame mashed, buckled, and sagged (A) from front-end collision and (B) from rear-end impact.

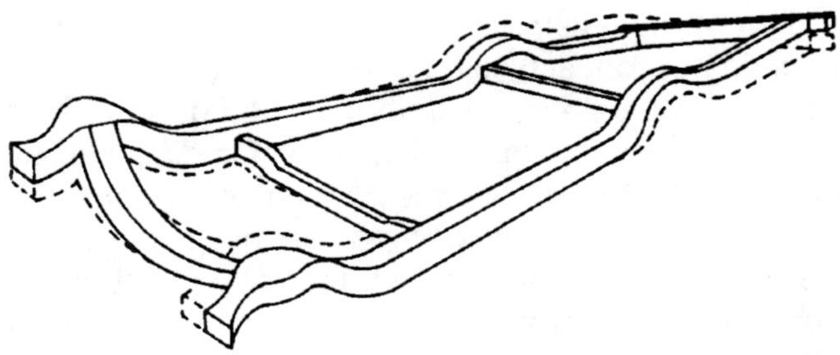

FIGURE 3–15 Typical frame twist condition.

or side member as the buckle's location will usually (but not always) indicate other types of damage.

For example, a buckle on the inner side of the frame rail or side member will indicate an inward sway condition. A buckle on the outer side of these components indicates possible outward sway. A buckle located on the upper surface of the rail or side member often indicates an upward sag condition, with the opposite applying to a buckle located at the lower side of the rail (downward sag).

Mash is simply an extensive or major buckle, resulting in any section or frame member of the vehicle being shorter than factory specifications (see Figure 3–12C). Mash is usually limited to forward of the cowl or rearward of the rear window. Doors might fit well and appear to be undisturbed, but wrinkles and severe distortion will be found in fenders, hood, and possibly frame horns. The frame generally will rise upward at the top of the wheel arch, causing the spring housing to collapse (Figure 3–14). With mash damage, there is very little vertical displacement of the bumper. The damage results from direct front or rear collisions.

When determining repair methods for severe mash conditions, the estimator or appraiser must remember that heating high-strength steels used in unitized construction to make pulls easier is highly restricted. Applying heat will weaken many of these steels.

On the other hand, cold straightening of deformed members can result in tearing or separation. While minor tears—3/4 inch or less in length—can be welded, the estimator must use sound judgment in determining whether a component should be repaired or replaced. If the structural part is kinked—the radius of the bend is greater than 90 degrees over a short distance—it should be replaced. If the structural part is bent—radius of the bend is less than 90 degrees—it may be possible to straighten it and still have full design strength. Simply straightening kinked structural parts will weaken the unibody structure. In the event of another crash, the weakened vehicle will not crush as designed. The air bags may not deploy at the proper time, and the vehicle's occupants could be injured.

DIAMOND

Diamond is a condition in which one side of the car has been moved to the rear or front causing the frame and/or body to be out of square. It will attain a figure similar to a parallelogram (see Figure 3–12D) and is caused by a hard impact on a corner or off-center from the front or rear. Visual indications are hood and trunk lid misalignment. Buckles might appear in the quarter panel near the rear wheelhousing or at the roof to quarter panel joint. Wrinkles and buckles probably will appear in the passenger compartment and/or trunk floor. There usually will be some mash and sag combined with the diamond.

A true diamond condition is most often associated with body-over-frame vehicles. One side rail of the frame moves forward or rearward in relation to the opposite rail. A diagonal cross measurement is made using a tram gauge to verify a diamond condition.

TWIST

Twist is the condition in which one corner of the car is higher than normal. The opposite corner might be lower than normal (see Figure 3–12E). Twist can happen when a car hits a curb or median strip at high speed. It is also common in rear-corner impacts. A careful inspection can reveal no apparent damage to the sheet metal; however, the real damage is hidden underneath. One corner of the car has been driven upward by the impact, and more than likely, the adjacent corner is twisted downward. If one corner of the car is sagging close to the ground as though a spring were weak, the car should be checked for twist. Twist can also occur as a result of a rollover accident.

A true twist condition can only occur in body-over-frame construction. One side rail of the frame moves vertically upward, while the opposite rail moves vertically downward (Figure 3–15). When viewed from the side, the two rails will appear to cross in the middle.

In a unitized vehicle, the front and rear side members do not connect, so this "true" twist condition cannot occur. However, a unibody damage condition closely resembling frame twist is reverse diagonal sag of the front and rear side members on both sides. For example, the right front side member has received upward sag damage, and the left rear side member has sustained downward sag. The left front side member would have downward sag with the right rear side member having an upward sag condition.

A distinction between frame twist and unibody twist is made because the repair method and related labor hours differ between the two. With a unitized vehicle, the pulling repair can be estimated as correcting a sag condition at each of four side members.

In a body-over-frame vehicle, the damage can be repaired with two pulls, one at the front end and one at the back end.

STRUT TOWER DAMAGE

Strut towers are used in MacPherson strut suspension systems found exclusively on unibody vehicles (Figure 3–16). The position of the strut tower is critical to proper wheel alignment. If *any* damage to the side member has occurred behind the strut tower assembly, such as sag, sidesway, buckle, or mash, the strut tower must be checked for alignment and, as needed, straightened during the pulling process.

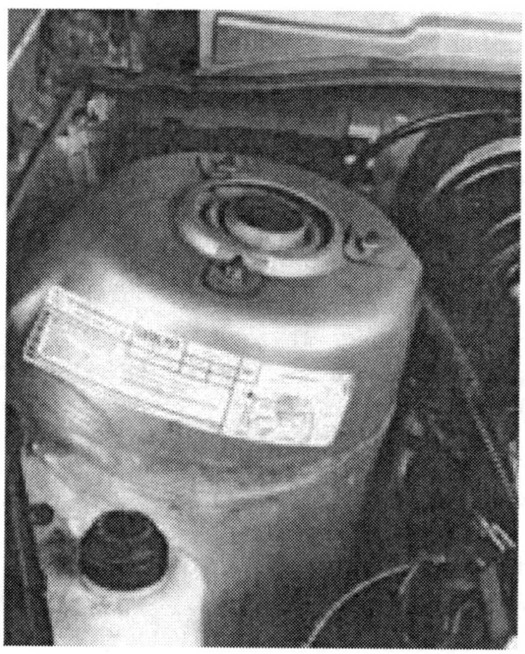

FIGURE 3–16 On vehicles equipped with MacPherson strut suspensions, the strut tower must almost always be checked for proper alignment.

3.4 IMPACT EFFECTS ON BOF VEHICLES

Figure 3–17 illustrates a perimeter frame with built-in collapsible sections. The circled areas indicate the softer sections of the frame, designed to absorb the major impact of a front- or rear-end collision. The body is attached to the frame by rubber mounts, which reduce the effects of shocks traveling from the frame to the body. Heavy impact can result in the body-to-frame connecting bolts and mounts bending or shearing. The result would be a gap between the frame and body. Also, depending on the magnitude and direction of the shock, the frame can receive damage while the body does not. Figure 3–18 shows full frame collapse sections found on Ford pickup trucks.

Frame deformation can be classified according to the following damage types discussed in the previous sections.

- Sidesway
- Sag
- Buckle or mash
- Diamond
- Twist

This listing also gives the order in which the damage types occur in moderate to major collisions involving BOF vehicles. As will be covered in Chapter 4, the most important rule in body/frame straightening and realignment is to repair in the reverse sequence of damage occurrence. This means on a BOF vehicle, the pulling or pushing of the damaged area must be done in the opposite direction of impact. The repair must be made in the reverse sequence that it happened.

Unfortunately, most collisions or accidents result in a mix of one or more of these damage problems. Sidesway and sag frequency occur almost simultaneously. Also, some of these collision solutions affect

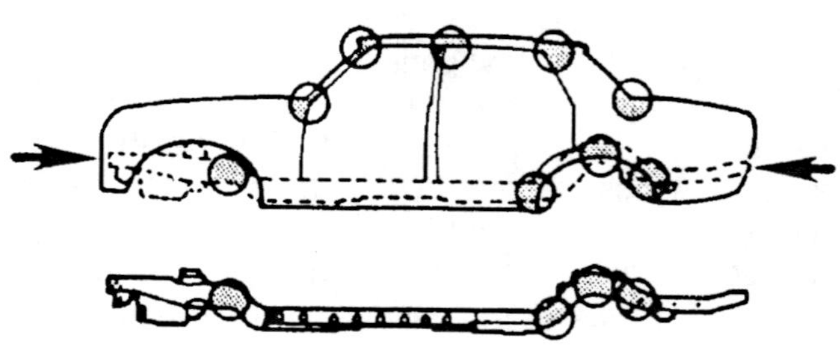

FIGURE 3–17 Typical perimeter frame collapsible sections.

the frame's cross members, especially the front member. In a rollover accident, for example, the front cross member on which the motor mounts are attached will be pulled or pushed out of shape because of the engine's weight. This will result in a sag of this cross member. While cross member damage is rather rare, it must be corrected, since cross members affect the handling of the vehicle.

A deformed frame can be inspected by comparing the space between the body rocker panel and the front and back of the frame, and by comparing the space between the front fender and the front and back of the wheel hub. To inspect front frame deformation, compare the left and right measurements from the rear hole of the front bumper installation to the front frame rail assembly.

Figures 3–19 through 3–22 show various body-over-frame vehicle impacts.

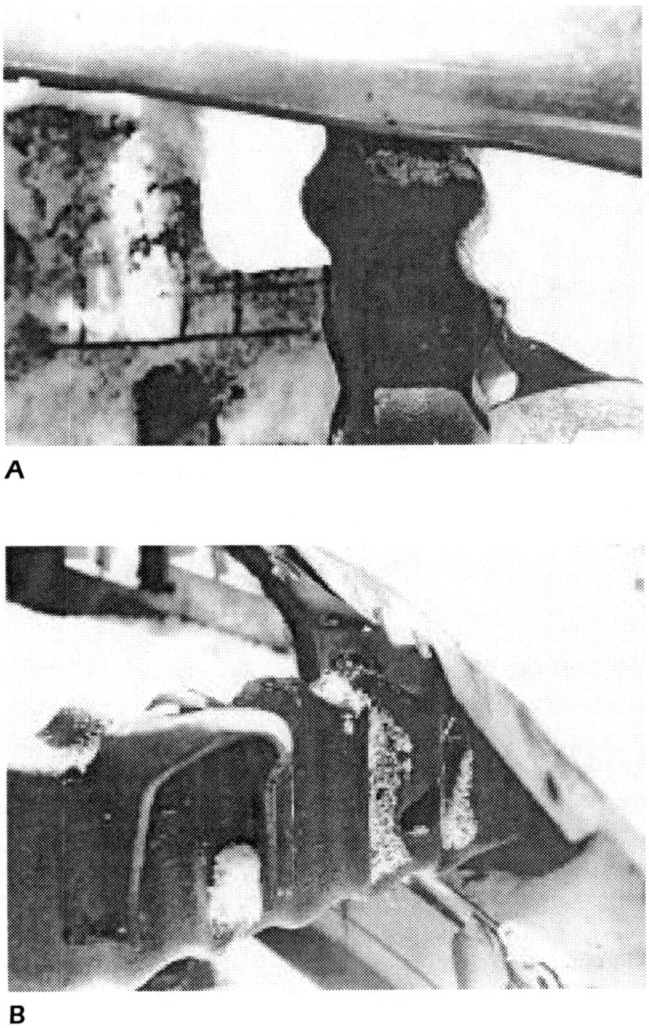

A

B

FIGURE 3–18 Full-frame collapse sections on the front of a Ford pickup truck: (A) top view (B) side view.

FIGURE 3–19 Ranger pickup front hit. Notice the buckled radiator support and inner panel.

FIGURE 3–20 Rear-hit Ranger pickup truck. Notice the frame damage forcing the rear axle housing out of alignment.

FIGURE 3–21 Side-hit F150. Damage to lower cowl has forced the door out of the top.

FIGURE 3–22 Ranger pickup rollover.

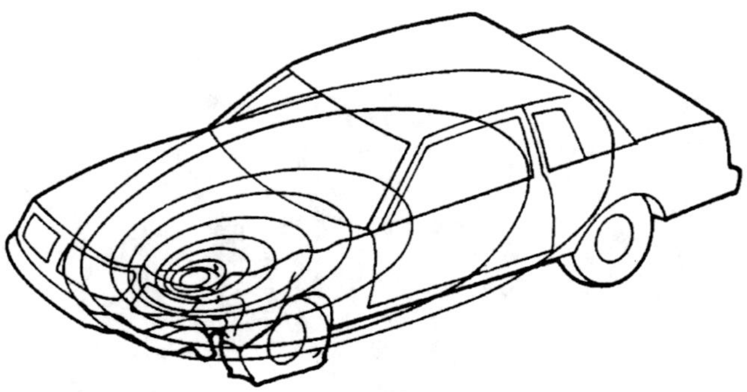

FIGURE 3–23 The cone concept is an ideal method of visualizing impact effect on unibody vehicles.

3.5 IMPACT EFFECTS ON UNIBODY VEHICLES

The damage that occurs to a unibody car as the result of an impact can best be described by using the cone concept (Figure 3–23). The unibody vehicle is designed to absorb a collision impact. When hit, the body folds and collapses as it absorbs the impact. As the force penetrates the structure, it is absorbed by an ever-increasing area of the unibody. This characteristic spreads the force until it is completely dissipated. Visualize the point of impact as the tip of the cone.

The centerline of the cone will point in the direction of impact. The depth and spread of the cone indicate the direction and area that the collision force traveled through the unibody. The tip of the cone is the primary damage area.

Since unibodies are structured entirely from the joining of pieces of thin sheet metal, the shock of a collision is absorbed by a large portion of the body shell. But the effects of the impact shock wave as it travels through the body structure is called *secondary damage*. Generally, this damage is toward the inner structure of the unibody or toward the opposite end or side of the vehicle (Figure 3–24). The lower body deflects away from the impact. The inertia of the upper body causes the upper body to deflect toward the impact.

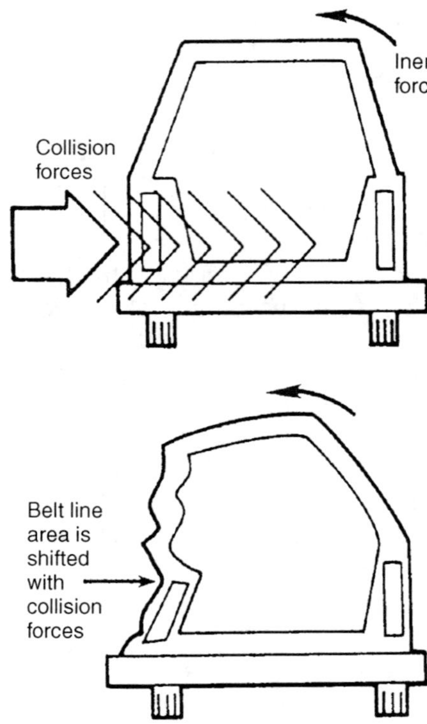

FIGURE 3–24 Example of secondary damage in unibody vehicles: the roof has shifted toward the side of impact due to the force of impact at belt line.

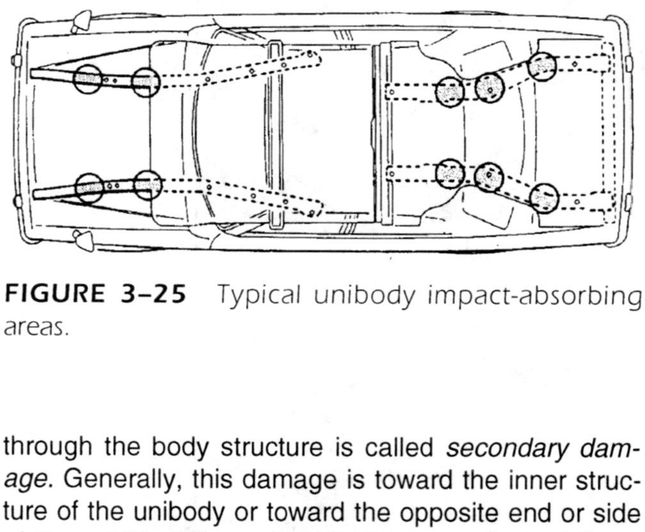

FIGURE 3–25 Typical unibody impact-absorbing areas.

To provide some control on secondary damage distortion and to give a much safer container or compartment for the passengers, a unibody vehicle is designed with crush zones or areas at the front and rear (Figure 3–25). These crush zones are engineered to collapse in a predetermined fashion; thus, the effects of the impact shock wave to the body structure are reduced as it travels through and is dissipated by the

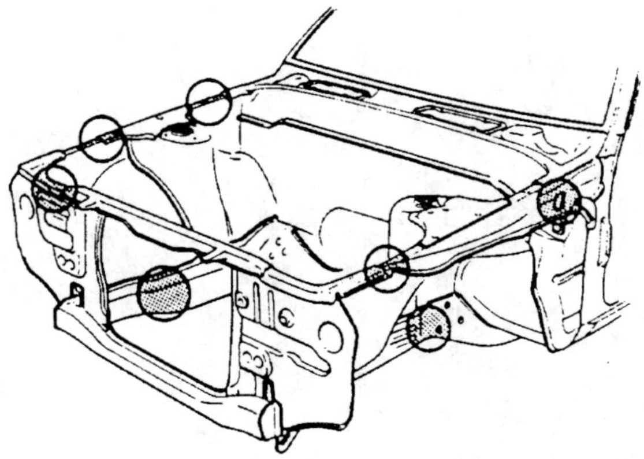

FIGURE 3–26 Typical unibody front crush zones.

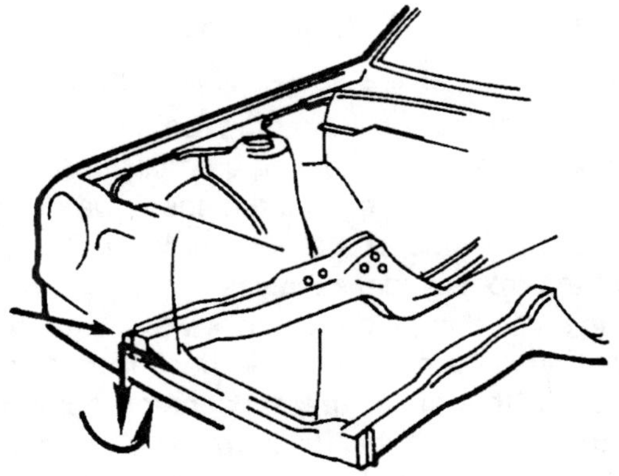

FIGURE 3–28 The lateral and vertical bending movement of a unibody vehicle.

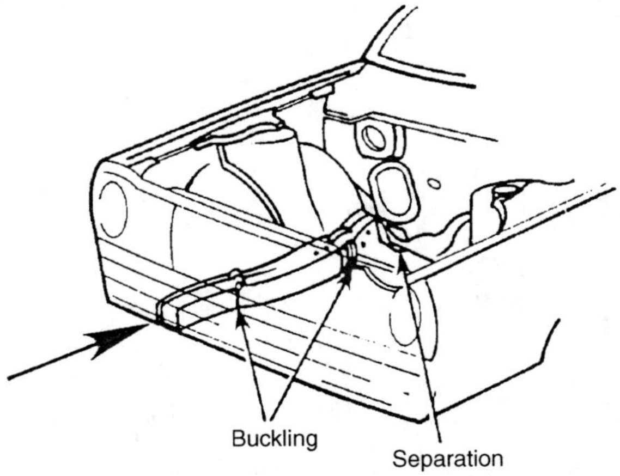

FIGURE 3–27 Buckling and separation action in a unibody vehicle.

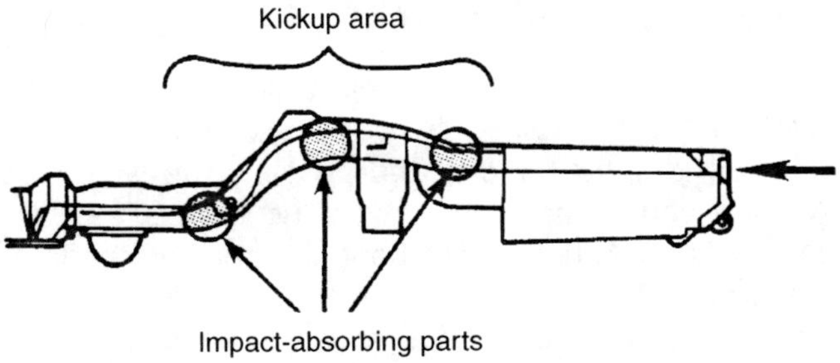

FIGURE 3–29 Rear side member impact-absorbing areas.

body structure. In other words, front impact shocks are absorbed by the front body and crush zones (Figure 3–26); rear shocks are absorbed by the rear body; and side shocks will be absorbed by the rocker panel, roof side frame, center pillar, and door.

UNIBODY FRONT IMPACTS

In the event of a minor impact, the bumper will be pushed back, bending the front side member, bumper stay, front fender, radiator support, radiator upper support, and hood lock brace.

If the impact force is greater, the front fender will contact the front door, the hood hinge will bend up to the cowl top, and the front side members will buckle into the front suspension cross member, causing it to bend (Figure 3–27). If the shock is great enough, the front fender apron and front body pillar (particularly the front door hinge upper installation area) will be bent, which will cause the front door to drop down. In addition, the front side members will buckle, the front suspension member will bend, and the dash panel and front floor pan will bend to absorb the shock.

If a frontal impact is received at an angle, the attachment point of the front side member becomes a turning axis and lateral as well as vertical bending occurs (Figure 3–28). Since the left and right front side members are connected through the front cross member, the shock from the impact is propagated from the point of impact to the front side member of the opposite side of the vehicle and causes deformation.

UNIBODY REAR IMPACTS

In rear-end collisions, the degree of damage will depend on factors such as the impact surface area, the impact speed, the impacting object, and the vehicle's weight. When the impact is comparatively small, the rear bumper, the back panel, the trunk lid, and the floor pan will be deformed, and the quarter panels will bulge out.

If the impact is severe enough, the quarter panels will collapse to the base of the roof panel, and on four-door vehicles, the center body pillar might bend. Impact energy is absorbed by the deformation of the above parts and by the deformation of the kickup of the rear side member (Figure 3–29).

UNIBODY SIDE IMPACTS

Generally, in severe side collisions, the door, front section, center body pillar, and even the floor will deform. When the front fender or quarter panel receives a large perpendicular impact, the shock wave extends to the opposite side of the vehicle. When the central area of the front fender receives an impact, the front wheel is pushed in, and the shock wave extends from the front suspension cross member to the front side member. In this case, the suspension components are damaged, and the front wheel alignment and wheelbase are changed. Since the steering equipment is usually affected by a side impact, the linkage and steering gear or rack can also be damaged.

UNIBODY TOP IMPACTS

When there is damage from falling objects, it not only involves damage to the roof panel but also to the roof side rail, quarter panels, and possibly the windows. When a vehicle has rolled over and the body pillars and roof panels have been bent, the opposite ends of the pillars will be damaged. Depending on the manner in which the vehicle rolled, the front or back section of the body will be damaged as well. In these cases, the extent of the damage can be determined by the deformation around the windows and doors.

Figures 3–30 through 3–33 show various unibody vehicle damage.

UNIBODY DAMAGE SEQUENCE

The damage sequence in a collision involving unibody vehicles is very similar to BOF vehicles except the entire shell is affected rather than a frame and a series of bolted-on panels. In unibody construction, a common damage sequence for front or rear collisions is as follows.

FIGURE 3–30 Unibody high front impact. A high impact bypasses the strength of the lower unibody and can do extensive damage to the engine compartment.

FIGURE 3–31 Unibody side hit. Notice the massive damage to the rocker panel and the roof. This hard type of hit shortens the wheelbase on the impact side.

FIGURE 3–32 Unibody rear hit. Buckles extended to under the rear set.

FIGURE 3–33 Unibody rollover.

Sag or Bending. In the first microseconds of impact, a shock wave attempts to shorten the structure, causing a lateral or vertical bending in the central structure. Most of the forces that broadcast impact shock to remote areas occur at this instant. Since the structure is still and springy, it tends to snap back to its original shape—at least momentarily. Bending is usually indicated by the height measurement being out of tolerance.

Mash or Crush. As the collision event continues, visible crushing occurs at the point of impact. Impact energy is absorbed in the deforming structure (helping to protect the passenger compartment). Remote areas might buckle, tear, or pull loose. Mash is indicated by the length measurement being out of tolerance. In a well-designed unibody structure, impact forces reaching the passenger compartment cause the side structure to bow out away from the passengers (never in), distorting side rails and door openings. Widening is indicated by out-of-tolerance width measurements.

Twist. Even if the initial impact is dead center, secondary impact can introduce torsional loads that cause a general twisting of the structure. Unibody structural twisting, like twisting of a conventional vehicle frame, is usually the last collision event. It is indicated by combinations of height and width measurements being out of tolerance.

Side impacts or less than dead-center front or rear impacts introduce sidesway damage into the collision sequence. When applicable, sidesway is the first type of damage sustained by the vehicle.

Like BOF vehicles, the puling sequence for realignment repairs is always last-in/first-out.

3.6 INERTIA DAMAGE

In moderate to major collisions, the weight and potential inertia force of heavy mechanical components, such as the engine, transmission, etc., can cause damage if these components break free. An engine that has broken free of its rubber mounts and pushed forward into radiator supports, front cross members, etc., or backward into the dash panel and cowl is the best example of inertia damage commonly encountered.

Passengers and luggage can cause secondary damage to the body as a result of inertia during a collision. The damage will vary depending on the position of the passengers and the severity of the impact. Parts with a high-damage frequency are the instrument panel, steering wheel, steering column, and seat backs. Luggage in the luggage compartment has also been known to cause damage to the body quarter panels.

3.7 BASIC SHEET METAL DAMAGE

The work of the estimator or appraiser involves a series of replace/repair decisions involving sheet metal damage. Is it most economical to replace the damaged panel with a new or salvaged part based on part cost and flat-rated labor? Or is repairing the damaged panel at a given labor rate per hour more economical? When light to moderate sheet metal damage occurs, repairing the existing panel is often the only logical choice.

Labor estimates must be accurate and fair. If the sheet metal repair option is chosen, insurance appraisers and collision repair facility estimators must understand the types of damage sustained by sheet metal, the repair problems they present, and the most efficient repair solutions available.

EFFECTS OF IMPACT FORCES

Like all other materials, sheet metal steels are composed of atoms that combine to form grain patterns within the steel. These grain patterns determine how the steel will react to force. A sheet metal's resistance to change caused by impact force has three properties:

- Elastic deformation
- Plastic deformation
- Work hardening

These properties are closely related to the steel's "yield point." As discussed in Chapter 2, this is the point where the steel will begin to tear and break as opposed to simply deforming.

Elastic Deformation. This is the steel's ability to stretch and return to its original shape (Figure 3–34). Sheet metal springback will be found in any area that is still relatively smooth, even though it has been carried out of position by buckles formed in adjoining areas. Many such areas will spring back into shape if they are released by relieving the distortion in the buckled areas that hold them out of place.

Plastic Deformation. Plastic deformation is the ability of metal to be bent or formed into different shapes. When metal is bent beyond its elastic limit, it will have a tendency to spring back, but not all the way to its original shape (Figure 3–35). This is because the grain structure has taken on a new set.

Both stretching and permanent plastic deformation occur in various areas of most damaged sheet metal panels. Permanent plastic deformation introduces permanent stress to the sheet metal. In other words, when a car is damaged in a collision, the stress sustained from the impact will remain unless it is taken

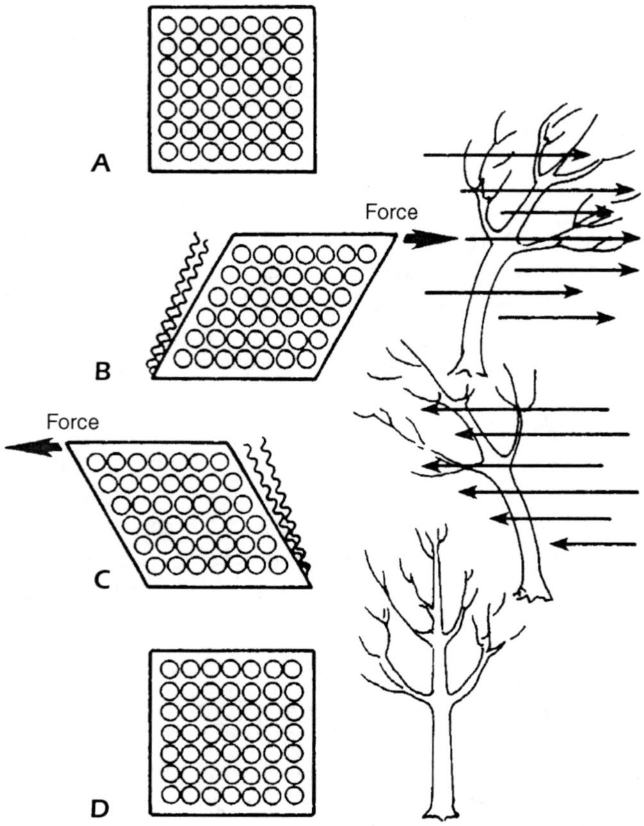

FIGURE 3–34 Elastic deformation is the metal's ability to bend and return to its original shape without assistance.

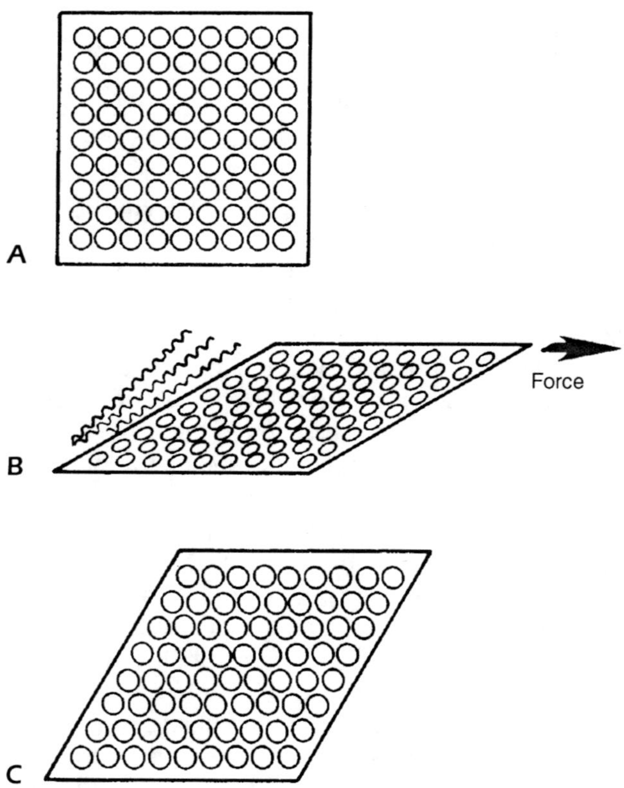

FIGURE 3–35 Plastic deformation occurs when the metal's grain structure is forced beyond the elastic limit and takes on a new set.

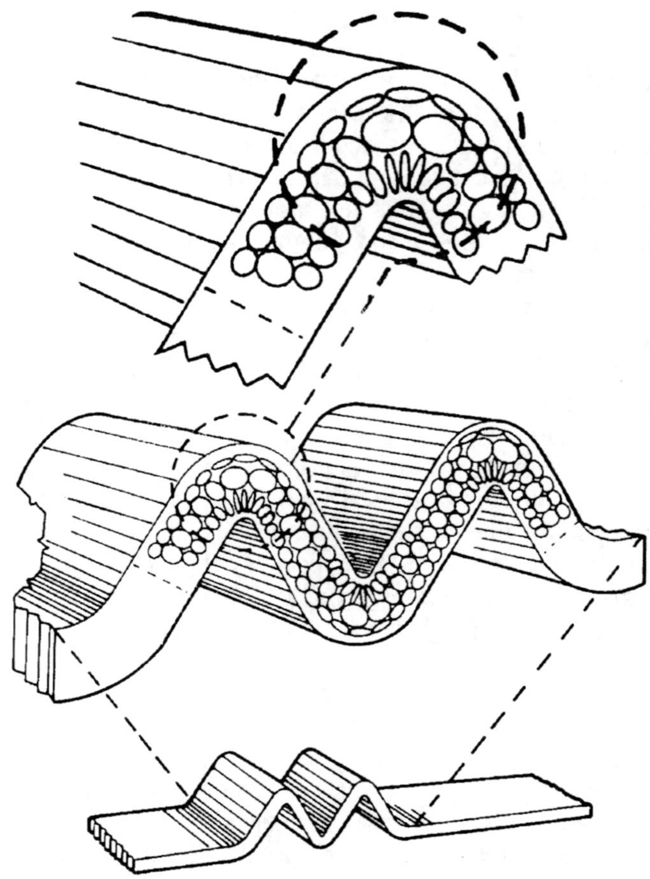

FIGURE 3–36 Grain realignment caused by bending is known as work hardening.

out. This is a condition in which there is an area with permanent stress surrounded by a neighboring area with elastic stress that cannot be removed. When repairing body panels with this type of damage, first remove the permanent stress that is restricting the elastic stress. When that is done, the elastic stress will disappear naturally, and the body panel will return to its original shape by itself.

Work Hardening. Work hardening is the upper limit of plastic deformation, causing the metal to become very hard in the area where it has been bent. For example, if the plastic deformation has been so great that the grain structure has been radically forced out of alignment, the metal becomes very hard and stiff (Figure 3–36). This increased hardness is called work hardening.

The importance of understanding how metal stiffens, making it stronger in areas that are bent or worked, cannot be overemphasized. It is the basis of practically all damage repair.

Some work hardness will be found in any undamaged body panel. It is the result of the original forming process (Figure 3–37). The bending caused by a collision adds still more work hardening in the areas affected. Sometimes, much more work hardening will be added by the cold working used by the body tech-

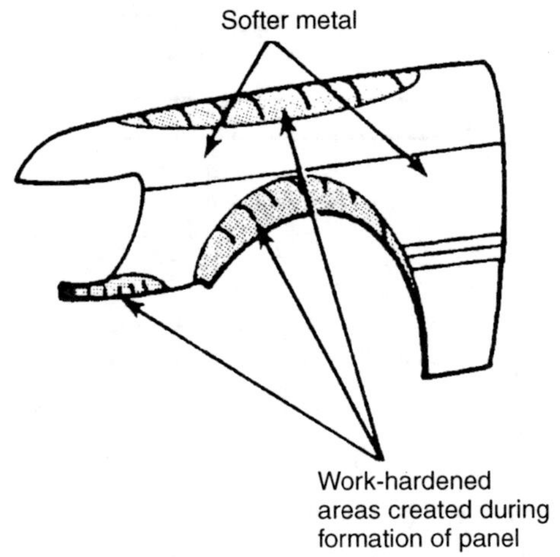

FIGURE 3–37 Examples of work-hardened areas created during manufacturing.

nician as he or she straightens the damaged area. If excessive work hardening is caused by working the metal improperly, the job will be made more difficult.

It is impossible to correct all the damage without creating some during straightening, but the point is to keep it to the absolute minimum. Do not create damage, repair it.

DIRECT DAMAGE

Direct sheet metal damage is very simple to define. It is usually visible as a gouge, tear, or scratch. It is the damage portion of the panel that came in direct contact with the object that caused the impact (Figure 3–38). Direct damage is usually only 10 to 15 percent of the total damage; however, it can be as much as 80 percent of the total damage if the impact was limited to a long gouge or crease. Repair options for direct damage are limited because the sheet metal used in today's cars is often too thin to be reworked. Straightening is also time consuming and usually not practical on areas containing direct damage. For these reasons, direct damage repairs usually involve the use of body fillers after all indirect damage has been repaired. Direct damage varies from collision to collision and job to job.

INDIRECT DAMAGE

As its name states, indirect damage is a result of direct damage. Indirect damage represents 80 to 90 percent of the total damage sustained by the sheet metal. Indirect damage does not vary from job to job. The types of buckles created in indirect damage are easily identified. The pressure forces in the damaged metal are predictable. So all indirect damage repairs are the same, only the tools used will vary due to the size, severity, and location of the damage.

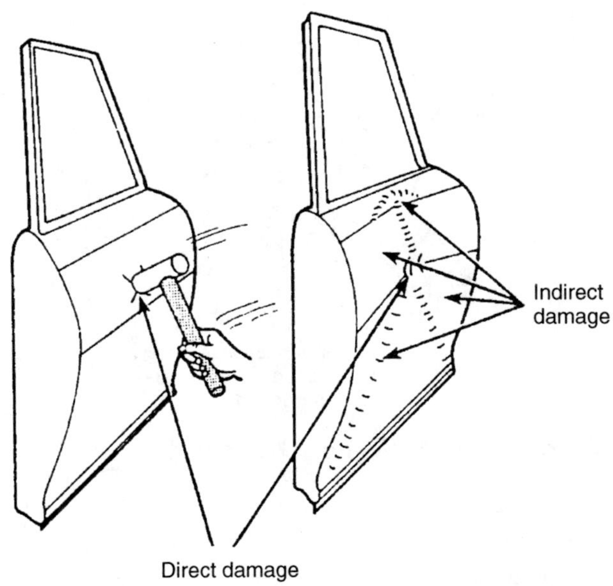

FIGURE 3–38 Examples of direct and indirect damage.

As mentioned in the previous section, buckles are a result of bending metal past its elastic limit. Beyond this point it will not return to its original shape. New work hardening and a new shape is formed. The work-hardened buckles will not allow the metal to return to its precollision or accident shape. The buckles in indirect damage are classified as follows:

1. Simple hinge
2. Collapsed hinge
3. Collapsed rolled
4. Simple rolled

The following sections will take a closer look at each of these classes.

Simple Hinge Buckles

The simple hinge buckle bends like a hinge equally along its entire length (Figure 3–39). The buckle usually causes little stretching or shrinking. If straightened incorrectly, however, it will cause considerable trouble for the technician. When severe, it should be "pulled" out rather than pushed out. Figure 3–39 is a drawing of a simple solid sheet metal hinge buckle showing—when it is created—how the top surface pulls, causing stretching on the upper surface, and how the bottom surface pushes together, causing shrinkage on the underside of the metal. If there is stretching on the top and shrinkage on the bottom, it stands to reason that somewhere in the middle there is an area that is unchanged. The correct repair procedures will result in straightening the metal and leaving the overall dimension the same. If corrected properly, the buckle reverses exactly as it was created. If not, additional damage is created to the adjacent metal and to the buckle. A simple hinge buckle always forms a "straight line" buckle.

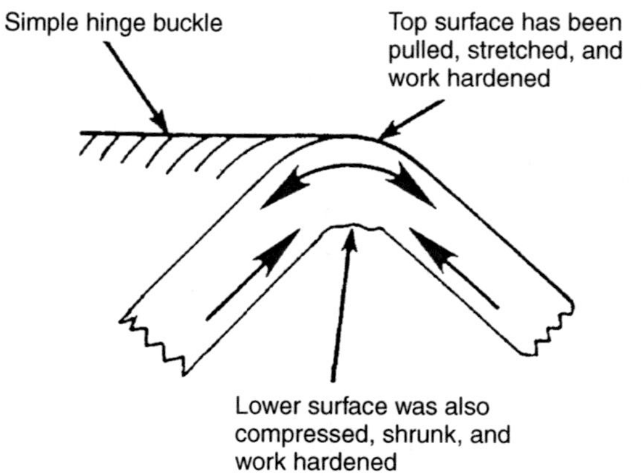

FIGURE 3–39 *Analysis of a hinge buckle.*

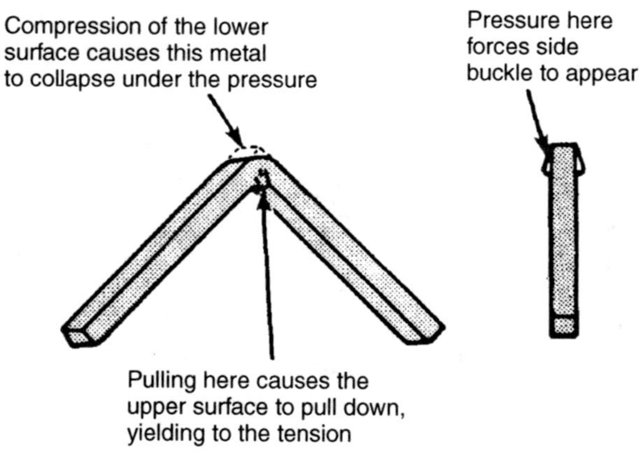

FIGURE 3–40 *A collapsed box section.*

Box Sections. The basic rule of metal bending applies to box sections as well as solid metal. There is a difference, however, in the results of bending (Figure 3–40). The box section has no strength in its center, and as a result, the upper (or top) sheet metal is pulled downward instead of stretching. Little or no stretching occurs. The lower (or bottom) sheet metal is under pressure from both sides and buckles easily. If not straightened with care, the top surface will buckle also. Severe overall shrinkage is the result. Unlike the buckle in the solid metal, the box section collapses on both sides, along with the bottom (and the top, when incorrectly straightened).

Collapsed Hinge Buckle

Sheet metal gives very little resistance to pressure forces exerted on its end. It has enormous resistance to a pulling force along its length. To demonstrate this, take a sheet of metal about one inch by eight inches in size and try to push something with it. It takes very little effort to bend the metal. Now secure

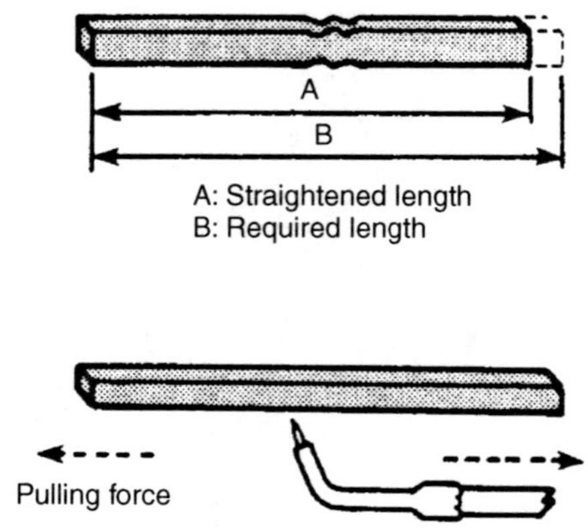

FIGURE 3–41 *(Top) Pulling without applying heat might not return the box section to the required length. (Bottom) Softening through heating might allow pulling to original length. DO NOT EXCEED HEATING LIMITS FOR HIGH-STRENGTH STEELS.*

the metal to a bench or table and pull. The metal strip will pull hundreds of pounds of weight without stretching or creating any damage whatsoever. This simply means that the top metal in the hinge buckle (see Figure 3–40) is relatively undamaged compared with the bottom metal. The pressure side of the buckle is shrunk severely, far more than if the box section were solid metal. This is called collapsing of a hinge buckle or a *collapsed box section.*

When the box section is straightened, further collapsing of the top surface could easily occur. Special effort with heat and pulling equipment is required to prevent it. Figure 3–41 shows the results of improper and proper corrective procedures.

WARNING: Know and never exceed heating limits for high-strength steels.

In the new unibody cars of today there are quite a number of complete boxed sections. Boxed structural rails, rocker panels, windshield pillars, center pillars, and roof rails are just a few examples. Some boxed sections like the door assembly are quite large. Any metal that is bent to form an angle is considered a box section. Late-model cars have a great number of ridges and flanges in them. These are all areas where work hardening is built into the panel and are considered *partial boxed sections* (Figure 3–42). Entire fenders can be thought of as partial boxed sections. As did the complete boxed section in Figure 3–40, the partial box also collapses. The results are the same. The name of the buckle is the same: collapsed hinge buckle. Straightening improperly will have the same results as with the complete box—overall shrinkage in dimensions.

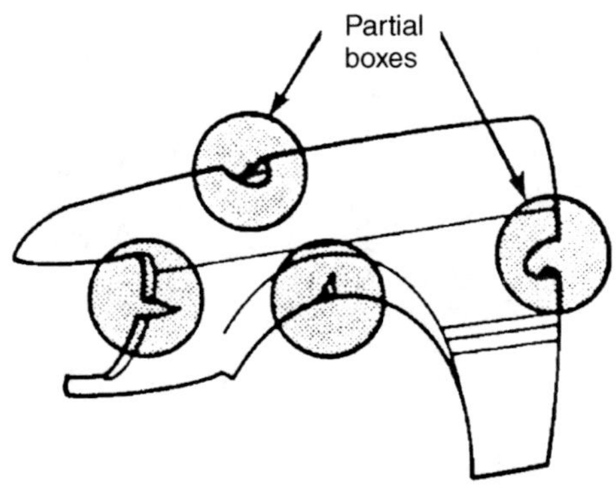

FIGURE 3–42 Examples of partial boxed sections used in unibody design.

Collapsed Rolled Buckle

When a hinge buckle crosses a panel, it not only shrinks any and all boxed or partially boxed sections, but it will also shrink any crowned surfaces it crosses. When this happens, a new buckle is formed. This buckle tries to turn the panel inside out and rolls along, increasing in length as it does. This increasing in length is the characteristic of the buckle. It is called a collapsed rolled buckle. Hinge-type buckles (collapsed or simple) increase in depth but not in length. Any buckles that are on crowned surfaces will shrink the metal. The collapsed rolled buckle is no exception. The shrinkage depends on the severity of the impact.

Simple Rolled Buckle

Two other buckles are also formed whenever a collapsed rolled buckle occurs. These two are adjacent to the collapsed roll. They are called simple rolled buckles. These are also shrinking type buckles because they are on the crown of the panel. The identification of the rolled-type buckles is not difficult. The collapsed and simple rolled buckles form an arrow design on the crown of the panel. At first glance, the fender in Figure 3–43 looks like it has only a single buckle crossing it vertically. Actually, it has five buckles, four types. Rolled buckles (simple and collapsed) occur only on crowned surfaces. This is because the crown is what causes them. If the metal was flat, it would bend like a hinge and a simple hinge buckle would occur. If the panel is crowned, the buckle crossing it tends to roll as it travels deep into the metal, not just because of the folding action on the surface, but because the metal shrinks within itself. Later, if the metal is laid flat (no folds), the shrinkage will still be there, within the metal. Collapsed rolled buckles always occur crosswise to all crowns they are in. The shrinkage they create is also in that direction. Simple rolled buckles shrink the metal as do collapsed rolled buckles, but in a different direction. The

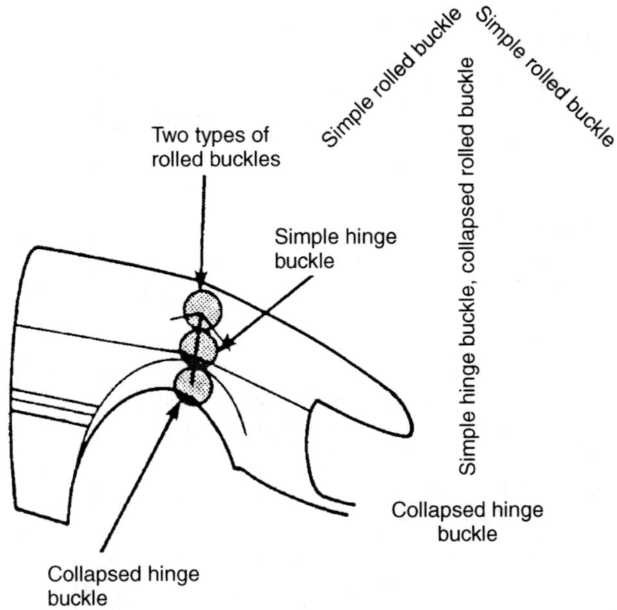

FIGURE 3–43 This fender has a combination of buckles making up indirect damage.

body technician should learn the four different kinds of buckles that are in the indirect damage area. The examples given here have been on only a few specific panels. Recognize the bulges that give clues to possible shrinkage somewhere. Be able to identify the buckles at a glance. Practice by examining many dents. Ask “Where is the direct damage?” and “What is the resulting indirect damage?” Find the work-hardened areas. Find the bent but undamaged areas. Look for the collapsed rolled buckle. Examine damage found in the repair shop or parking lot. Make note of the buckles found, and try to figure out a procedure to correct the damage.

PRESSURE FORCES

The terms *pressure* and *tension* are descriptive words that are commonly used to describe the conditions in metal after damage. These conditions are often described as *high spots* and *low spots.* The main thing is to understand their meanings. There are pressures and tensions (stresses and strains) within the metal before any damage occurs. All crowns, for instance, are under pressure. This “pressure,” however, is not the type of force being discussed here—the forces that occur (or start) at the time of the damage. Metal that is forced up has a new pressure applied to it. The pressure is being held there by the work-hardened buckles. If they were to suddenly disappear, the metal would return to its original shape. Consider the metal as being without pressure or tension before it was damaged when evaluating the changes that have taken place and the corrections that must be made.

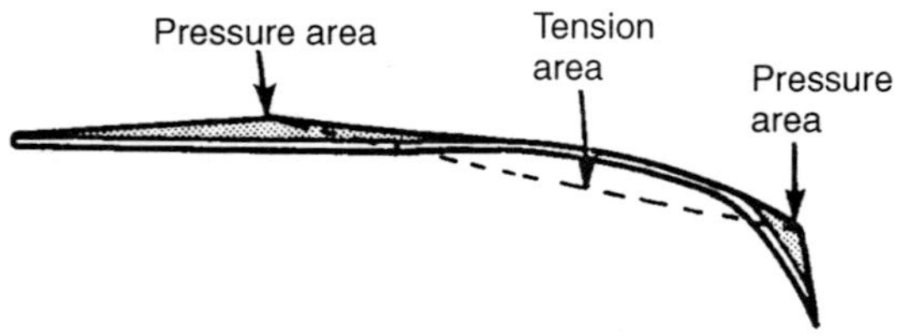

FIGURE 3–44 High and low areas in a damaged combination-type panel.

The metal that is pushed up is called a *pressure area.* Areas that are pushed down are called *tension areas.* The drawing in Figure 3–44 shows a typical cross section of a crowned panel that is damaged. The movement of the metal is obvious.

All panels are crowned to some degree. A highly crowned panel is called a *high crowned.* A nearly flat panel is called a *low crowned.* When a low-crowned panel is damaged, it is pulled into the center of the damage. This pulling brings the panel below its original level. Damaged areas that are below the normal level are called tension areas. Any area of the panel that moves above the original level is called a pressure area. There are three types of outer panels:

- Single crown
- Combination crown
- Double crown

The important thing about this information is that the repair procedure and the application of tools are determined by whether the area is under tension or pressure. A hammer should never strike in a tension area; the dolly should never strike the underside of a pressure area. Power hookups are determined by the direction of pressure forces. No plastic filing can be done when there are pressure areas present.

Single Crown Panels

Figure 3–45 shows a drawing of a *single-crowned panel.* This panel is flat in one direction (left to right) and crowned in another (90 degrees or crosswise). The damage shown in the drawing has tension in one direction and pressure in another. Look at the damage from the flat side (the side view of the drawing); it appears similar to that in Figure 3–46. Look at it from the end view; it appears similar to that of Figure 3–47.

This means there is a three-dimensional effect on all pressure and tension areas. An area that is low is accompanied by a high (or pressure) area adjacent to it. This is true on all crowned areas. All damage to the crowned areas of a panel must be straightened first rather than just filled with plastic.

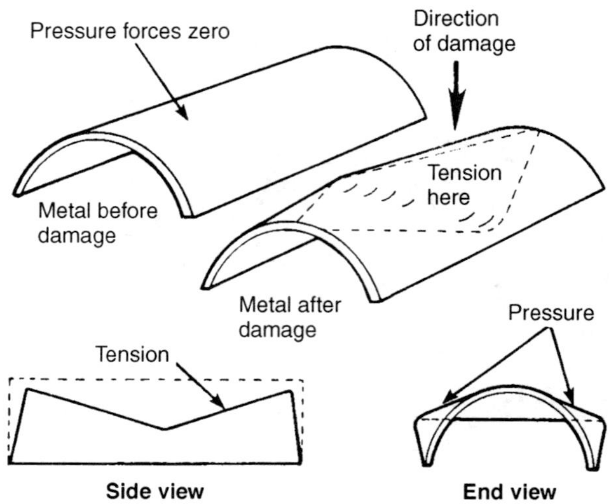

FIGURE 3–45 High and low spots created in a damaged crown panel.

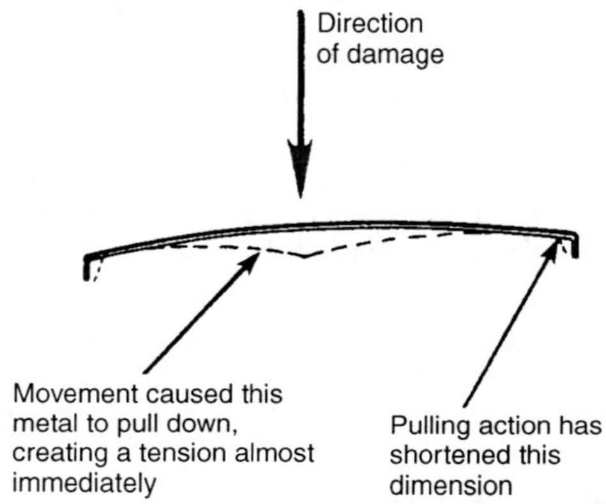

FIGURE 3–46 Tension area in a damaged low-crown area.

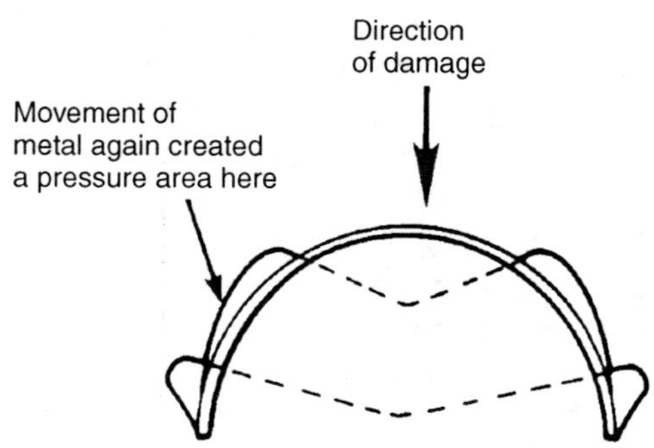

FIGURE 3–47 Pressure areas in a damaged high-crown panel.

"Parking lot" dings are a good example of pressure and tension areas (Figure 3–48). The impact creates a shallow tension area surrounded by a ridge or pressure area. The ridge must be filed level with the panel, and the low area must be leveled with the plastic body filler.

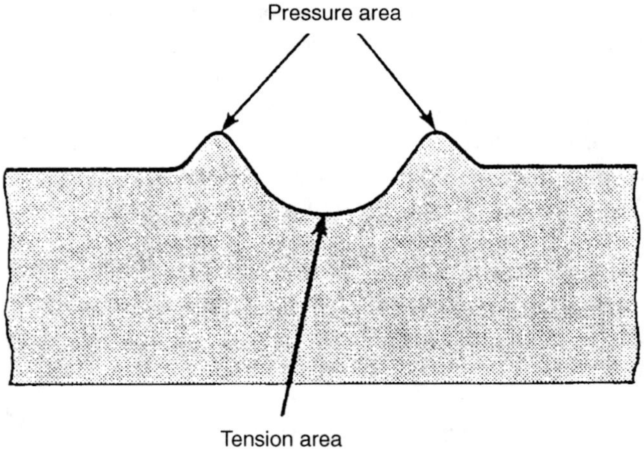

FIGURE 3–48 Pressure and tension areas in a sheet metal ding.

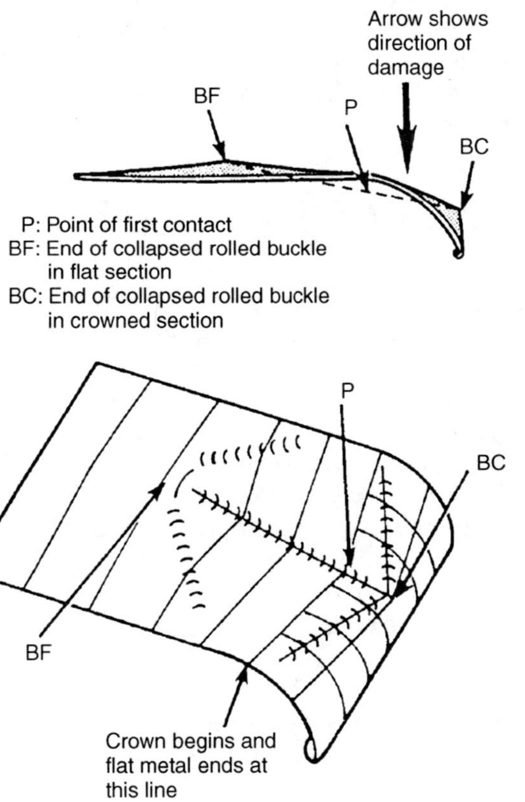

FIGURE 3–49 Work-hardened areas in a damaged combination panel.

Combination Crown Panels

The shifting of the pressure areas on a combination panel is shown in Figure 3–49. The direction of damage shown is from above and almost straight down, yet there are two collapsed rolled buckles (P to BF and P to BC) that are of different lengths. This is because the crown is stronger than the flatter area and resisted the pressure forces more. Actually, during the damage, the same force was applied to both sides of

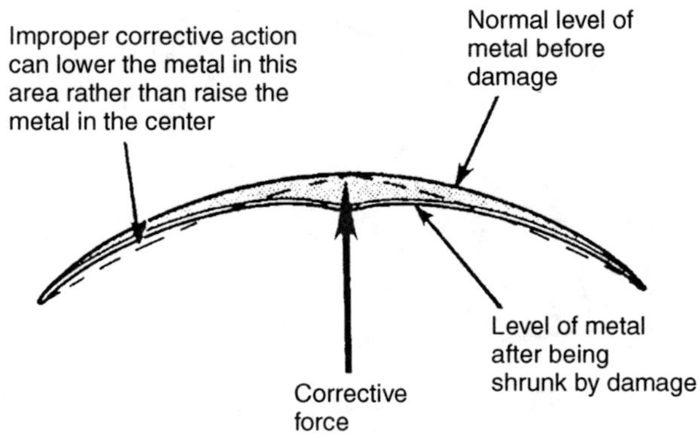

FIGURE 3–50 A shrunken panel must be stretched to return it to its original contour.

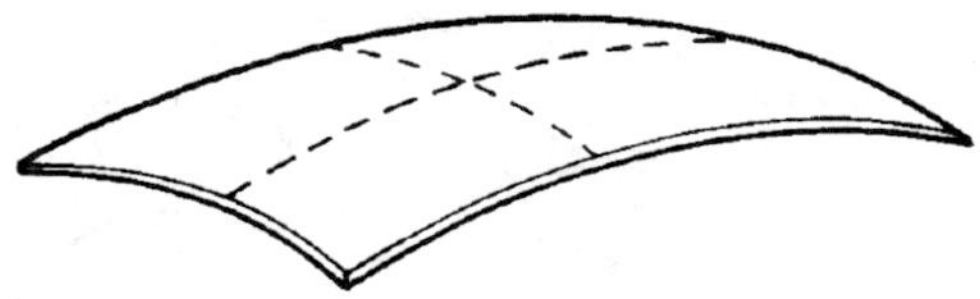

FIGURE 3–51 Example of a double-crowned panel.

the arrow (at P), yet the metal to the left has a greater area of damage. The significance here is that if the novice body technician merely "drives" the metal upward to correct it, he or she would further damage the flatter area of the panel. The flat area would yield to corrective forces (blows), and the greatest resistance area (P to BC) would remain intact. The correction here is to unroll the buckle (P to BC) because it is the "key" to opening up the metal in the flatter area, and it is the area of greatest resistance.

If a crowned panel has a shrunken area on it (caused by either welding, improper hammer and dolly techniques, or the results of a buckle in the crown), the level of that area will be lower than the normal level yet have no pressure areas present. Whenever a low area on a crown is not accompanied by a nearby pressure area, the low area is shrunken and can be corrected only by stretching. To attempt straightening by picking up the low area will only lower the adjacent metal, as shown in Figure 3–50. A damaged crown panel will always have some pressure areas present somewhere, unless it is damaged from underneath. In this case, the metal will be pulled inward, creating the opposite condition from those described in Figures 3–46 and 3–47.

Double Crown Panels

For simplicity, the various buckles were discussed as they occurred on panels that are curved or crowned in one direction and flat in another direction. Most panels are reasonably close to this type of construction. Some, however, are crowned in both directions (Figure 3–51), called *double crowns.*

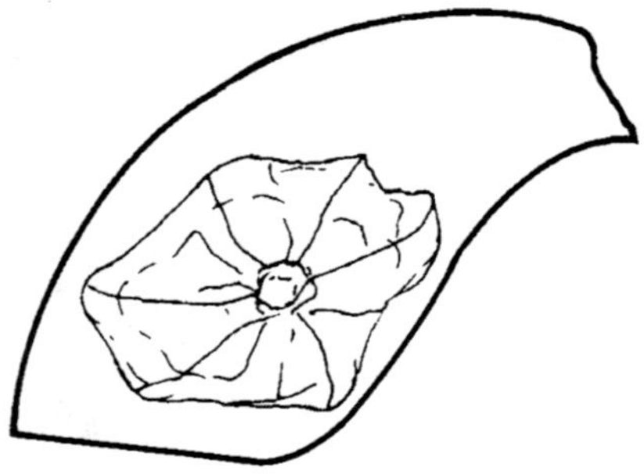

FIGURE 3–52 Collapsed rolled buckles in a double-crown panel.

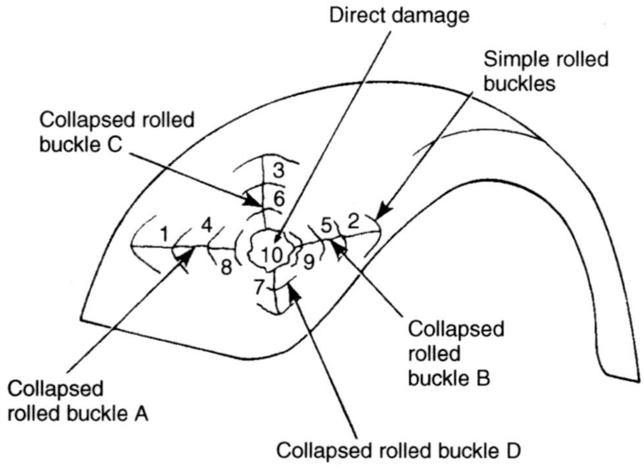

FIGURE 3–53 The procedure for bumping out a collapsed rolled buckle.

Rolled buckles occur on crowned surfaces and roll (or travel) toward the nearest flat area. In the case of a panel that has a double-crowned surface, the rolled buckles will travel normally in all directions from the point of impact. The collapsed rolled buckles spread out from the impact point like spokes of a wagon wheel, the hub being the initial point of impact. The damage shown in Figure 3–52 is typical of an impact on that type of panel.

Determining the direction of damage is extremely important to accurately assessing overall damage. Collision damage must be reversed exactly as it occurred. This is true of frame and shell damage discussed earlier in this chapter. It is also true of simpler sheet metal repairs.

Visual inspection of simple sheet metal damage can usually determine damage direction, but sometimes it becomes complicated by overlapping conditions. Several helpful clues to establishing damage direction are listed in the following sections.

Collapsed Rolled Buckles. These buckles always move away from the point of first contact. When two or three of them are present, it becomes easier. Where they all converge (like spokes of a wheel to the hub) is the point of first contact. It is also usually approximately 90 degrees straight out from the collapsed rolled buckle (Figure 3–53).

Direct Damage Marks. The direction of damage always ends at the mark created by the impact object. Sometimes there are no marks, however.

Pressure Forces. The point of first contact is always in a tension area. The metal moves to the pressure area. The side with the greatest pressure shift of metal will indicate which way the metal traveled. This method is used in combination with the other clues.

By visually reversing the accident, studying each buckle, and unfolding each work-hardened section of metal, the appraiser or estimator can determine how the work will progress and the actual amount of time needed for the repair. As you will see in the following chapters, there are general labor estimating rules that can be applied to sheet metal work. But accurately applying these rules of thumb is based on a thorough understanding of the damage types covered in this chapter.

REVIEW QUESTIONS

1. An off-center hit causes more damage than an impact along the vehicle's center of gravity.
   True
   False
2. Technician A says that inertia damage can be found in the roof distortion of a front-hit unibody. Technician B says that a unibody vehicle is designed to absorb the energy of a collision. Who is right?
   a. technician A
   b. technician B
   c. both A and B
   d. neither A nor B
3. Frame damage can be described as
   a. sag.
   b. mash.
   c. sidesway.
   d. any of the above.
4. Technician A says that mash affects datum. Technician B says that sidesway affects datum. Who is right?
   a. technician A
   b. technician B
   c. both A and B
   d. neither A nor B
5. Diamond is found on full-frame vehicles only.
   True
   False

6. Technician A says that work hardening is found in body lines. Technician B says that direct damage is usually the largest amount of total damage. Who is right?
   a. technician A
   b. technician B
   c. both A and B
   d. neither A nor B
7. A nearly flat panel has _____ crown.
8. Technician A says that rolled buckles happen only in low-crown panels. Technician B says that indirect damage does not vary from job to job. Who is right?
   a. technician A
   b. technician B
   c. both A and B
   d. neither A nor B
9. A collapsed box section should be straightened with heat.
   True
   False
10. Technician A straightens a low area on a crown by picking. Technician B straightens a low area on a crown by stretching. Who is right?
    a. technician A
    b. technician B
    c. both A and B
    d. neither A nor B

Chapter 4

Auto Collision Repair Methods

Objectives

After reading this chapter, you should be able to:

- **List the procedures involved in frame straightening.**
- **Explain the purpose and procedures for stress relieving.**
- **Describe how to remove and install panels.**
- **Identify the methods of repairing cosmetic panel damage.**
- **Explain the materials and methods used in refinishing.**

Key Terms

Stress
Stress relief
Hookup
Tie-down
Resistance spot weld
ARC braze
Hammer on dolly
Hammer off dolly
Stretch
Shrink
Body filler
Primer
Surfacer
Sealer
Base coat/clear coat
Tricoat

An accurate, thoroughly written damage report will account for all labor and materials required to restore a damaged vehicle to its precollision state. To accomplish this, estimators or damage appraisers should be familiar with repair procedures. They must understand the process of pulling a damaged frame back into alignment, the tasks involved in removing and installing a damaged panel, the steps involved in preparing a panel for refinishing, and the various kinds of paint systems employed today. This chapter will give an overview of these major repair areas.

4.1 REALIGNING THE VEHICLE

Realigning a frame requires following three basic guidelines. Follow these three principles to assure that misalignment and damage will be corrected with a minimum of metal working and without further damage to the structure:

1. Measure and diagnose the damage (explained in Chapter 5).
2. Plan the repair sequence. Make a damage diagram of the vehicle. Draw the tie-downs and hookups. Think through the repair, reversing the damage.
3. Repair the damage following the plan . . . first in, last out.

ANCHORING THE VEHICLE

A body-over-frame vehicle can be anchored by placing a suitable plug hook in one of the fixture holes in

FIGURE 4–1 This pickup truck is tied down with a chain wrapped around the frame. The frame cannot move down under tension because the hard wood blocks hold it at the proper height.

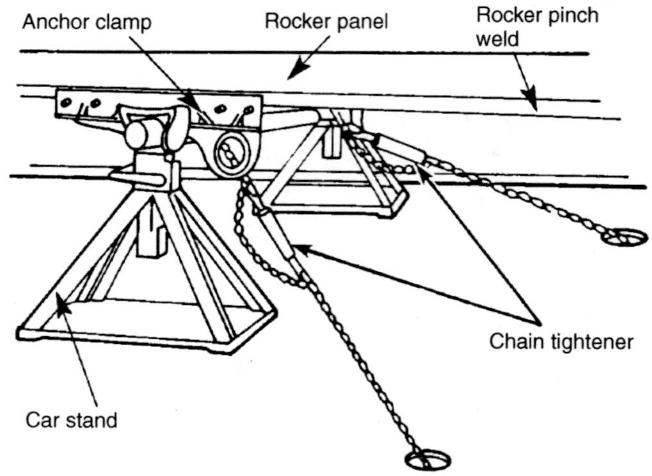

FIGURE 4–2 Pinch weld clamps hold a unibody to the frame machine. A clamp is placed at each corner of the center section.

the bottom of a frame rail. Blocking is used to keep the hook in line with the frame rail and to maintain frame rail height. Identical hookups should be made on both sides of the vehicle. If this type of tie-down is not possible, chains can be wrapped at the cross member and frame rail junction. The chains are tied to the floor or frame machine. In many cases, proper tie-downs will be difficult to place because brake lines, gas lines, body parts, or brackets are in the way. Properly placed tie-downs will help correct height and in some cases, sidesway (Figure 4–1).

A unibody car is anchored by attaching pinch weld clamps and a crossbar as shown in Figure 4–2.

The unitized body has made multiple anchoring a must. At least four anchors are required, one on each of the body clamps. Depending on the vehicle construction, additional anchoring might be required.

ATTACHING CLAMPS

After the vehicle is anchored, clamps are attached to the panel(s) to be pulled (Figure 4–3). There are many cases in which a clamp cannot be fastened to the exact area of deformation. In such a case, a piece of steel can be temporarily welded to the section. After repairs are completed, the temporary piece of steel is removed.

Most damaged panels must be clamped in several places and pulled from several points simultaneously Figure 4–4. This requires the use of a bench or rack with a multiple-pull capability. Multiple-pull systems have the ability to exert a great deal of control over any pulling, improving the precision with which a pull can be made. They also eliminate the need for disconnecting and moving the power posts in order to change the direction of the pull.

The use of multiple-pull points also reduces considerably the amount of force required at any single point, thus reducing the risk of tearing the thin, lightweight metal. Due to the design of today's cars, in many cases there simply is not enough material available in any one place to secure an attachment to transmit sufficient force to complete a repair. Therefore, the pull load must be distributed through several attaching points.

4.2 STRESS RELIEVING

The basic objective of the pull procedure is to work the damaged metal back to its original shape and, just as importantly, to its original state. Restoring damaged panels to their original state requires relieving stresses that were created in the metal during the collision. Metal has a "memory," or elastic property, because it "knows" its original state and will return to that condition if the stress caused by the accident is relieved.

FIGURE 4–3 Multiple hookups spread out the force of the pull, making it less likely to tear out the metal. (Courtesy of Chief Automotive Systems.)

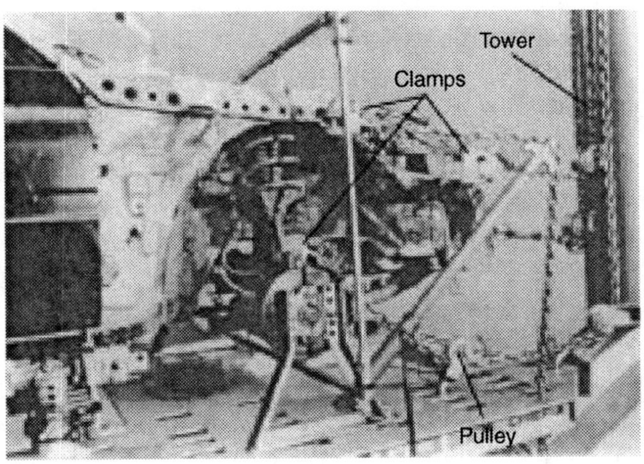

FIGURE 4–4 This picture shows two forward pulls, one angle pull, and one down pull.

Unbent metal (Figure 4–5A) contains layers of grain or molecules all in a relatively relaxed state. As a piece of metal is bent (Figure 4–5B), these grains become slightly distorted, introducing stress. If a piece of metal is flexible enough, the grain will return to its original state once pressure is released. If the metal is bent too far, as in a collision, the grain on the outside of the bend is severely distorted by tension, while the grain of the inside is distorted by compression (Figure 4–5C).

These forces actually cause the grain to shift, creating a grain structure that is harder and less flexible than the predamaged structure. Stresses become locked into the metal.

If an attempt is made to pull the metal back into shape without relieving the stresses, the metal will either tear or take on a weakened shape like that shown in Figure 4–5D. The shape is close to the original but the grain structure remains distorted and new areas of distortion are formed.

When stresses are not removed from the metal, metal fatigue and cracking can eventually occur in the weakened area, or in the event of a second similar collision, less force might be required to cause the same or greater damage, possibly jeopardizing the occupants of the vehicle.

There are two methods of relieving stress: spring hammering and applying heat. Spring hammering is done with a body hammer and a spoon or dolly

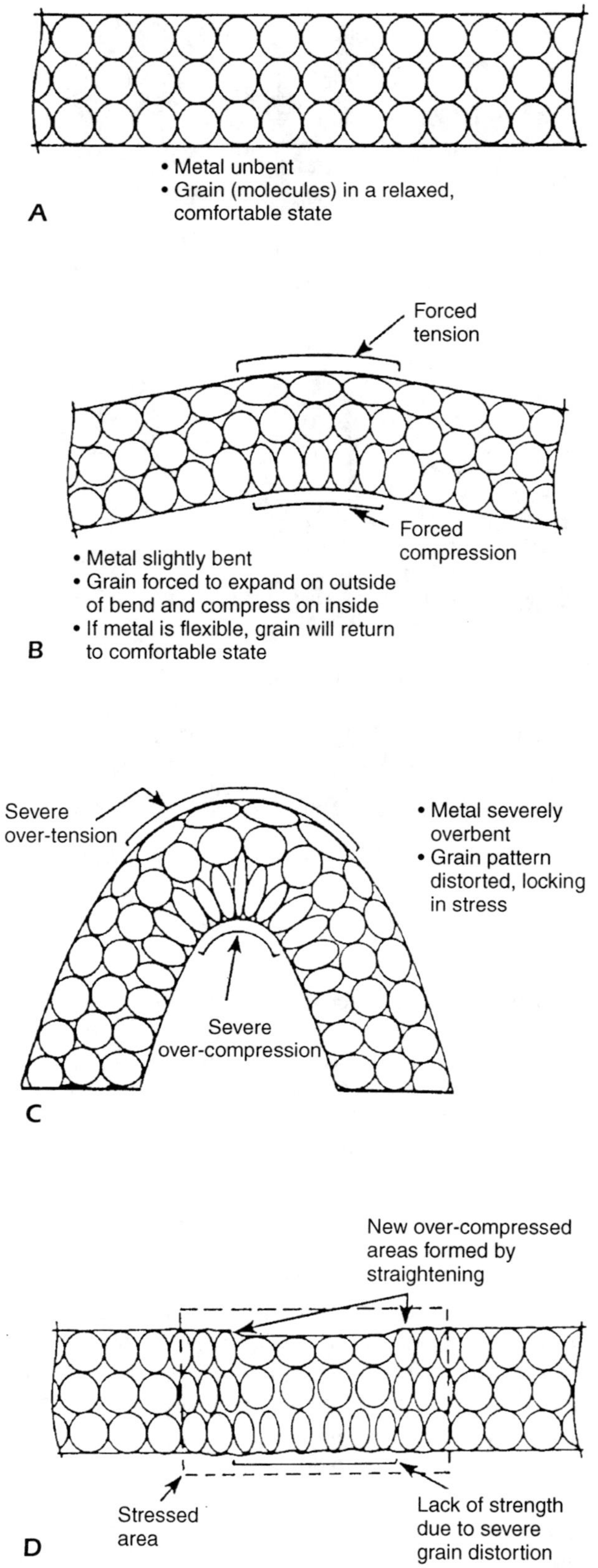

FIGURE 4–5 Condition of metal: (A) unbent, (B) slightly bent, (C) overbent, and (D) straightened (without relieving stress).

(Figure 4–6). While pull tension is applied to the panel, the technician places the spoon on the panel near the work-hardened area. Then, the technician will lightly and rapidly strike the spoon, while moving the spoon around the buckled area. The shock of the hammering will cause the metal grain structure to return to its original state. As the stressed metal relaxes, additional pull tension can be applied. Gradually the deformed metal will return to its original shape and state. This type of stress relieving is called "cold working" the metal.

Badly deformed metal sometimes will not respond to spring hammering. Then, heat must be applied (Figure 4–7) to excite the metal grains before spring hammering. Heat causes the deformed grain structure to loosen its "lock" on the metal. However, heat must be applied very cautiously. Too much heat can destroy the grain structure, causing the metal to soften and lose strength. The car manufacturer's recommendations for stress relieving must always be followed to the letter when applying heat. Some manufacturers prohibit the application of heat to certain high-strength rails and reinforcements. Other manufacturers allow high-strength steel to be heated to not more than 1,000° F for not more than three minutes. Always check and follow the manufacturer's recommendations.

PULLING THE PANEL

Most structural panels are designed to deform and collapse in such a way as to absorb the impact force, transferring as little secondary damage to adjacent panels as possible. For that reason, each panel must be pulled separately, starting with the last panel to be damaged and working back to the point of contact.

Each pull is made slowly and in stages. Measurements must be taken often to ensure that the metal is moving in the right direction and is not being stretched. The proper repair method is to pull, then spring hammer under tension. The spring hammering releases the stress.

ALIGNING FRONT END DAMAGE

In unibodies, the damage correction sequence is length, center line, and then height. Work from the center of the vehicle outward. If twist is present, it must be corrected first. Block and hold the undamaged corners. Push up on the low corner and pull down on the high corner.

The general repair method for front end damage is best covered by going through a typical example of front end collision damage (Figure 4–8). The repair process will include the replacement of a front cross member, front fender apron, and side member on one side, as well as repairs to the front fender, front fender

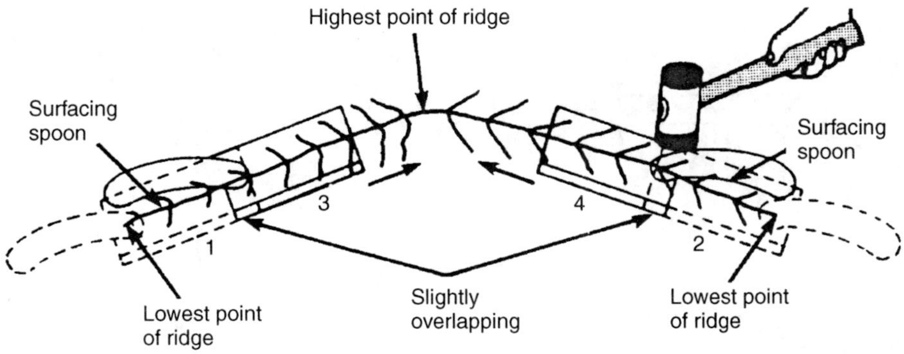

FIGURE 4–6 Spring hammering a ridge with hammer and spoon.

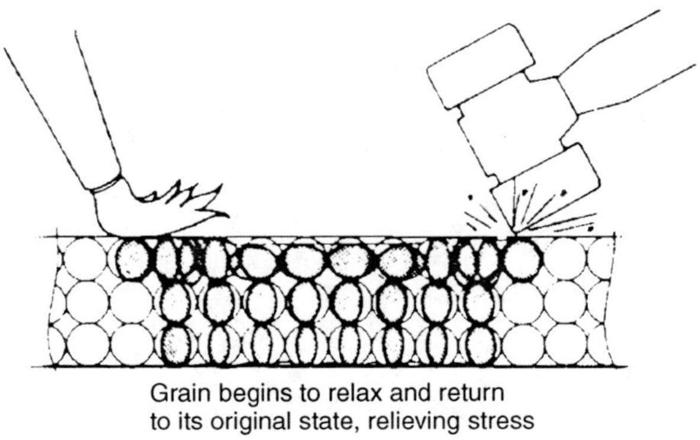

FIGURE 4–7 Methods of returning metal to a relaxed state.

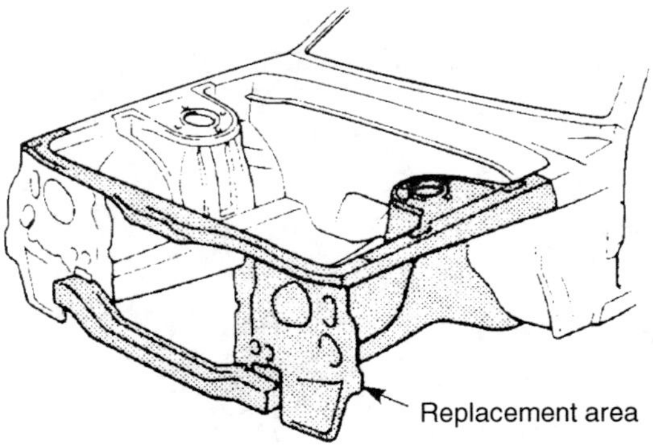

FIGURE 4–9 The shaded components must be replaced.

FIGURE 4–8 Front-hit unibody.

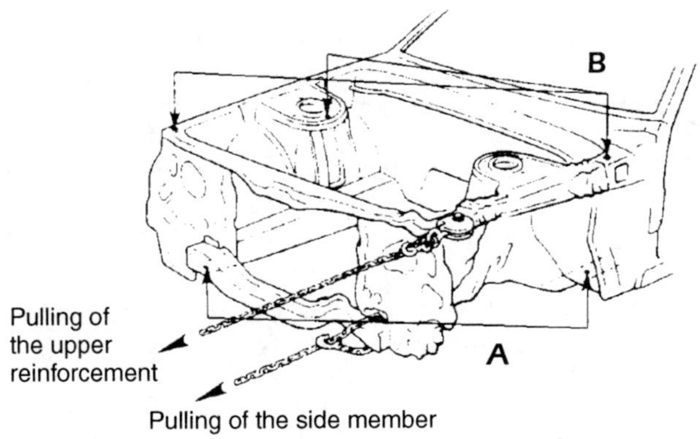

FIGURE 4–10 Checking the front dimensions as the damaged panels are being pulled.

apron, and side member on the opposite side. The repair procedure begins by restoring the right fender apron and side member to their predamaged condition and repairing the support structure on the replacement (left) side (Figure 4–9).

First, the side member on the replacement side is pulled in the direction opposite to the impact direction. Then the fender apron and side member on the repair side are repaired, and at the same time the front fender apron and side member installation areas on the replacement side are repaired. There are many cases in which the entire fender apron or side member on the repair side is deflected left or right only. Since there is practically no warping in the lengthwise direction, repairs involve measuring the diagonal dimensions A and B, as shown in Figure 4–10, and correcting that distance while keeping an eye on the repair condition. The operation can be done efficiently if

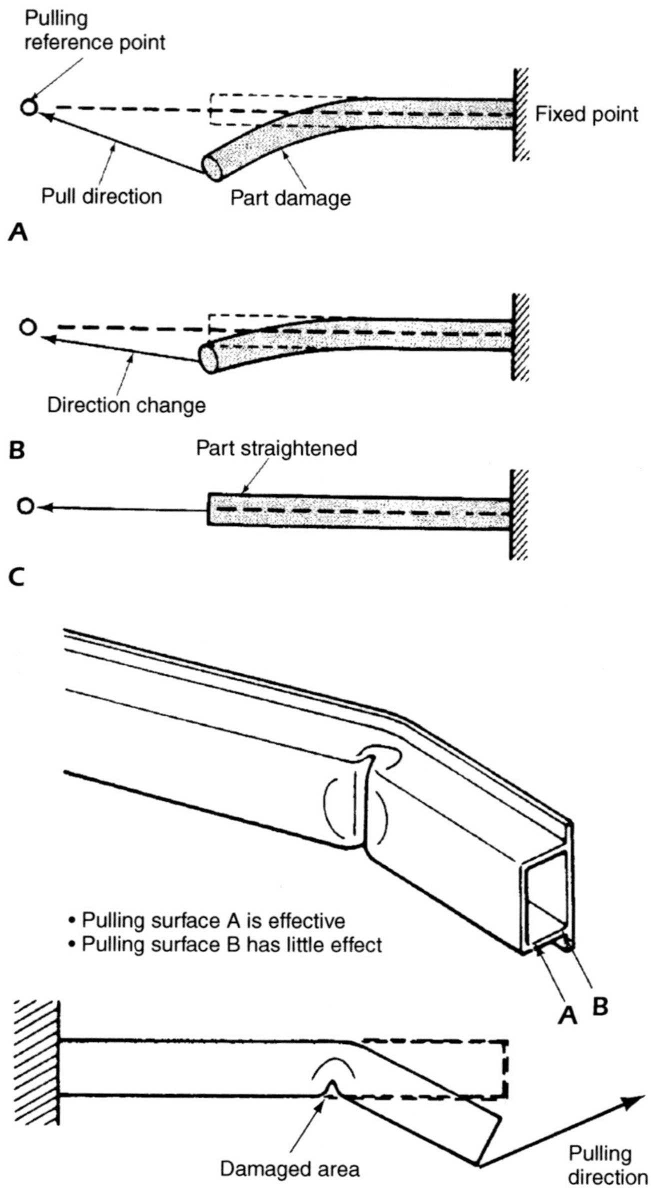

FIGURE 4–11 These diagrams show how to position the hookup, anchor, and pull to correct sidesway.

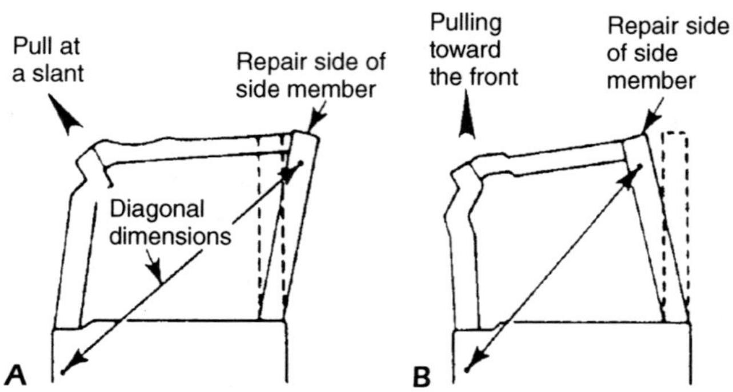

FIGURE 4–12 (A) Pull at a slant if the side member is pushed out. (B) Pull straight ahead if the side member is pushed in.

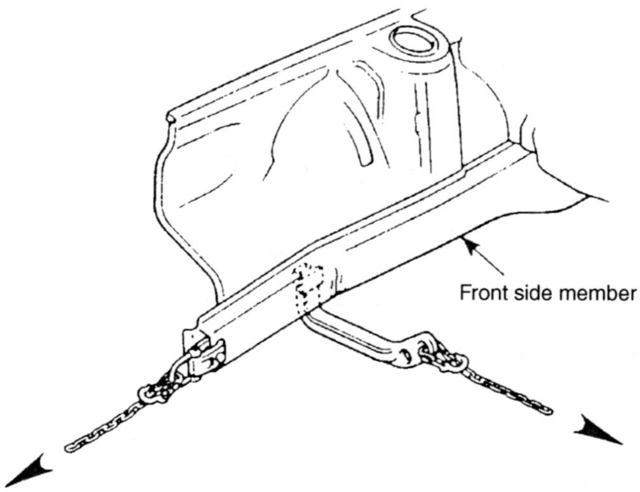

FIGURE 4–13 Front side member bending repairs.

the fender apron upper reinforcement is pulled at the same time as the side member.

The type of pull shown in Figure 4–11A works quite well in restoring rail length and correcting sidesway. Be sure to hook up as shown in Figure 4–11A. In this way, the mashed side of the rail is lengthened and the stretched side is not.

If the side member on the side being repaired is deflected toward the outside, the pull is slanted toward the front, keeping an eye on the diagonal dimensions (Figure 4–12). If it is deflected toward the inside, it must be pulled toward the front.

If there is severe bending damage to the side member on the repair side, it may be necessary to separate the front cross member and radiator upper support at the point where the diagonal dimension is correct and repair them separately. Grip the inside broken face of the side member, and while pulling it forward, pull the broken piece from the inside or push it from the outside (Figure 4–13). After repairing the bent portion, match up the dimensions to the standard diagonal dimensions.

On the replacement side, the main repairs are near the dash panel and the cowl panel. A severe impact will also damage the front body pillar (the door will fit poorly in this case). Gripping the front edge of the side member of the fender apron and pulling will repair the damaged dash and cowl panels but will not repair the major damage to the front body pillar or the dash panel (Figure 4–14A). In this case, the fender apron and side member must be cut away near the installa-

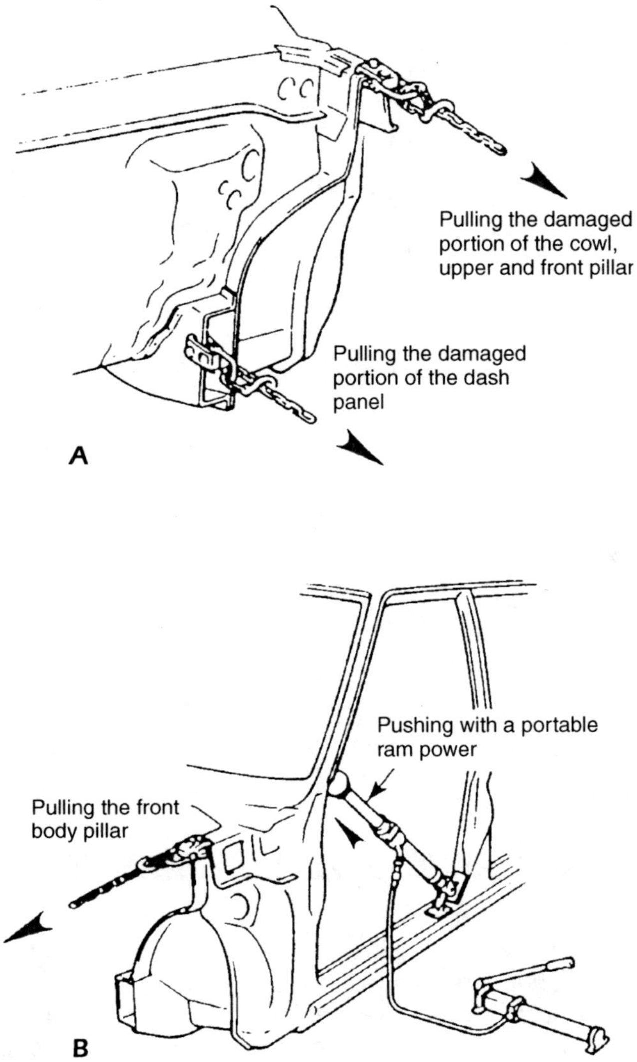

FIGURE 4–14 Front pillar and dash panel repairs. Do not cut off structural panels before they are straightened.

tion area, a clamp must be installed near the major panel damage, and the damaged panel must be pulled into alignment. Also, at the same time that the pillar is being pulled forward, pushing can be done from the interior side with a power ram (Figure 4–14B).

During body aligning, confirmation of the degree of restoration is made by measuring the dimensions, but the reference hole in the underbody front floor reinforcement and the rear of the front fender installation hole are the standard reference points. Therefore, it is important if it can be confirmed at the time of the diagnosis that the damage does not extend to these areas (Figure 4–15). If the impact to the front side member structure on rear wheel drive vehicles (Figure 4–16), is severe, there is a tendency for it to take the shape shown in the figure. The height of the standard measuring point might be distorted. Further, the front side member used in front wheel drive vehicles has a reference point in the rear that has a tendency to be deflected upward when damaged (Figure 4–17).

To correct lateral bending damage to the front caused by side impact, frame straightening equipment is necessary. The clamping point (Figure 4–18) receiving the greatest force is point B, which must be clamped securely. If point C is not secured, point A cannot be pulled. If point C does not have a good clamping location, there is a method for fixing the body sill at four points with underbody clamps.

To pull a MacPherson strut tower into position, a pull plate is attached to the vehicle and a tower pull chain is connected (Figure 4–19). Most MacPherson adapter plates can also be used on frame horns, hinge mounts, or similar locations. If both towers are tipped left or right, they can be repositioned by mounting adapter plates to both towers and installing a strap to make the pull. After the pull is made, a dimension check should be performed using a MacPherson strut measuring gauge.

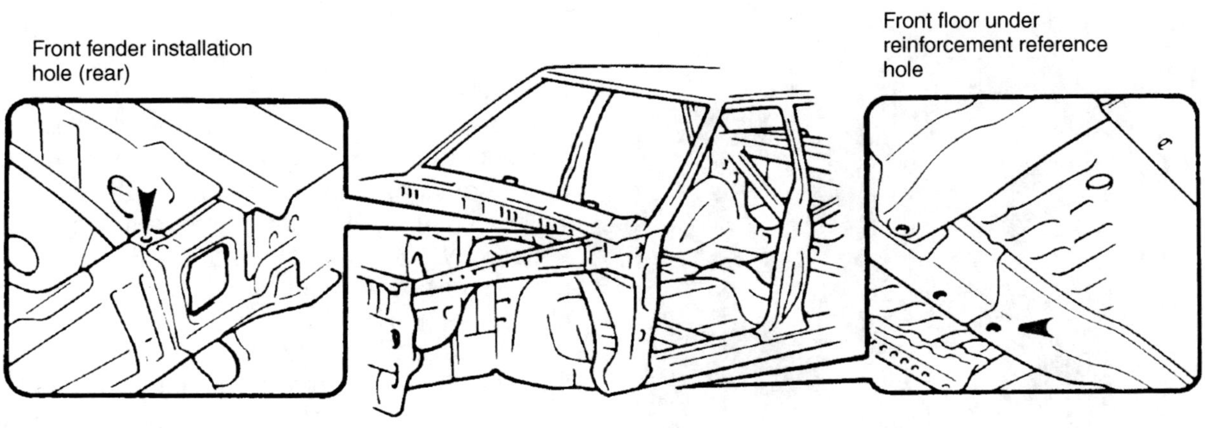

FIGURE 4–15 Standard measuring points.

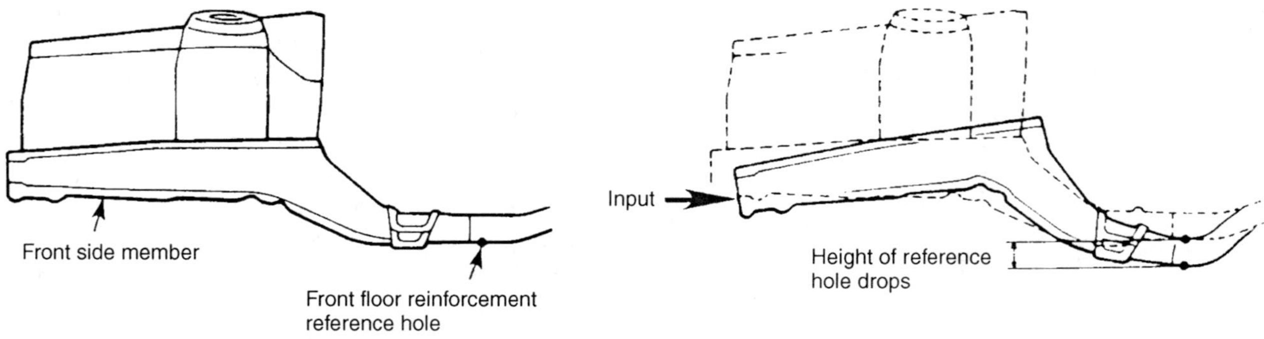

FIGURE 4–16 Typical FR vehicle front side member damage.

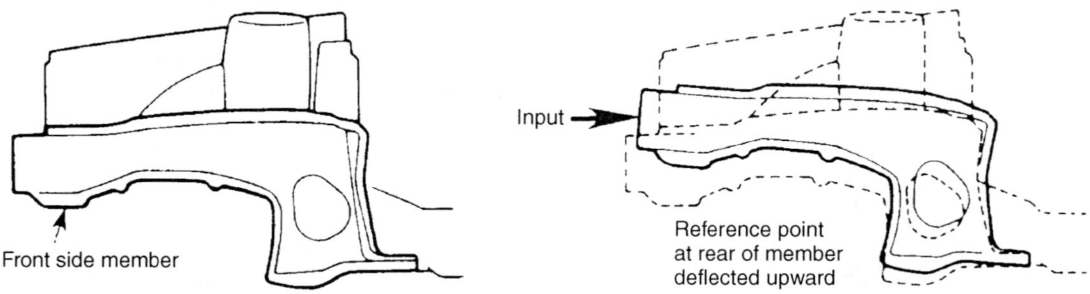

FIGURE 4–17 Typical FF vehicle front side member damage.

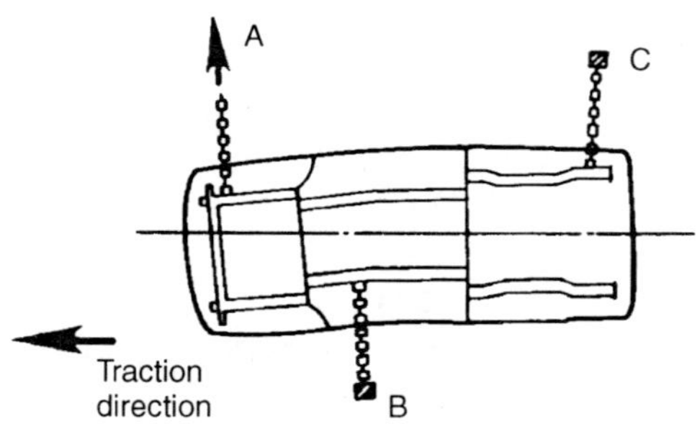

FIGURE 4–18 Correcting lateral bending damage.

FIGURE 4–19 MacPherson strut tower pull hookup.

REAR DAMAGE VEHICLE REPAIRS

Since panel construction of the rear body is more complex when compared to the front body, the path of propagation taken by the collision forces is also more complex, and the damage can be more extensive. Therefore, the damage diagnosis must be done accurately. In most cases the bumper is hit during rear-end collisions, and the impact force will usually propagate through the rear ends of the rear side members or nearby panels and cause damage to the kickup area (Figure 4–20). Next, the wheelhousings will deform, causing the entire quarter panel to move forward, causing clearance problems between other components. If the impact is severe enough, it will have an effect on the roof, door panels, and center body pillar.

Rear-end damage is repaired by attaching clamps or hooks to the rear portion of the rear side member, rear floor pan, or quarter panel. The pull is made while measuring the dimensions of each part of the underbody and while determining the degree of repairs necessary by the conditions of panel fit and clearances (Figure 4–21).

When a rear side member has been pushed into the wheelhousing, creating a clearance problem between the quarter panel and rear door, pulling the rear side member will correct the problem. The quarter panel does not have to be pulled unless the panel has been permanently deformed. Sometimes the wheel-

FIGURE 4–20 Typical rear end damage.

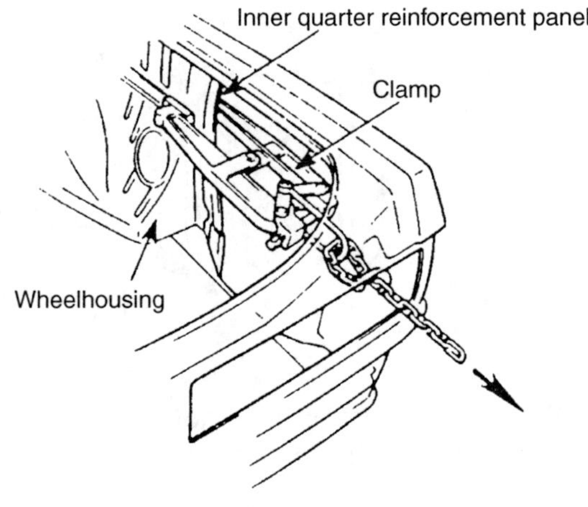

FIGURE 4–22 Rear quarter panel repair.

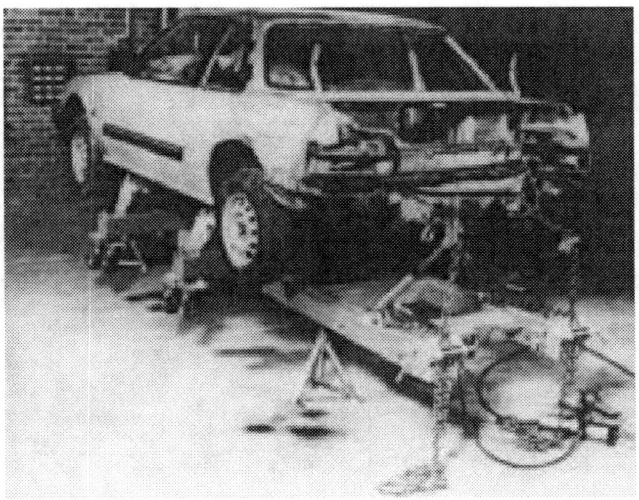

FIGURE 4–21 Typical rear damage repair pulling setup. (Courtesy of Hein-Werner.)

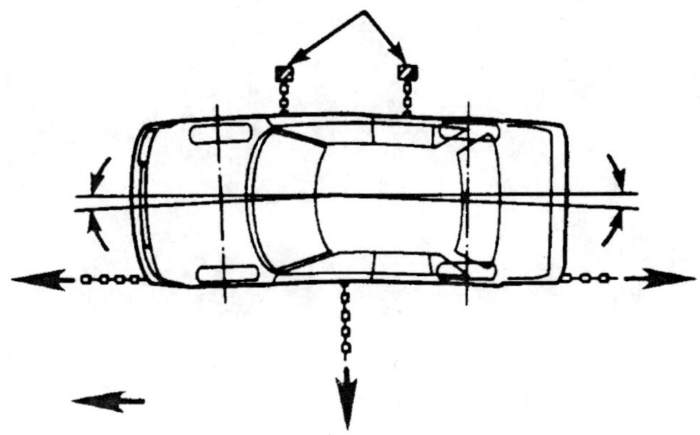

FIGURE 4–23 For side damage repairs, it might be necessary to pull in three directions. (Courtesy of Nissan Motor Corp.)

housing or the roof side inner panel must be clamped and pulled along with the rear side member to properly maintain the clearances with the door panel (Figure 4–22).

Where there is rear structure twist from a front hit, the rear lower structure is clamped to the repair bench. The preliminary pulling will restore some of the lower alignment points, and clamps can be repositioned to preserve the alignment. Moving forward with subsequent pulls, the alignment and number of anchoring points will, of course, move right along with them.

SIDE DAMAGE ALIGNING

If there is a severe impact to the center of the rocker panel, the floor panel will deform and the entire body will take on a curved shape like a banana. This type of damage is called a "banana hit." To align this type of damage, use a method similar to straightening a piece of bent wire. The two ends of the body are pulled apart and the side that is caved in is to be pulled outward (three-way pulling—Figure 4–23).

Anchoring unitized body vehicles for side pulls can be very difficult due to the limited chain hookup areas. Figure 4–24 shows an anchoring method. When a side pull is made on the end of a vehicle, the center section can be anchored by passing a chain around the pinch weld clamp and hooking to the edge of the bench. Tension/anchoring can be applied by attaching the pull chain to a pinch weld clamp.

It is advisable to make an end-to-end stretch pull whenever pulling outward on the center section of a vehicle (Figure 4–25). If pulling high on the body, it will be necessary to tie the vehicle down on the opposite side (Figure 4–26). Pulling outward on the center section of the car can also be done with the portable beam and knee attached to the pulling tower or ram.

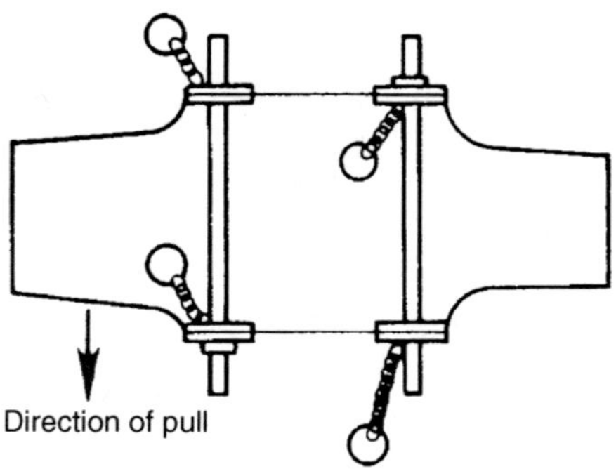

FIGURE 4–24 Anchoring a side pull. (Courtesy of Blackhawk Division, Applied Power Inc.)

FIGURE 4–25 Making an end-to-end stretch pull (Courtesy of Bee Line Co., Bettendorf, IA.)

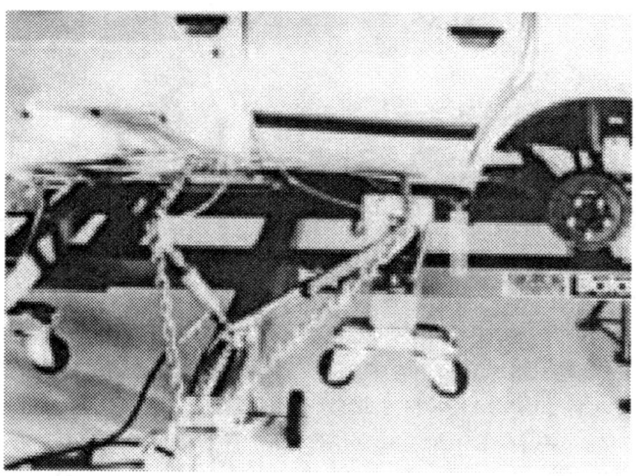

FIGURE 4–26 Pulling outward on the center section. (Courtesy of DatalinerNicator.)

Damaged body-over-frame vehicles are more difficult to pull and align when compared to unibody vehicles (Figure 4–27). Full frames can have diamond, twist, mash, sag, or sidesway, and the rails themselves can be twisted. The presence of gas lines, brake lines, body panels, and gas tanks makes it difficult to hold or pull the frame (Figure 4–28). Often

FIGURE 4–27 Side-hit pickup truck with considerable frame damage. A tough fix.

FIGURE 4–28 Blocking and hold-downs on a pickup truck.

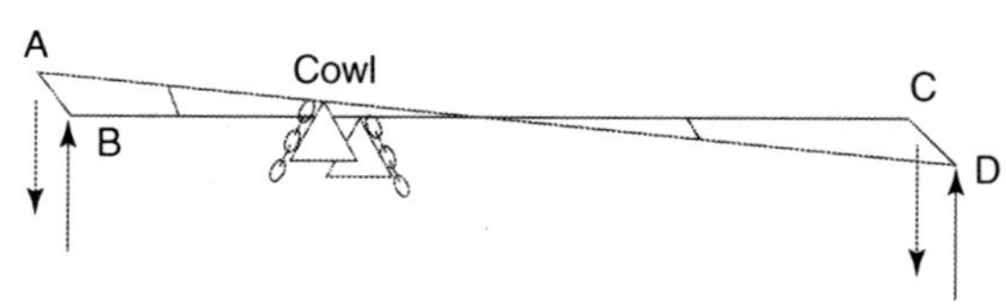

FIGURE 4–29 Setup to correct twist. Block and hold cowl. Pull down at A and C, push up at B and D.

these obstructions will need to be removed. Remember that the hold-downs must always be stronger than the pull. If the pull is stronger than the hold-down, the frame will be damaged even more. Work from the center of the vehicle outward when making repairs. If there is twist, remove it first.

- Twist: Block and hold down frame under the cowl. Pull down on high sides and push up on low sides (Figure 4–29).
- Mash: Hold. Heat if allowed by manufacturer. Pull.

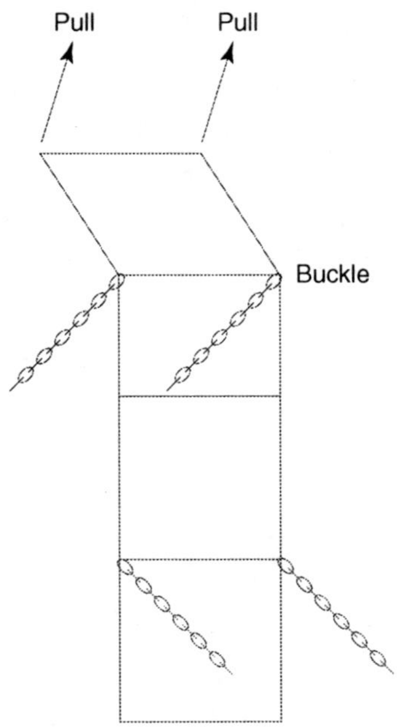

FIGURE 4–30 Setup to correct sidesway. Be sure that the hold-downs are stronger than the pull.

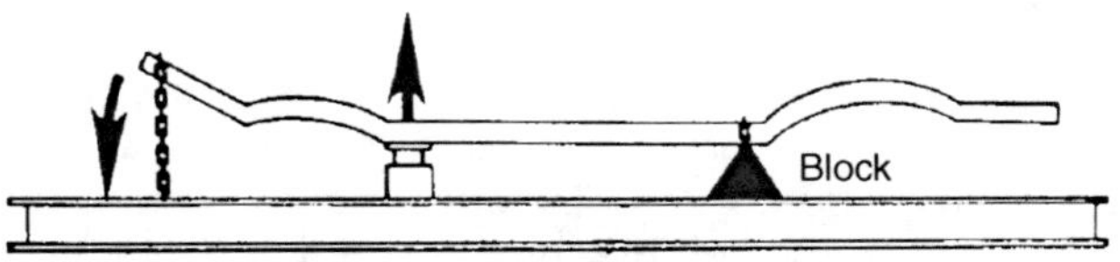

FIGURE 4–31 Setup for sag and kickup.

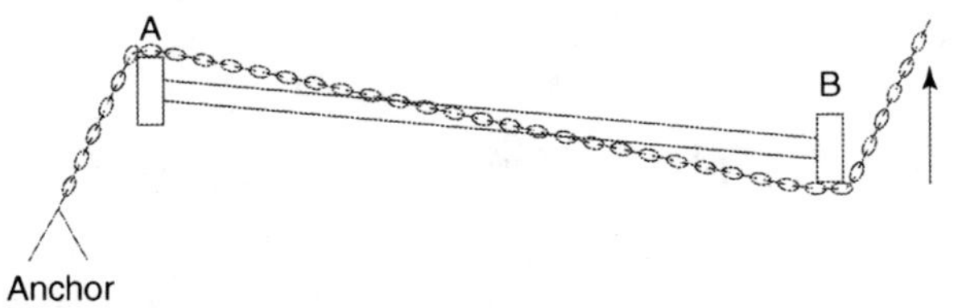

FIGURE 4–32 Setup for out-of-level at the rear of the pickup truck. Rail A is high, rail B is low. Position the chain over A and under B. Anchor the chain and pull up. Rail B is pulled up and rail A is pulled down.

- Sidesway: Block and hold undamaged portion of frame behind the buckle. Pull in the proper direction (Figure 4–30).
- Sag or kickup: Block and hold down portion of frame at the proper height. Pull up or down as needed (Figure 4–31).
- Out-of-level: Anchor chain. Wrap the chain over the high rail and under the low rail. Pull up on the low (Figure 4–32).
- Diamond: Block and hold undamaged side rail. Pull damaged side. Correct this last (Figure 4–33).

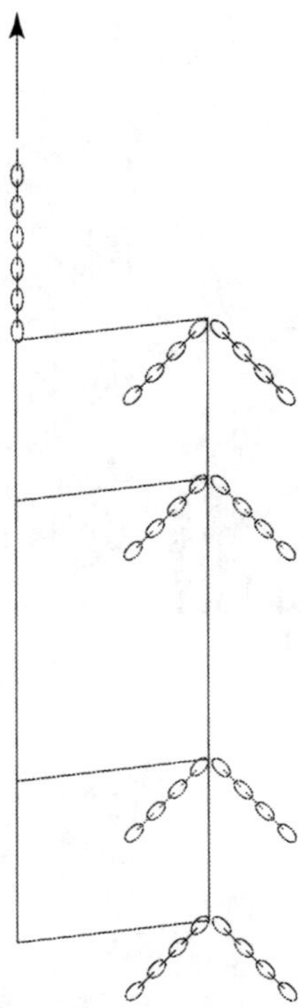

FIGURE 4–33 Setup for diamond correction.

4.3 REMOVING STRUCTURAL PANELS

Once kinked structural parts are pulled to within 3 mm of proper dimensions, they can be removed. Structural body panels are joined together in the factory by spot welding. Therefore, removing panels mainly involves the separation of spot welds. Spot welds can be drilled out, blown out with a plasma torch, chiseled out, or ground out with a highspeed grinding wheel. The best method for removing a spot-welded panel is determined by the number and arrangement of mating panels and the accessibility of the weld.

Drilling out the numerous spot welds in a panel can be tedious. To make the job easier, a spot removing drill is used with an integral clamping mechanism (Figure 4–34). The special rounded bit is forced into the weld by hand pressure.

The removal of spot welds using a plasma torch cutter (Figure 4–35) is much faster. The plasma torch works a little like an acetylene torch. Where spot welds are removed using a plasma torch, a hole is blown in all the thicknesses of metal at the same time. Obviously, the use of a plasma torch does not preserve the integrity of the underlying panels.

A high-speed grinding wheel can also be used to separate spot-welded panels (Figure 4–36). This technique is used only when the weld is not accessible

FIGURE 4–34 A spot weld removing tool.

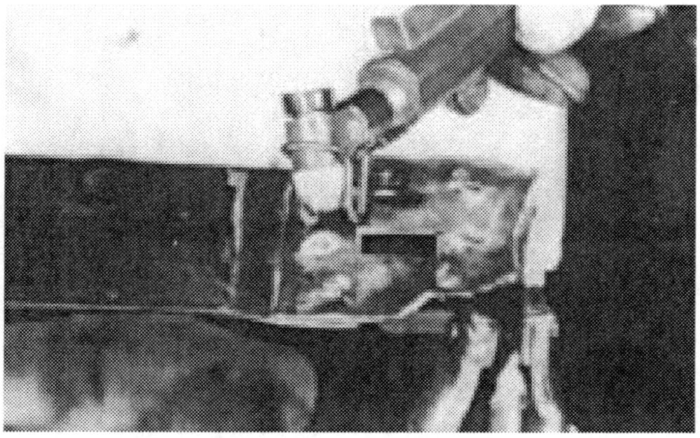

FIGURE 4–35 Removing spot welds with a plasma torch.

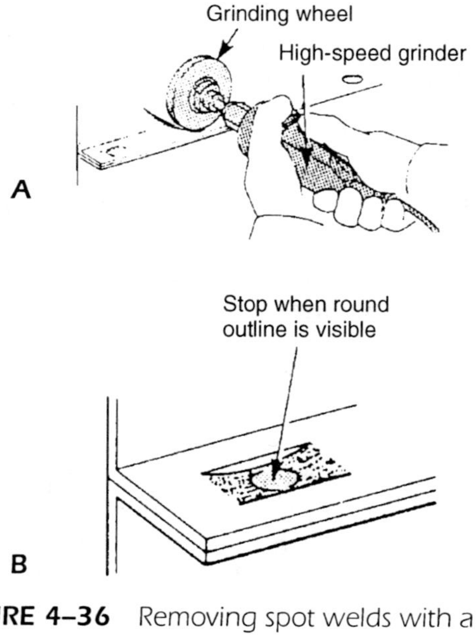

FIGURE 4–36 Removing spot welds with a grinder.

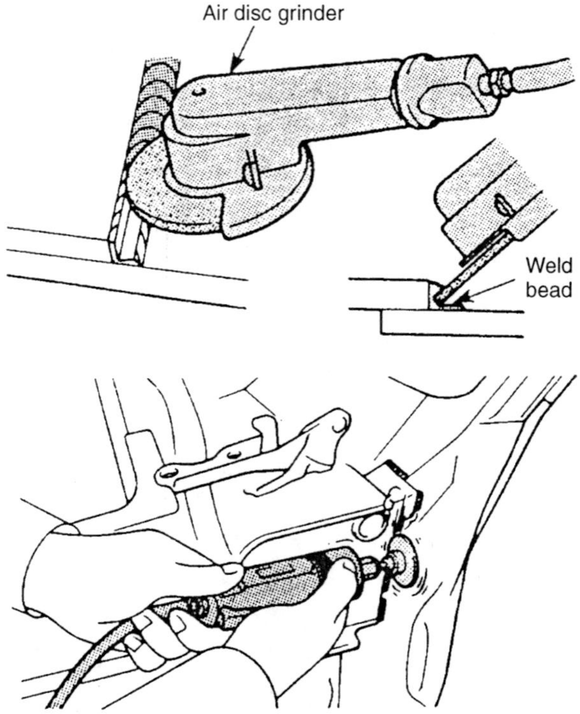

FIGURE 4–37 Removing continuous weld with disc grinder.

with a drill, where the replacement panel is on top, or where a plug weld (from a previous repair) is too large to be drilled out.

After the spot welds have been drilled out, blown out, or ground down, a chisel is driven between the panels to separate them.

SEPARATING CONTINUOUS WELDS

In some vehicles, panels are joined by continuous MIG welding. Since the welding bead is long, a grinding wheel or high-speed grind is used to separate the panels, as shown in Figure 4–37. After grinding through the weld, a hammer and chisel are used to separate the panels.

SEPARATING BRAZED AREAS

Brazing is used at the ends of outer panels or at the joints of the roof and body pillars to improve the finish quality and to seal the body. Generally, separation of brazed areas is done by melting the brazing metal with an oxyacetylene or propane torch and scraping the brazing metal away. However, in areas where arc brazing is used, the fusion temperature of the brazing metal is higher than with ordinary brazing, and melting the brazing metal would result in damaging the panels underneath. Therefore, areas that are arc brazed are normally separated by grinding. Ordinary brazing can be distinguished from arc brazing by the color of the brazing metal. Ordinary brazed areas are the color of brass, but arc-brazed areas are a reddish copper color.

4.4 INSTALLING NEW PANELS

As stated earlier, exterior sheet metal panels are attached with either fasteners or welds. The fastener method of installing panels is simple and fast. It is a matter of bolting the new panel in place and adjusting the fit. Inspect and measure the adjacent or adjoining panels for proper alignment before tightening the bolts. If necessary, straighten any adjacent panel that is out of alignment. One indication of misalignment or damaged inner structure is that the bolt holes do not line up with the bolt holes in the new panel. Also, tapered gaps between panels indicate misalignment.

Welding replacement panels requires care in preparation and alignment. The following procedure is typical of many panel replacement operations. Nonstructural panels, such as door skins, can be bonded rather than welded if allowed by the OEM.

VEHICLE PREPARATION

After removal of the damaged panels, these steps are taken to prepare the vehicle for the installation of the new panels:

1. Sanding off the primer from the plug welding areas (Figure 4–38).
2. Removing dirt, rust, paint, sealers, zinc coatings, and so on from the joint surfaces.
3. Applying weld-through primer to bare metal to preserve corrosion protection.
4. Removing paint and undercoating from the backsides of the panel joining surfaces on parts that will be plug welded during installation.
5. Smoothing the dents and bumps in the mating flanges with a hammer and dolly (Figure 4–39).

REPLACEMENT PANEL PREPARATION

Since all new parts are coated with a primer, the primer must be removed from the mating flanges to allow the welding current to flow properly during welding. To prevent excessive metal removal, use a nylon abrasive wheel instead of a grinder. Drill 5/16-inch holes for plug welds. The following steps are taken to prepare the new panel for welding:

FIGURE 4–38 Removing paint with an abrasive wheel does not remove metal.

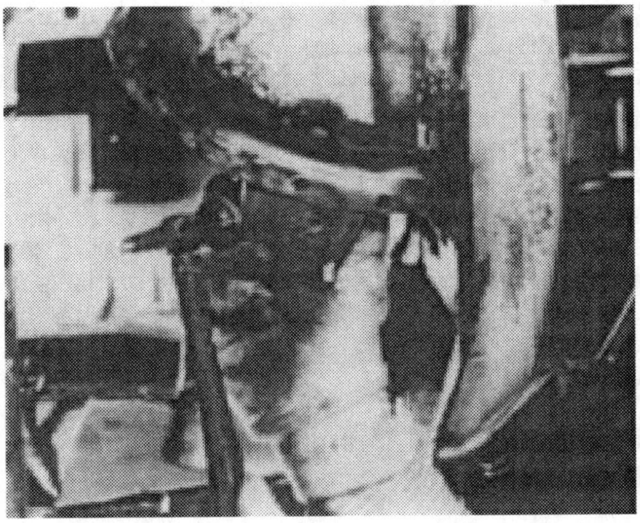

FIGURE 4–39 Smoothing panel flange with hammer and dolly.

1. Using a nylon wheel to remove the paint from both sides of the welding area.
2. Applying weld-through primer (as an anti-rust treatment) to the welding surfaces where the paint film was removed.
3. Making holes for plug welding with a punch or drill. Body repair manuals for each type of vehicle specify the number of plug welding holes. Often, more holes are required than the number of spot welds used on the factory

assembly line. The holes must be a specific diameter and spaced a specific distance apart.

4. If the new panel is sectioned to overlap any of the existing panels, the new panel must be rough-cut to size, using an air saw or a plasma arc cutting tool. The edges should overlap the portion of the panel remaining in the sectioning area on the body by 3/4 to 1 inch. If the overlap portion is too large, it will make matching the position of the panel more difficult during temporary installation.

POSITIONING NEW PANELS

Aligning new parts with the existing body is a very important step in repairing vehicles with major damage. Improperly aligned panels will affect both the appearance and the driveability of the repaired vehicle. Basically, there are two methods of positioning body panels. One way is to use dimension measuring instruments to determine the installation position, and the other is to determine the position by the relationship between the new part and the surrounding panels. The dimensional accuracy of structural panels making up the engine compartment or underbody parts such as the fender aprons or the front side members and rear side members has a direct effect on wheel alignment and driving characteristics. Therefore, when replacing structural panels in unibody vehicles, the dimensional measurement positioning method is used because it is more accurate. There is also a definite relationship between the fit of the new and old parts and the finish appearance, so whether structural or cosmetic panels are being replaced, the emphasis is on proper fit. Of course, it is desirable to use both methods together and, therefore, assure the accuracy and the finish necessary for a high-quality vehicle repair.

Positioning by Dimensional Measurement Methods

When one is using a dedicated bench system of measurement or a mechanical universal system (Figure 4–40), the vehicle must be properly positioned on the bench before the new panel can be correctly aligned. This will usually have already been done in order to pull and straighten damaged panels that do not require replacing. All straightening must be done before replacing panels. Otherwise, proper alignment of the new panels will be impossible.

So, when using a bench system for panel replacement at a factory joint, place the fixtures or gauges on the bench in their correct locations and tie them to the bench. Then the new panel is placed and visually inspected for alignment with adjacent panels. Any necessary adjustments are made, the panel is clamped in place, and the panel is welded into place.

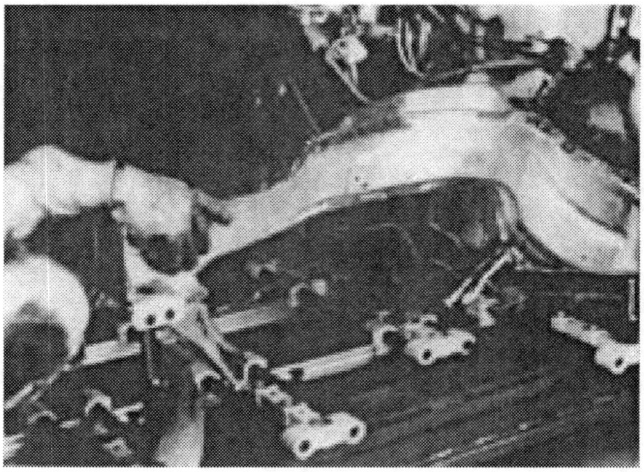

FIGURE 4–40 Aligning a new panel on universal bench.

Positioning by Visual Inspection

Nonstructural outer panels can sometimes be visually aligned with adjacent panels without the precise measurements necessary in replacing structural panels. This is true of both mechanically fastened panels and welded panels. The emphasis here is on appearance. Body lines must be aligned and gaps between panels must be even, not tapered.

4.5 WELDING

Most structural and cosmetic panels are joined together on a unibody vehicle by one or more welding processes. In the repair process, there are basically three welding processes used to replace damaged panels:

1. *Fusion Welding.* In fusion welding, pieces of metal are heated to the melting point, jointed together with a filler rod, and allowed to cool. Of the several fusion welding processes, MIG welding is the most popular in auto repair shops. MIG welding may be used on steel or aluminum. A Tungsten Inert Gas (TIG) welder may also be used on steel or aluminum. Although in a collision repair facility its primary use would be on aluminum.
2. *Pressure Welding.* In this process, pieces of metal are heated to a softened state by electrodes, pressure is applied, and the metal

pieces are joined. Of the various types of pressure welding, electric resistance welding is used in auto body repair.
3. *Arc Brazing.* In a brazing process, metal with a melting point lower than that of the base metals being joined is melted over the joint without melting the pieces of metal.

Figure 4–41 shows areas of a unibody vehicle where these three welding processes are employed. MIG welding, resistance welding, and brazing are explained in the following paragraphs.

MIG WELDING

Metal inert gas (MIG) welding (Figure 4–42) became a necessity in body shops when auto manufacturers began using thin gauge, high-strength, low-alloy (HSLA) steels. Carmakers insisted that the only correct way to weld HSLA and other thin gauge steel was with MIG (or a similar gas metal arc welding [GMAW] system). MIG welding is also applicable for other welding tasks and is much quicker to use than the oxyacetylene welding once commonly used on heavy, body-over-frame vehicles. Welding a rear quarter panel with an oxyacetylene welder averages about 4 hours. A MIG welder can do the same job in about 40 minutes.

The MIG welding method (Figure 4–43) uses a welding wire that is fed automatically at a constant speed as an electrode. A short arc is generated between the base metal and the wire, and the resulting heat from the arc melts the welding wire and joins the base metals together. Since the wire is fed automatically at a constant rate, this method is also called semiautomatic arc welding. During the welding process, either inert or active gas shields the weld from the atmosphere and prevents oxidation of the base metal. The type of inert or active gas used depends on the base material to be welded. For most steel welds, 75 percent argon and 25 percent carbon dioxide (CO_2) is used. With aluminum, either pure argon gas or a mixture of argon and helium is used, depending on the alloy and the thickness of the material. Stainless steel is welded by using argon gas with a little oxygen (between 4 and 5 percent) added.

MIG welding uses the short circuit arc method, which is a unique method of depositing molten drops of metal onto the base metal. The short circuit arc method uses very thin welding rods, a low current, and low voltage. With this technique the amount of heat introduced into the panels is kept to a minimum, and penetration of the base metal is quite shallow.

In summary, the MIG process (Figure 4–44) works like this:

- At the weld site the wire undergoes a split-second sequence of short circuiting, burnback, and arcing.
- Each sequence produces a short arc transfer of a minute drop of electrode metal from the tip of the wire to the weld puddle.
- A gas curtain or shield surrounds the wire electrode. This gas shield prevents contamination from the atmosphere and helps to stabilize the arc.
- The continuously fed electrode wire contacts the work and sets up a short circuit, and resistance heats the wire and the weld site.
- As the heating continues, the wire begins to melt and thin out or neck down.
- Increasing resistance in the neck accelerates the heating in this area.
- The molten neck burns through, depositing a puddle on the workpiece and starting the arc.
- The arc tends to flatten the puddle and burn back the electrode.
- With the arc gap at its widest, it cools, allowing the wire feed to move the electrode closer to the work.
- The short end starts to heat up again, enough to further flatten the puddle but not enough to keep the electrode from recontacting the workpiece. This extinguishes the arc, reestablishes the short circuit, and restarts the process.
- This complete cycle occurs automatically at a frequency ranging from 50 to 200 cycles a second.

Basic MIG Welding Techniques

As shown in Figure 4–45, there are six basic welding techniques employed with MIG equipment.

- *Tack Weld.* The tack weld is a relatively small, temporary MIG spot weld that is used instead of a clamp or sheet metal screw to tack and hold the panels in place while proceeding to make a permanent weld.
- *Continuous Weld.* In a continuous weld, an uninterrupted seam or bead is laid down in a slow, steady, ongoing movement.
- *Plug Weld.* For a plug weld, a hole is drilled or punched through the outside piece (or pieces), the arc is directed through the hole to penetrate the inside piece, and the hole is filled with molten metal. The plug weld is the body shop alternative to the OEM resistance spot welds made at the factory, because it can be used anywhere in the body structure that the factory used a resistance spot weld. Its use is not restricted. It has ample strength for welding load-bearing structural

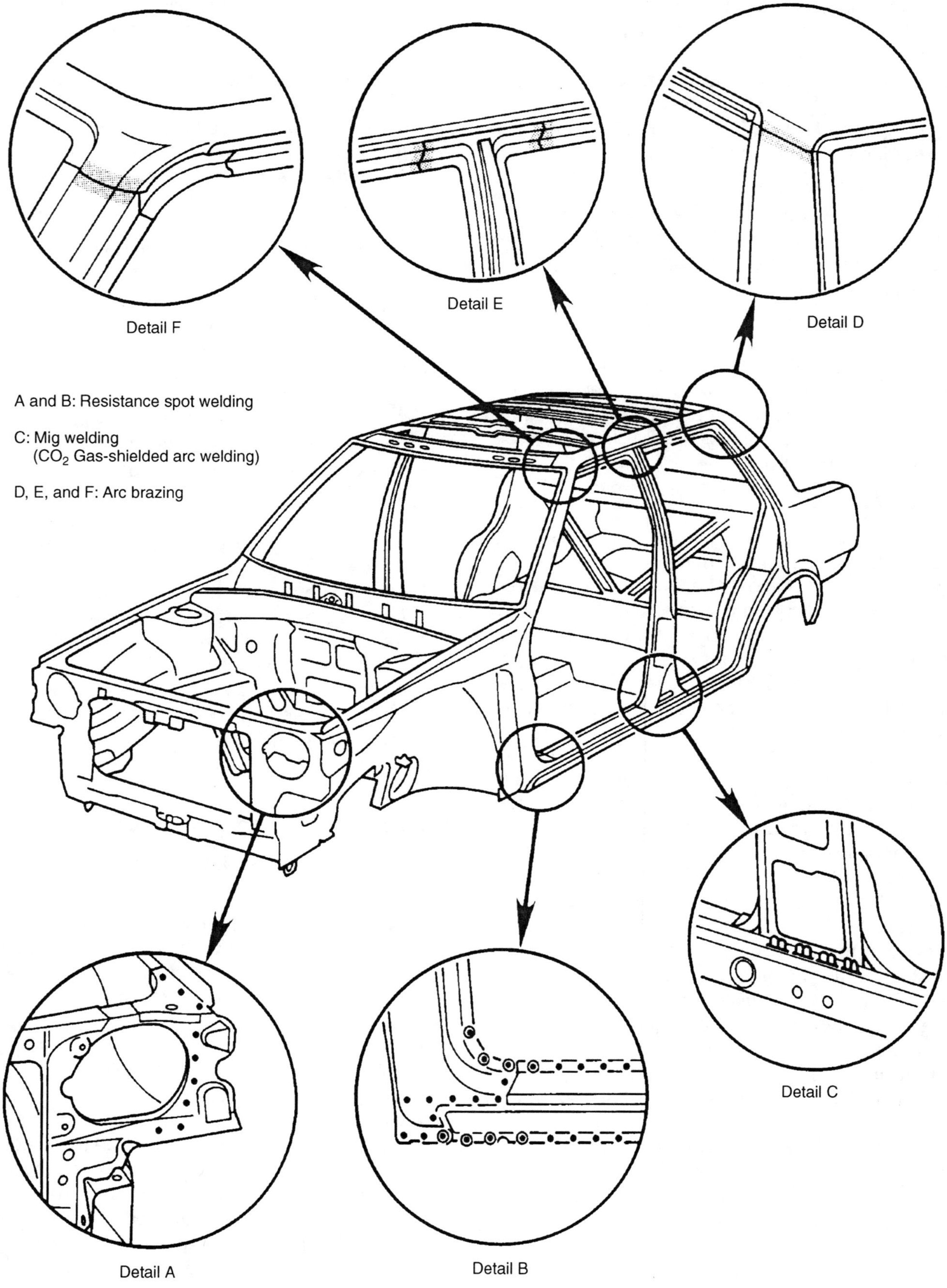

FIGURE 4–41 Body welding on a production vehicle. (Courtesy of Nissan Motor Corp.)

FIGURE 4–42 MIG welding is an important unibody repair skill.

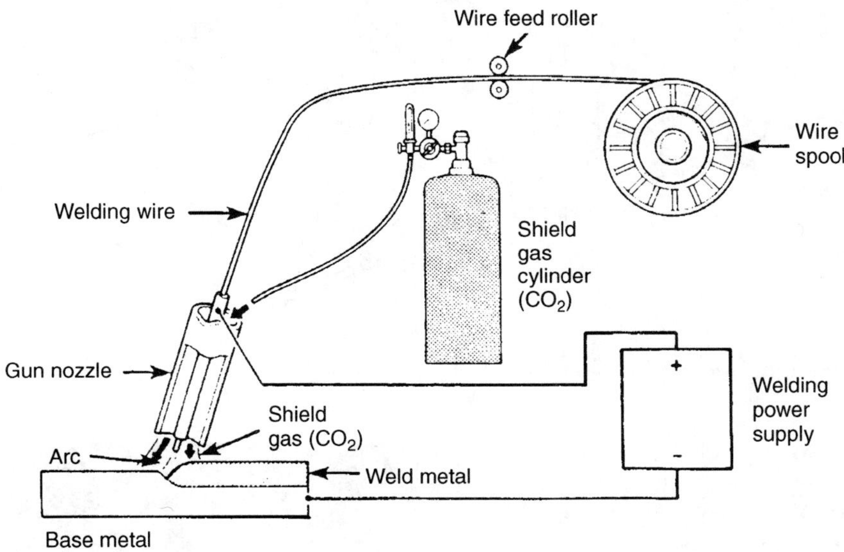

FIGURE 4–43 The principle of MIG welding.

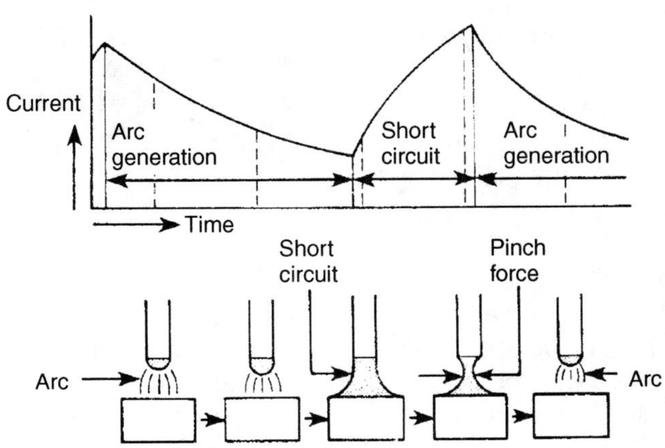

FIGURE 4–44 How the short arc method operates.

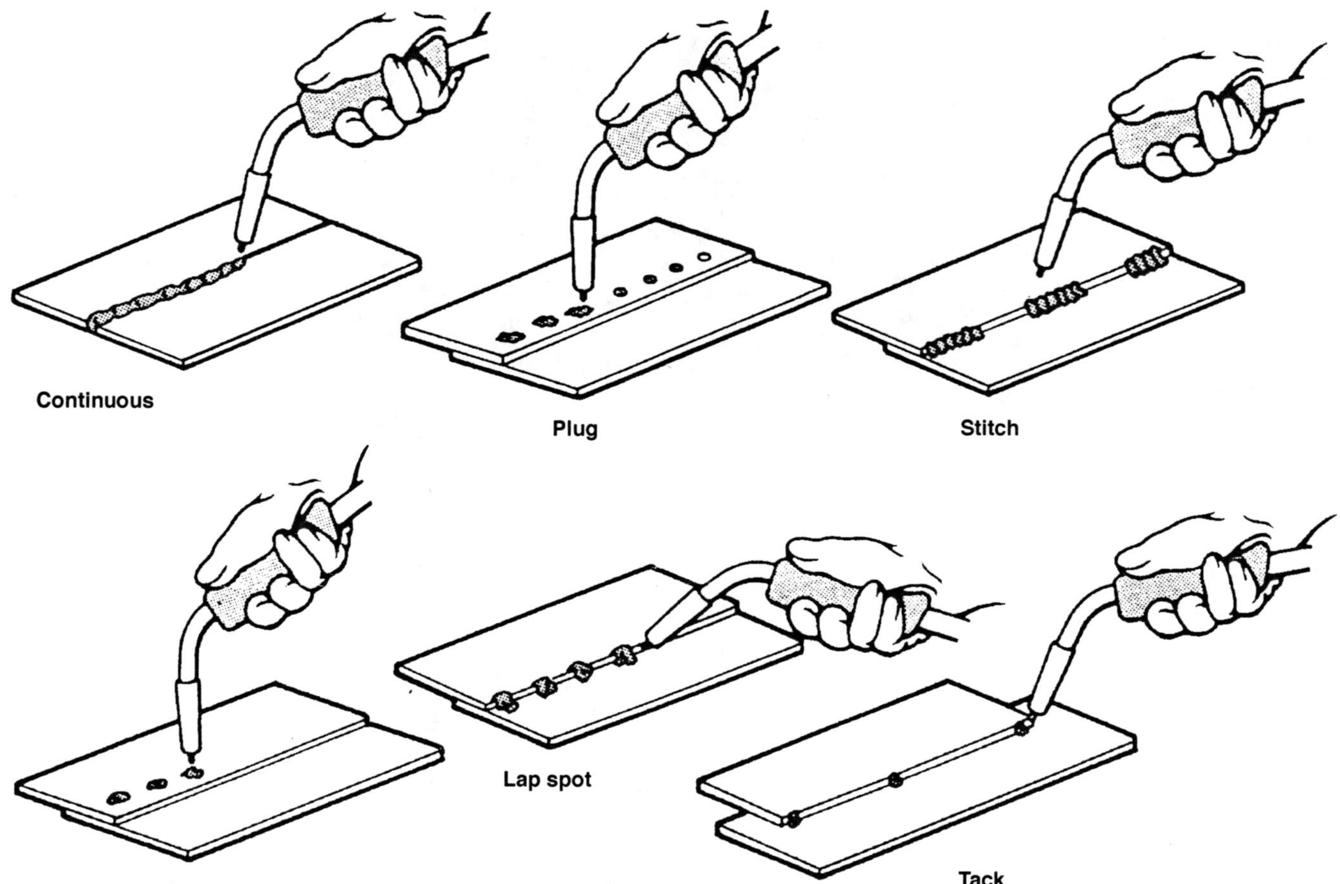

FIGURE 4–45 Basic welding techniques.

members. It can also be used on cosmetic body skins and other thin-gauge sheet metal, too.

- *Spot Weld.* In a MIG spot weld, the arc is directed to penetrate both pieces of metal, while triggering a timed impulse of wire feed.

  For MIG spot welding, a special welding nozzle must replace the standard nozzle. Once this gun is in place and with the spot timing, welding heat, and backburn time set for the given situation, the spot nozzle is held against the weld site and the gun triggered. For a very brief period of time, the timed pulses of wire feed and welding current are activated and the arc melts through the outer layer and penetrates the inner layer (Figure 4–46). Then the automatic shutoff goes into action, and no matter how long the trigger is squeezed, nothing will happen. However, when the trigger is released and then squeezed again, the next spot pulse is obtained.
- *Stitch Weld.* A stitch weld is a series of connecting or overlapping MIG spot welds that create a continuous seam (Figure 4–47). The standard nozzle is used, not the spot nozzle. A stitch weld is a combination of the spot welding and continuous welding techniques. The automatic timer on the MIG machine is set so that the welder pulses and shuts off with automatic regularity. The arc-off period allows the last spot to cool slightly and start to solidify before the next spot is deposited, meaning less distortion and less melt-through or burn-through. These characteristics make the stitch weld preferable to the continuous weld for working thinner-gauge cosmetic panels (Figure 4–48).

FIGURE 4–46 Spot welding with a MIG welding machine.

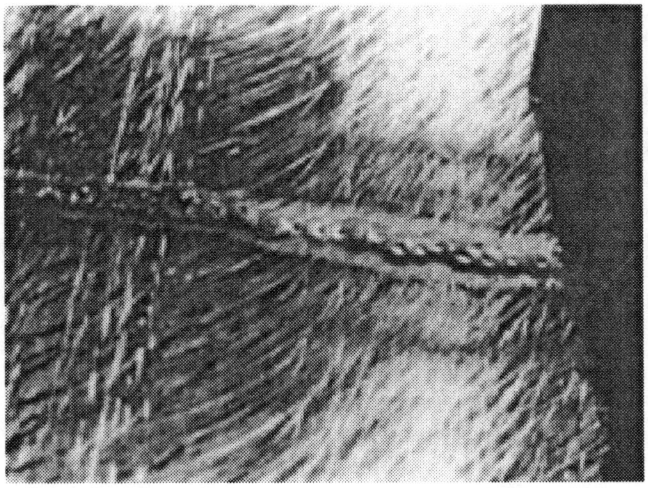

FIGURE 4–47 Results of combining the spot-welding process with the continuous welding gun technique and gun travel.

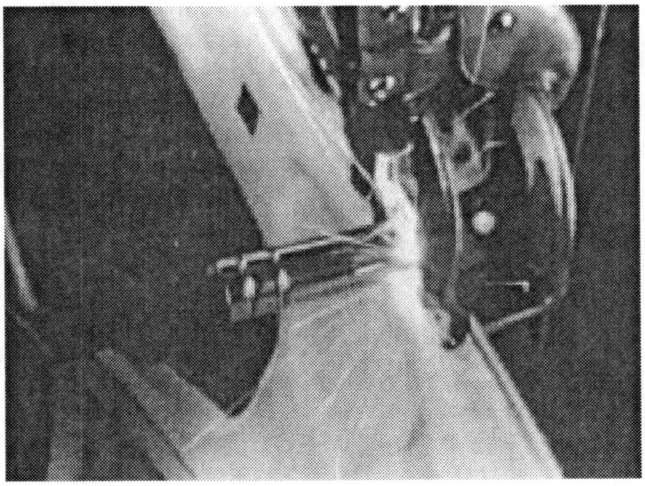

FIGURE 4–48 During one brief, timed pulse of wire feed and welding current, the arc melts through the outer layer and penetrates the inner layer.

Aluminum Welding

Aluminum frames can be welded using a MIG or TIG machine. The MIG welder needs to have different shielding gas and electrode wire as well as a modified wire feed system. The high-frequency current generated by a TIG welder requires that all electronic components be removed from the vehicle prior to welding. Before any technician attempts to weld an aluminum frame or part, he should have instruction and considerable practice. Aluminum does not change color when heated. Heat also travels faster and spreads further in aluminum when compared to steel. The backhand travel method and rapid movement are recommended when welding aluminum.

RESISTANCE SPOT WELDING

Resistance spot welding is the welding process used by automobile manufacturers to make many of the assembly line welds on unibody cars. It is estimated that between 90 and 95 percent of all factory welds in a unibody structure are spot welds. Since resistance spot welding is also specified by a growing number of automobile manufacturers for repair welding of their vehicles, the repair specialist must know how to use a resistance spot welding gun.

The squeeze-type resistance spot welder is ideal for welding many of the unibody's thin-gauge sections that require good weld strength without distortion.

Resistance spot welding has several advantages over MIG welding:

- Reduces welding costs.
- No consumable filler wire, rod, or gas required.
- Clean; no smoke or fumes.
- Allows use of weld through conductive zinc primers to restore corrosion protection to repair joints.
- Duplicates OEM factory weld appearance.
- Eliminates need for grinding of welds.
- Fast; weld times of one second or less make strong welds on HSS and HSLA steels as well as mild steels with a very small heated zone, eliminating distortion of metal.

How Resistance Spot Welding Works

Resistance spot welding relies on the resistance heat generated by low-voltage electric current flowing through two pieces of metal held together, under pressure, by the squeeze force of the welding electrodes. Thus, the three important factors in the operation of resistance spot welding are:

- *Pressurization.* The mechanical welding bond between two pieces of sheet metal is directly related to the amount of force exerted on the sheet metal by the welding tips. As the tips squeeze the sheet metal together, an electrical current flows from the tips through the base metal, causing the metal to melt and fuse together.
- *Current Flow.* When pressure is applied to the metal, a high electric current flows through the electrodes and through the two pieces of metal. The temperature rises rapidly at the joined portion of the metal where the resistance is greatest (Figure 4–49). If the current continues to flow, the metal melts and fuses together.
- *Holding.* If the current flow is stopped, the melted portion begins to cool and forms a round flat bead of solidified metal (nugget). This structure becomes very dense due to the pressurization force, and its subsequent mechanical bonding is excellent.

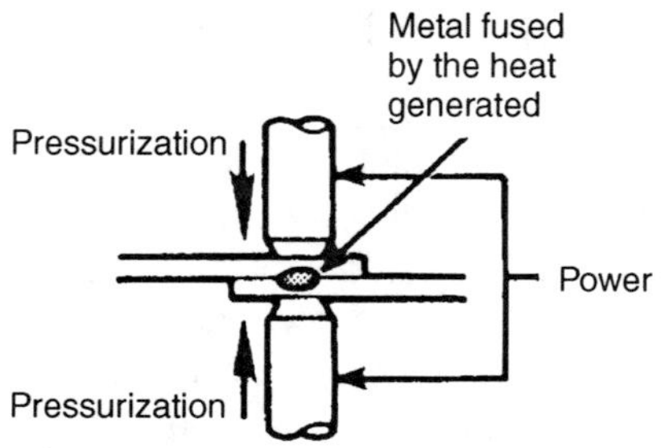

FIGURE 4–49 The resistance of the metal to the flow of electricity causes the metal to melt and fuse together.

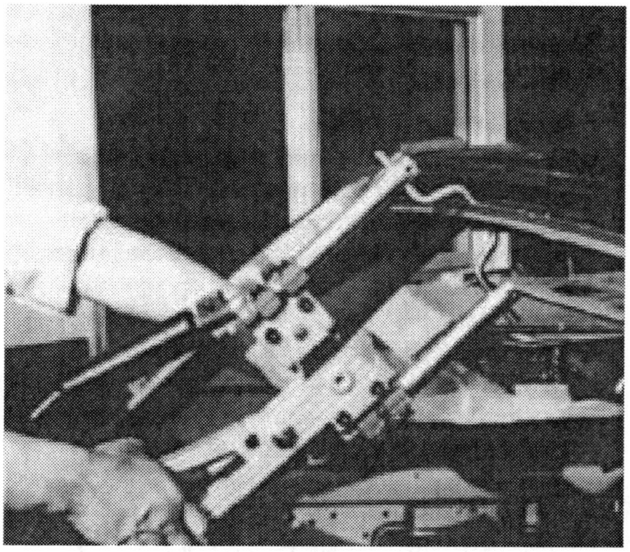

FIGURE 4–50 Resistance spot welding.

The following preparations must be made before spot welding.

1. The panel flanges or mating surfaces must be reshaped if they are damaged. A hammer and dolly are used to straighten flanges.
2. Paint, rust, dust, and all other contaminants on the mating surfaces must be removed so that there is metal-to-metal contact and uninterrupted current flow.
3. The mating surfaces must be coated with an anticorrosion material before welding.

To perform a spot weld with a resistance spot welder, the technician holds the welding gun and positions it so that the welder arm electrodes contact the body parts to be welded. The technician then uses the squeeze mechanism to apply weld force to both sides of the metal being welded. As force is applied and maintained on the metal, the force mechanism initiates an electrical signal to the welder control that switches on the flow of weld current for a preset time and then switches it off. Since the weld time is usually less than 1 second, the entire process is very fast.

Generally, the number of replacement welds should be equal to the number of OEM welds. However, some car manufacturers specify using 30 percent more replacement welds, due to the smaller capacity of spot welding equipment available to repair shops (Figure 4–50).

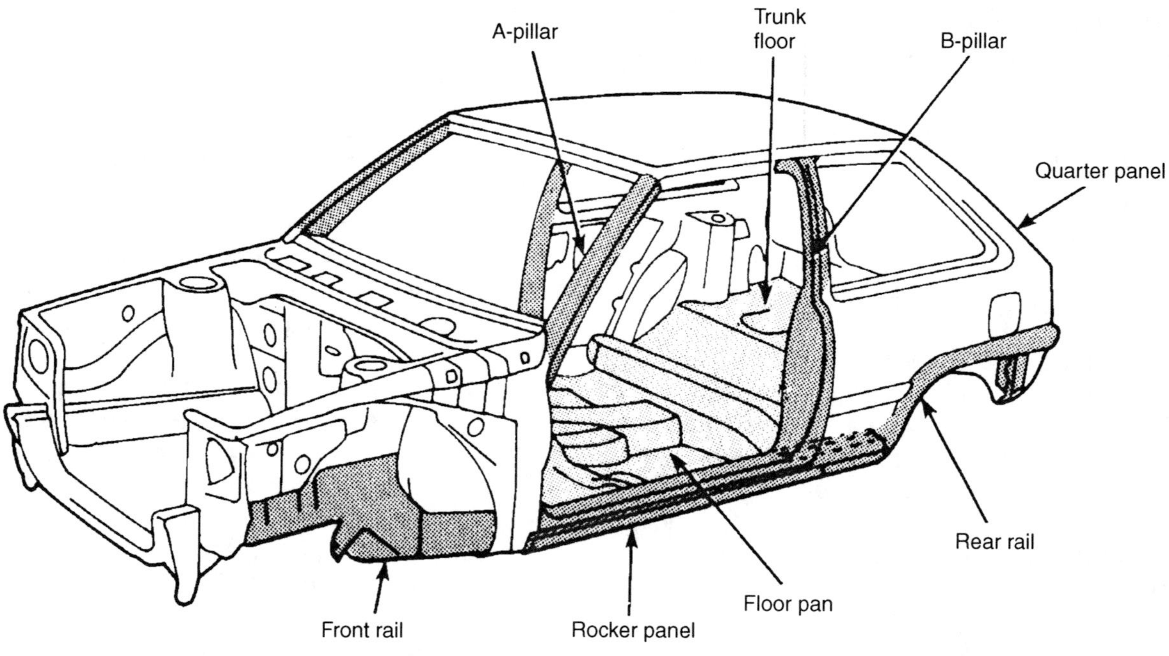

FIGURE 4–51 Commonly sectioned structural parts.

ARC BRAZING

Automobile assembly plants use arc brazing to join the roof and quarter panels together. Arc brazing uses the same principles as MIG welding. However, argon is used with brazing metal instead of the welding wire. Since the amount of heat applied to the base metal is low, overheating is minimized, and there is little distortion to warpage of the base metal. When compared to brazing—depositing brass on the base metal—arc brazing shortens both the time for making the weld and finishing. Also, there is no danger of lead poisoning that is associated with soft brazing with lead.

REPLACING STRUCTURAL PARTS

Structural parts such as radiator support, front upper frame rails, and aprons are usually replaced as a complete assembly. Front lower frame rails, rocker panels, center pillars, and rear frame rails are often sectioned (Figure 4–51). Sectioning, when done properly, is an acceptable method of repair. A sectioned structural part will protect the passengers as well as an undamaged part. To section a structural panel, the damaged part is pulled to within 3 mm of the specified value. The splice joint location can be determined by checking the manufacturer's body repair manual or looking at an estimating guide (Figure 4–52). The damaged panel and the replacement panel are cut at the required location (Figure 4–53). The panels are prepped for welding. The panels are fitted together and proper dimensions are verified. Then the panels are welded together. After welding, the weld joint is treated to prevent corrosion.

Open or U-shaped frames can also be spliced. The damaged frame is pulled to meet specifications. The damaged frame and replacement frame are accurately cut. A sleeve is made from metal the same thickness as the frame. The sleeve is twelve inches long, overlapping the splice joint by six inches on either side. The replacement rail is fitted to the frame with a gap about the thickness of the rail. The sleeve is then clamped in place on the inside of the rail. Holes can be drilled in the sleeve to get around rivets or bolts. The sleeve is MIG-welded in place; then the splice joint is welded. The welds are ground down and corrosion prevention treated. Cracked full frame rails can be welded in the same manner. Non-repairable welded cross members can also be replaced.

REPLACING NON-STRUCTURAL PARTS

Non-structural parts can be welded together—quarter panel, roof and rear body panel—or bolted on—hood, deck lid, door, and fender (Figure 4–54). When the repair time of the damaged part is 75 percent of the cost to replace the part, replacement is the best choice. Figure 4–55 shows a side-hit Taurus wagon. The door and quarter panels are to be replaced. The quarter glass is cut out and saved. The quarter panel is removed by drilling out spot welds and rough cutting with a hacksaw. The right rear door is unbolted; the hinges remain on the center pillar (Figure 4–56). Butt joints will be made in the pillars in front of and to

SIDE

Refinish Wheelhouse Panel & Rails Complete
Right 1.5
Left 1.5
Refinish Wheelhouse Panel .... R/L 1.0
Refinish Upper Side Rail .... R/L .5
Refinish Front Ext. Panel .... R/L .5
Refinish Side Rail Assy. .... R/L 1.0
Section Frame Rail Complete
Right .... #6.5
Left .... #6.5

#Section Approximately 12' on Drivers Side & 14' on Passenger Side from Curve of Frame Rail at Cowl Area, w/Front Apron Assy Removed Deduct 1.0

NOTE: For sectioning methods, refer to TECH-COR information Bulletin #04-88. See disclaimer on Page P3 concerning labor times.

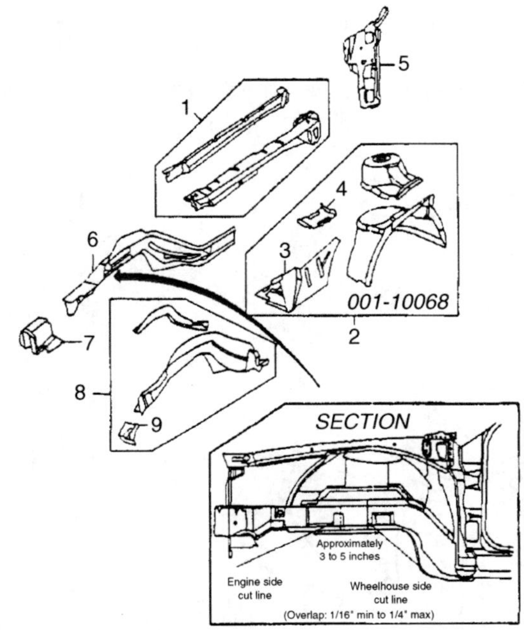

Part	Year/Side	Part No.	Time	Price
1 **Rail, Side Upper**				
HSLA	R/L	20250194-5	**s 2.0**	179.00
2 **Panel, Wheelhouse**	82-91 R	20693766	**s #8.0**	232.00
	L	20681979	**s #8.0**	d223.00
	92 R	20693766	**s #8.0**	232.00
	L	10221444	**s #8.0**	232.00
	93 R	10261844	**s #8.0**	232.00
	L	10254189	**s #8.0**	232.00
	94-96 R	10261844	**s #8.0**	232.00
	L	10254189	**s #8.0**	232.00

#w/Necessary Bolted Parts Removed, Includes R&R Side Upper Rail, w/Radiator Support Removed Deduct 1.0

FIGURE 4–52 Section location from estimate guide.

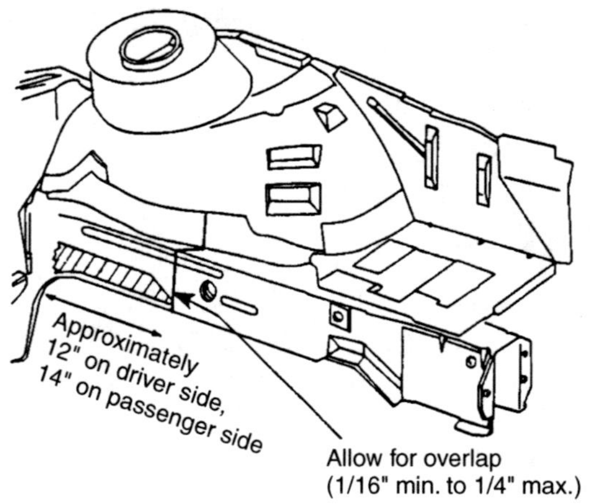

FIGURE 4–53 Lower front frame rail sectional location.

the rear of the quarter window. To make an exact fitup the new quarter panel is rough cut, prepped, and fitted into place. The replacement pillars overlap the existing pillars. The pillars are then cut with a hacksaw at the overlap. This allows a joint gap the thickness of a saw blade. The pillars are MIG stitch-welded together. The panel flanges are plug welded (Figure 4–57).

Bonding

Some damaged non-structural panels can be replaced by bonding if allowed by the OEM. These panels are usually door skins. Tests of properly bonded replacement panels have shown them to be as crashworthy as welded replacement panels. Advantages of bonding are: no warpage from welding heat and no corrosion hot spot from welding. The bonding also provides a good seal. The following steps are used to bond a replacement panel:

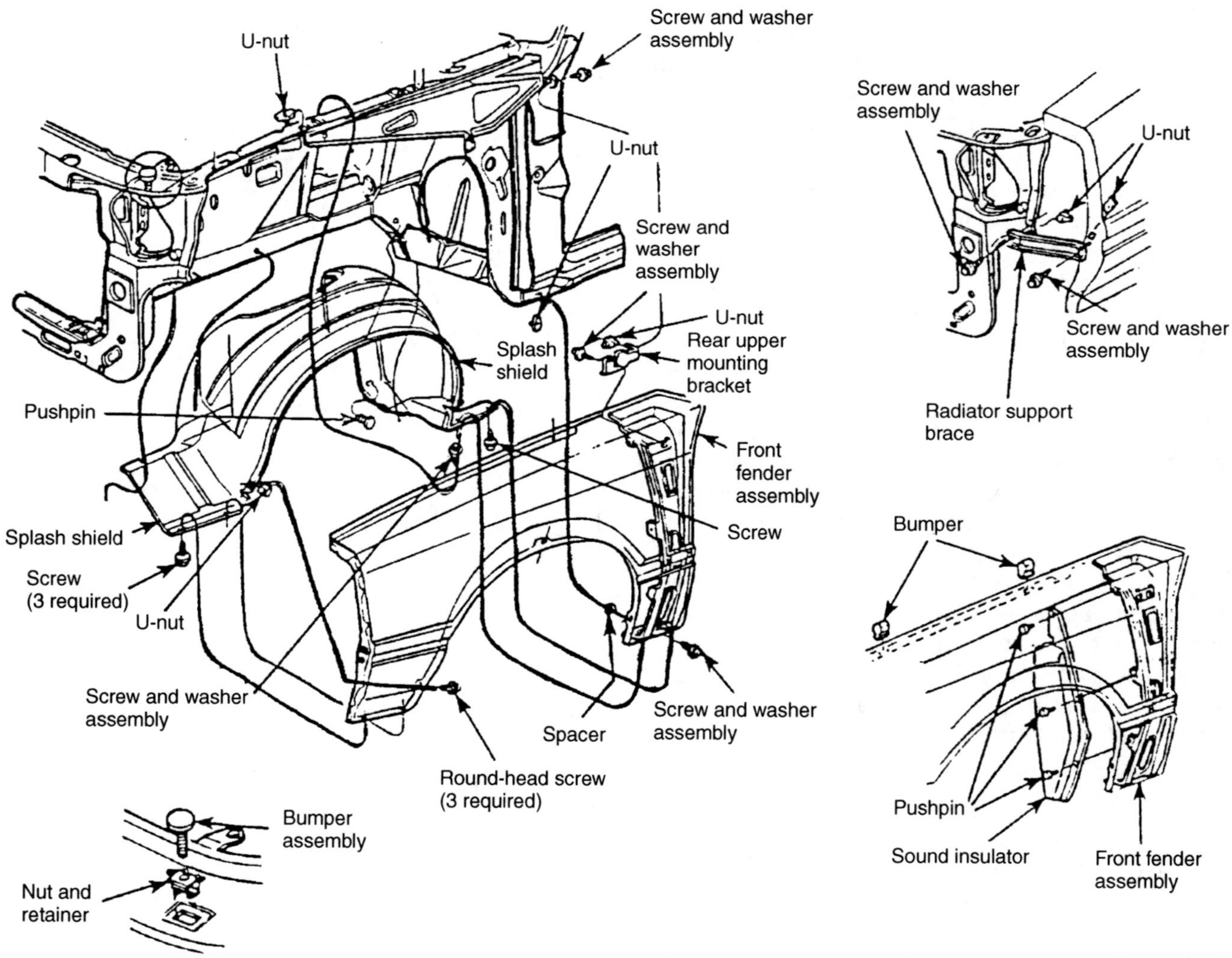

FIGURE 4–54 Typical front fender assembly and splash shield.

FIGURE 4–55 Collision-damaged door and quarter panel.

FIGURE 4–56 Vehicle with damaged panels removed.

1. Remove the damaged panel. On the door skin grind hem flange.
2. Use a hammer and dolly to straighten the flanges.
3. Use a grinder to remove coatings from all surfaces to be bonded.
4. Test fit the replacement panel. Make sure the alignment is correct.

FIGURE 4–57 New panel welded in place.

FIGURE 4–58 Metal adhesive bonding.

5. Apply 3/8-inch bead of bonding material to the edges of the replacement panel (Figure 4–58).
6. Position the replacement panel on the vehicle. Once positioned, do not pull away. If movement is needed, slide the panel.
7. Clamp the panel with even pressure.
8. Allow bonding material to cure.
9. Remove the clamps. Grind off the excess material.

Non-structural bolted-on parts are usually trimmed before they are installed on the vehicle. Trimming or edging involves sanding and priming the part. Then the areas that will be inaccessible once the part is installed are sprayed with color and if needed, clear coat. The trimmed part is aligned and bolted into

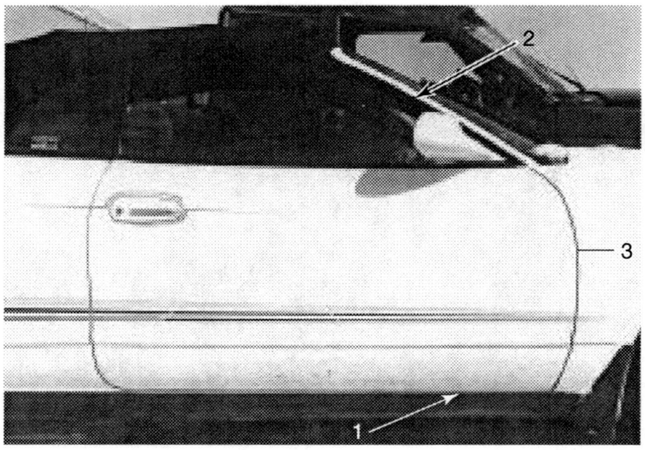

FIGURE 4–59 Sequence to use when checking a door for fit.

FIGURE 4–60 Sequence to use when checking a hood for fit.

place. On doors, the gap at the bottom of the door is checked first. When this is even, the gap at the windshield pillar is checked and adjusted. The door is checked for proper operation. Then the gap at the fender is checked (Figure 4–59). To adjust a hood, first X-check the engine compartment. If the X-check measurements are within specifications, the hood and fenders can be adjusted to align properly. Install the hood and look at the gap between the cowl vent panel or windshield and the hood. Adjust the hood to make this gap even. When the gap is even, the hood is mounted straight. Check for an even gap between the hood and the header panel or grille. Hood height can be adjusted by moving the hood stops up or down. Check the hood for proper latching by shutting it. Adjust the latch if needed (Figure 4–60). To check a fender, install and check the gap between the fender and door, then the gap between the hood and fender (Figure 4–61). Bumpers should be level with even gaps.

FIGURE 4–61 Sequence to use when checking a fender for fit.

4.6 METALWORKING

The buckles and creases in a dented panel can be unlocked in a variety of ways. On panels where the backside of the panel is accessible, hammers and dollies or spoons are used for the initial roughing out. On sealed panels or in areas where the backside of the panel is difficult to reach, slide hammers, picks, and welded studs can be used to reverse the damage.

Damage in an exterior panel can be locked in by indirect damage to structural panels or inner reinforcements. Before this type of damage is corrected, structural damage must be repaired by pushing or pulling the damaged understructure into alignment. Usually when this is necessary, it is best to maintain the hydraulic push or pull until the exterior panels have been straightened using one of the techniques described in the following paragraphs.

Metalworking usually consists of three or four steps:

1. Roughing out the damage and realigning the panel
2. Grinding off the paint
3. Metal bumping and heat shrinking to restore contour
4. Metal finishing

ROUGHING-OUT STAGE

The repair procedure begins with the *rough-out stage.* rough-out is not usually completed on a damaged panel until 80 to 90 percent of the paint can be removed without turning the grinding disc on its edge. Roughing out a panel uses the metalworking techniques discussed later to return the panel close to its original shape.

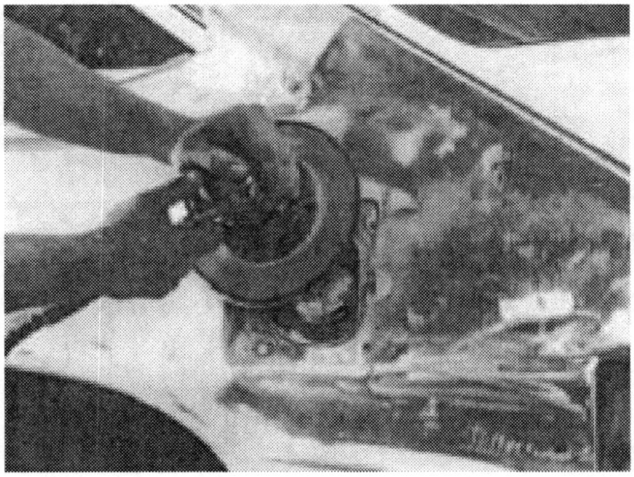

FIGURE 4–62 Removing old paint finish with a grinder.

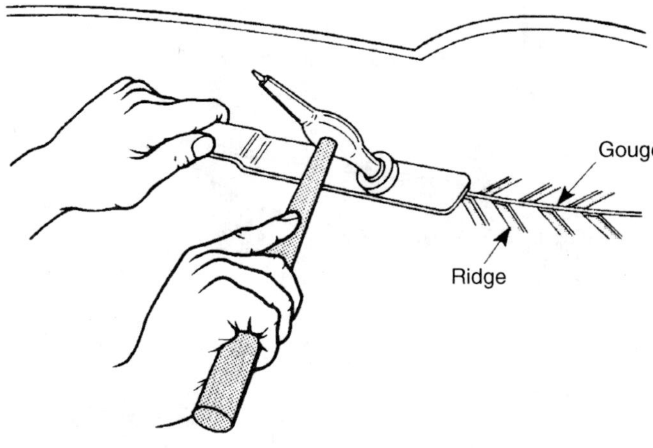

FIGURE 4–63 Lowering a ridge with a spoon and hammer.

REMOVING THE PAINT

After the initial roughing out, many sheet metal repairs require removing the paint finish. Usually this is done with a disc sander or grinder (Figure 4–62). Some body technicians use a 7-inch diameter disc and an electric sander that operates at 4000 rpm or more. Using a sander that is too fast might overheat the metal, causing warpage and additional damage to the metal. Grit sizes range from #16 to #60. Use #24 or #36 for grinding away metal and the higher grits for removing file marks and polishing metal. Two types of backing pads are used. An inflexible pad is used for removing metal and a softer backup pad should be used when removing paint or polishing metal. The softer pad allows the disc to "roll" and give with the metal.

BUMPING DENTS WITH SPOONS

Spoons can be used in a number of ways, depending on their design. Spoons can be used to pry out dents, certain kinds can be struck with a hammer to drive out dents, and others are used as a dolly in hard-to-reach areas. Some are even designed to be used in place of a hammer.

Spring hammering is done with a hammer and a dinging spoon. The dinging spoon is the one most commonly used on sheet metal. It is lightweight and has a low crown. When used, it is held firmly against the high ridge or crease. The spoon is then struck with a ball peen or bumping hammer (Figure 4–63). The force of the blow is distributed by the spoon over a large area of the crease or ridge. This reduces the likelihood of stretching the metal. Firm pressure on the spoon must be maintained when spring hammering. It must never be allowed to bounce. Part of the

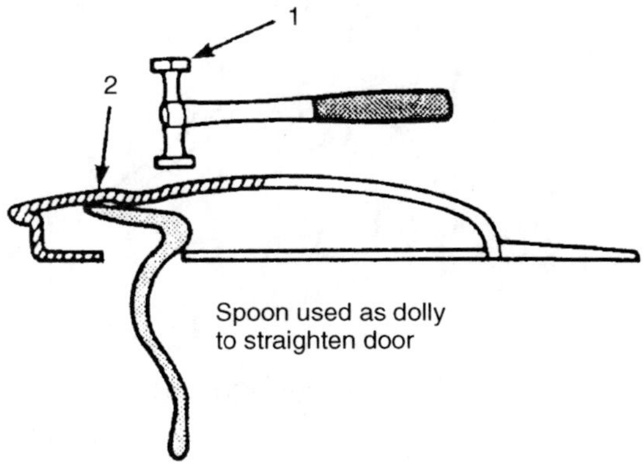

FIGURE 4–64 Using a spoon as a dolly.

corrective force is the pressure of the spoon. Begin at the ends of a ridge (hinge buckle) and work toward the high point on the ridge, alternating from side to side.

Slappers are sometimes used instead of hammers. They can be driven down harder and more often without damaging the panel. They can be used with a dolly. Hammer blows on top of a panel can be a corrective force only when they are placed on pressure (high) areas of the damage. A bumping file can also be used to "slap" down ridges. It has a serrated surface that shrinks the stretched metal.

Spoons can be used to back up the hammer or in combination with a slapper spoon. With a long body spoon, the technician can often reach into places inaccessible to a hammer or dolly. Pressure can be applied to tension areas with the spoon, while pressure or high areas are bumped down (Figure 4–64).

Spoons can also be used to pry metal up in the rough-out stage or to drive deep dents out. Figure 4–65 shows a double end lower back and quarter-panel spoon being used to pry out a dent in a door panel. The door is supported on blocks of wood to

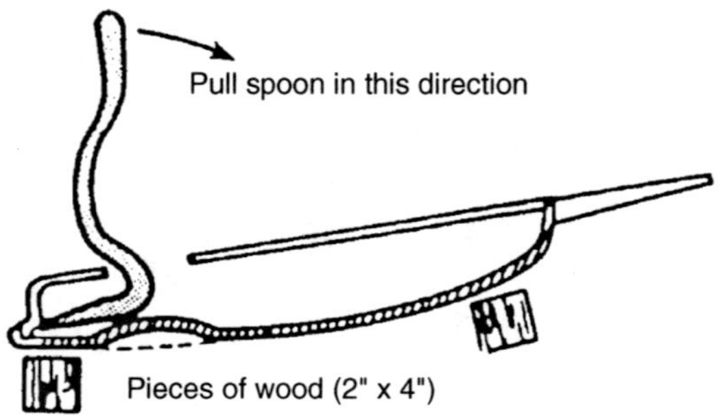

FIGURE 4–65 Using a spoon to pry a dent in a door panel.

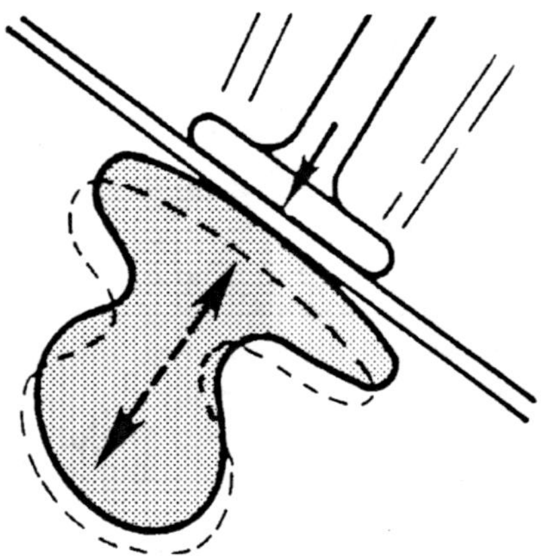

FIGURE 4–66 The dolly rebounds after the panel is struck with a body hammer.

provide clearance for the door panel to move. Once the dent is roughed out by prying with a spoon or dolly, the spoon or dolly can be used with a body hammer to finish the area.

BUMPING DENTS WITH DOLLIES

The dolly block is used throughout the repair process. In the rough-out phase, it is used as an impact tool. The underside of the metal can be hit with the dolly to raise low areas and to unroll buckles. The dolly is also used as a backing block for the hammer. The contour of the dolly must fit the contour of the underside of the damaged area or additional damage will result. As each hammer blow is struck, the dolly tends to rebound slightly from its position against the metal, creating somewhat of a secondary lifting action (Figure 4–66). The amount of rebound force is determined by the pressure exerted against the dolly and by the size and shape of the dolly face in relation to the shape of the metal. A greater lifting action is also obtained when the hammer blows are closest to the dolly. The further away the blows are from the dolly, the less the metal is moved.

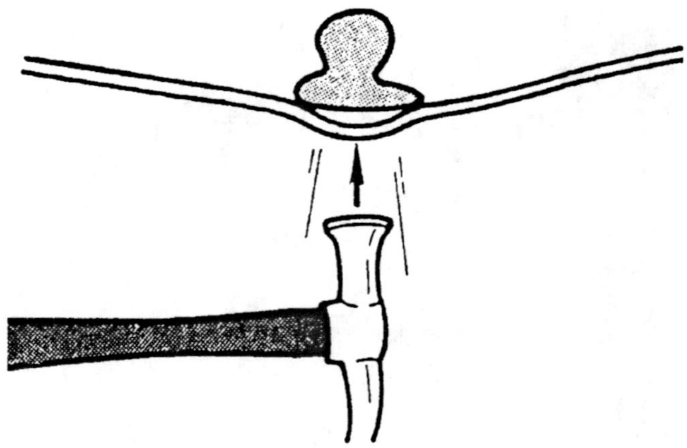

FIGURE 4–67 Hammer-on-dolly technique.

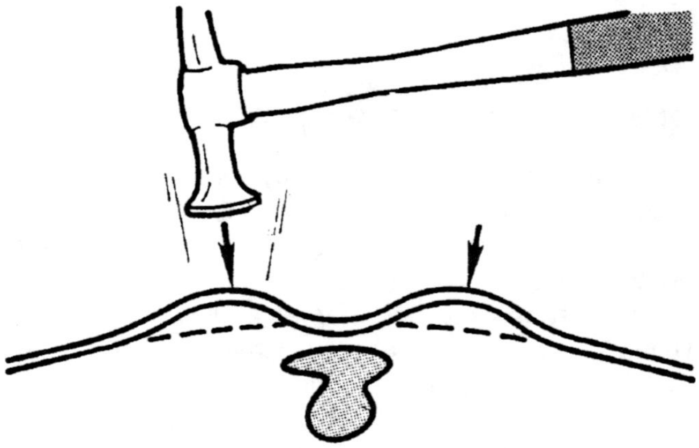

FIGURE 4–68 Hammer-off-dolly technique.

HAMMER-ON-DOLLY METHOD

Hammer-on-dolly repairs are used to smooth small, shallow dents and bulges and to stretch metal so that it can return to its original shape. This occurs usually on crowns and occasionally on flat panels. A bulge is flattened by placing a dolly against the backside of the panel directly behind the bulge and using a hammer from the front side (Figure 4–67). There will be a slight rebound as a result of the hammer hitting the dolly. The dolly will hit the backside of the panel, and as the force of the dolly pressing against the panel is increased, the flattening action will also increase.

HAMMER-OFF-DOLLY METHOD

The hammer-off-dolly method is used to straighten metal just before the finishing stage. In it the hammer does not actually strike the dolly. The dolly is held under the lowest area, and the hammer hits the high area nearby (Figure 4–68). The dolly will rebound

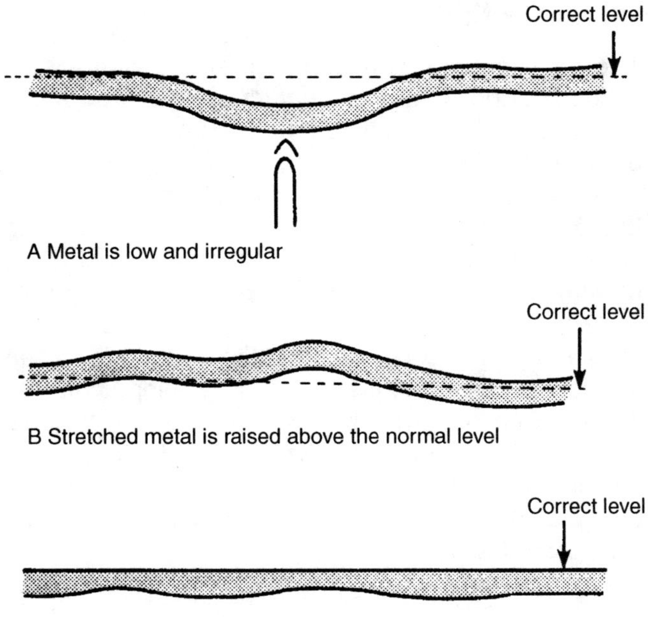

FIGURE 4–69 Raising a dent with a pick.

from the shock of the hammer and strike the underside of the panel. The dolly, like the hammer, is designed to correct damage in that it must strike in only tension areas to be effective as an impact tool.

The hammer-off-dolly method is used mostly on flat or low-crown panels. These panels are soft in comparison with crowned panels.

PICKING DENTS

There are several methods of picking up metal with the use of a pointed (not necessarily sharp) tool. Pick hammers, long picking tools, the edge of a dolly, and even a scratch awl can be used. When picking up a small dent with a striking tool, it is better to use several light blows rather than one or two hard blows. After an area has been picked up, a file or grinder can be used to level the damaged area. Figure 4–69 shows a low spot being raised with a pick. A certain amount of stretching has occurred; so, the metal must be filed after the low spot is eliminated. Hard filing will heat the metal causing it to shrink partially back to shape.

Long picking tools can also be used to pry up metal in areas that cannot be reached with a dolly or spoon. A car door is a good example. A pick can sometimes be inserted through a drainage hole or a hole drilled behind the door gasket. This eliminates the need to remove the inside door trim or to drill holes in the outer panel for pulling the dent.

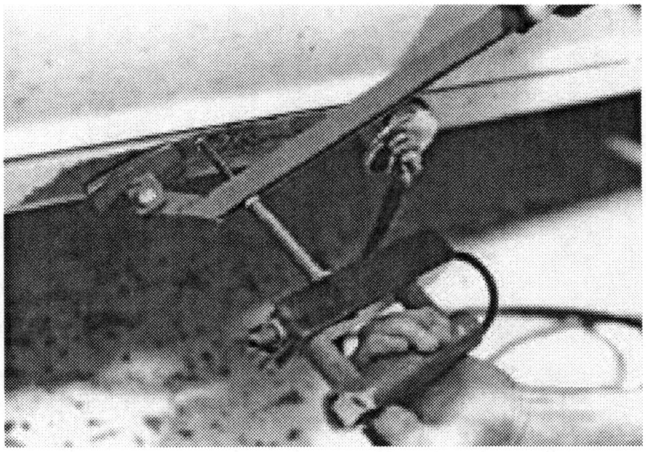

FIGURE 4–70 Pulling a dent with a dent puller.

PULLING DENTS

Dents can be pulled out with a number of tools: suction cups, pull rods, dent pullers, even a sheet metal screw and vise grips. By far, though, the most common method of pulling dents is with a dent puller. Figure 4–70 shows one type of dent puller in use. The technician places the dent puller shaft on the bare metal dent, then presses the trigger. The shaft is welded to the dent. A leverage handle is used to pry up on the shaft, raising the dent. The shaft is twisted to release the weld. If a portion of the dent remains the shaft can be rewelded.

Access to the inside of many panels is blocked by welded inner panels. Window mechanisms often interfere when a door panel is damaged. Using a dent puller, the technician can often repair a simple dent in less time than would be required to make the disassembly necessary to start the repair by other means.

Before car manufacturers began using galvanized sheet metal, dent pullers with screw tips were commonly used to pull dents. This type of dent puller is used as follows: After scrutinizing the dent to determine the point and angle of the initial impact, the technician should drill or punch holes 1 inch apart along the length of the initial crease. Beginning at the lowest point of damage, the screw tip on the dent puller is threaded into the first hole. The handle of the dent puller is held in one hand and the weight is slid forcefully against the handle (Figure 4–71). Moving from hole to hole, the dent or crease is gradually worked out. The surface must be worked as close to the finished contour as possible without pulling the metal beyond the original surface. High spots are tapped down with a body hammer to relieve pressure in ridge areas. The holes must be welded shut. Low spots can be filled with body filler.

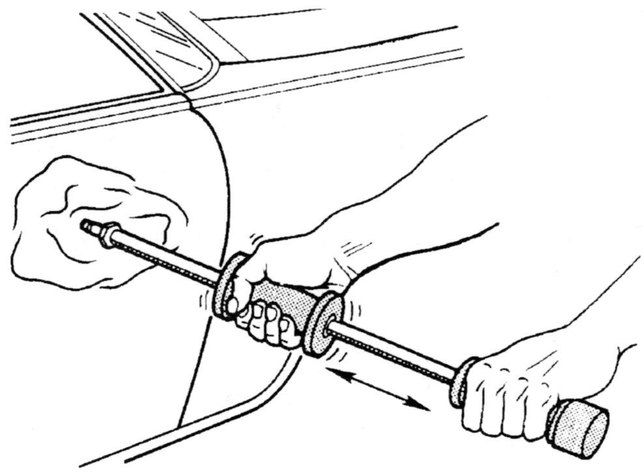

FIGURE 4–71 Removing a dent with a dent puller.

Pulling with Studs

The most popular way to pull dents is with a stud welded to the dent. Pulling with spot-welded studs avoids drilling or punching through the metal and undercoating, potentially a corrosion creating technique.

A specially designed spot welder fuses a metal pin to the dent area (Figure 4–72A). Fusing the pin to the panel takes only a fraction of a second. Then, the pin or pins can be pulled with a dent puller (Figure 4–72B). Several pins can be welded in a crease and all pins pulled at the same time (Figure 4–72C). When the dent is corrected, the pins can be snipped off with cutters and ground flush with the panel (Figure 4–72D). The whole process is very quick with no damage to the panel. It is especially useful in pulling out small door dings where the backside of the panel is inaccessible.

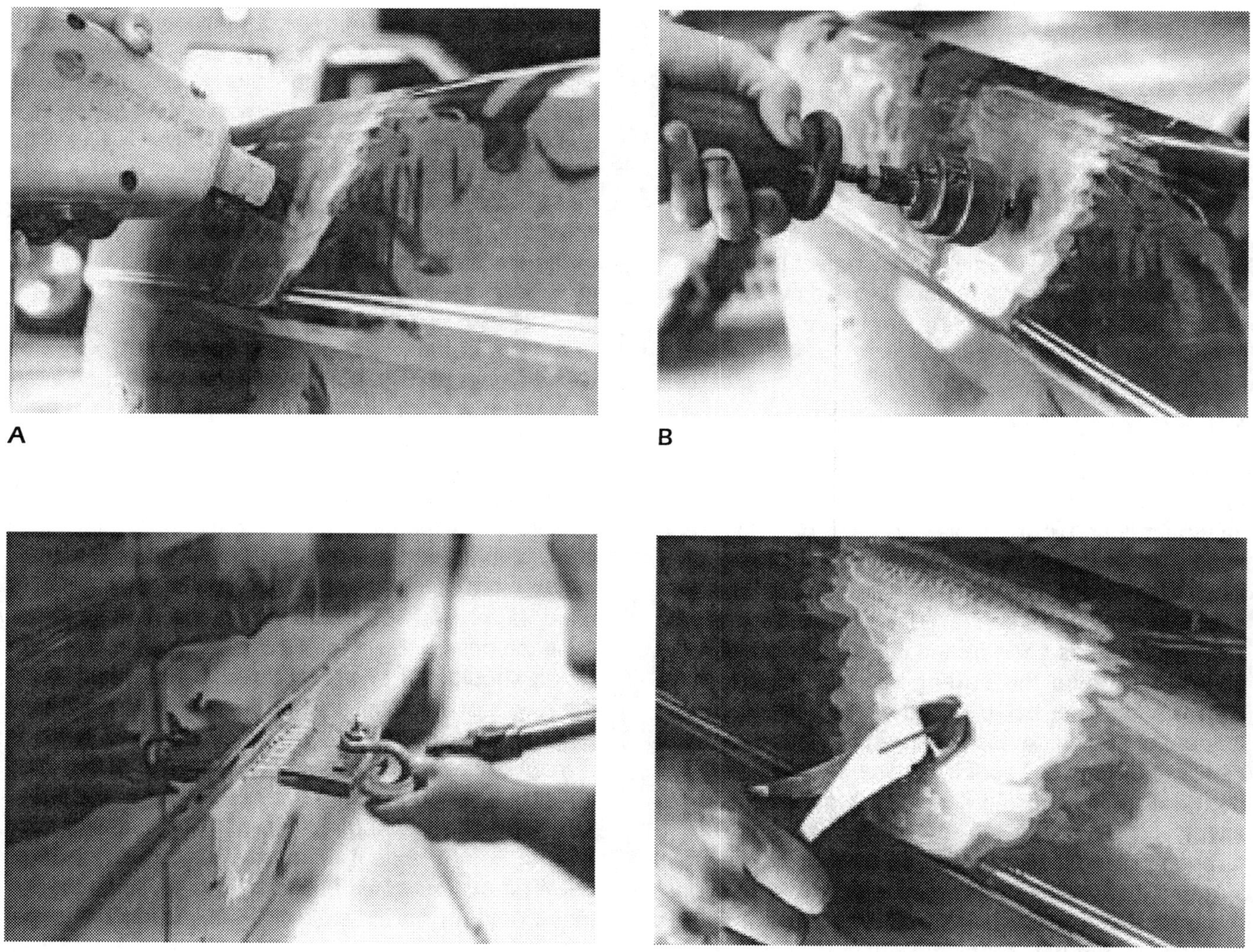

A B C D

FIGURE 4–72 Pulling a dent with stud-welded pins: (A) welding the stud to the dented panel; (B) pulling the dent with dent puller; (C) pulling several pins at one time; and (D) cutting off the welded pins.

4.7 SHRINKING METAL

When metal is severely damaged in a collision, it is often stretched in the badly buckled areas. These same areas are also sometimes stretched slightly during the straightening process. Most of the stretched metal will be found along the ridges, channels, and buckles in the direct damage area.

When an area of metal is stretched, the grains of metal are moved further away from each other. The metal is thinned and work hardened. Shrinking is the method used to bring the molecules back to their original position and thus to restore the metal to its proper contour and thickness. The object of shrinking is to remove the stretched metal without disturbing the relatively undamaged elastic metal in the surrounding area.

A variety of pieces of welding equipment can be used to heat metal for shrinking. Attachments are available for spot and MIG welding equipment as well as stud welders to transform them into shrinking equipment. But the most commonly used tool is the oxyacetylene torch with a #1 or #2 tip.

Shrinking Operation with a Gas Torch

To shrink an area with a torch, a small spot of the stretched area or bulge is heated to a cherry red. The "shrink" is placed in the highest spot of the stretched area, then in the next highest spot, and so on, until the attached area has been shrunk back to its proper position (Figure 4–73).

As the heat from the torch is applied to a small spot in the panel, the heated metal expands. The cooler metal surrounding the hot spot resists the expansion forces. As the temperature increases, the heated metal becomes softer. This soft metal piles up and forms a bulge in the hot spot.

The metal usually bulges up instead of down because the top of the metal is heated first. When it starts to bulge, the rest of the metal in the hot spot follows.

When the spot has been heated, a body hammer is used to tap around the spot with several sharp hammer blows. This will drive the molecules of metal closer together.

During this procedure, it is not necessary to support the metal with the dolly unless the metal collapses. If it is necessary to support the panel, the dolly should only be held lightly under the metal. As soon as the redness disappears, off-dolly blows and light, on-dolly blows are used to smooth and level the area around the hot spot.

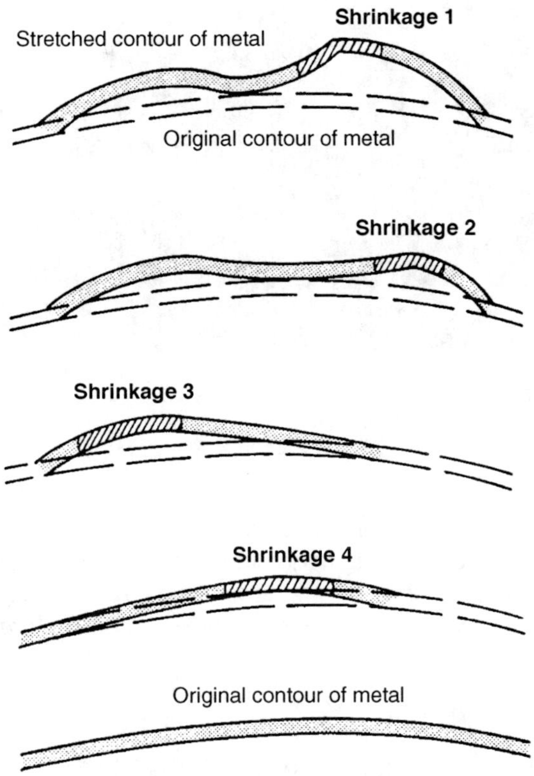

FIGURE 4–73 Shrinking stretched metal usually requires more than one hot spot. The highest spot is always heated first.

Once the redness has disappeared and the area has been dollied smooth, the "shrink" can be cooled with a wet rag or sponge. When this is done, a greater degree of contraction occurs, and a slight amount of distortion could result. The warpage should get straightened before the next "shrink" is attempted.

It is not uncommon to find that an area has been overshrunk when the "shrink" has completely cooled. When the area has been overshrunk, the metal in the area of the last "shrink" is usually collapsed or pulled flat and sometimes the metal surrounding the "shrink" area can even be pulled out of the proper contour. Overshrinking is corrected by using hard, hammer-on-dolly blows to stretch the last "shrink." The last shrunken area is usually the direct cause of overshrinking.

Aluminum panels, such as hoods and deck lids, should be repaired if the damage is light and near the edge of the panel. There are some differences between aluminum and steel, one being that aluminum is softer and more easily stretched than steel. Since aluminum spreads heat more rapidly, an aluminum panel is easily warped by the heat of grinding. The hammer-on-dolly technique should be avoided. The buckles in aluminum will be more difficult to remove than steel. Use a body spoon to spring hammer buckles. Picking

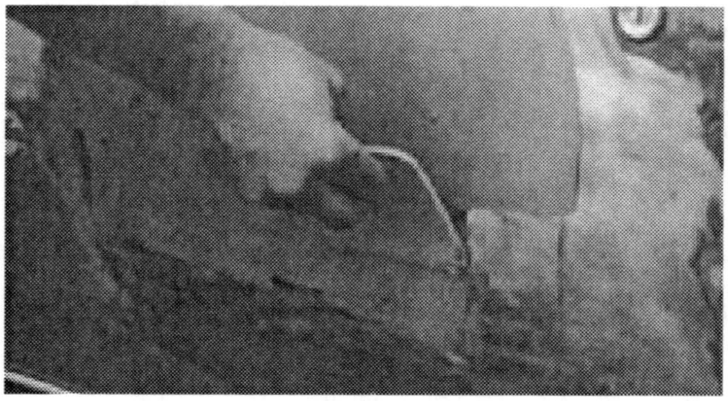

FIGURE 4–74 Applying filler with a plastic spreader.

with a blunt end pick hammer may be used to raise small dents. Careful, light filing can remove the picks. Remove paint with 80 grit paper on a DA. The repaired area should be sprayed with epoxy primer before applying body filler.

4.8 APPLYING BODY FILLER

After the metalwork is complete, plastic body filler must be applied to fill in the remaining depressions in the panel and to create a smooth, level surface. There are several different kinds of body filler on the market. The better fillers go on smoothly, dry tack-free without pinholing, and are easy to file and sand. Some have fiberglass fibers or metal particles to make them stronger and harder.

Plastic fillers must be mixed with a hardener, or catalyst. The hardener causes the filler to undergo a chemical change that causes the filler to become very hard. The hardening process occurs quickly after the hardener is mixed with the filler, so it must be applied quickly (Figure 4–74).

After fifteen to twenty minutes, the filler will become hard enough that it can be filed (Figure 4–75). A file called a "cheese grater" is used to shape and level the filler. Filing the filler at this stage greatly reduces the time and effort involved in sanding the filler.

After the filler has thoroughly hardened, the file marks are removed using a coarse grit sandpaper, followed by medium and fine grit papers until the filler is level with the contour of the panel and is smooth and free of scratches (Figure 4–76).

After cleaning the filler with compressed air and a tack cloth, a layer of two-part spot putty is often applied to fill in small sand scratches and pinholes. This ensures that the repair area will be smooth and prevents hardener bleed through on clear coats.

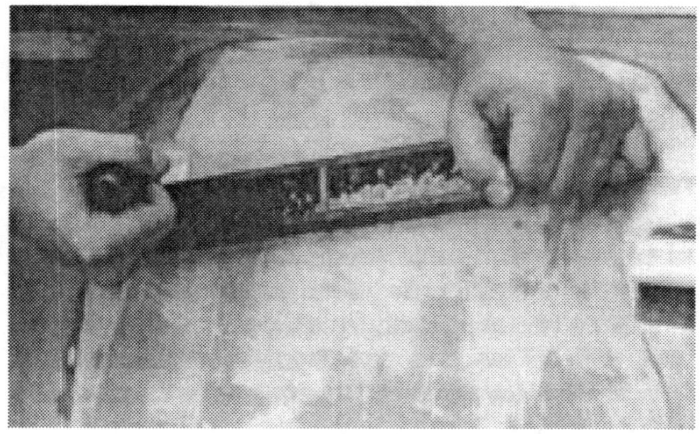

A

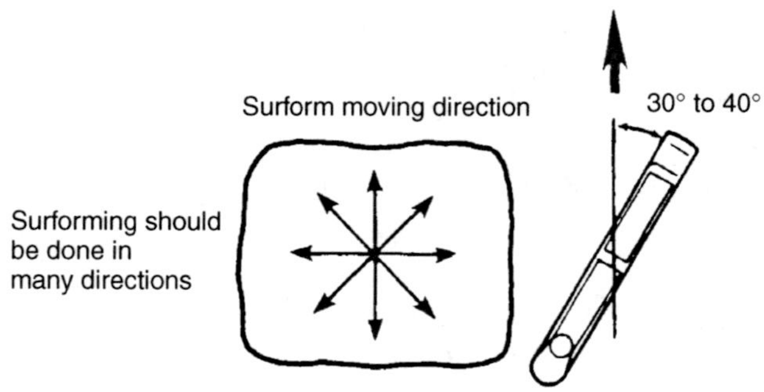

B

FIGURE 4–75 Grating the semi-hard plastic.

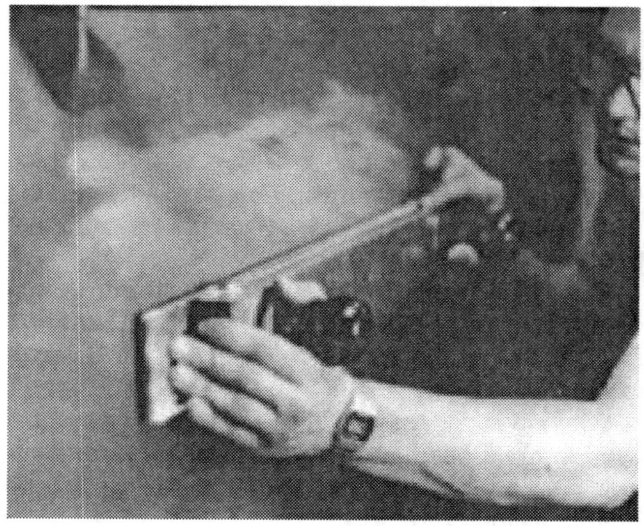

FIGURE 4–76 Sanding the plastic filler with a sanding board.

4.9 SURFACE PREPARATION

After the damaged panels have been either straightened or replaced, the vehicle must be prepared for refinishing. There are several steps involved in preparing a panel for refinishing.

1. *Washing.* It is important to thoroughly wash all repair panels prior to refinishing. Otherwise, dirt nibs will be visible through the new paint film.
2. *Dewaxing.* Any foreign substance on the repair panel will result in poor adhesion of the new paint. Many waxes also contain silicone. A condition called fisheye occurs in the paint if the silicone is not removed. A wax and grease remover (Figure 4–77) is applied to the repair area with a cloth. A second clean cloth is used to wipe the dissolved waxes off the panel.
3. *Stripping.* If the old finish is badly damaged or deteriorated, the paint should be removed. Either chemical paint removers or sandblasting can be used to quickly strip a panel.
4. *Sanding.* Usually the old finish is sound enough that it does not have to be removed. However, blemishes, such as nicks or scratches, must be sanded smooth. Broken paint edges must be *featheredged* to a wide taper (Figure 4–78); otherwise, the edges of the old finish will be visible as ridges under the new paint film. The panels to be refinished must also be *scuff sanded.* Scuff sanding removes the hard surface gloss of the old finish, enhancing the adhesion of the new finish.
5. *Metal Conditioning.* Bare metal very quickly begins to rust. Microscopic rust particles must be neutralized with an application of metal conditioner; otherwise, the surface rust will eventually destroy the bond between the new finish and the metal substrate, resulting in peeling. Metal conditioner also *etches* the metal, a process which chemically encourages the bonding of paint to metal. Soon after treating the bare metal with conditioner, the metal should be sprayed with a primer to seal it against further rust formation.
6. *Masking.* Before spraying any primer or topcoats, nearby glass and trim should be masked with tape and paper (Figure 4–79). Adjacent panels should also be masked to avoid overspray.
7. *Cleaning.* The repair area must be given a final cleaning just prior to application of the new finish. Dust should be blown out of the corners and crevices with compressed air. Then, the panels should be wiped down with a tack rag.

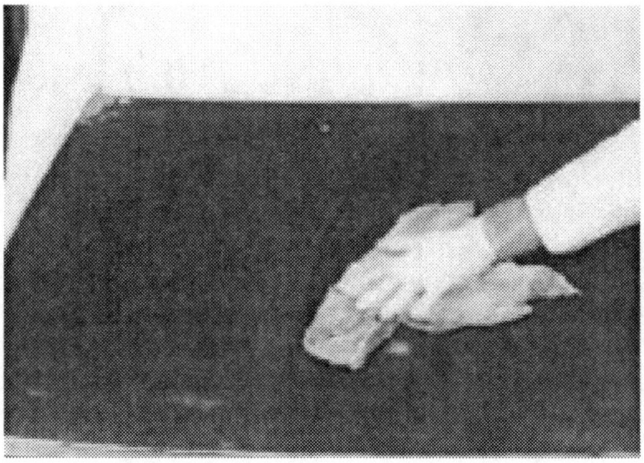

FIGURE 4–77 Applying a wax and grease remover. (Courtesy of DuPont Co.)

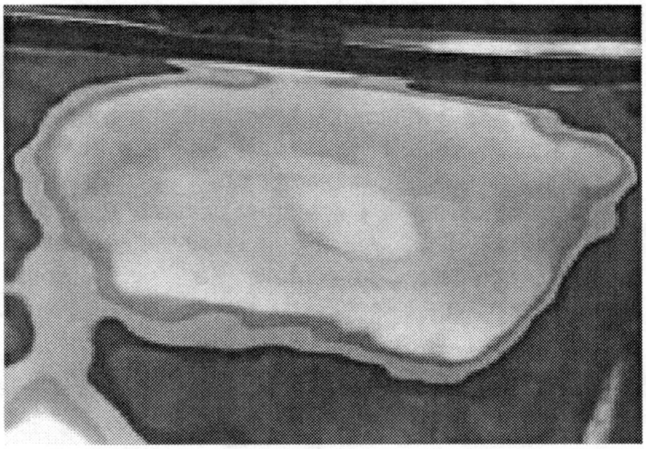

FIGURE 4–78 Feather edge of paint layers around the body filler.

FIGURE 4–79 Masking has been applied to all glass, trim, and exterior accessories.

4.10 APPLYING THE FINISH

After the metal damage has been repaired and the panel surfaces have been properly prepared, the new finish can be applied. The painter will choose a paint system that is compatible with the original or existing finish. Paint systems contain several refinishing materials that are designed to be compatible. Some of the materials classify as *undercoats;* others are called *topcoats.*

Undercoats and topcoats are made of two types of paints: lacquer paint or enamel paint. Lacquer paints dry and physically harden by the evaporation of solvents. As the solvents evaporate, the binder and pigments left behind combine into a solid state.

Enamels on the other hand cure in two stages; the first stage is evaporation of the solvents and the second stage is chemical oxidation of the binder. Oxidation of the binder occurs as the binder combines with the oxygen in the air.

Lacquer will soften and dissolve if solvent is applied to it; enamels, however, are insoluble after the oxidation process is completed. An enamel paint will also dry to a glossy finish; lacquer paints on the other hand dry to a dull finish and must be rubbed with a polishing compound to achieve the desired gloss. Lacquer paints are rarely used today.

UNDERCOAT PRODUCTS

Proper undercoating is the foundation for an attractive, durable topcoat. If the undercoat is not correctly applied, the topcoat appearance will suffer and might even crack or peel. There are three general or basic types of undercoats:

- Primer
- Surfacer
- Sealer

Primers. By definition, a primer is generally the first coat in any finishing system (Figure 4–80). It is designed to prepare the bare substrate to accept and hold the color topcoat. Primers will also produce a corrosion-resistant foundation. Primers generally do not fill surface imperfections and often do not require sanding.

The two part *epoxy primers* are probably the most versatile and valuable primer products on the market. An epoxy primer system consists of the primer itself and an activator (or catalyst), which must be mixed together. Epoxy primers restore corrosion resistance to a repaired panel.

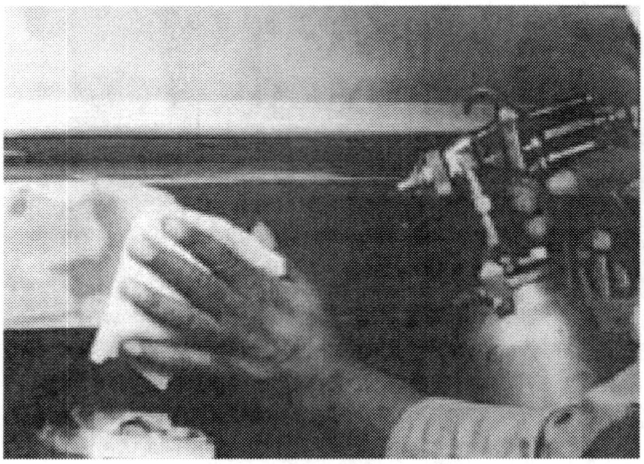

FIGURE 4–80 Spraying primer over a repaired panel.

Zinc chromate primer is a product that has been around for many years. It is a primer designed mainly for adhesion to aluminum and for use where dissimilar metals will come in contact with each other (such as in a truck body where steel ribs might be used to support aluminum sides and top). One thin coat of zinc chromate will prevent an electrolysis reaction between dissimilar metals, which causes rapid corrosion.

Zinc weld-through primer is a zinc-rich material that is frequently used to protect welded joints against corrosion. The zinc in the primer will provide galvanic protection in the weld zone.

When corrosion resistance is very important *self-etching primer* is used. The reducer in a self-etching primer contains phosphoric acid that "bites" into the surface to provide superior adhesion to steel, aluminum, rigid and flexible plastics, and certain kinds of zinc-coated metal. The bond is so strong, in fact, that when a self-etching primer has been applied to bare metal it is extremely difficult to remove without grinding it off.

Chip-resistant paints are specially formulated for use on lower body sections that are prone to gravel and stone impact damage. They give improved resistance against corrosion and help to reduce drumming noise.

Adhesion to plastic panels that have come from the factory unprimed can be improved by the use of a *plastic primer.*

Another ready-to-spray primer product designed specifically for use with basecoat/clear coat OEM finishes is the so-called *adhesion promoter* or *mild coat primer.* This product is a water-clear primer with good durability and excellent adhesion to the very hard clear coats. It is recommended that it be applied beyond the repaired area on a spot repair before any

FIGURE 4–81 Appling a polyester primer-surfacer with gravity feed spray gun.

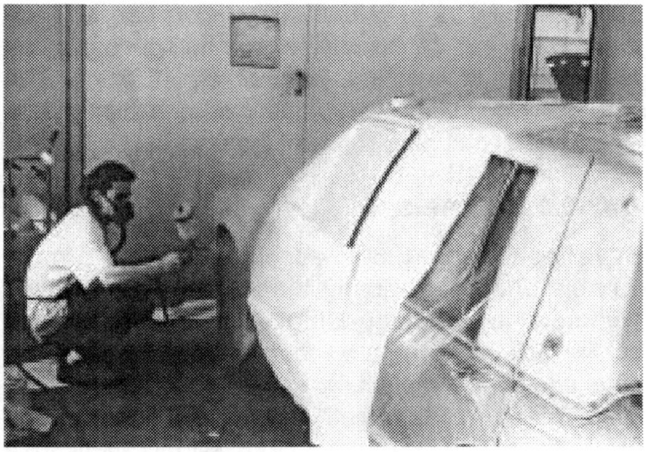

FIGURE 4–82 Sealer is sprayed on this replacement fender prior to topcoating.

other primer is used. Its purpose is to provide a surface to which a blend edge can adhere.

Surfacers. Surfacers are used to build up feathered-edged areas, to fill in grinder marks and sand scratches, and to smooth over other rough areas in the base metal and existing finish. The high-build capability of a surfacer reduces the time spent sanding the repair surfaces. Surfaces also sand easily to a smooth and level surface (Figure 4–81).

Sealers. Sealers are sprayed over a primer or surfacer or a sanded existing finish. They are used in automotive refinishing to promote adhesion between the topcoat and the repair surface, to prevent bleed-through of the old topcoat, and to provide uniform holdout. Some sealers also provide sufficient build to fill shallow scratches in the old finish (Figure 4–82).

TOPCOAT PRODUCTS

There are a variety of topcoats used today. Some are solid colors, some are metallic colors (mixed with flakes of aluminum or mica), and others are transparent. All topcoats are either an enamel paint or a lacquer paint. However, different formulations of the two basic paint bases have produced many different paint products with specific characteristics. Some of the more commonly used paints are described in the following paragraphs.

Acrylic Lacquers

Lacquer paint was first used as an automotive finish in 1924. The introduction of nitrocellulose lacquer in that year enabled cars to be completely finished in only one day, rather than the forty to fifty days required by earlier finishing methods. Prior to 1929 cars were hand-painted with many layers of lampblack pigment and varnish and then hand-rubbed with water and pumice. Nitrocellulose lacquer could be applied with a spray gun and air-dried very quickly.

Acrylic lacquer was introduced in 1956 and quickly replaced nitrocellulose lacquers. Acrylic lacquer dries to a high gloss and maintains its color much longer than nitrocellulose lacquer.

Lacquers are physically hardened and cured by evaporation. As the solvents evaporate, the binder and pigment left behind combine into a solid state. Lacquers are quick to surface dry but take a very long time to dry all the way through. After curing for several hours, lacquer topcoats usually must be compounded to bring out the gloss. Most acrylic lacquer basecoat/clear coat finishes allow a lacquer basecoat to be clear coated with an enamel topcoat. This helps eliminate the need for compounding or polishing to bring out the gloss.

Enamel Paints

Alkyd resin enamel was introduced in 1929. Unlike the nitrocellulose lacquers available at that time, alkyd enamel dries to a high gloss without the need for compounding. Alkyd enamels are no longer used in OEM applications. However, many refinishers still select this enamel because it provides an economical two-coat system with high initial gloss that hides surface imperfections better than newer enamels. This is especially true when a synthetic enamel is reduced with hardener. The synthetic is applied the same as without hardeners, but the drying time is shortened when the hardener has been added. The finish can be two-toned when it is dry enough to tape and generally can be recoated at any time without fear of lifting. The hardener also makes the synthetic enamel highly wrinkle resistant, though it does not improve durability.

Acrylic Enamels

Developed in the 1960s, acrylic enamels are more durable and faster drying than alkyd enamels and eliminate the compounding step associated with acrylic lacquer. Generally recommended for panel repairs and overall repainting, some acrylic enamels are used for spot repair of both lacquer and enamel original finishes. Acrylic enamel costs slightly more than synthetic enamels but is less expensive than other finishing systems available today.

Acrylic enamel is fast drying and is about 50 percent more durable than synthetic enamel. It is also easy to apply, produces a deep luster and high gloss, and has excellent color and gloss retention. Two-toning or recoating is possible within the first six hours of drying. During this period, the solvent evaporates from the film, leaving the finish nontacky and dry to the touch.

For the next six to forty-eight hours, however, most acrylic enamels enter the hard-curing stage. Recoating must be done before this period because the enamel will no longer partially dissolve and melt in with the new coat.

Polyurethane Enamels

Ordinary polyurethane enamel provides significantly greater durability than alkyd or synthetic enamels, acrylic enamels, and acrylic enamels with hardeners.

The polyurethane enamel—made with a polyester resin—dries by evaporation of the solvents and by chemical cross-linking between the two principal base components to cure the paint film. A high-gloss, extremely durable, chemical- and solvent-resistant finish results. Without an accelerator, dry time to tape is about sixteen hours. With an accelerator, dry time to tape can be reduced to approximately three to five hours.

Acrylic Urethane Enamels

A recent development in the urethane family is the acrylic urethane enamel (Figure 4–83). It is formulated with an acrylic resin that is much more durable than any other finish. Because of the characteristics of this resin, an acrylic urethane enamel offers higher gloss retention and, ultimately, a much longer life. Recent manufacturer tests have shown that this finish retains its gloss from 50 percent to 100 percent longer than ordinary polyurethane enamels.

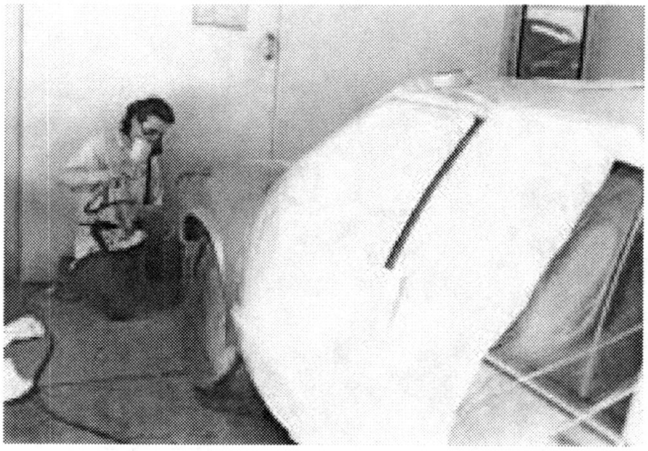

FIGURE 4–83 Basecoat application of an acrylic urethane basecoat/clear coat. Note supplied air respirator.

The acrylic urethane enamel also dries tack-free faster than polyurethane enamels and is more resistant to film degradation caused by the sun's ultraviolet (UV) rays. When compared to metallic polyurethane colors, acrylic urethane enamel metallic colors have twice the durability. The urethane's durability also offers exceptional resistance to stone chipping and does not break down when exposed to weather, gasoline, chemicals, dirt, and road grime. It virtually eliminates sand scratch swelling.

Basecoat/Clear Coat Finishes

In the past few years, OEMs around the world have increasingly adopted basecoat/clear coat systems as the finish of choice for new cars rolling off assembly lines. The technology for basecoat/clear coat finishes was developed in Europe. The durability and popularity of these finishes prompted Japanese and American automobile manufacturers to begin offering them, too. Today, more than 90 percent of all domestic and import cars feature OEM basecoat/clear coat finishes.

When the basecoat/clear coat system was first introduced, it utilized either an acrylic lacquer or polyurethane enamel clear coat over an acrylic lacquer basecoat. Now all manufacturers are using an acrylic urethane basecoat/clear coat system.

Basecoat/clear coat finishes (Figure 4–84) are made up of a highly pigmented color coat, followed by a clear, weather-resistant unpigmented urethane clear coat applied on top. The color coat can be solid or it can be metallic. The unpigmented topcoat acts as a barrier, preventing damaging ultraviolet rays, acid rain, and other pollutants from attacking the pigments in the basecoat. The result is a longer-lasting, high-

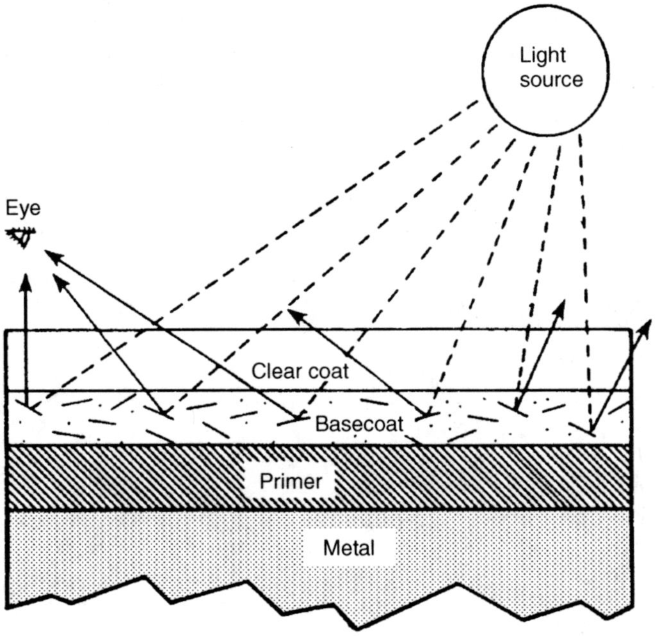

FIGURE 4–84 Metallic basecoat/clear coat paint system.

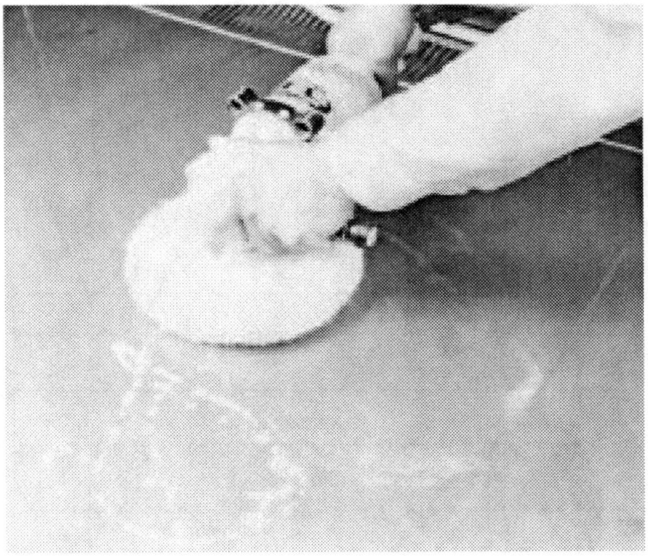

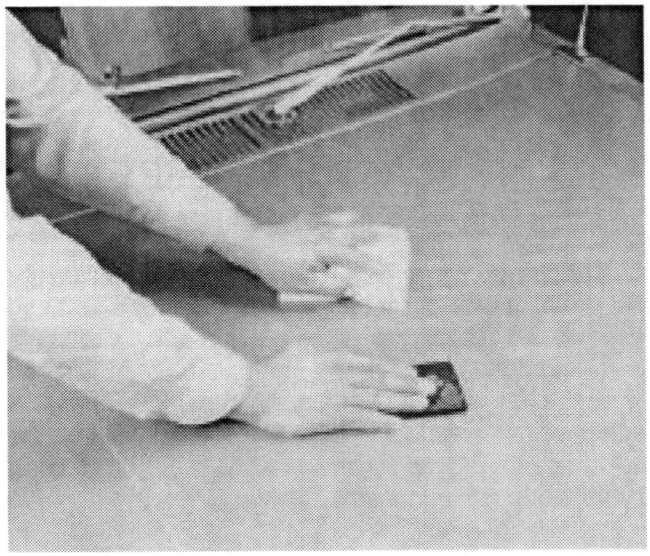

FIGURE 4–85 The clear coat can be sanded and buffed to remove defects. (Courtesy of 3M.)

gloss finish with a deeper color than is possible with single-stage paint systems. Figure 4–85 shows how a clear coat can be buffed.

Identifying cars that are clear coated is easy. Looking at the vehicle identification code and the color chip book is a quick way to find out if the car has the basecoat/clear coat system. If the code has been removed or destroyed, sanding a small spot in a concealed area of the vehicle to be finished, using a fine sandpaper, can help determine the type of finish. If the dust is white, the car has a basecoat/clear coat finish. If the dust is the color of the car, it does not.

Three-Coat Finishes

One of the newest OEM coating developments is the three-coat finish (also referred to as three-stage and tricoat finishes). Currently the rage in Europe, three coats are growing in domestic popularity on upscale foreign and U.S. makes. Most domestic and import manufacturers have added three-coat finishes to their lineups for future years.

These unusual new finishes consist of three distinct layers that produce a pearlescent appearance: a basecoat, midcoat or interference coat, and a clear coat (Figure 4–86). Unlike other coating systems such as metallics and some micas, which change the value of a color (lightness/darkness) when viewed from different angles, the new three coats actually change the hue (color) as the angle of view changes. Thus a three-coat finish may look red viewed from straight on and blue when seen from the side. The effect is similar to that of a thin layer of oil floating on water.

In a three-coat finish the interference coat, or midcoat, is the layer contributing the most to the final appearance of the color. Particles making up the midcoat color can be designed to reflect, absorb, and refract differing amounts of the light striking them. Each particle in the midcoat color is made from a thin flake of mica coated with titanium dioxide and/or iron oxide. Changing the amount, or color of the mica flake's coating drastically alters the color of the finish when viewed from straight on or from an angle.

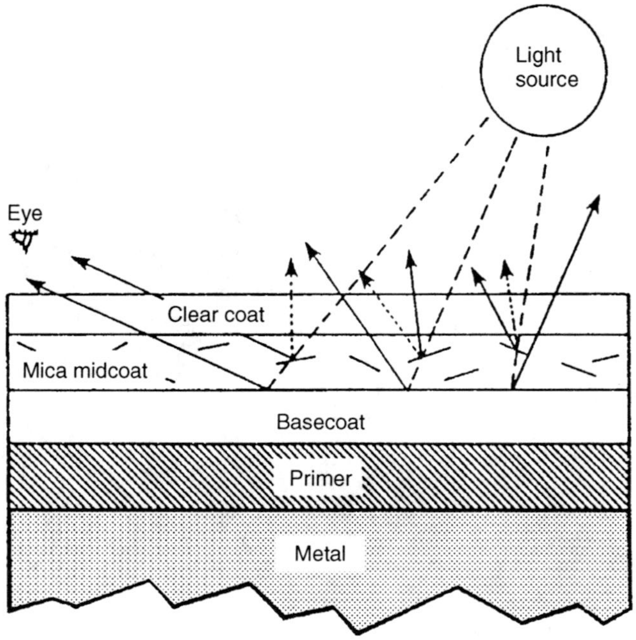

FIGURE 4–86 Metallic tricoat paint system.

4.11 PLASTIC REPAIRS

In Chapter 2, some basic guidelines were given for determining whether a plastic part should be repaired or replaced. Once the decision is made to repair a part, one of two basic repair methods must be chosen to do the job. The part can be repaired using chemical adhesives or plastic welding. Table 4–1 indicates the best repair systems for the plastics most often used by the automotive industry.

CHEMICAL ADHESIVE BONDING TECHNIQUES

The main advantage adhesives have over welding is that not all plastics can be welded. Only thermoplastics soften when heated.

Repair of Minor Cuts and Cracks

Adhesives are ordinarily used for repairing minor cuts and cracks in plastic parts. Table 4–2 outlines the basic repair procedure used for repairing minor cuts and cracks.

Repair of Gouges, Tears, and Punctures

The procedure for repairs of gouges, tears, and punctures is somewhat more involved than the previous one but requires no special skills or tools (Figure 4–87).

TABLE 4–1: PLASTIC PARTS REPAIR SYSTEMS

KEY

AR Adhesive repair **S** Anerobic (instant) adhesive
FGR Fiberglass repair **PC** Patching compound
W Welding

ISO Code	Name	Repair System
ABS	Acrylonitrile-butadiene-styrene (hard)	W, S, FGR
ABS/PVC	ABS/Vinyl (soft)	PC, W
EPI II or TPO	Ethylene propylene	AR, W
PA	Nylon	S, FGR, W
PC	Lexan	W
PE	Polyethylene	W
PP	Polypropylene	W
PPO	Noryl	FGR, W
PS	Polystyrene	S
PUR, RIM, or RRIM	Thermoset polyurethane	AR, W
PVC	Polyvinyl chloride	PC, W
SAN	Styrene acrylonitrile	W
SMC	Sheet Molded Compound (polyester)	AR
TPR	Thermoplastic rubber	AR, W
TPUR	Thermoplastic polyurethane	AR, W
UP	Polyester (Fiberglass)	FGR

PLASTIC WELDING

Although most types of plastics can be repaired with adhesive materials, hot-air welding is usually preferred for thermoplastics because of its speed and ease. A plastic welder only takes a couple of minutes to heat up to operating temperatures and, once hot, can weld at speeds ranging from five to thirty inches per minute, depending on the application. Best of all, one can go directly to and from welding to sanding and painting without waiting for the repair to cure (although it might be necessary to allow a few minutes for the welded plastic to cool).

Many adhesives do require mixing and can take anywhere from thirty minutes to several hours to cure, but heat can be used to shorten the curing time. This is not necessarily a disadvantage because repair technicians can always find something else to do while the adhesive cures. As for surface preparation, both welding and adhesives require some grinding, sanding, or trimming for good adhesion. More surface preparation is generally needed with adhesives, however, because the edges of the damaged area often have to be featheredged before the adhesive is applied. Some adhesives also require reinforcing or support patches behind the damaged area, making the repair a several-step process. With plastic welding,

TABLE 4–2: ADHESIVE REPAIR SYSTEMS

1. Clean damaged area.
2. Grind with #36 grit disc.

Most plastics sand normally (dust).	Polyolefins (polyethylene, polypropylene, and similar materials) melt and smear when sanded.
3. Featheredge with #180.	3. Featheredge with #180.
	3A. Apply adhesion promoter.
4. Apply adhesive repair material.	4. Apply adhesive repair material.
5. Sand with #180 and #240.	5. Sand with #180 and #240.
	5A. Apply adhesion promoter.
6. Apply flexible filler.	6. Apply flexible filler.
7. Sand with #320.	7. Sand with #320.
	7A. Apply adhesion promoter.
8. Apply flexible sealer.	8. Apply flexible sealer.
9. Scuff with #320.	9. Scuff with #320.
	9A. Apply adhesion promoter if sealer is scuffed through.
10. Apply color.	10. Apply color.

however, there is no need to feather out the edges. A V-groove made along the cracked or damaged area is generally all that is needed for a good weld. Remember, however, plastic welding *cannot* be used on thermosetting plastic parts.

Although only in existence for a relatively short period of time, plastic welding has become very popular with the auto body repair industry. The basic plastic welding procedure is as follows:

1. Wash the damaged part with soap and water.
2. Wipe or blow dry.
3. Clean the damaged part with plastic cleaner.
4. Align and tape the broken or split sections with aluminum body tape.
5. V-groove at least 50 percent of the way through the panel if a two-sided weld, 75 percent of the way through if a one-sided weld.

FIGURE 4–87 Repair of plastic bumper with adhesives.

FIGURE 4–88 Priming a plastic panel.

6. Use a slow-speed grinder with a #60 or #80 grit disc to remove the paint from around the damaged area and blow dust free.
7. To start the weld after setting to the proper temperature, *slowly* feed the rod into the melt tube.
8. Apply a light pressure to the rod and slowly force it *out* and into the grooved area.
9. As the rod melts, very slowly start to move the torch tip in the direction of the intended weld. Overlap the edges of the groove with the melted plastic while progressing forward.
10. After completing the weld, use the flat "shoe" part of the weld torch tip to smooth out the weld area.

REFINISHING PLASTICS

The many different types of plastics used in automobile bodies (Figure 4–88) have different refinishing requirements, so the first step in plastics refinishing is to determine the plastic used to form the part. Looking on the back of the part for the International Organization for Standardization Codes is the best place to start. The Code List will then identify appropriate refinish systems for use on that particular part. Table 2–1 in Chapter 2 lists the various kinds of plastics used in today's cars, their abbreviations, and their applications.

With the exception of PVC, most exterior plastics can be topcoated using most acrylic lacquers, acrylic enamels (catalyzed or uncatalyzed), and acrylic enamel basecoat/clear coat systems. Semirigid plastics require the addition of a "flex agent" to the paint system. The additive is required because semirigid plastics expand, contract, and bend more easily than other substrates. The flex agent will keep the paint film flexible so it can accommodate the movement of the substrate without cracking.

Plastic Primers. Some plastic auto body replacement parts come from the factory already primed, while others are delivered unprimed. If the parts are factory primed, no additional priming is necessary. If they are not, both rigid and semirigid plastics can benefit from the use of a special plastic primer, primer-sealer, or vinyl wash-primer to improve paint adhesion. TPO, in particular, has an extremely slick, waxy surface that makes it difficult for the topcoat to form a strong bond to the substrate unless a primer is used. Adhesion to ABS also is greatly enhanced by priming.

Unpainted rigid and semirigid plastic parts should be solvent cleaned to remove mold release agents (typically silicones) before a primer is applied. Like topcoat systems, some undercoat products require the addition of a flexible additive when used on semirigid plastic parts.

The use of plastic parts in vehicle construction is expected to continue to increase. Fenders and outer door skins, bumpers and hoods, and other plastic parts are predicted to become more and more common. In the future, knowing how to recognize plastics and how to prepare and refinish them will become, for many shops, an everyday activity.

REVIEW QUESTIONS

1. The initials LIFO mean ____________________.
2. Technician A relieves stress with tension and heat. Technician B relieves stress with tension and spring hammering. Who is right?
   a. technician A
   b. technician B
   c. both A and B
   d. neither A nor B

3. A unibody vehicle is anchored to the frame machine with pinch weld clamps.
   True
   False
4. Technician A cuts off all damaged structural parts before puling. Technician B pulls the damaged structural parts to within one-eighth inch before removing them. Who is right?
   a. technician A
   b. technician B
   c. both A and B
   d. neither A nor B
5. Factory spot welds can be removed with a
   a. drill.
   b. plasma cutter.
   c. both A and B
   d. neither A nor B
6. Technician A welds a replacement door skin. Technician B bonds a replacement door skin. Who is right?
   a. technician A
   b. technician B
   c. both A and B
   d. neither A nor B
7. A plug weld is made by using a 5/16-inch hole.
   True
   False
8. Technician A uses plug welds to replace factory spot welds. Technician B uses an oxyacetylene welder to weld HSLA. Who is right?
   a. technician A
   b. technician B
   c. both A and B
   d. neither A nor B
9. In basecoat/clear coat finishes, the basecoat is always a metallic color.
   True
   False
10. Technician A applies body filler over sanded paint. Technician B says that primer must be applied to bare metal before the surfacer. Who is right?
    a. technician A
    b. technician B
    c. both A and B
    d. neither A nor B

Chapter 5

Making a Damage Appraisal

Objectives

After reading this chapter, you should be able to:

- **List the tools and equipment used in estimating.**
- **Explain why a routine sequence of inspections is important.**
- **Explain the "zone" method of damage analysis.**
- **Point out tell-tale signs of secondary damage.**
- **Explain how to measure body alignment using a tram gauge.**
- **Inspect a vehicle for damage to mechanicals, accessories, trim, upholstery, and exterior finish.**

Key Terms

Zone 1 (primary damage)
Zone 2 (secondary damage)
Zone 3 (mechanical components)
Zone 4 (passenger compartment)
Zone 5 (exterior components)
Datum
Centerline
Length
X check

Chapters 2, 3, and 4 explained vehicle construction, the types of damage a vehicle can sustain in a collision, and typical repair processes. With that background of understanding, the estimator (body shop personnel or adjustor) can approach a damaged vehicle with confidence in his or her ability to identify collision damage and to write an accurate damage report of the repair costs. However, it is very important that the estimator have certain tools and equipment and follow a set sequence of inspections when analyzing a damaged vehicle. Particularly on a heavily damaged vehicle, the damage analysis can be a complex process. Unless an orderly inspection procedure is followed, the written estimate will most likely have errors and oversights.

Many times damaged vehicles are evaluated by the estimator in the shop area. There are many hazards in a collision repair shop; use good common sense to avoid injury. If you need to evaluate a car in the shop, follow these guidelines:

1. Wear safety glasses at all times to protect your eyes.
2. Do not look at a MIG weld arc.
3. Do not go into an area where paint is sprayed unless you are wearing at least a cartridge respirator. If isocyanates are sprayed, you need a supplied air respirator.
4. Do not crawl under a raised car unless it is supported on jack stands or locked on a lift.

5.1 THE ESTIMATOR'S TOOLS

The estimator needs several tools when making an analysis of damage (Figure 5–1). Here is a list of the basic equipment required to complete a thorough damage analysis:

FIGURE 5–1 Carefully record damaged panels when making a damage analysis.

1. The estimator will need some method of recording information. Depending on the extent of damage, the estimator can simply use a notepad and pencil, or the damage can be verbally recorded on a tape recorder and transcribed later. Those with access to a computerized estimating program can use computer-generated exploded parts illustrations and estimating worksheets or even enter parts codes into a hand-held computer.
2. The estimator needs manufacturer's parts lists and specification sheets to assist in the identification of parts and in the measuring of critical dimensions.
3. Measuring tools are essential to accurately analyzing damage. A tram gauge and a twelve-foot steel tape with both metric and fractional-inch (SAE) dimensions are necessary. Most manufacturer's specifications are given in the metric system.
4. The estimator must have at his or her disposal a lift or jack and support stands to raise the vehicle. In all but the very light collision damage, it is necessary to perform underbody visual inspections and measurements.
5. The estimator should have at hand common mechanics hand tools, such as wrenches (both metric and SAE), screwdrivers, and pliers. These are sometimes necessary for disassembly and removal of parts on a vehicle that might be obstructing visual or measurement checks. Stools, creepers, trouble lights, and extension cords should also be accessible in the estimating area.

5.2 A SEQUENCE OF INSPECTIONS

In addition to knowledge and tools, the estimator should have a systematic method of inspection. A set pattern or sequence of checks is extremely important. The damage incurred during a collision can be very complex, especially in a heavily damaged vehicle. If a careless, random method of inspection is used, the estimating process can be very confusing and errors are the inevitable result.

An orderly sequence of inspections will minimize the possibility of overlooking parts that are damaged or must be removed and reinstalled when making repairs. Following a predetermined system of analysis will also help the estimator to avoid duplicating parts and will assist him or her in spotting overlap and included operations. With a predetermined sequence of inspections each time an estimate is written, a thorough, accurate estimate can be written, minimizing the likelihood of a supplement or add-on.

ZONE INSPECTIONS

This chapter describes the "zone" method of damage analysis. The zone concept was developed by American car manufacturers and the Inter-Industry Conference on Auto Collision Repair (I-CAR). This system arbitrarily divides the vehicle into five logical areas of concern:

- Zone 1, the immediate impact area; note that this is the area containing the primary damage.
- Zone 2, remaining body damage defined as secondary damage.
- Zone 3, mechanical components. Any damage to drive train and to accessories.
- Zone 4, passenger compartment. Includes all damage to the compartment, including trim, lights, accessories, controls, and finish.
- Zone 5, the exterior components and trim.

The damaged vehicle is carefully inspected zone by zone and the damage recorded in that order.

When using the zone concept or any other sequential method of inspection, these guidelines should be followed:

1. Work from the front of the vehicle to the rear (or from the rear to the front in the case of a rear-end collision).
2. Work from the outside in. List the damage to the exterior, cosmetic panels first and then the damage to underlying structural panels and attachments.
3. List major subassemblies first, and then list smaller components and attachments that are not included in the overall subassembly.

As the estimator inspects the vehicle for design, he or she must also be on the lookout for non-OEM parts. These parts will not be listed in the collision estimating guides. One must also be on the lookout for signs of previous damage. Dents, dings, and rust that are obviously not related to the current damage are not going to be covered by the insurance company's appraisal of the repair costs. Remember that the insurance company is only obligated to restore the vehicle to its precollision condition. Repair to any other damage will have to be paid for by the vehicle owner. It is also a good idea to make a list of personal items found in the vehicle. Although this is not part of the damage analysis, it helps to avoid potential conflicts if an item is "missing" later.

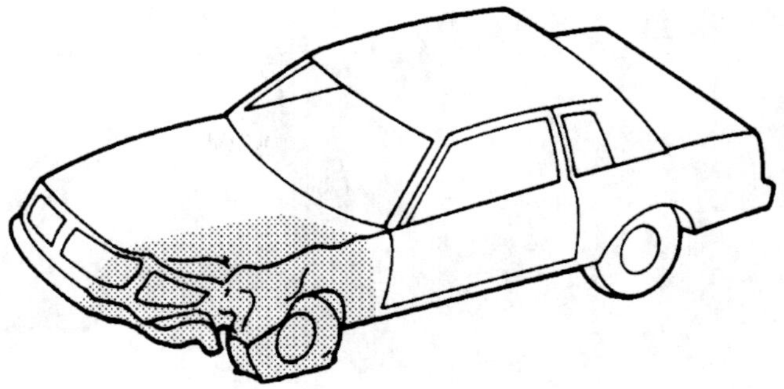

FIGURE 5–2 Zone 1 includes all primary body damage present near the immediate point of impact.

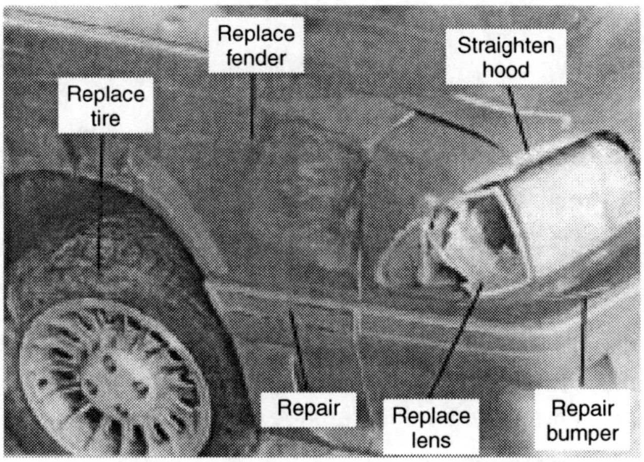

FIGURE 5–3 Primary damage.

5.3 ZONE 1 PRIMARY DAMAGE

The first step in the systematic zone inspection is to visually check and then list the primary body damage at the immediate point of impact (Figure 5–2). The area that includes primary damage will vary, depending on the vehicle construction, force and angle of impact, and other factors. For the most part, primary damage results in crumpled panels and cracked or broken components (Figure 5–3). Primary damage is visible and does not require measuring.

Begin the Zone 1 inspection by examining cosmetic outer panels, attaching plastic panels, glass, chrome, and structural metal beneath the outer panels. In the case of frontal impact, the area to inspect would probably include (and there could be more):

- Bumper system
- Grille
- Hood
- Fenders
- Lights
- Glass
- Doors
- Wheels
- Fluid leakage

Should the point of impact occur at the rear of the vehicle, look for such damage as:

- Bumper system
- Quarter panels
- Deck lid
- Lights
- Glass
- Wheels
- Fluid leakage

If the impact point is on the side of the car, check such items as:

- Doors
- Roof
- Glass

FIGURE 5–4 Check underneath the vehicle for hidden damage.

- Pillars
- Floor pans
- Supports
- Fluid leakage

After listing the damage incurred by outer panels and components, raise the vehicle (Figure 5–4) and check the extent of the damage to:

- Underbody panels
- Component supports such as the engine cradle
- Structural supports
- Rails

While examining the damaged area, look for clues to see the extent of the following;

- Split seams
- Flange damage
- Popped spot welds
- Distorted metal

Pay close attention to structural members. Since the strength of the vehicle depends on the whole structure, all small splits, tears, or popped welds must be properly repaired if the vehicle is to come back and perform as it was designed.

Working from the front to the rear and from the outside in, list all primary damage on the estimate form or worksheet. Refer to the collision estimating guides, manufacturer's parts lists, or computer-generated parts illustrations to determine the correct part name and to locate panel seams. Also consult the manuals to make sure that all parts are accounted for. It is very easy to overlook small parts such as braces or extensions.

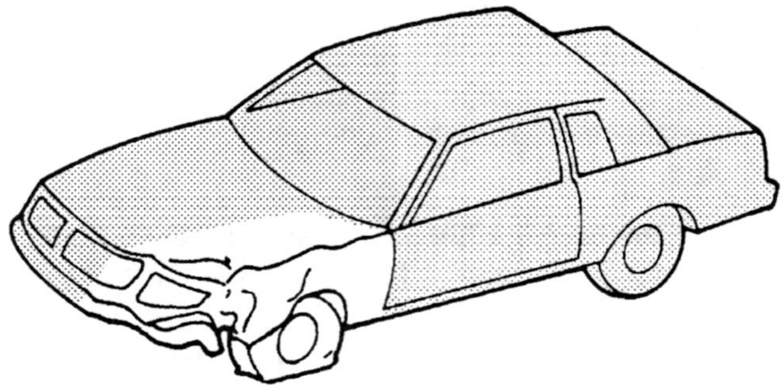

FIGURE 5–5 Zone 2 includes the secondary damage that might have occurred in the rest of the vehicle body.

5.4 ZONE 2 SECONDARY DAMAGE

Secondary damage occurs outside Zone 1 and some distance away from the point of impact (Figure 5–5). As mentioned in Chapter 3, secondary damage is caused by the force of the collision as it moves through the vehicle. That is, the collision forces travel from the impact area to adjacent areas of the body, and the impact force is absorbed as it travels to the neighboring panels. These forces travel over a wide area, and any part of the car can be affected.

The distance the forces travel through the car and the extent of secondary damage are determined by the force of the impact, the angle of the force, and the strength of the members absorbing the collision forces. As mentioned in Chapters 2, 3, and 4, many unibody components are designed to collapse and absorb collision forces in order to protect the vehicle occupants. Secondary damage is often found in these areas.

Secondary damage can also be caused by the mass of the power train and rear axle. As the vehicle comes to a sudden halt due to a collision, the mass and forward motion of the heavy mechanical components place strain on the mounting points and supporting structures. The adjoining metal can buckle or tear, and welds can pop. Always inspect suspension, axle, engine, and transmission-mounting points carefully.

Secondary damage is not always easily detectable. There are often visible signs of it, but usually secondary damage analysis depends on measurements.

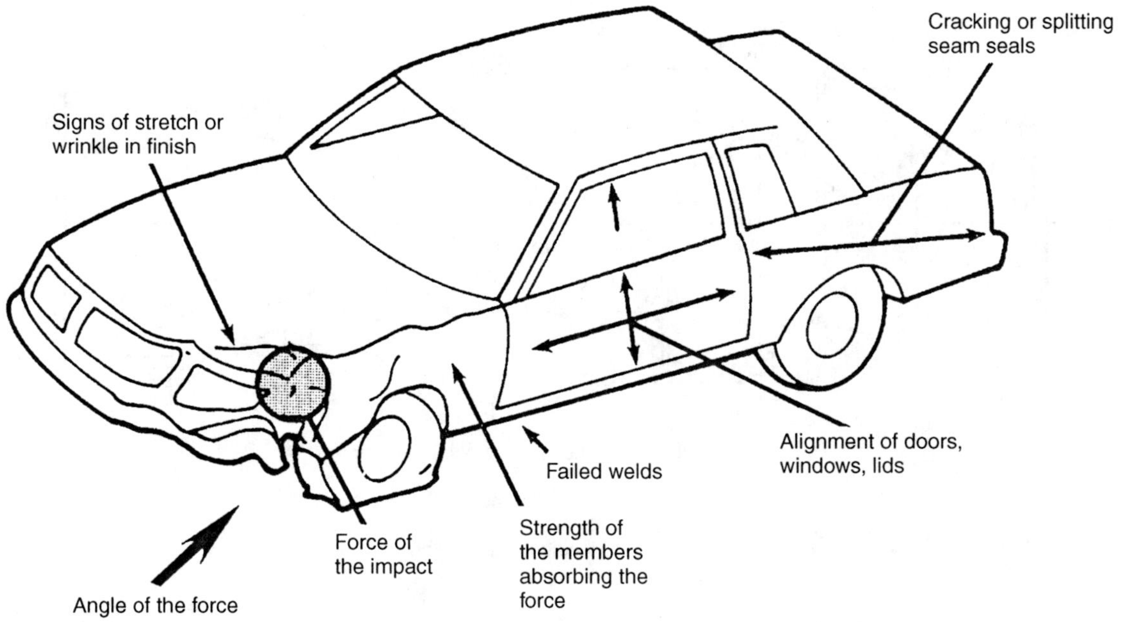

FIGURE 5–6 Signs of secondary damage.

VISUAL SIGNS OF SECONDARY DAMAGE

There are several visual clues (Figure 5–6) that should alert the estimator to secondary, hidden damage:

- Buckled panels
- Wrinkles or stretch marks in the paint
- Misaligned panel gaps
- Split seam seals
- Failed welds

These points can give clues as to where to look for secondary damage. For example, look for signs of panel movement and irregular gaps in the fit of fenders, hood, and doors (Figure 5–7). A vehicle damaged in a front-end collision might sustain secondary damage to the rear of the car so that the deck lid or hatch will not open or close.

On hard front end hits, look for an increase in the gap between the upper windshield pillar and the upper front corner area of the door window frame. Compare the right and left sides. Gapping here is a sign that cowl movements pushed the pillar upward, and the roof might have been damaged.

Look for buckles in exterior panels. Buckling of the roof panel directly above the center post is common on a hard hit. If the vehicle has a sunroof, check for buckles at the corners of the opening. Buckled exterior panels are a sign that secondary damage has occurred in underlying structural panels.

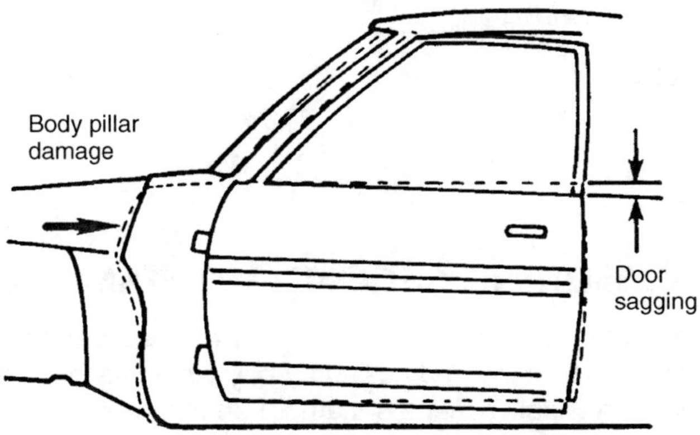

FIGURE 5–7 Misalignment of panel-to-panel gaps indicates secondary damage to underlying structural panels.

Check the dogleg over the rear wheel well and behind the rear doors for splits and buckles. Also check the quarter panel just below the rear quarter window pillar for buckling, a sign that the rear body rails might be kinked.

Open the hood and trunk, and look for wrinkled paint or splits in the sealer that covers welded joints. Look at the welds that join the panels. Strain and tearing pull the metal around the welds and break paint loose.

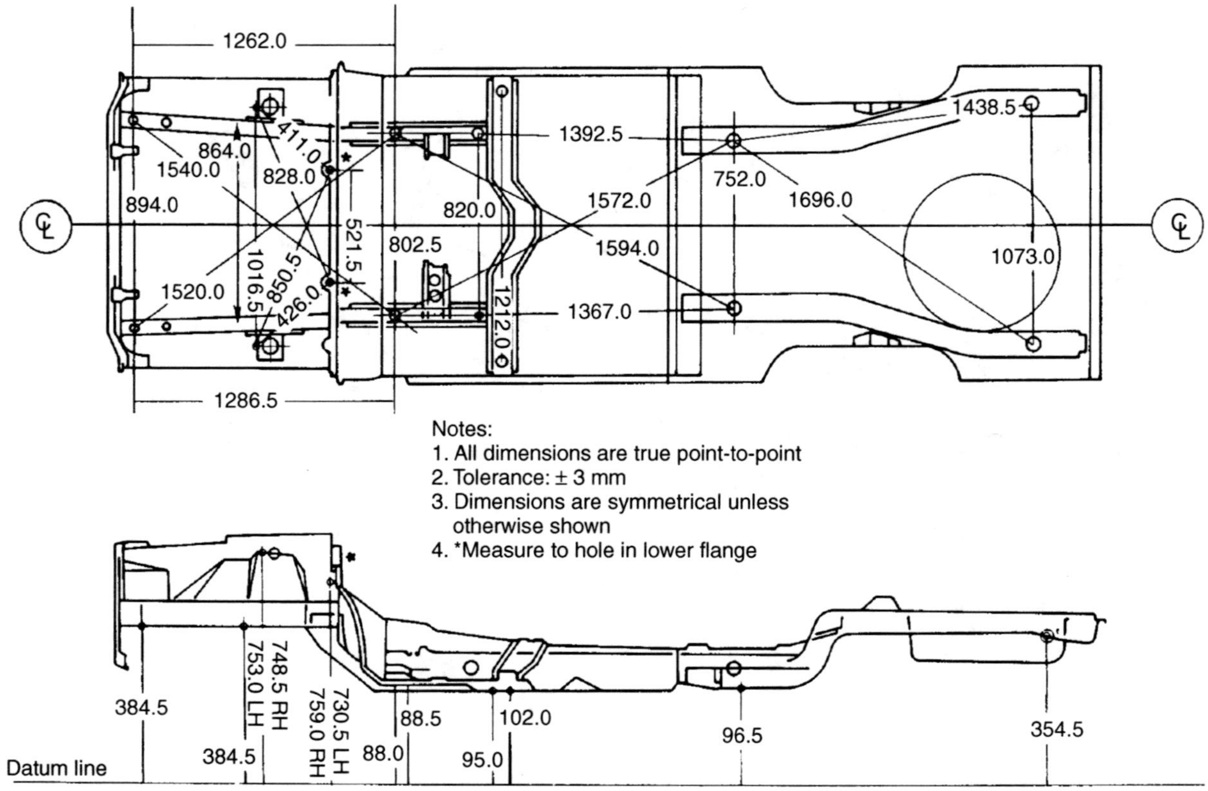

FIGURE 5–8 Underbody dimensions and specifications taken from a typical manufacturer's service manual. (Courtesy of Mitchell International.)

MEASURING SECONDARY DAMAGE

Accurate damage assessment can be made at specific points on the body using a body dimensions chart (this is a widely accepted method). In the body dimensions chart, measurements are based on the diagonal line measuring method. (Figure 5–8 shows a typical unibody car dimensions chart, while Figure 5–9 illustrates one for a conventional frame vehicle.) Dimensions for engine compartment body, door, trunks, and so on, are recorded on the charts.

Body dimensions are most often given in metric. Scales and gauges used to measure panel alignment should have metric scales.

Measurements can be taken using a steel measuring tape (Figure 5–10) or a tram gauge (Figure 5–11). The tram gauge measures one dimension at a time. Each dimension must be recorded and must be cross-checked from two additional control points—at least one being a diagonal measurement. The best areas to select for tram gauge measurements are the attachment points for suspension and mechanical components, since these are critical to alignment.

Body dimensions are measured from *control points* on the vehicle. Most control points are actually holes in the vehicle structure, and dimensions are center-to-center distances (Figure 5–12). Control point holes are frequently larger in diameter than the tram gauge tip. To measure accurately with the tram gauge (when the holes are the same diameter), measure like-edge to like-edge. A few manufacturers' specification books give the measurement based on the gauge's bar length. Always check the method used for specification measurements.

When the holes are not the same size, they will usually be the same type of hole: round, square, oblong, and so on. In this case, to find the center-to-center measurement, measure inside edge to inside edge, then outside edge to outside edge (Figure 5–13). Add the results of the two measurements and divide by 2. For example, two round holes, one being 1/2 inch in diameter, the other 1–1/2 inches in diameter, have an inside measurement of 30 inches and an outside measurement of 32 inches. The center-to-center dimension is 30" + 32" ÷ 2 = 31" (Figure 5–14). The 31 inches is the dimension for the tram gauge. It should center at

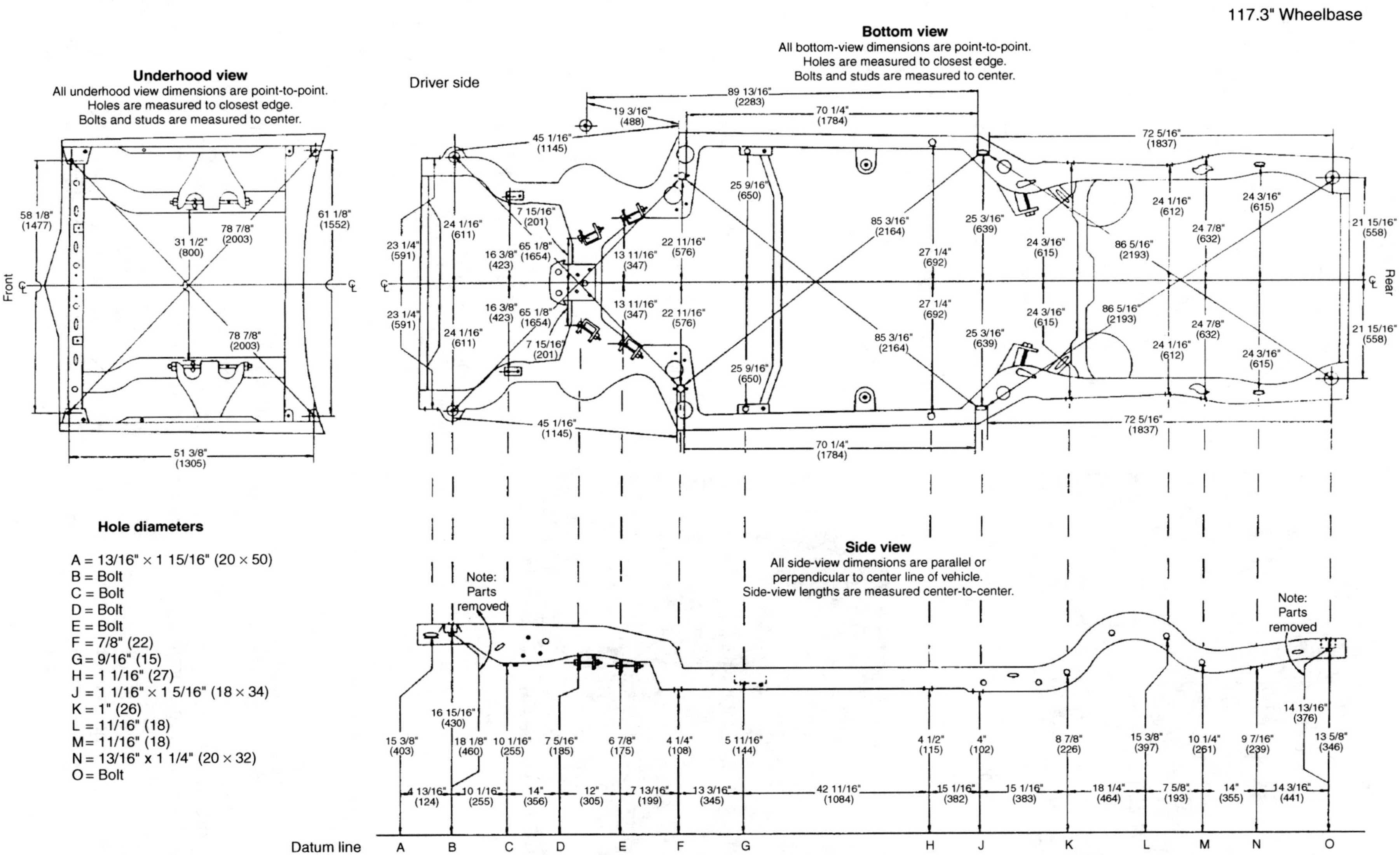

FIGURE 5–9 Underbody dimensions and specs for body-over-frame type of vehicle. (Courtesy of Hein-Werner.)

FIGURE 5–10 Using a tape to measure a hood opening.

FIGURE 5–11 Using a tram gauge to measure a hood opening.

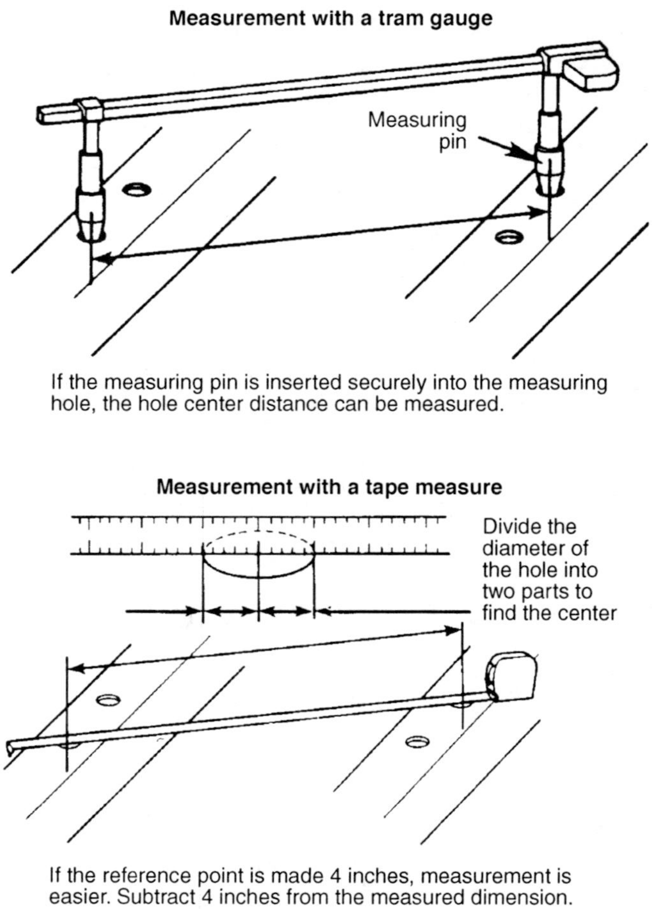

FIGURE 5–12 Measuring the distance between hole centers.

the smallest holes and the other end should center at the larger hole. If not, there is damage to the unibody.

In using a tram gauge for measuring, the manufacturer's specifications for the vehicle are needed to accurately assess the damage. If the manufacturer's specifications are not available, use an undamaged vehicle of the same make and body style as a source for correct factory dimensions. Frequently, if only one side of a vehicle is damaged, it is possible to take measurements on the undamaged side that can be set on the tram gauge and applied to the damaged side for a comparison measurement.

UPPER BODY DIMENSIONING

In addition to underbody frame dimensions, upper body damage can also be determined by the use of tram gauges and a steel measuring tape. Their use is basically the same as when doing an underbody evaluation. Manufacturers furnish specifications on the most important upper body control points (Figure 5–15).

Measuring the Front Body

When the front body has been damaged in a collision, measure the front sheet metal to determine the extent of the damage. Even if only the one side of the body received the impact, the other side will usually be damaged also. Therefore, the extent of deformation must be checked. Figure 5–16 shows the typical front body control points, which can be checked against the manufacturer's body dimensions diagram.

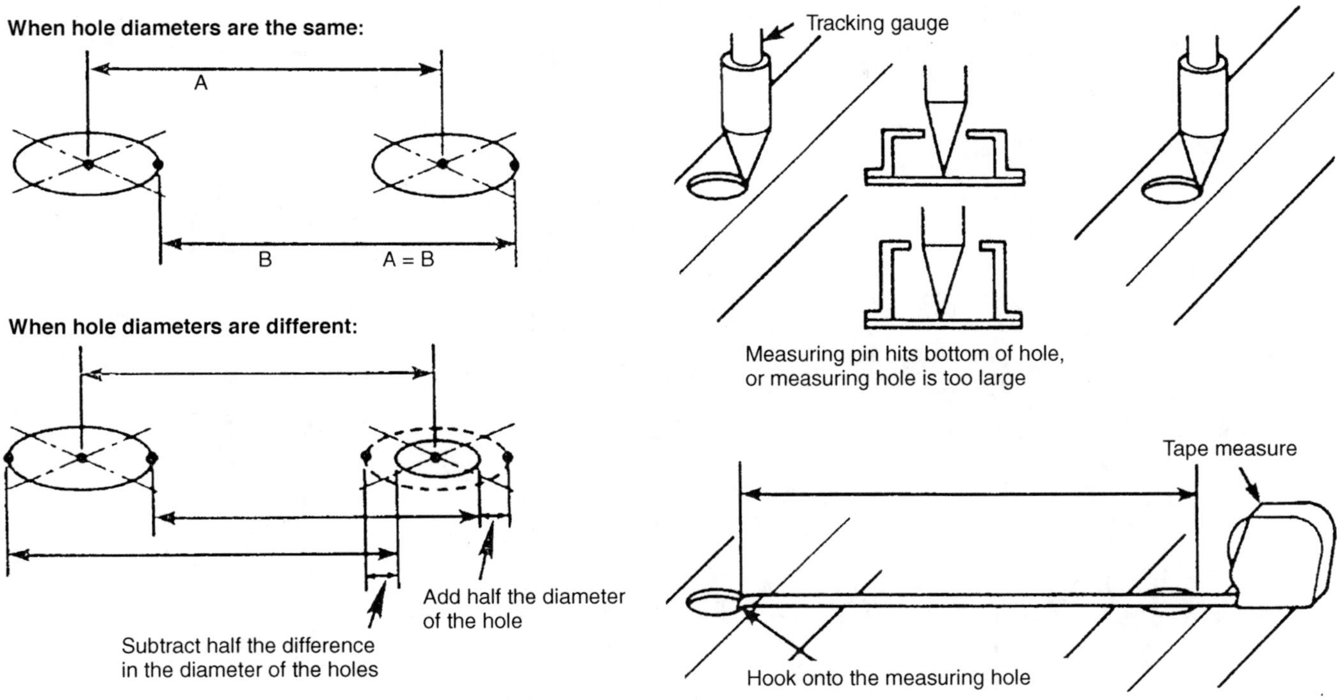

FIGURE 5–13 Measuring the edge-to-edge distance between control points.

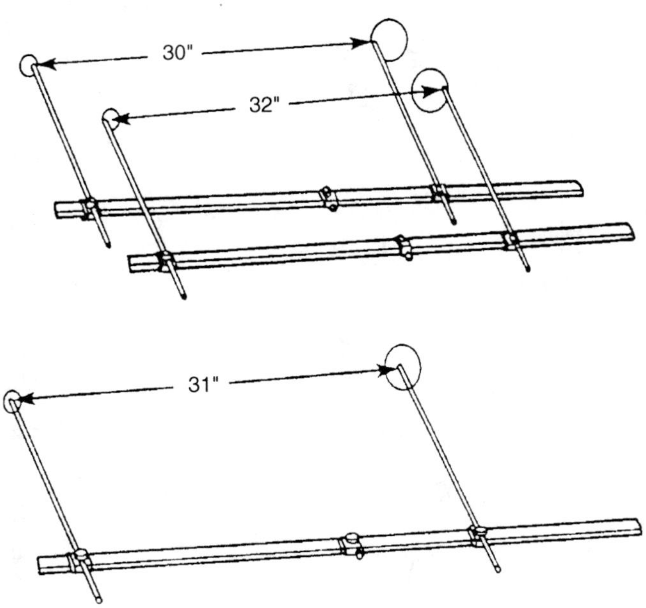

FIGURE 5–14 Measuring center-to-center distance between different-size holes.

Be sure to check which measurements are symmetrical. Symmetrical means that the measurements are equal from the centerline. In some cases, measuring points are asymmetrical. A vehicle with asymmetrical measuring points under the hood will not X-check as equal. In this case, set the tram to the specified length for each leg of the X.

When checking front end dimensioning, the best areas to select for the tram gauge measurements are the attachment points for suspension and mechanical components, since these are critical to proper alignment. Each dimension should be checked from two additional reference points with at least one reference point being a diagonal measurement. The longer the dimension, the more accurate the measurement. For example, a measurement from the lower cowl area to the front mount of the engine cradle is a better gauge than a measurement from a lower cowl area to another lower cowl area, because the longer dimension takes in a larger area of the vehicle. The use of two or more measurements from each control point assures greater accuracy and helps to identify the extent and direction of any panel damage.

Measuring the Body Side Panel

Any deformation of the body side structure can be ascertained by the appearance and irregularities in the door when it is opened and closed. Depending on where the deformation is located, attention should be given to possible water leakage. Thus, accurate measurements must be taken. The tracking tram gauge is primarily used to measure the body side panel (Figure 5–17).

Warping can generally be detected if the left-to-right symmetry of the body (part installation holes or reference holes) is used for measuring diagonal lines,

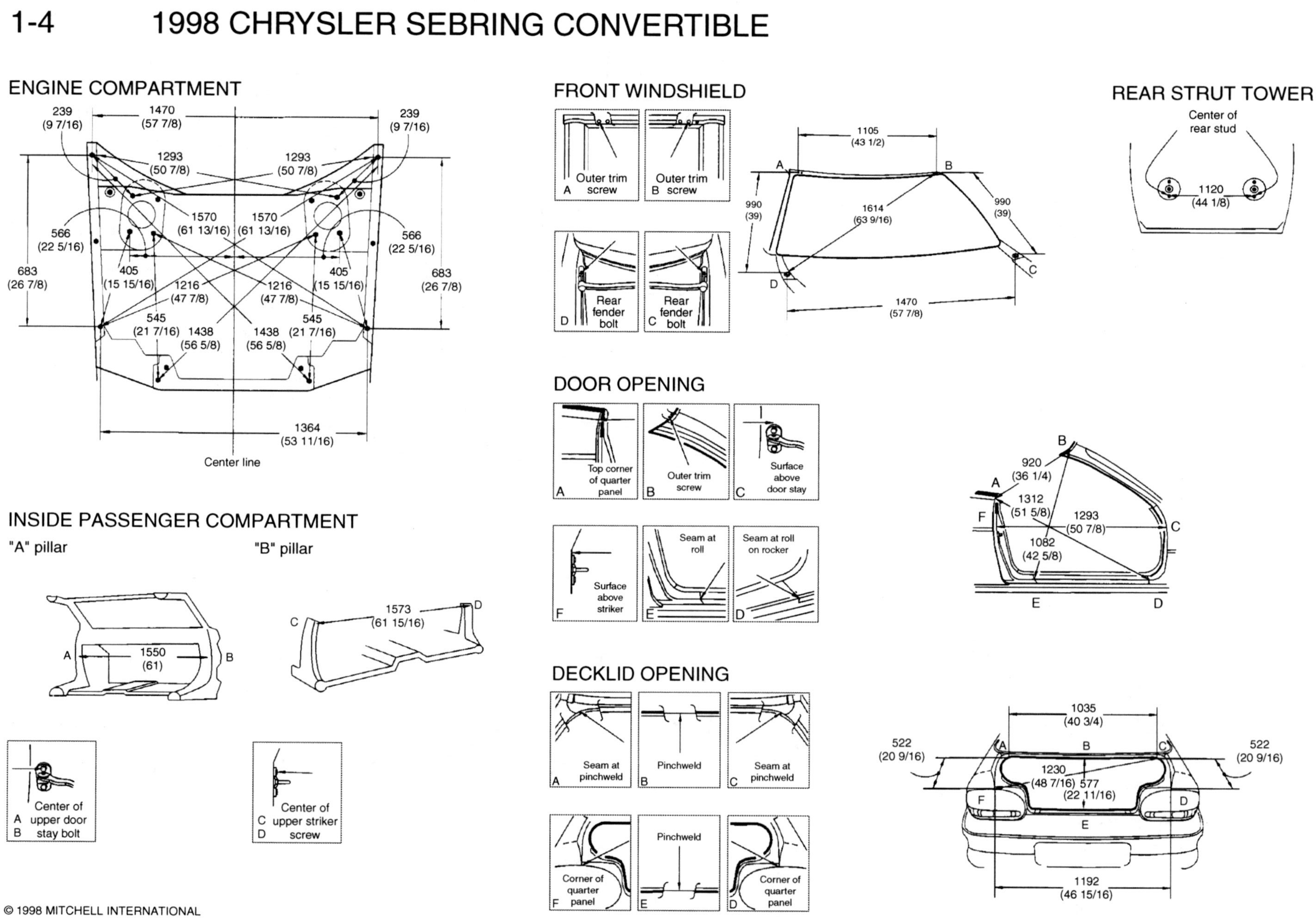

FIGURE 5–15 Upper body dimensions and specifications. (Courtesy of Mitchell International.)

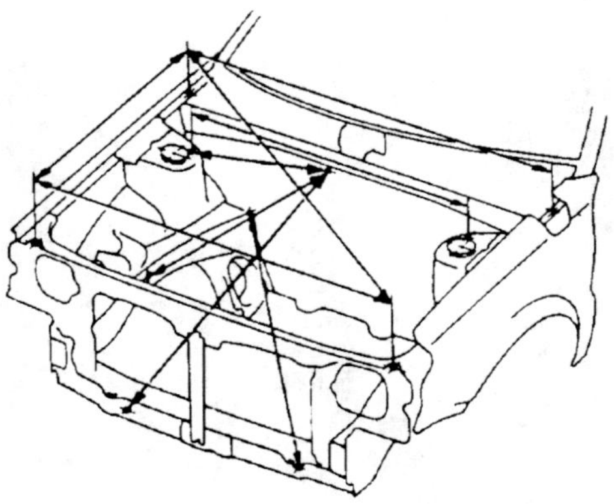

FIGURE 5–16 Typical front body measurement points.

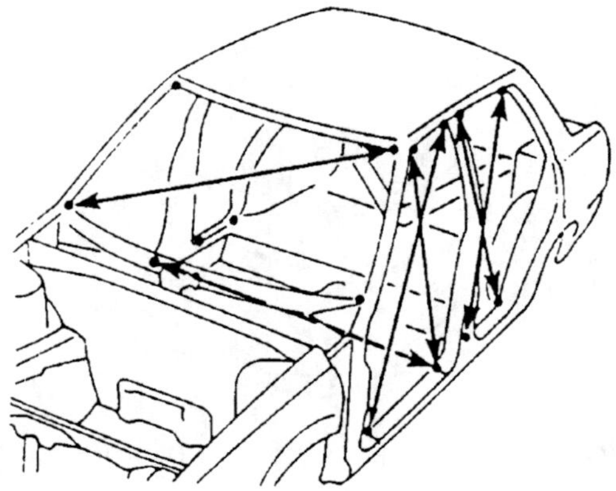

FIGURE 5–17 Typical body side panel measurements.

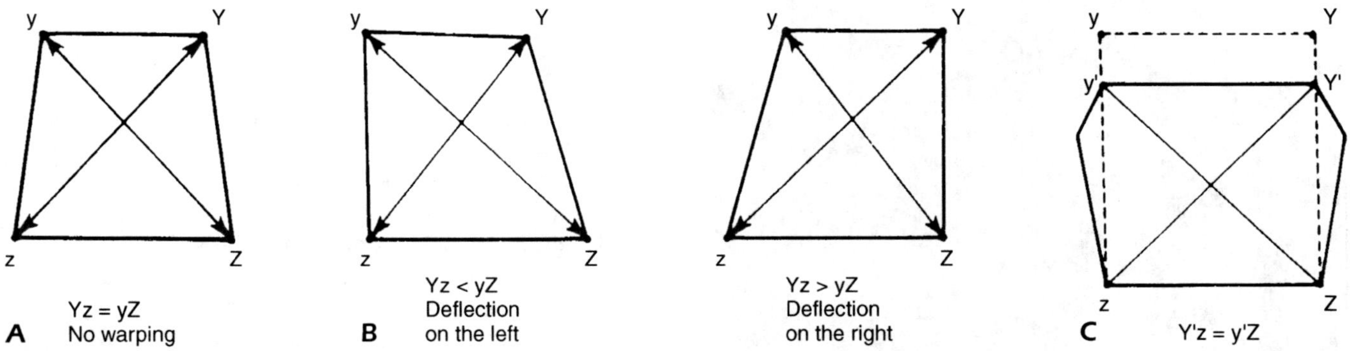

FIGURE 5–18 Measuring body side panels with the diagonal line measurement method.

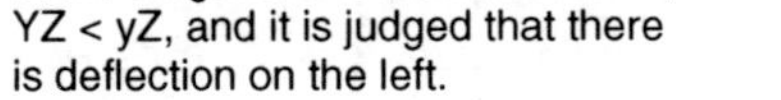

FIGURE 5–19 Comparison of length dimensions on body side panel.

as shown in Figure 5–18A. Use this measuring method if the data on the engine compartment and underbody are missing, if there is no data available in the body dimensions chart, or if the vehicle has been severely damaged in a rollover.

The diagonal line measurement method is not adequate when inspecting damage to both sides of the vehicle or in the case of twisting, since the left-to-right difference in the diagonal lines cannot be measured (Figure 5–18B). If deformation is the same on the left and the right, left-to-right difference will not be apparent (Figure 5–18C).

In Figure 5–19, the measurement and comparison of the left and right lengths between yz and YZ will give an even better indication of damage conditions (this method should be used in conjunction with the diagonal line measurement method). This method can be applied where there are parts that are symmetrical on the left and right sides.

Measuring of the Rear Body

Any deformation of the rear body can be roughly estimated by appearance and irregularities (for example, catching) evident when the trunk lid is opened and closed. Because of the location of the deformation and the possibility of water leakage, it is essential to take accurate measurements (Figure 5–20). Furthermore, any wrinkle in the rear floor is usually due to buckling of the rear side member. Thus, measure the rear body together with the underbody. In this way, the straightening work can be performed effectively.

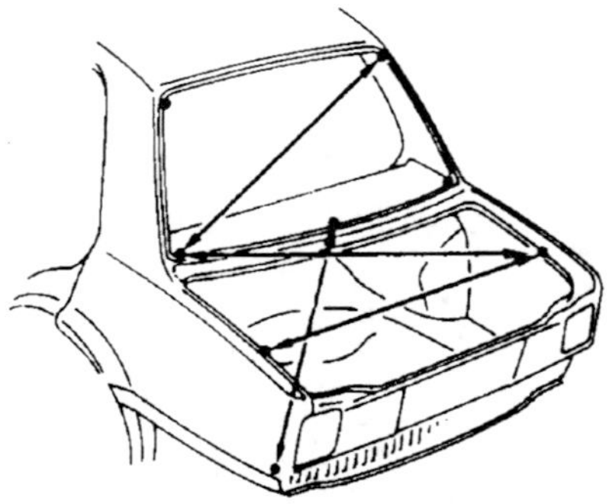

FIGURE 5–20 Typical rear body measurement points.

FIGURE 5–21 Tram gauge used to measure over obstructions.

When using a tram gauge, be sure to keep the following pointers in mind:

- Measurements are made to fixed points on the vehicle such as bolts, plugs, or holes.
- A point-to-point measure is the direct actual measurement between two points.
- The tram bar should be parallel to the car body. This might require the pointers on the tram bar to be set at different lengths.
- Long pointers can be installed to work around obstructions (Figure 5–21).
- Some body dimension manuals show dimensions in bar length. Other dimension books show dimensions in point-to-point lengths. Some manuals use both. The important point is that the technician must use the same method of measurement as the dimension manual employs or there is a very good chance of introducing an error.

FIGURE 5–22 Tram gauge used to measure rail length.

- Make all measurements on the damaged vehicle at the points specified in the body dimensions manual (Figure 5–22). The amount of damage can usually be determined by subtracting the actual measurement from the specified measurement. However, the actual amount of panel movement, or misalignment, is not important. In the estimating process, any misalignment is a sign that damage has occurred and time for repair must be included in the damage analysis.

Gauging

Gauging is used to determine the location and degree of frame damage. The vehicle is evaluated for the following:

Length
Centerline
Datum

Length is simply how long a fame rail is. Centerline is an imaginary line down the center of the vehicle that divides the vehicle into equal left and right halves (Figure 5–23). Datum is an imaginary plane underneath the vehicle from which all height measurements are taken (Figure 5–24). Vehicles can have the following types of frame damage:

- *Twist.* In the vehicle center section, the rocker panels are out of level, disrupting datum (Figure 5–25).
- *Sidesway.* Side impact deflects rails, disrupting the centerline (Figure 5–26).
- *Mash.* Impact shortens the rail length (Figure 5–27).

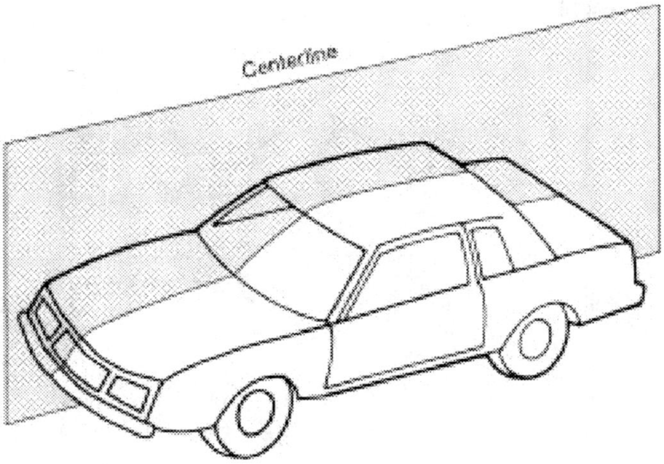

FIGURE 5–23 Centerline.

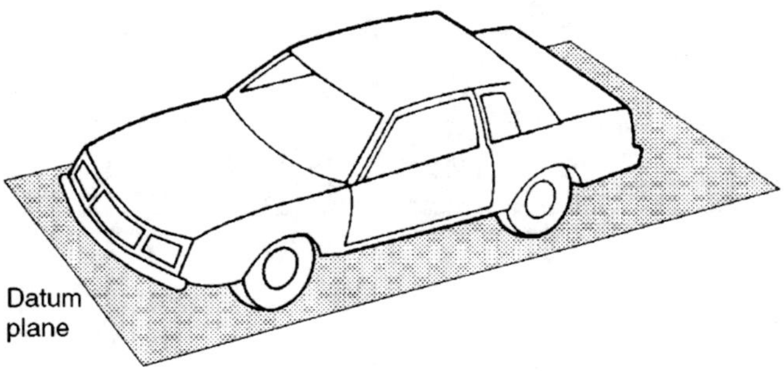

FIGURE 5–24 Datum.

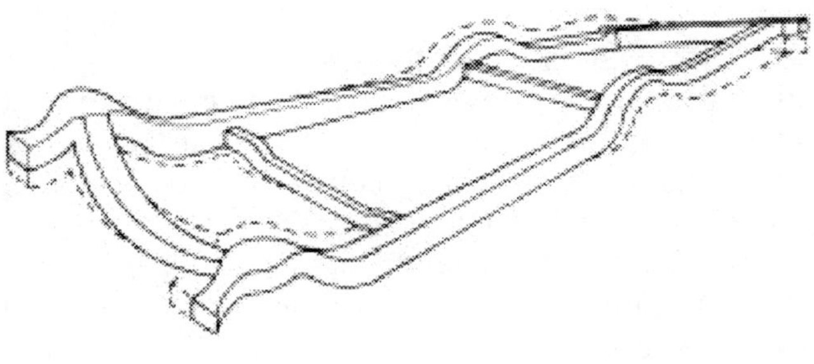

FIGURE 5–25 Twist.

- *Sag or Kickup.* Impact forces the rail up or down, disrupting datum (Figure 5–28).
- *Diamond.* Found only on full-frame vehicles, impact on one frame rail drives it back relative to the undamaged rail (Figure 5–29).

There are three devices to gauge a vehicle for damage: Hanging gauges, mechanical gauges, and computer.

When using hanging gauges, consult a frame dimension manual for locations and heights at which to hang the gauges (Figure 5–30A and B). Use a tram

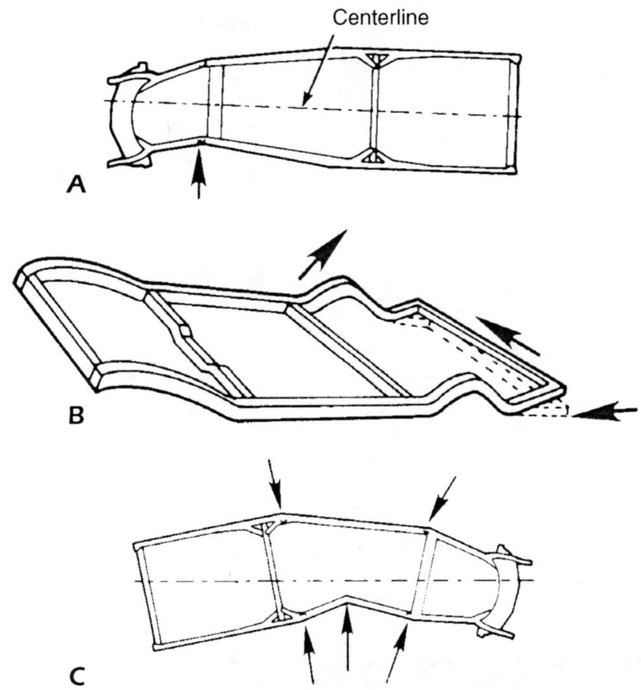

FIGURE 5–26 Various types of sidesway.

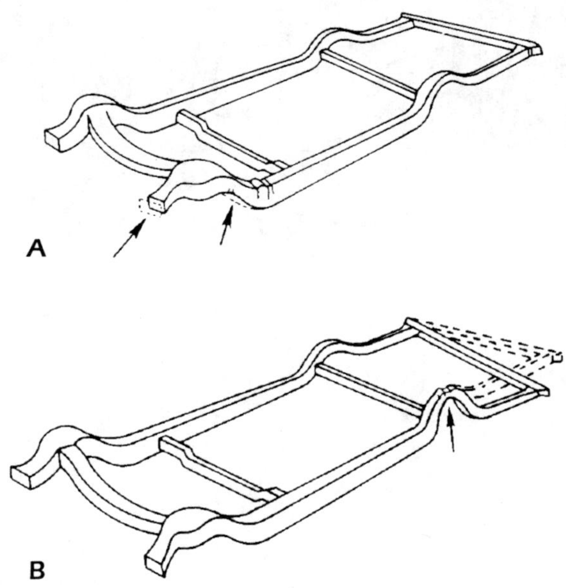

FIGURE 5–27 Mash damage also includes other types of damage.

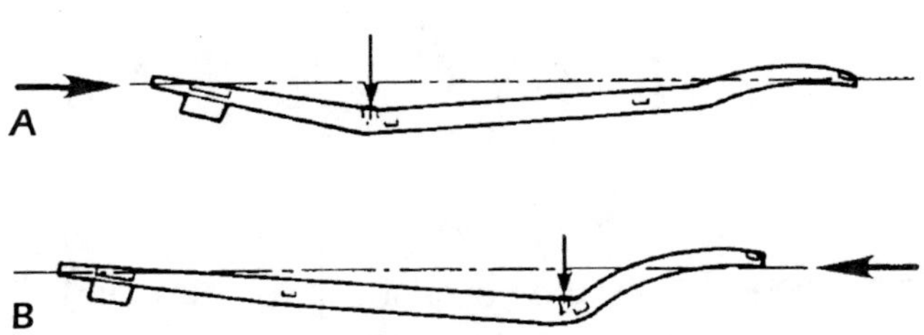

FIGURE 5–28 Sag (A) front impact; (B) rear impact.

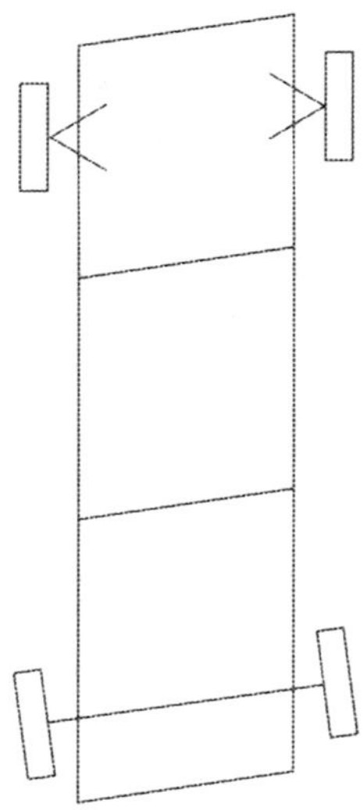

FIGURE 5–29 Diamond.

A

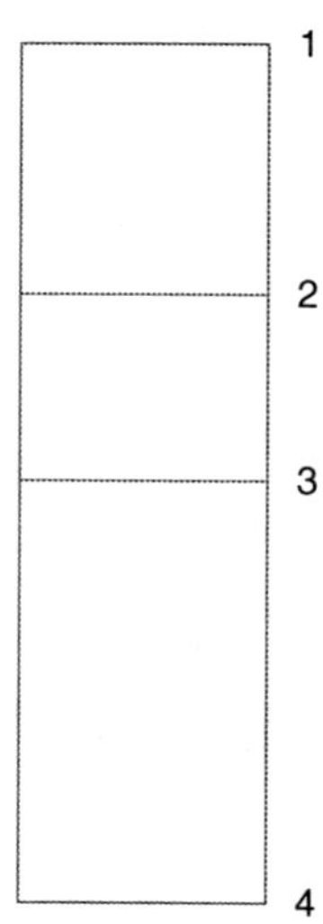

B

FIGURE 5–30 Hanging gauges: (A) one gauge is visible; (B) gauge locations of truck.

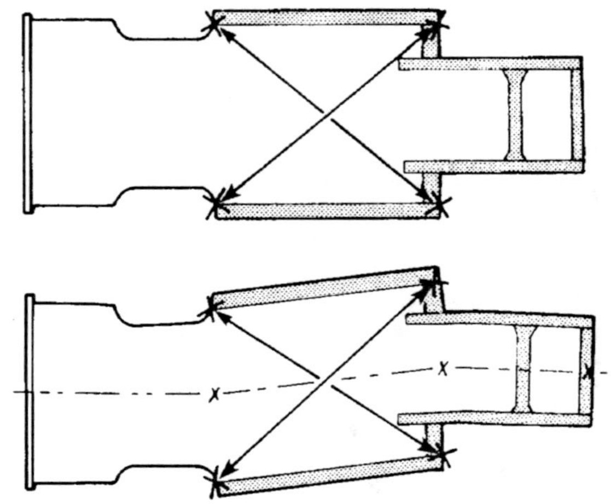

FIGURE 5–31 Unequal diagonal measurements show diamond.

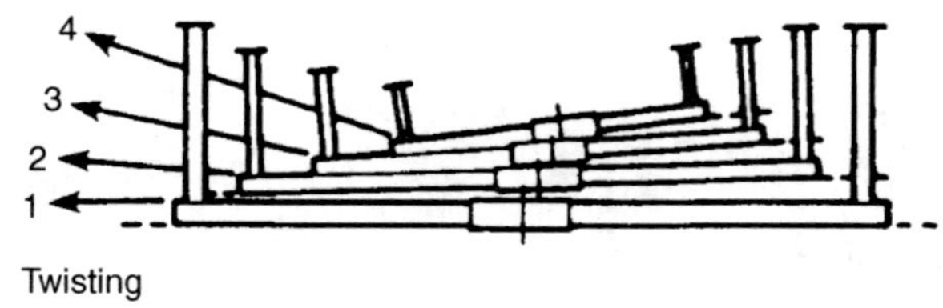

FIGURE 5–32 If self-centering gauges read like this, the body or frame has twist damage.

gauge to measure length. The center section of the vehicle is measured first. On full-frame vehicles, check for diamond. Use the tram gauge to X-check under the cab for diamond. Do not use the cab mount bolts; use holes or rivets on the rails. If there is at least a half-inch or 6 mm difference in the X-check, then diamond damage is present (Figure 5–31). All of the remaining tests are performed on both unibodies and full frames. Hang a gauge under the cowl and one under the rear seat or in the back of the cab on a pickup truck. Stand at least ten feet away from the vehicle and check to see if the gauges are level. If they are not, there is twist damage (Figure 5–32). If they are level, there is not any twist. Next, hang a gauge at the front of the vehicle, usually under the radiator support. Hang the last gauge at the rear, attached to the frame rails. If the gauges are all level to each other, there isn't any datum problem with the vehicle. Gauges not level to the two center gauges indicate a sag or kickup (Figure 5–33A and B). Look through one eye to sight the centerline pins. If they are in a straight line, there is no sidesway. Centerline pin misalignment indicates sidesway (Figure 5–34). Check for mash by using a tram gauge to measure rail

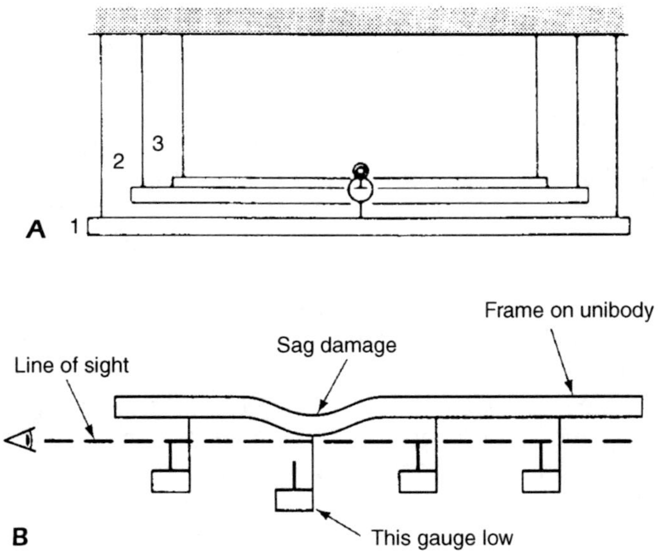

FIGURE 5–33 Gauges indicate sag (A) front view, (B) side view.

FIGURE 5–34 Gauges indicate sidesway.

FIGURE 5–35 This mechanical measuring system uses pointers attached to a bridge.

FIGURE 5–36 Computer display of vehicle dimensions.

length. Make a drawing of the underbody and indicate any damage.

A mechanical measuring system usually consists of a bridge for framework and measuring pointers and pins. To set up the mechanical system, consult the manual for locations to set pointers on the bridge (Figure 5–35). In some cases, the pointers are adjustable in height. In other cases, there are several types of pointers and the proper length pointer is specified. The bridge is assembled under the vehicle. If the pointers do not align at the proper location on the car, there is damage. All types of unibody damage, mash, sidesway, sag, and twist can be read from the pointers. As in a gauge system, the technician should make a drawing of the vehicle and damaged areas.

The computerized measuring system has the vehicle's dimension information stored in the computer software (Figure 5–36). The technician enters the type of vehicle and the computer monitor indicates the targets to use. These targets are placed at the specified locations on the frame. A laser bridge reads the target location and sends the information to the computer. The computer compares the actual measurements to the specified measurements. Any differences are indicated on the computer screen, and a printout can be generated as well. Other computer systems use ultrasonic emitters or mechanical arms instead of laser.

5.5 ZONE 3 MECHANICAL DAMAGE

After checking for primary and secondary damage to the vehicle body, the estimator should next turn his or her attention to Zone 3—the mechanical components of the vehicle (Figure 5–37). If the vehicle was damaged in a front collision, check under the hood for damage to the radiator, fan, power steering pump,

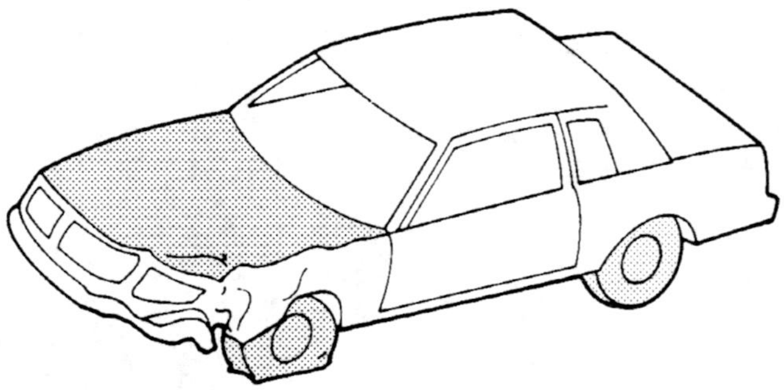

FIGURE 5–37 Zone 3 includes all mechanical components of a vehicle.

FIGURE 5–38 Checking for mechanical damage under the hood.

air-conditioning components, alternator, battery, fuel evaporative canister, windshield washer reservoir, and other mechanical and electrical components. Look for fluid leaks, misaligned pulley belts, dislocated hoses and wires, and dents and cracks as signs of damage (Figure 5–38).

Depending on the severity of the collision, the engine and transmission might also be damaged. Always start the engine if possible and allow it to warm to normal operating temperatures. With the vehicle's wheel lifted off the floor, run the engine in each gear. Listen for any unusual noises; check for smoothness of shifting and clutch operation if the vehicle has a manual transmission. Observe any binding in the throttle linkage, clutch engagement, and gear linkage.

Turn on the air conditioner and make sure it works properly. Check the dash lights and gauges for charging condition, oil pressure, etc. An illuminated CHECK ENGINE light, or its equivalent, is also an indication of mechanical or electrical problems under the hood.

More and more new cars are being equipped with computerized engine control systems that have self-diagnostic capabilities. The self-diagnostic circuits in the computer system are programmed to output trouble codes during certain operating conditions. The earlier self-diagnostic systems required the use of a voltmeter, electronic scan tool, or some other diagnostic equipment to obtain and display the trouble codes. Some newer cars display the codes on the instrument panel or on a small CRT screen. Instructions for initiating the self-diagnostics and for interpreting the trouble codes can be found in the manufacturer's service manuals. The trouble codes indicate specific systems where a fault of some sort is present. This information can be very helpful in pinpointing and verifying damage. Be aware, however, that most computer systems store trouble codes in a memory, and the fault indicated by a trouble code might have existed prior to the collision. If there is not sufficient reason to believe that the mechanical/electrical fault is the result of the collision, do not list it on the estimate form. Remember that the vehicle owner is only entitled to have his or her vehicle restored to its precollision condition—not to have preexisting conditions corrected. Any additional work or improvement must be approved and paid for by the customer.

Mechanical damage is sometimes the result of secondary damage rather than direct collision forces. The engine and transmission have a tremendous mass and in a collision can be thrown forward as much as six inches, causing damage to attached accessories and other underhood components. Because the power train will return almost to its original position, secondary damage might not be immediately noticeable. Check for damage to motor mounts, misaligned pulleys and drive belts, and loose hoses and linkages.

After checking under the hood, jack up the vehicle and support it with jack stands. Roll under the vehicle and check for bent steering and suspension components, kinked brake hoses, and leaking brake and fuel lines and fittings. Check for leakage from the engine, transmission, differential, steering rack or gearbox, and struts. Turn the steering wheel from stop to stop and check for binding or noise. Rotate the wheels and check them for runout, cuts, scratches, and bruises. Lower the vehicle onto its tires and adjust the steering so that the wheels are in the straight-ahead position. Measure the vehicle from the front wheel hub to the rear wheel hub. The measurement should be identical for left and right sides of the vehicle. If not, the steering and/or suspension components are damaged.

Wheel alignment can be quickly tested by doing a jounce-rebound test.

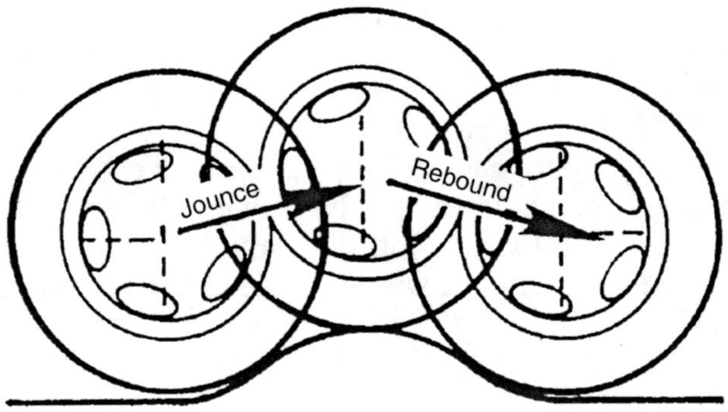

FIGURE 5–39 *Jounce and rebound motions.*

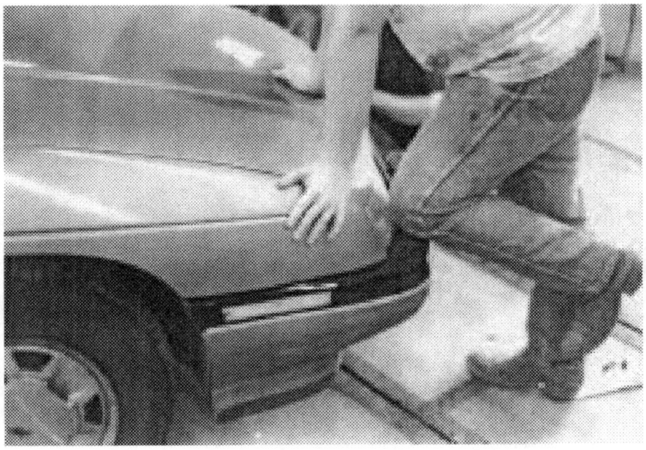

FIGURE 5–40 *This simple jounce-rebound test will check steering arm or gear alignment.*

- Jounce (Figure 5–39) is the motion caused by a wheel going over a bump and compressing the spring. During jounce, the wheel moves up toward the chassis. Jounce can be simulated for in-shop testing by pushing down on the car by sitting on the fender. The car must be jounced equally on both sides.
- Rebound is the motion caused by a wheel going into a dip, or returning from a jounce and extending the spring. During rebound, the wheel moves down away from the chassis. Rebound can be simulated for in-shop testing by lifting up on the fender. The car must be lifted equally on both sides.

This jounce-rebound check (Figure 5–40) will determine if there is some misalignment to the rack-and-pinion gear.

For a quick check, unlock the steering wheel and see if it moves during the jounce and/or rebound. For a more careful check, employ a pointer and a piece of chalk. Use the chalk to make a reference mark on the tire tread and place the pointer on the same line as the chalk mark. Jounce and rebound the suspension system a few times while someone watches the chalk mark and the pointer. If the chalk mark on the wheel moves more than one tread width (or 3/32 inch) left or right, the chances are that there is a steering arm or gear out of alignment. Repeat this procedure for the opposite tire.

Another alignment test measures cornering angle. The cornering angle check evaluates the proper relationship of the two front wheels as they are turned through a steering arc. To measure cornering angle, one wheel is turned on a turn plate or *protractor* a given amount; the amount of rotation of the opposite wheel is measured on a similar turn plate or protractor.

The process is then repeated with the opposite wheel. The results are compared right to left to determine if the two front wheels are rotating through the same arc.

During a cornering angle check, the left front wheel should be turned out 20 degrees. The right wheel rotation would be measured. The readings will usually not match those of the right wheel. The right wheel should turn “in” the same amount or about 2 degrees less. The difference accounts for the turning radius difference between the inside and outside wheels during cornering.

The process is repeated with the right wheel; the right wheel is turned 20 degrees “out.” The movement of the left wheel is measured on the protractor or turn plate. The left wheel should turn “in” the same amount, or about 2 degrees less.

By design, a vehicle might use a different turning radius from one side to the other. If in doubt, refer to the manufacturer’s specifications. If these measurements do not repeat within 2 degrees, damage to the steering arms or gear is indicated. Cornering angle measurements are especially useful in determining whether improper toe conditions are caused by poor wheel alignment or damaged suspension components.

A check of the camber angle will indicate suspension damage. The following camber checks can be made to diagnose the condition of a strut and can be measured easily with a camber gauge (Figure 5–41). One is called a jounce-rebound camber measurement (Figure 5–42A) and can be made by loading the suspension in a similar fashion to jounce-rebound toe change and measuring the camber angle from an individual wheel.

The suspension is then unloaded as in the jounce-rebound toe check and a second camber reading of the same wheel is made (Figure 5–42B). The two readings are compared; these readings should not differ more than 3 degrees on a MacPherson strut–type suspension. If they do, side-to-side strut damage is probably the reason.

Front-to-rear bending of the strut can be checked by making a swing camber measurement (Figure 5–43). Turn the front wheel full right and take a camber reading. Then turn the wheels full left and take another camber reading. If the readings vary by more than 6 degrees, the strut suspension is probably bent front to rear.

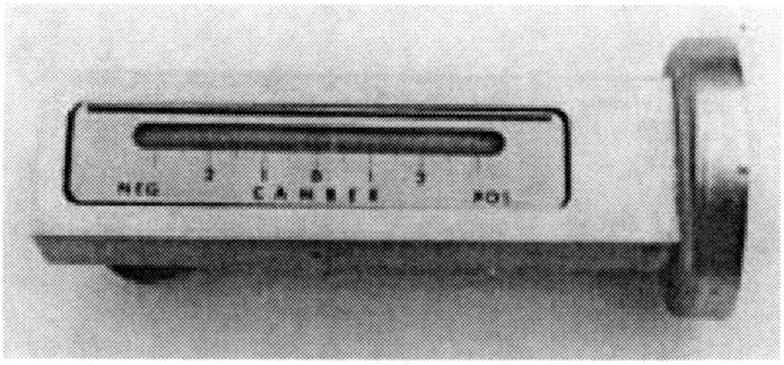

FIGURE 5–41 A camber gauge.

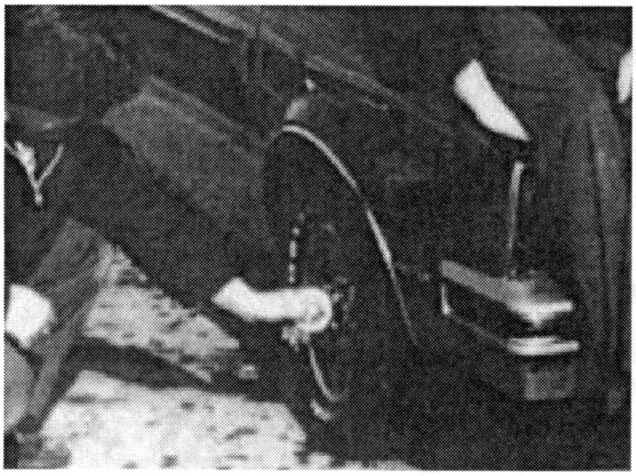

A

B

FIGURE 5–42 Jounce and camber checks: (A) loading the suspension and (B) unloading the suspension.

A

B

FIGURE 5–43 Swing camber measurements: (A) full left and (B) full right.

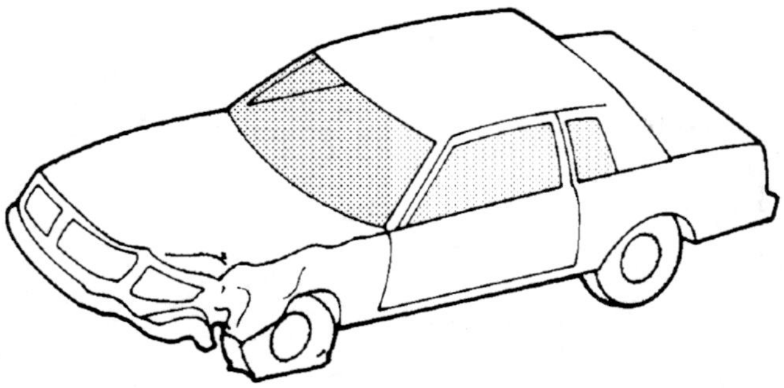

FIGURE 5–44 Zone 4 includes interior trim and components.

FIGURE 5–45 Deployed passenger-side air bag.

When making the camber measurements, it is not necessary to have the vehicle on a level surface. Neither is the actual camber angle important. Only the difference between camber angle readings is needed to determine whether the suspension has been damaged. Be sure to take the camber readings from the same spot on the wheels.

5.6 ZONE 4 PASSENGER COMPARTMENT

Damage to the passenger compartment (Figure 5–44) can be the result of direct collision forces, such as in a side hit. Damage to interior trim and accessories also results from the momentum of passengers and objects within the compartment.

Begin the inspection with the dash. If the collision caused damage to the cowl or door pillars, damage might have occurred to the instrument panel, heating core and ducts, radio, electronic control modules, airbags, etc. Check the operation of any components that were not checked during the Zone 3 inspection (Figure 5–45).

Check the steering wheel for damage. Check its mounting hardware, tilt and telescopic features, horn, headlight, turn signal switches, ignition key, and steering wheel lock. Point the wheels straight ahead and make sure the steering wheel is centered. If the steering wheel is the energy-absorbing type, make sure it is not collapsed.

Check for damage to knobs, levers, dash glass, and trim. Open and close, and lock the glove compartment to verify that it has not been knocked out of alignment. Check the brake pedal for distortion, binding, soft pedal, etc. Pull the carpet back and check the floor and toe pans for popped rivets and split seams.

FIGURE 5–46 Checking window for proper operation.

Check the seats for damage. In a front-end collision, the body mass of the passenger strapped into the seat can cause damage to the seat frame adjuster tracks and mounting hardware. In a rear-end collision, the hinge points of the seat can be damaged. Check the seat adjustments for full range of movement.

Check the doors. Arm rests, trim panels, and inner door panels can be damaged by the occupant's momentum. If the collision was a side hit, door locks and window regulators will probably be damaged. Even in a front-end collision, the mass of the window glass can cause damage to the window track and regulator. Operate the window fully down and back up to check for binding or interference (Figure 5–46). Lower the window 1–1/2 inches and make sure that the window is parallel with the door frame. Make sure that all accessories such as electronic locks and antitheft systems, master window and door lock controls, and remote mirror controls work properly.

Check the passenger restrain system. If the car has a passive restraint system, make sure the belt engages and disengages fully and without binding or hesitation. Check seat belt latches for safe operation. Make sure that the lap and shoulder belts in nonpassive systems buckle and unbuckle easily. Make sure that retractors, D-rings, and anchor plates are not damaged. Some seat belts have a tension-sensitive label. If the seat belt is worn in a crash and if the tension on the seat belt exceeds the design limit, the label tears. Seat belts with torn labels must be replaced. The label is visible by drawing the belt completely out of the retainer.

Also make a list of non-OEM accessories in the vehicle—such as CB radios, tape players, and stereo speakers.

5.7 ZONE 5 EXTERIOR TRIM AND FINISH

After thoroughly inspecting the body, mechanicals, and interior upholstery and accessories, walk around the vehicle once more and list damage done to trim, mouldings, vinyl roof material, wheel covers, side markers, and other body accessories (Figure 5–47).

FIGURE 5–47 *Zone 5 includes all exterior trim, accessories, and finish.*

Turn on the lights and check the headlights, taillights, turn signal indicator lights, and flashes. Light filaments often break due to the jar of the collision impact, especially if the lights are on when the collision occurs.

If the bumpers were not checked in the Zone 1 or Zone 2 inspections, they should be inspected now. Check for cracked fascia and dust covers, impacted or leaking energy absorbers, and torn rubber isolators.

Carefully inspect the paint condition. Make a note of which panels must be refinished, and list any special considerations, such as clear coating, flexible plastic parts, and surface rust. Light panel damage might only require spot painting. Other repairs might necessitate painting one or more panels. In either case, time will be needed to blend and color match the new finish with the old finish remaining on the undamaged panels. A heavily damaged car or one with an old, weathered finish might require a whole-car paint job.

It is also important to check the finish for prior damage. Repair of dents, dings, scratches, and paint defects present on the vehicle prior to the collision will not be covered by the insurance company's settlement. The customer must pay for any repair that improves the precollision condition of the car.

5.8 FIRE, FLOOD, AND VANDALISM DAMAGE

Fire in an engine compartment will melt wiring and plastic parts, such as batteries and reservoirs, and damge heat-sensitive components, such as gaskets and diaphragms. Paint might bubble or be discolored by heat and smoke. Metal panels might buckle.

Flood damage can cause extensive damage if the engine and other mechanical and electrical parts are submerged. Water can enter the engine through the carburetor or throttle body and through the dipstick. Water can also enter transmissions and differentials through vent holes. Engines and transmissions must be drained, flushed, and filled before starting and test driving the vehicle. Submersion in salt water is especially corrosive and often requires teardown of the power train. Vehicles submerged below the belt line are often totaled.

Damage due to vandalism is much easier to determine. Broken window and locks, distortion caused by prying, missing parts, and torn wiring and mounting areas are very visible. Test drive the vehicle and operate all accessories before completing the damaged appraisal.

REVIEW QUESTIONS

1. At minimum, how many gauges need to be hung to check for sidesway?
   a. one
   b. two
   c. three
   d. four

2. Technician A checks a pickup truck for diamond by X-checking under the cab. Technician B does not check a unibody vehicle for diamond. Who is right?
   a. technician A
   b. technician B
   c. both A and B
   d. neither A nor B
3. Sidesway is the distortion of ________________.
4. Technician A X-checks a hood opening with a tape measure. Technician B X- checks a hood opening with a tram gauge. Who is right?
   a. technician A
   b. technician B
   c. both A and B
   d. neither A nor B
5. On a symmetrical unibody vehicle, if one leg of an X-check is 1181 mm and the other leg is 1179 mm, is there a problem?
   a. Yes
   b. No
6. Technician A determines the center-to-center distance between same-sized holes by measuring the near side of one to the far side of the other. Technician B does the same on different-sized holes. Who is right?
   a. technician A
   b. technician B
   c. both A and B
   d. neither A nor B
7. In which zone would you find mechanical damage?
   a. Zone 1
   b. Zone 2
   c. Zone 3
   d. Zone 4
8. Technician A says a bent McPherson strut can change the camber of a wheel. Technician B says that four gauges should be hung when gauging a unibody vehicle. Who is right?
   a. technician A
   b. technician B
   c. both A and B
   d. neither A nor B
9. A computer measuring system is more accurate than a gauge system.
   True
   False
10. On a unibody vehicle, technician A checks for damage in the center section first. Technician B says that damaged full frames can be checked with a tape measure only. Who is right?
    a. technician A
    b. technician B
    c. both A and B
    d. neither A nor B

Chapter 6

Estimating Labor and Parts Costs

Objectives

After reading this chapter, you should be able to:

- **Use the collision estimating guides to determine labor and parts costs.**
- **Determine whether a panel should be repaired or replaced.**
- **List operations that are not included in the collision estimating guides' operation times.**
- **Name the three major types of operations for which labor times are given in the collision estimating guides.**
- **Explain how to adjust operation times for overlap, included operations, and access time.**
- **Explain how betterment reduces the amount of the parts costs that an insurance company will cover.**

Key Terms

R&I
R&R
O/H
Sectioning
Overlap
Betterment

After the estimator finishes his or her visible damage analysis, the estimator must decide whether individual panels and components will be repaired or replaced. To do that, he or she must compare the cost of replacement parts and labor with the labor involved in repairing the damage. Much of the information needed to make this comparison is found in the collision estimating guides.

The collision estimating guides contain labor times for many repair operations, including refinishing. They also contain current list prices for most OEM parts. Figure 6–1 illustrates the information given in a typical grouping of components. The rest of this chapter and the following chapters will explain how to interpret the information given in the collision estimating guides and how to use the information to write an estimate of the repair costs.

6.1 REPAIR OR REPLACE?

When making the decision to repair or replace a damaged panel, several factors must be considered. first what is the extent of the damage? As explained in chapter 3, all metal will bend to a degree and still return to its original shape. This tendency is called *elastic deformation.*

However, when a piece of metal is bent beyond a certain point (the elastic deformation limit), it will permanently deform, or buckle. The grain structure of the metal will slip and realign, taking on a new arrangement or set. The grain along the top of the bend will stretch, and the grain along the underside of the bend will compress and shrink (Figure 6–2). Stretched and compressed areas are said to be *work-hardened.* The realigned grain structure becomes hard and stiff. As explained in Chapter 4, heat may need to be applied to the buckle to soften the metal before it can be hammered or pulled back into shape. The estimator must recognize work-hardened areas in a panel. Panels with excessive amounts of work hardening might

FRONT BUMPER

Use Procedure Explanations 1 and 28 with the following text.

STD, GT, GTU, INDY

Refinish Bumper Cover **2.2**
R&I Bumper Assy .... **.7**
O/H Bumper Assy (Includes R&I) .... ***#2.2***
#Includes R&R Park/Signal & Side Marker Lamps & Front Deflector
Add to O/H Bumper Assy w/GTU or Indy .... **.6**
NOTE: All Parts in this section are included in overhaul unless noted otherwise.

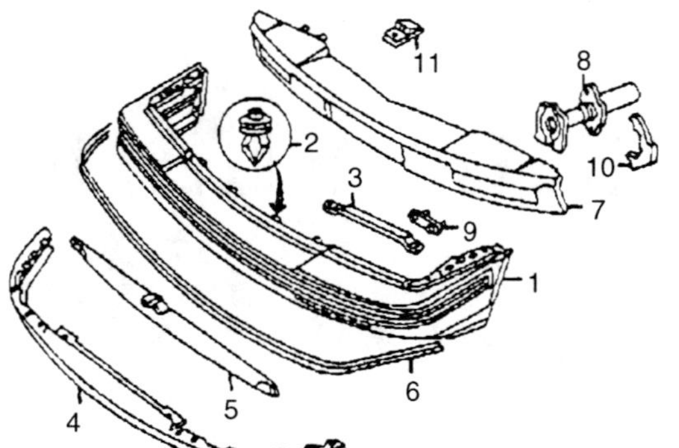

1 **Cover, Bumper (P)**				
Standard, GT	87-92	16509230	***#2.2***	225.00
	93-96	16520556	***#2.2***	392.00
GTU, Indy¶		16509230	***#2.8***	225.00
(P) Paint to Match				
¶Order by Application				
#Includes R&R Park/Signal & Side Marker Lamps & Front Deflector				
U-Nut, Cover End		12338002		.44
2 **Retainer, Push-In (Each)**		14093088		.45
3 **Support, Cover**	87-92 R/L	10049954		7.18
	93-96 R/L	22577070-1		4.15
Bracket, License	87-94	14083146	**.3**	18.41
	95-96	22598649	**.3**	14.25
4 **Deflector, Cover**				
Std, GT		10134430	**.3**	74.00
GTU, Indy-See Ground Effects Section				
5 **Deflector, Radiator Air**		14093081	**.2**	16.75
6 **Moulding, Cover**				
1987-90				
2.0L Eng				
Bright		14103493	**IOH**	32.00
Black		14103494	**IOH**	35.00
2.2L Eng		10116145	**IOH**	38.25
2.8L Eng				
Black		14093080	**IOH**	31.25
Red		10115261	**IOH**	38.25
3.1L Eng				
Black		10116144	**IOH**	38.25
Red		10116143	**IOH**	38.25
Charcoal (Indy)		12507536	**IOH**	119.00
1991-92 (a)		10173550	**IOH**	38.25
1993-96-Serviced w/Bumper Cover				
(a) Black listed, Order by Application				
7 **Bar, Impact**		22578820	**IOH**	145.00
Bumper, Bar	R/L	14083150-49	**IOH**	3.64
8 **Absorber, Impact**	87-91 R/L	22074508	***#.3***	49.02
	92-96 R/L	22149565	***#.3***	70.59
#w/Bumper Removed, Not Included in O/H				
9 **Plate, Stud (2/Side)**	R/L	14083135	**IOH**	2.05
10 **Shim, Absorber¶**	R/L	10051921	***#.3***	3.45
¶As Required				
#R&R Absorber & Shim Complete, w/Bumper Assy Removed, Not Included in O/H				
11 **Pad, Tow**	87-91	10077223	***#.2***	10.95
#Not Included in O/H				

GTZ & Z26

Refinish Bumper Cover **2.4**
R&I Bumper Assy .... **.7**
O/H Bumper Assy (Includes R&I) .... ***#2.8***
#Includes R&R Park/Signal & Side Marker Lamps
NOTE: All Parts in this section are included in overhaul unless noted otherwise.

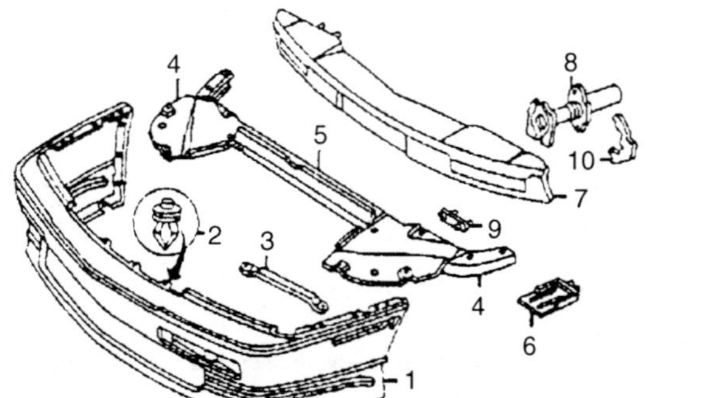

1 **Cover, Bumper (P)**		16515071	***#2.8***	446.00
(P) Paint to Match				
#Includes R&R Park/Signal & Side Marker Lamps				
2 **Retainer, Push-In (Each)**				
1990-91		10077283		.59
1992-96		14093088		.45
3 **Support, Cover**				
Steel	R/L	10049954		7.18
Plastic	R/L	22577070-1		4.15
Bracket, License		10115284	**.3**	17.58
Deflector, Lower				
4 Outer	R/L	2257642617	***#.2***	16.75
5 Center		10108701	***#.3***	16.75
#R&R Complete .5				
6 **Extension, Deflector**	R/L	10119386-7		3.95
7 **Bar, Impact**		22578820	**IOH**	145.00
Bumper, Bar	R/L	14083150-49	**IOH**	3.64
8 **Absorber, Impact**	90-91 R/L	22074508	***#.3***	49.02
	92-96 R/L	22149565	***#.3***	70.59
#w/Bumper Removed, Not Included in O/H				
9 **Plate, Stud (2/Side)**	R/L	14083135	**IOH**	2.05
10 **Shim, Absorber¶**	R/L	10051921	***#.3***	3.45
¶As Required				
#R&R Absorber & Shim Complete, w/Bumper Assy Removed, Not Included in O/H				

FIGURE 6–1 Information contained in Mitchell Collision Estimating Guide. Look for paint time, part number, part name, labor time, part price. (Courtesy of Mitchell International.)

require excessive labor to repair; in that case, the buckled panels should be replaced.

Sometimes the type of metal used in the buckled panel will determine whether or not the panel is repairable. Panels made of mild carbon steel are easier to shrink or stretch back into shape than are panels made of high-strength, low-alloy steels. Some high-strength steels lose their grain orientation, or strength, when subjected to the heat involved in shrinking or stretching operations. Therefore, damaged panels

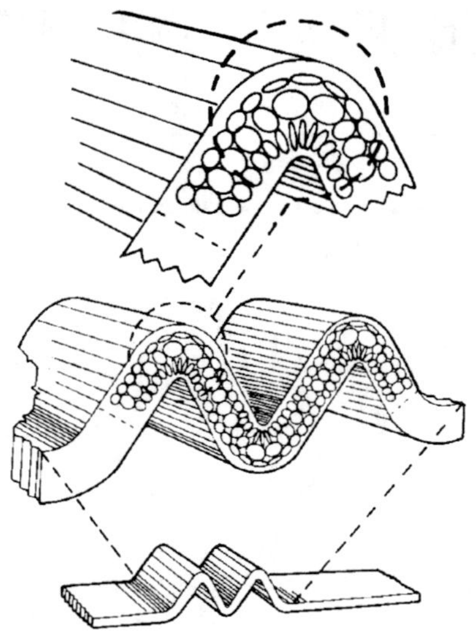

FIGURE 6–2 Grain structure disruption in a work-hardened buckle.

that are formed from high-strength steels are more often replaced than panels formed from softer, carbon steels.

Information regarding parts made with high-strength steel is not always available from the vehicle manufacturer. When such information is provided, the publishers of the collision estimating guides will identify the high-strength steel components with the abbreviation HSS or HSLA following the part name (Figure 6–3). The designation should alert the estimator that conventional repair methods might not be applicable to that particular part and that replacement or sectioning of the damaged part is the better solution.

Another factor affecting the repairability of a panel is its construction. In many sections of a car, the outer panels are welded to the inner panels to create a strong box structure. High-strength steel reinforcements are often placed inside the boxed structures. These reinforcements restrict sectioning and welding

QUARTER PANEL

Use Procedure Explanations 20 and 28 with the following text.

SHEET METAL

Operation		
Refinish Quarter Outside	R/L	**2.4**
Add to Edge Panel	R/L	**.5**
Add for Pillar	R/L	**.5**
Refinish Fuel Door		**.5**
Refinish Quarter Rear Extensions	R/L	**.5**
Upper	R/L	**.3**
Lower	R/L	**.3**
R&R Both Quarter Panels		**30.2**
R&R Roof & Both Quarter Panels		
w/Sunroof		**44.7**
w/o Sunroof		**44.2**
R&R Rear Body Panel & Both Quarter Panels		**31.7**
R&R Roof, Both Quarter Panels & Rear Body Panel		
w/Sunroof		**46.2**
w/o Sunroof		**45.7**
Add to R&R Quarter Panel		
To Drill for Installation of Cladding	R/L	**.3**
To R&I Black Glass		**.5**

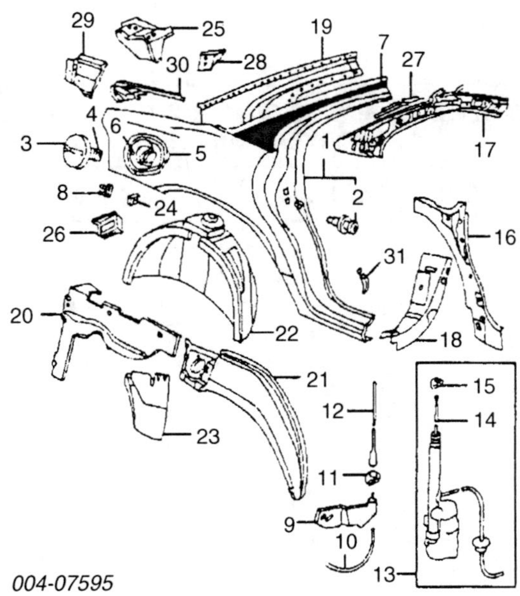

Part		Part No.	Labor	Price
1 **Panel Assy, Outer Quarter (HSLA)**				
Right	93	4646186	**#17.5**	125.00
	94-97	4646768	**#17.5**	135.00
#Includes R&R Fuel Door, Fuel Filler Tube Reinf & Antenna Assy., Deduct 4.4 w/Roof Removed, Deduction Includes Common Surfaces & Included Operations				
Left	93	4646187	**#16.5**	125.00
	94-97	4646769	**#16.5**	135.00
#Deduct 4.4 w/Roof Removed, Deduction Includes Common Surfaces & Included Operations				
2 **Striker, Door Interlock¶**	R/L	4696937		2.85
¶Included w/Outer Quarter Panel Assy				
3 **Door, Fuel**				
To 3-15-95		4583784	**.3**	15.50
From 3-15-95		4769163	**.3**	16.25
4 **Spring, Fuel Door**	93	4646925		3.00
	94-96	4756019		2.60
5 **Reinf, Fuel Filler Tube**		4646530		16.50
6 **Bumper, Fuel Door**		4416406		3.70
7 **Applique, Quarter Upper**	R	4856064	**#.2**	67.50
	L	4773733	**#.2**	72.50
#Included in R&R Back Window				
8 **Plate, Filler Door Mtg**		4696543		3.10
9 **Base, Manual Antenna**		4608017	**#.5**	8.20
#Includes R&R Antenna Mast & Grommet, Does Not Include Replace Cable				
10 **Cable, Antenna**	93-95	4608021		18.75
	96	4759128		20.50
	97	N.A.		N.A.
11 **Grommet, Antenna**		4608149		1.60
12 **Mast, Manual Antenna**		4608146	**.1**	8.35
13 **Antenna Assy, Power**	93-95	4760528	**.5**	86.00
	96	4760059		64.00
	97	4760528		86.00
14 **Mast, Power Antenna¶**		4740369	**.4**	24.25
15 **Grommet, Antenna¶**		4608149		1.60

FIGURE 6–3 Listing of HSLA part in estimating guide. (Courtesy of Mitchell International.)

techniques. It is important that the estimator stay abreast of construction techniques so that he or she is aware of the reinforcements' location.

Sometimes there is no access to the backside of the damaged panel. This prohibits the use of hammer and dolly, spoons, or picks. If the damage is too extensive to be pulled out by a slide hammer and draw pins, the panel probably should be replaced.

The major deciding factor, of course, is cost. There comes a point when it is more economical to replace a panel than to repair it. The commonly accepted rule of thumb is this: if the repair costs exceed 75 percent of the cost of replacement, the damaged parts should be replaced.

FRONT LAMPS

HEADLAMP

R&I/R&R Does Not Include Aim Lamps			.4	
R&I Headlamp Assy	R/L		#.4	
#Includes R&I Grille & Park/Side Marker Lamps, R&R Both Sides .6				

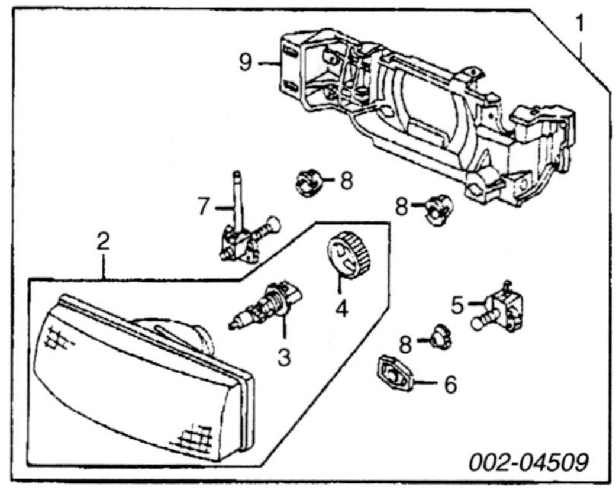

1 **Lamp Assy**	R/L	F1CZ 13008 A-B	**#.4**	183.47
2 **Lens Assy¶**	R/L	F1CZ 13007 A-B	**#.4**	89.63
#Includes R&R Park/Side Marker Lamps & R&I Grille, R&R Both Sides .6				
3 **Bulb¶ (a)**	R/L	E5LY 13N021 A	**.2**	20.70
4 **Retainer, Bulb¶ (a)**	R/L	E4LY 13N019 A		3.08
(a) Included w/Lens Assy				
5 **Screw, Adjusting¶**	R/L	F1CZ 13032 B		12.80
6 **Clip, Screw Retaining¶**	R/L	F1CZ 13A147 B		2.70
7 **Bar, Adjusting¶**	R/L	F1CZ 13032 A		12.80
8 **Bushing, Retainer¶ (6)**	R/L	F1CZ 13A147 A		5.23
9 **Housing, Headlamp¶**	R/L	F1CZ 13100-1 A	**#.7**	29.72
¶Included w/Lamp Assy				
#Includes R&I Grille Assy & R&I/R&R Park/Side Marker Lamp, R&R Both Sides 1.2				
Module, Daytime Running				
To 10-90		N.A.	**#.3**	N.A.
From 10-90		N.A.	**#.3**	N.A.
Bracket, Daytime Running		N.A.	**#.3**	N.A.
#R&R Complete				

FIGURE 6–4 Operation times in estimate guide for headlight. (Courtesy of Mitchell International.)

TABLE 6–1: FRACTIONAL HOUR-TO-MINUTE CONVERSION CHART

Tenths of an Hour	Minutes
.1	6
.2	12
.3	18
.4	24
.5	30
.6	36
.7	42
.8	48
.9	54
1.0	60

6.2 LABOR TIMES

When the decision has been made to replace a damaged panel, the estimator will find labor allowance times listed in the crash guides (Figure 6–4). The operation times are listed in hours and tenths of an hour. For example, the labor time allotted for removing and replacing a headlight assembly might be 0.4 hours, which is equivalent to 24 minutes. A labor time of 3.5 hours is equal to 3-1/2 hours. Table 6–1 shows the minute equivalent of each tenth of an hour.

Labor times are based on information supplied by vehicle manufacturers and time studies. The manufacturers arrive at the times by performing each operation under normal shop conditions. The average mechanic, working under average conditions and following procedures outlined in the manufacturer's service manuals, should be able to complete the operation within the time allotted. The labor time studies include time for gathering tools and materials in preparation for the operation.

The labor times suggested by the vehicle manufacturer are for the removal and replacement of new, undamaged OEM parts from new, undamaged vehicles. Sometimes the manufacturers' operation times are proven by field experience to be unrealistic when performed on damaged vehicles. Therefore, the publishers of the collision estimating guides are constantly collecting data from the collision repair industry, comparing the quoted times with actual performance in the field. Time studies are performed in an actual body shop environment to validate the operation times listed in the collision estimating guides. These time studies centers are performed throughout the country and cover a multiplicity of operations in a variety of environments. These "hands-on" experiences, in addition to industry surveys and technical input from collision repair facilities, are used to develop the labor allowance illustrated in the collision estimating guides.

ADDITIONAL TIME ALLOWANCES

Operation times are based on new, undamaged parts installed as an individual operation. These times apply only to standard OEM stock items. The times do not apply to vehicles with equipment other than that supplied by the manufacturer as standard or regular production options. If other parts or equipment must be replaced, the operation times should be adjusted to compensate for the increased labor. For example, if the decision is made to replace a damaged part with an aftermarket or salvage part, additional time might be justified for adjusting and aligning the part to the car.

Additional time may also be required for operations not included in the quoted times. There are a variety of vehicle conditions and a wide range of collision damages that simply cannot be anticipated by the publishers of the collision estimating guides. Therefore, the following items are not taken into consideration when formulating the operation times:

- *Access Time.* Remove extensively damaged parts by cutting, pushing, pulling, etc.
- *Anticorrosion Rust-Resistant Material.* Remove or apply weldable zinc primers, wax, petroleum based coating, undercoating or any type of added conditioning.
- *Broken Glass Cleanup*
- *Detail.* Clean vehicle to preaccident condition.
- *Electronic Components.*
- Time to remove and install as necessary; includes wiring and/or wiring harness and computer module.
- Time to reset memory code function when battery has been disconnected to perform repairs.
- *Frame Setup*
- *Free Up Parts.* Time necessary to free up parts frozen by rust or corrosion.
- *Measure and Identify.* Structural damage to unibody vehicles.
- *Plug and Finish Holes.* Time to plug and finish unneeded holes on parts being installed.
- *Repair or Align.* Parts adjacent to parts being replaced.

16 FRONT/REAR OR BACK DOOR

IMPORTANT REMINDER: Labor times for reinforcements are with outer panels removed.

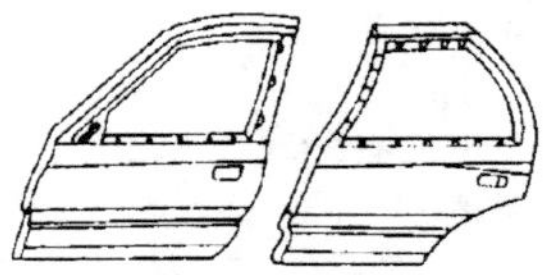

CEG28

DOOR SHELL R&R

INCLUDED OPERATIONS

- Remove and install trim panel
- Remove and install or replace:
  Lock cylinder
- Replace or transfer parts attached to door (example: outside handle, glass, run channels, regulator, vent assembly, weatherstrip)
- Replace caulking for standard factory application if necessary
- Replace clip type moulding for base model vehicle (example: window surround moulding, belt moulding)

NOT INCLUDED OPERATIONS

- Refinish door panel
- Replace channels attached to glass; add .2 hour for lower and .3 hour for upper and lower
- Replace sound deadening
- Remove and install or replace:
  Rear view mirror
  Weld-on type hinges
  Hinge anchor plates
  Decals (example: EPA, body identification, caution/danger)
- Recode lock cylinder
- Remove and install adhesive exterior trim; add to clean and retape
- Replace new adhesive exterior trim; deduct one-half of R&R time
- Install stripes, decals, transfers or overlays
- Drill holes for installing exterior trim
- Broken glass clean up

FIGURE 6–5 Included, not included operations listed in P pages. (Courtesy of Mitchell International.)

- *Rework Parts.* To fit a particular year or model (example: cutting holes for lamps, modifying a radiator support).
- *Tar and Grease.* Removal of these or any other materials that would interfere with operation.
- *Transfer Time.* For welded or riveted brackets, braces or reinforcements from old part to new part.

Operation times and the actual repair operations covered are further defined in the p pages of the collision estimating guide (Figure 6–5). It is important to refer to all pages when writing an estimate. These pages explain which operations are included or not included in an operation time. They might also specify adjustments to the operation times based on the vehicle's accessories and trim packages (Figure 6–6). Other notes give information on sectioning, pricing, ordering parts, and refinishing.

7 FRONT FENDER

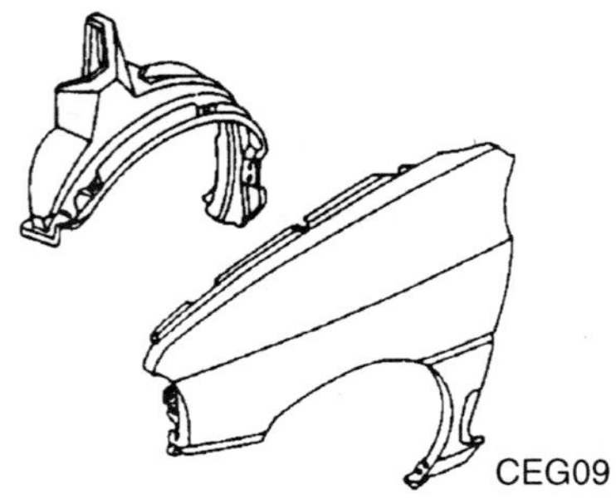

FRONT FENDER R&R

INCLUDED OPERATIONS

- Remove and install or replace:
  - Headlamp assembly if attached to fender
  - Cornering lamp if so equipped
  - Side marker lamp if so equipped
  - Turn indicator lamp if so equipped
  - Parts attached to fender except those listed in not included section
- Replace clip type moulding for base model vehicle

NOT INCLUDED OPERATIONS

- Refinish front fender
- Aim lamps
- Remove and install front bumper
- Remove and install or replace:
  - Front header panel
  - Skirt or liner
  - Wheel
  - Antenna
  - Rocker moulding
- Remove and install adhesive exterior trim; add to clean and retape
- Replace new adhesive exterior trim; deduct one-half of R&R time
- Install stripes, decals, transfers or overlays
- Drill holes for installing exterior trim
- Cut holes for installing lamps

FIGURE 6–6 Mustang rocker panel replacement times. (Courtesy of Mitchell International.)

FRONT LAMPS

HEADLAMP

R&R Does Not Include Aim Lamps		.4	
R&I Headlamp Assy	R/L	.3	

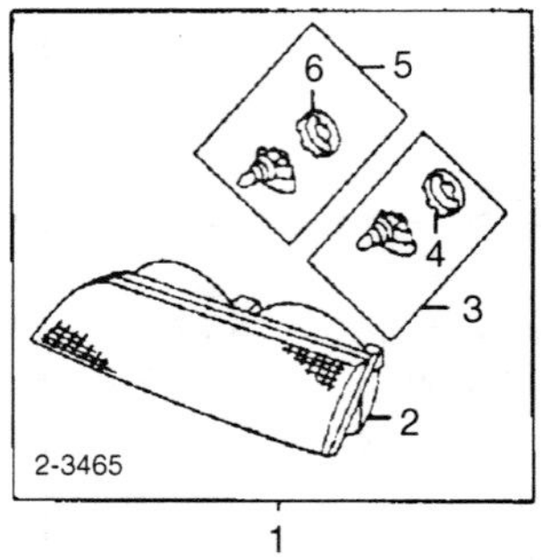

Part	Side	Part No.	Time	Price
1 **Headlamp Assy**	R/L	F2AZ 13008 A-B	**#.3**	152.50
#Included in R&R Front Panel				
2 **Lens¶**	R	F2AZ 13007 A	**.4**	66.84
	L	F2AZ 13007 B	**.4**	71.37
3 **Bulb Assy, Low Beam¶**	R/L	E9SZ 13N021 A	**#.3**	11.61
#Time is to R&R One, Each Additional .1				
4 **Retainer, Bulb¶ (a)**	R/L	E9SZ 13N019 A		3.95
(a) Included w/Low Beam Bulb Assy				
5 **Bulb Assy, High Beam¶**	R/L	E9SZ 13N021 B	**#.3**	11.61
#Time is to R&R One, Each Additional .1				
6 **Retainer, Bulb¶ (a)**	R/L	E9SZ 13N019 B		3.95
¶Included w/Headlamp Assy				
(a) Included w/High Beam Bulb Assy				
Screw, Adjust				
Pivot	R/L	E9SZ 13032 C		8.32
Horizontal	R/L	E9SZ 13032 A		5.92
Vertical	R/L	E9SZ 13032 B		8.32
Clip, Headlamp Mtg (6)	R/L	E9DZ 13N129 A		3.78

FIGURE 6–7 An example of R&R time. (Courtesy of Mitchell International.)

6.3 LABOR OPERATIONS

Repair operations fall into three categories in collision estimating guides: remove and replace (R&R), remove and install (R&I), and overhaul (O/H). In addition to these major categories, operation time is also given for drilling holes and for refinishing. However, unless specifically stated in the text or footnotes, the times given in the collision estimating guides are for R&R. It is important that the estimator understands which category each specific repair falls into and that he or she estimates the repair accordingly.

REMOVE AND REPLACE

Remove and replace (R&R) times should be used when a damaged part must be replaced with a new part. R&R times (Figure 6–7) allow for the damaged part or assembly to be removed, for bolted- or clipped-on parts to be removed and transferred to the new part, for installation of the new part or assembly on the vehicle, and for alignment of the part or assembly (except when noted otherwise in the text).

The key to properly estimating R&R times is understanding which repair operations are covered by R&R times and which are not. For example, replacement of an apron panel and rail typically includes:

- All necessary welding
- R&I of floor mats insulation, and trim
- Remove and replace caulking

Listed operation times for apron panel and rail replacement typically do not include:

- Setup of the vehicle on a frame machine and diagnosis of the damage
- R&I of all bolted-on parts
- Wheel alignment
- R&I of body sheet metal
- Refinishing

- Applying undercoating, sound-deadening material, and anticorrosion protection
- Removal of adjacent panels
- Pulling and straightening
- Aim headlamps
- Replace decals

If the nonincluded operations are necessary in the replacement procedure, labor times must be entered separately on the estimate for those operations. On the other hand, the estimator must be careful not to list operations that are already included in the R&R times.

Subassemblies such as a door or bumper include time for removal and reinstallation of individual components, which should not be listed separately on the estimate. For example, the operation time for replacing a car door includes time for transferring the handles, locks, window glass, and other hardware. Operation time is also listed for these components individually; however, separate R&I for these parts should not be included on the estimate, since the labor for these operations has already been figured into the time for door R&R.

Additional information regarding included and not-included operations will be discussed in Chapters 7, 8, and 9. Those chapters explain how to estimate damage to specific sections of a vehicle using the collision estimating guides.

REMOVE AND INSTALL

Sometimes it is necessary to remove a part that is undamaged in order to repair an adjacent panel or part. After the repair is made, the part is simply reinstalled. This operation is called remove and install (R&I). For example, a vehicle driven into a ditch might sustain damage to the wheels and frame. To repair this damage, the undamaged bumper, hood, and fenders might have to be removed to gain access to the undercarriage. These undamaged parts must then be reinstalled and aligned after the repairs are complete. R&I time (Figure 6–8) is used for the removal and reinstallation of these parts.

R&I times also include time for aligning and adjusting the part or assembly, except when noted otherwise in the text or procedural pages of the collision estimating guides. R&I times do not include repairing or aligning parts adjacent to the parts being removed and installed. R&I times also exclude specific operations or parts that are specified in the text as not being part of the removal and installation procedure. R&I time is also used for the removal and installation of a damaged part so that it can be repaired away from the car. It might be easier to repair a damaged door, for example, by first removing it from the car.

FRONT BUMPER

Use Procedure Explanations 1 and 28 with the following text.

Operation		Time
Refinish Impact Bar		**.9**
Refinish Bar Extension	R/L	**.5**
Refinish Front Cover		**1.0**
Refinish Lower Flare	R/L	**.5**
Refinish Front Air Deflector		.9
R&I Lower Fender Flare	R/L	**.2**
R&I Bumper Assy		**#1.0**
#Includes R&I Grille		
O/H Bumper Assy (Includes R&I)		**#1.6**
#Includes R&I/R&R Front Air Deflector, Park/Signal Lamps & R&I Grille		
Add to O/H Bumper Assy w/Side Mouldings		**3**

NOTE: All parts in this section are included in overhaul unless noted otherwise.

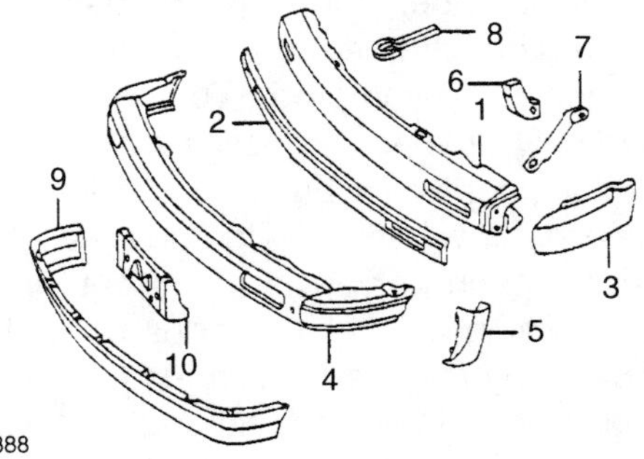

001-12388

Part	Year/Side	Part No.	Time	Price
1 **Bar, Impact**				
Chevrolet				
Dark Gray		15647596	**IOH**	123.00
Body Color (P)				
w/Appearance Pkg	94-96★	15647597	**IOH**	123.00
	97★	12389356	**IOH**	118.00
w/o Appearance Pkg	94-96★	15959830	**IOH**	118.00
	97★	12389359	**IOH**	123.00
GMC	94-96	15984100	**IOH**	118.00
	97	12389360	**IOH**	118.00
(P) Paint to Match				
2 **Strip, Impact**				
Chevrolet				
w/Appearance Pkg		15972385	**.3**	74.50
3 **Extension, Bar¶**				
w/Appearance Pkg	R/L	12383120-19	**.6**	86.00
w/o Appearance Pkg	R/L	12383112-1	**.6**	40.75
¶Black listed, Order by Application				
4 **Cover Front**				
GMC				
SL				
1994		15972306	**#1.6**	76.75
1995-97 (P)		12549246	**#1.6**	217.00
SLS (P)		12549246	**#1.6**	217.00
SLE		15993850	**#1.9**	316.00
(P) Paint to Match				
#Includes R&R Front Air Deflector, Park/Signal Lamps & R&I Grille				
5 **Flare, Lower**				
Highrider, ZR2 Pkg				
Chevrolet	R/L	15672032-1	**#.2**	50.25
GMC¶	R/L	15672036-5	**#.2**	50.25
¶Black listed, Order by Application				
#w/Fender Removed, Not Included in O/H				
6 **Bracket, Impact Bar-See Frame Section**				
7 **Brace, Impact Bar**				
w/Highrider, ZR2 Pkg	R/L	15653994-3	**#.2**	5.77
w/o Highrider, ZR2 Pkg	R/L	15647600-599	**#.2**	3.07
#R&R Complete				
8 **Hook, Tow**	R/L	15661184-3	**.2**	25.25
9 **Deflector, Front Air¶**				
2 WD				
w/Fog Lamps	94-95	15658467	**#1.2**	37.00
	96-97	15998332	**#.12**	31.25

FIGURE 6–8 An example of R&I time. (Courtesy of Mitchell International.)

OVERHAUL

Overhaul (O/H) times allow for the removal of a subassembly such as a bumper or suspension assembly; disassembly, inspection, and replacement of damaged parts; and reinstallation and adjustment of the subassembly on the vehicle.

Overhaul time is actually a shortcut for the estimator. The collision estimating guides list overhaul time most often in connection with bumpers (Figure 6–9). It is very common for a car involved in a collision to have bumper damage. But, in most cases, many of the individual parts can be salvaged, so that an overhaul of the assembly is in order. To correctly estimate the disassembly, inspection, replacement, reassembly, and adjustment of a bumper and its individual parts, an estimator would have to add up all the operation times for each part, plus figure out overlap for each one. Collision estimating guides do that for the estimator by listing the overhaul time.

The Mitchell and Motor collision estimating guides differ somewhat in the way overhaul times are given. In the Mitchell guides, the Bumper and Front Suspension sections designate items contained in the overhaul time with the abbreviation IOH (Figure 6-9). Other bumper or suspension parts that are part of the overhaul procedure but can also be removed and replaced separately are given separate time allowances. In the Motor collision estimating guides all parts, even those that would be included in an overhaul of the assembly, are given individual time allowances. The estimator or appraiser must not list the items separately if they are included in overhaul times on the estimate.

SECTIONING

Another operation that is published in the collision estimating manuals is sectioning. Oftentimes it is less expensive and time-consuming to section a damaged panel and weld part of a new panel in place than to replace the damaged panel at the factory seams. The R&R time for replacing a right quarter panel on a 1997 Chevy Malibu is 13.0 hours. However, to section the quarter panel at the belt line is flat rated at 11.0 hours (Figure 6–10). Other body parts that are given sectioning time in the crash manuals are rocker panels and side pillars.

The collision estimating guides do not always provide flat rate time for sectioning. When the sectioning of a panel is allowed by the OEM and the sectioning time is not listed in the collision estimating guide, the estimator must determine the labor allowance. The allowance must include time to:

FRONT BUMPER

Use Procedure Explanations 1 and 28 with the following text.

Refinish Bumper Cover	**2.9**
Refinish Bumper Cover Applique	**.5**
R&I Cover Assy	**1.0**
O/H Bumper Assy (Includes R&I)	**#2.4**
#Includes R&I/R&R Headlamp Assemblies & Side Marker Lamps, Lower Shield & Deflector, Loosen & Pull Back Fender Liner	
Add to O/H Bumper Assy	
w/Fog Lamps	**.2**
w/Grille	**.2**

NOTE 1: Bumper cover is constructed of Thermoplastic Olefin-TPO.
NOTE 2: All Parts in this section are included in overhaul unless noted otherwise.

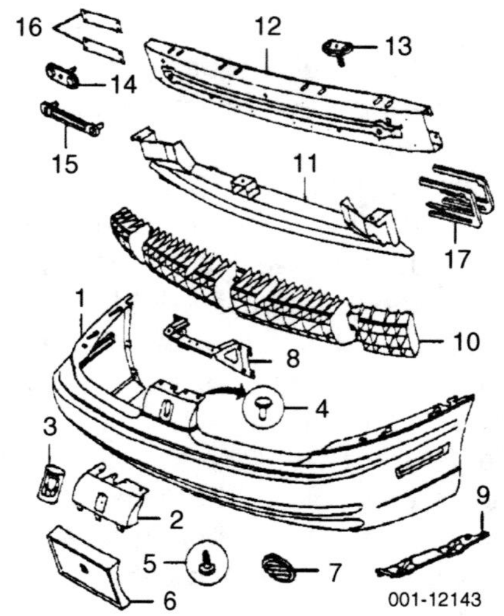

1 **Cover, Bumper**				
SL, SL1, SW1		21110670	**#1.4**	227.17
SL2, SW2 (P)		21110424	**#1.4**	201.92
#Includes R&I/R&R Side Marker Lamps, Lower Shield & Deflector, Loosen & Pull Back Fender Liner				
2 **Applique, Bumper Cover (P)**				
SL, SL1, SW1		21110425	**#.3**	9.44
(P) Paint to Match				
3 **Emblem, Cover**		21111139	**#.3**	6.05
#R&R Complete				
4 **Retainer, Upr Cover (3)**		21030249		.18
5 **Retainer, Lwr Cover (7)**		21030249		.18
6 **Brkt, License**		21108512	**.3**	10.63
7 **Insert, Bumper Cover**	R/L	21095869	**.2**	2.67
8 **Brkt, Bumper Cover**		21110752		15.11
9 **Support, Cover Outer**	R/L	21108498-9		4.71
10 **Absorber, Impact**		21110631	**IOH**	44.09
11 **Grille, Lwr Cover**		21108505	**#1.4**	20.00
#Includes R&I Front Bumper Cover				
12 **Bar, Impact**	96	21111006	**IOH**	124.13
	97	21111006	**IOH**	124.13
13 **Plate, Bar Stud**				
Center (3)		21095374	**IOH**	1.13
Plate, Bar Stud				
14 Upper	R/L	21075310	**IOH**	1.83
15 Lower	R/L	21075311	**IOH**	5.00
16 **Insulator, Impact Bar**	R/L	21075206		.37
17 **Shim, Impact Bar (2/Side)**	R/L	21077302		5.37

FIGURE 6–9 An example of overhaul time. (Courtesy of Mitchell International.)

QUARTER PANEL

Use Procedure Explanations 20 and 28 with the following text.

SHEET METAL/EXTERIOR TRIM

Operation	Side	Time
Refinish Quarter Outside	R/L	**2.1**
Add for Pillar	R/L	**.5**
Add to Edge Panel	R/L	**.5**
Refinish Fuel Door		**#.3**
#Included in Refinish Quarter Panel		
Refinish Mudguard	R/L	**.4**
R&R Both Quarters		**21.9**
R&R Roof & Both Quarters		
w/Sunroof		**35.3**
w/o Sunroof		**34.3**
R&R Rear Body Panel, Extensions & Both Quarters		**27.3**
R&R Roof, Both Quarter Panels & Rear Body Panel & Extensions		
w/Sunroof		**40.7**
w/o Sunroof		**39.7**
Section Quarter Panel	R	**#11.0**
	L	**#11.5**
Section Both Quarter Panels		**#20.8**
Section Both Quarter Panels, R&R		
Rear Body Panel & Extensions		**#26.2**
#Horizontal Cut Below Back Glass & Vertically into Trunk Opening		
Add to R&R Quarter Panel		
W/Mudguard	R/L	**.2**
To R&I Back Glass		**.3**

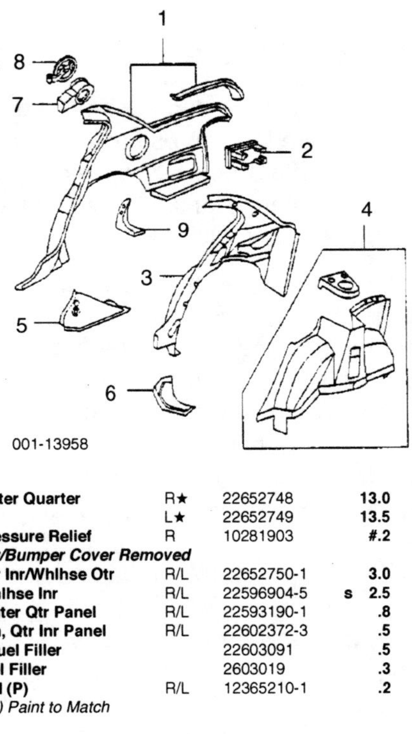

Part	Side	Part No.	Time	Price
1 **Panel, Outer Quarter**	R★	22652748	**13.0**	343.00
	L★	22652749	**13.5**	343.00
2 **Valve, Pressure Relief**	R	10281903	**#.2**	9.00
#w/Bumper Cover Removed				
3 **Panel, Qtr Inr/Whlhse Otr**	R/L	22652750-1	**3.0**	144.00
4 **Panel, Whlhse Inr**	R/L	22596904-5	s **2.5**	99.00
5 **Brace, Outer Qtr Panel**	R/L	22593190-1	**.8**	12.99
6 **Extension, Qtr Inr Panel**	R/L	22602372-3	**.5**	2.50
7 **Pocket, Fuel Filler**		22603091	**.5**	6.80
8 **Door, Fuel Filler**		2603019	**.3**	32.25
9 **Mudguard (P)**	R/L	12365210-1	**.2**	10.00
(P) Paint to Match				

FIGURE 6–10 An example of sectioning time. (Courtesy of Mitchell International.)

- R&I any obstructions
- Measure
- Cut
- Align
- Weld or join panels
- Finish with body filler as needed
- R&R caulking

FRONT FENDER

Operation	Time
Refinish Outside	**2.0**
Add to Edge Fender	**.5**
Add for Inner Panel	**1.0**
R&I Fender Assembly	**2.5**
Add to Fender R&R	
To Drill for & Install	
One Moulding	**.3**
Each Additional Moulding	**.2**
HSLA - High Strength Low Alloy Steel	

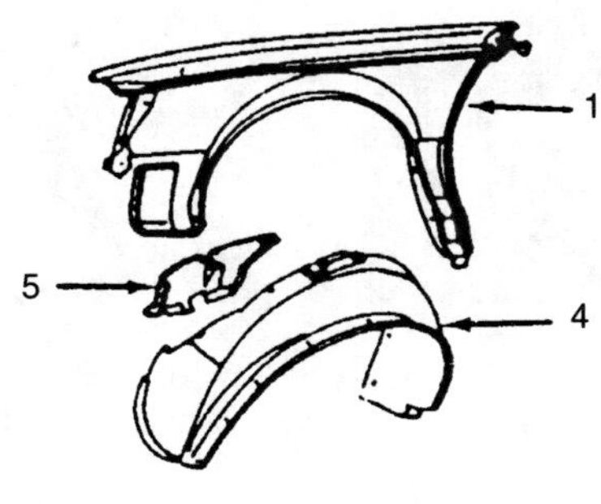

FIGURE 6–11 An example of drilling time. (Courtesy of Mitchell International.)

OTHER REPAIR OPERATIONS

The collision estimating guides also contain labor allowances for drilling operations and painting. The allowances for drilling holes are for drilling round holes only. If square holes or holes of another shape are required, they should be estimated separately. Figure 6–11 shows an example of drilling time given in a collision manual.

6.4 ADJUSTMENTS TO OPERATION TIMES

In order to write an accurate estimate of the labor involved in repairing a damaged vehicle, the estimator must do more than list the operation times given in the collision estimating guides on the estimate sheet. There are several factors that might or might not result in a reduction of the given time allowances. These are overlap, included operations, and access time.

OVERLAP

When repair operations overlap, reduction to one or more of the labor times must be considered. Overlap occurs when the replacement of one component duplicates some labor operations required to replace an adjacent component. For example, when replacing a

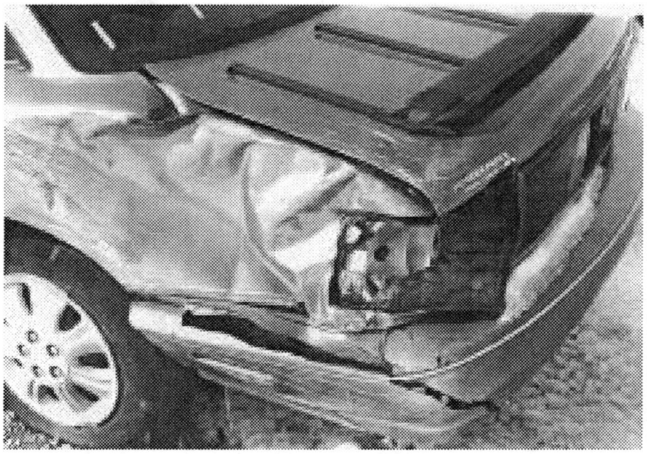

FIGURE 6–12 *Time for overlapping seams—quarter panel and rear body panel—must be deducted.*

quarter panel and a rear body panel on the same vehicle, that area where these two components join is considered overlap (Figure 6–12). Where a labor overlap condition exists, less time is required to replace adjoining components collectively than is required when they are replaced individually.

Overlap deductions must be agreed upon by the collision repair facility estimator and the insurance appraiser. Guidelines for figuring overlap are given in Chapters 7, 8, and 9.

INCLUDED OPERATIONS

Another labor cost reduction is noted as included operations. These are jobs that can be performed individually but are also part of another operation. As an example, when replacing a door, the suggested time for the door replacement would include the replacement of all parts attached to the door, except for the ornamentation. It would be impossible to replace the door without transferring these parts. Consequently, the time involved in transferring these parts would be "included operations" and should be disregarded because the times for the individual items are already included in the door replacement time.

ACCESS TIME

A third way in which operation times can be reduced is by the removal of adjacent parts with another repair operation. If prior removal of a part or panel creates easier access for a second operation, the flat rate listed in the collision estimating guides for the second operation must be reduced proportionately. For example, the R&R time for a rear body panel on a 1994 Laser is 5.5 hours. However, that time includes the removal of adjacent parts to provide access to the cover. If prior repair operations have already removed adjacent parts, making the cover more accessible, the R&I time should be reduced accordingly. The reduction for access time is expressed in Figure 6–13.

Other parts that may require reduction to operation times if access has been created by another operation are:

- Radiator
- Radiator hoses
- Fan blades, clutch, and pulley or electric motor
- Fan shroud
- Crankshaft pulley
- Water pump
- Oil pan
- Vibration damper
- Alternator, power steering, A/C, and AIR pump belts

ADDITIONS TO OPERATION TIMES

The following individual operations are additional factors to be considered by the estimator when writing an estimate. Such added time is usually negotiated between the shop estimator and the insurance company's appraiser or the customer.

- Setup of the vehicle on a frame machine and damage diagnosis. Setup times suggested by the frame machine manufacturer are used as guidelines by the insurance industry.
- Pushing, pulling, cutting, and so on to remove collision-damaged parts
- Straightening or aligning related parts
- Removing undercoating, tar, grease, and similar materials
- Repairing rust or corrosion damage to adjacent parts
- Freeing up rusted or frozen parts
- Drilling for ornamentation or the fabrication of mounting holes when holes are not provided in the new part being installed
- Filling or plugging unnecessary holes in new parts
- Repairing damaged replacement parts prior to installation
- Checking suspension and steering alignment
- Removing shattered glass
- Rebuilding, reconditioning, and installing aftermarket components. (Parts replacement time does not include refinishing time.)
- Applying sound-deadening material, undercoating, and caulking, and painting inner areas

REAR BODY

Use Procedure Explanations 26 and 28 with the following text.

SHEET METAL

Refinish Rear Body Panel Outside		1.5
Add for Edge & Inside		.8
Refinish Rear Floor Pan		1.0
Section Rear Floor Side Member	R/L	#4.5

#Time is to Cut at Dotted Line in Illustration, w/Rear Body Panel Removed, Deduct 1.5 w/Rear Floor Pan Removed

NOTE: For sectioning methods, refer to Manufacturer's Body Repair Manual.

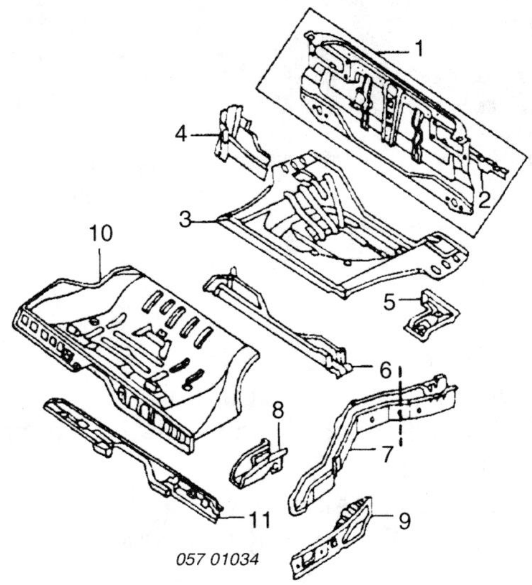

1 **Panel Assy, Rear Body**					
FWD	90-91	MB549808	**#5.5**	320.00
	92-94	MB835815	**#5.5**	345.00
AWD		MB593326	**#5.5**	310.00
#Deduct 2.0 for Each Quarter Panel Removed, Deduction Includes Common Surfaces & Included Operations				
2 **Brkt, Upr Bumper Face¶**		MB549809	**1.0**	111.00
¶Included w/Rear Body Panel Assy				
3 **Pan, Rear Floor**				
FWD		MB549723	**#10.0**	205.00
AWD		MB593314	**#10.0**	140.00
#w/Rear Body Panel & Bolted Parts & Assemblies Removed, w/Each Sidemember Removed Deduct 1.5, w/Each Inner Wheelhouse Removed Deduct .5, Includes R&R Suspension Lateral Rod Bracket & Jacking Reinforcement				
4 **Brkt, Susp Lateral Rod**				
FWD		MB549735	**2.0**	59.00
5 **Reinforcement, Jacking**				
FWD		MB593348	**1.0**	30.00
Crossmember, Rear Floor				
Rear End Point				
AWD		MB593316	**5.5**	151.00
6 **Crossmember, Rear Floor**				
Front End Point				
FWD		MB549736	**3.5**	82.50
AWD		MB593315	**3.5**	90.00
7 **Sidemember, Rear Floor**				
FWD	90-91 R	MB549738	**s #6.5**	260.00
	L	MB549737	**s #6.5**	270.00
	92-94 R	MB835812	**s #6.5**	325.00
	L	MB835811	**s #6.5**	345.00
AWD	R	MB593318	**s #6.5**	370.00
	L	MB593317	**s #6.5**	410.00
#w/Rear Body Panel Removed				
Extension Sidemember Front				
AWD	R/L	MB593372-1	**3.0**	86.00
8 **Bracket, Trailing Arm**				
FWD	R/L	MB593358-7	**1.5**	39.25
9 **Sill, Rear Floor Side**	R/L	MB593356-5		122.00
10 **Floor, Rear Seat**		MB549724	**#14.5**	265.00
#Includes R&R Rear Seat Floor Crossmember				
11 **Crossmember, Rear Seat Floor**				
FWD		MB593353	**3.0**	66.00
AWD		MB593313	**3.0**	56.50

FIGURE 6–13 Deductions to labor times must be made when a previous operation makes access to an operation easier. (Courtesy of Mitchell International.)

- Replacing rust-resistant anticorrosion material
- R&I of the main computer module when excessive temperatures (above 176°F) are necessary in repair operations. Consult the manufacturer's service manuals for recommended removal procedures.
- R&I of wheel or hubcap locks
- Replacing such accessories as trailer hitches, sunroofs, and fender flares
- Correction to previous repairs or damage

Another job often overlooked in estimating is aiming the headlights. Any repairs or replacement of front end parts should include a headlight alignment. Remember that the estimator's job is to define and determine the cost of every repair operation needed to restore the vehicle to its precollision operation, safety, and value.

6.5 ESTIMATING FRAME STRAIGHTENING LABOR

Because there are many variables in frame repair, extent of damage, types of equipment, and the attachment of parts, no specific labor times will be listed. Each collision repair facility must determine a policy for labor times. This policy should take into account the amount of time it takes to do the following:

1. *Mount the Vehicle on the Frame Repair Equipment.* In some bench systems, the vehicle is raised on a lift and the bench is moved underneath. Then the vehicle is lowered onto it. Other bench systems are assembled underneath the vehicle and the

vehicle is gradually raised in the assembly process. In drive-on frame racks, the vehicle is driven or pulled onto the machine. In a floor pot system, the vehicle is tied to the floor or for unibodies, a pinch weld clamp apparatus is bolted to the vehicle and then tied to the floor. Severely damaged vehicles—that is, those that will not roll because of suspension damage—will require extra time to mount.

2. *Measure Damage.* To properly plan and make the pulls, the frame damage must be measured. Hanging gauges and mechanical or computerized measurements may be used. Each of these methods may require a different amount of time. The estimator should determine how long it takes to set up the equipment, then read and record the data. The amount of time may vary between unibody and full-frame vehicles.
3. *Access Time.* In order to make pulls on damaged panels, undamaged parts may need to be removed. This labor is called access time. For example, on a front corner hit on a unibody with repairable radiator support damage, the fender on the non-impact side may be undamaged. However, in repairing the radiator support, the non-impact side will need to be held in place. The undamaged fender will need to be removed so a clamp can be attached to the radiator support. On a pickup truck, the frame needs to be tied down to hold it in place while pulls are made on the damage. In some cases, the gas tank may need to be drained and removed to allow the frame to be wrapped with a chain. Any access time will need to be added to the frame repair time. In many cases, the collision estimating guide will list the time for R&R parts. If there isn't any time listed, the estimator will need to allow time to remove and replace the parts. Rough pulls may need to be made to allow access to bolts or components.
4. *Align the Frame.* The time to align the frame is determined by how many attachments or hookups are required. Extensive frame damage consisting of multiple types of damage will require more hookups or time than simple damage consisting of a single type of frame damage. In some cases, a single hookup can correct two types of damage, such as sidesway and sag, at once. Each hookup should allow time for the following:
   - Plan the pull
   - Install clamps
   - Attach chains
   - Position towers
   - Apply tension with pulling
   - Relieve panel stress by vibration or spring hammering
   - Release tension
   - Check alignment by measuring
   - Remove chains
   - Remove clamps

FIGURE 6–14 The damage to this quarter panel is repairable. The estimator must calculate the appropriate judgment time to repair.

In many cases, the estimator will need to negotiate with the insurance company to determine the labor judgment time allowed to straighten a frame.

In summary, the estimator needs to know the shop equipment for mounting, measuring, and pulling. It will take considerable experience to be able to estimate the amount of time needed for access and aligning.

6.6 ESTIMATING PANEL STRAIGHTENING LABOR

Labor to straighten damaged panels is known as judgment time (Figure 6–14). The calculation of judgment time for a steel panel requires the estimator to allow for the following:

- Examine the panel.
- Plan a repair sequence.
- Rough out the damage.
- Remove the paint by grinding.
- Shrink if required.
- Mix body filler.
- Apply body filler.
- Cure body filler.
- Sand body filler.

Material cost per hour of straightening work is best derived by carefully recording the material expenses and the time involved in straightening work done at a shop over a period of weeks or months. Then, an average cost of materials per labor hour for that shop can be accurately determined.

After determining the labor and materials cost for straightening a damaged panel, the total cost of repair should be compared with the expense of replacing the damaged panel. As stated earlier, if the straightening costs exceed 75 percent of the cost of labor and parts for a new panel, the panel should be replaced, especially on late-model vehicles. A panel damaged to such an extent is difficult to restore to a like-new appearance. The slightly higher cost of using a replacement panel is offset by obvious safety, appearance, and quality considerations.

6.7 ESTIMATING PLASTIC REPAIR

Plastic—urethane, SMC, PC, ABS, FRP—is frequently used by most automobile manufacturers. Common usages are header panels, bumper fascia, fenders, grilles, and ground effects.

Damage to plastic can be easily and inexpensively repaired when the proper techniques are employed. Some types of plastic, PP, PC, and ABS may be welded. Repairs also can be made using compatible adhesives and fillers, the same materials used in the fabrication of the panel. Properly made, plastic repairs are strong and durable.

For welded repairs, the labor allowance should include time to:

- Examine the damage.
- Plan the repair.
- Warm up the welder.
- Align.
- Clean the part.
- Weld the damage.

Judgment times for bonded plastic repair should allow time for the following:

- Examine the damage.
- Plan the repair.
- Clean the panel.
- Scuff sand.
- Align.
- Mix adhesive/filler.
- Apply adhesive/filler.
- Cure adhesive/filler.
- Sand adhesive/filler.

If reinforcement such as fiberglass mat or urethane backing is needed, allow additional time.

ROOF

Use Procedure Explanations 17 and 28 with the following text.

Refinish Roof Outside				
2 Door			**3.5**	
4 Door Sedan			**4.0**	
4 Door Wagon			**5.0**	
R&R Headliner				
2 Door			**3.5**	
4 Door Sedan			**4.5**	
4 Door Wagon			**5.2**	
Add to R&R Roof				
w/Luggage Rack			**1.8**	
w/Standard Vinyl Cover			**8.0**	
w/Tucked Vinyl Cover			**9.0**	
Panel, Roof				
2 Door				
w/o Hatch Roof				
w/o Sunroof		20607943	**15.0**	d231.00
w/Sunroof		20607946	**18.0**	d545.00
w/Hatch Roof		★20607882	**18.0**	d0.00
4 Door Sedan		20145627	**15.5**	d270.00
4 Door Wagon		20307351	**18.0**	d456.00
Moulding, Inside Garnish (P)				
Side				
2 Door	R/L	20540409	**.2**	d0.00
4 Door Sedan	R	20540344	**.2**	N.A.
	L	20540345	**.2**	d27.75
4 Door Wagon				
Front	R	20399136	**.2**	d0.00
	L	20399137	**.2**	N.A.
Rear	R	20398792	**.2**	d0.00
	L	20398793	**.2**	N.A.
Rear Header (Wagon)		20398814	**.2**	d0.00

(P) Paint to Match

FIGURE 6–15 An example of OEM part prices in an estimating guide. (Courtesy of Mitchell International.)

6.8 PARTS PRICES

Having decided to replace the damaged part, the estimator must decide whether to use new parts from the original equipment manufacturer (OEM) or whether an aftermarket or salvage part will suffice.

OEM PARTS

Car manufacturers keep to a fairly regular schedule of price increases. The collision estimating guides are updated on a regular basis to reflect the most recent price changes. Thus, the parts prices quoted in the collision estimating guides are very reliable. However, actual retail prices might vary between localities and between dealers.

If a manufacturer makes an unscheduled price change just before the collision estimating guides go to press, a special section will be included in the back of the collision estimating guide to list the price changes. The list prices in the guide should be amended accordingly until the next edition of the collision estimating guide is published.

An example of price information as contained in the collision estimating guides is given in Figure 6–15.

When a price is prefixed by the letter *d,* it indicates that this part has been discontinued by the vehicle manufacturer. The given price is the last available price. When a price is prefixed by the letter *A* in the Motor guides or by a black box in the Mitchell guides, it indicates that this part is included in the General Motors Target Marketing program and does not have a manufacturer's suggested list price. The price shown following the letter *A* or the box symbol is simply an approximation. Such prices are to be regarded as a guide only. The actual price, which may be higher or lower than the listed price, can be determined by calling a local GM dealer. Part prices do not include the cost of state and local taxes, bolts, rivets, screws, nuts, washers, clips, fasteners, body repair materials, and refinishing materials, unless otherwise noted in the collision estimating guides.

The collision estimating guides provide two sources of glass prices: the manufacturer's and the National Auto Glass Specifications (NAGS). The collision estimating guide text lists prices for clear, shaded, and tinted glass (Figure 6–16). Where applicable, it will also specify whether the glass has antenna wires or a heating element. The estimator can determine the type of glass needed by reading the label on the glass (Figure 6–17).

Glass suppliers often give the collision repair facility a hefty discount on glass prices—up to 60 percent. The insurance companies also discount the prices on their appraisals but at a smaller percentage, leaving a reasonable profit margin for the shop.

For example, if a windshield costs $350.00 and the discount was 40 percent, the insurance appraiser might reduce the list price by 20 percent and list the cost of the glass as $280.00 (350 – 20%). That would leave the insurance company and the collision repair facility with a profit margin of 20 percent. Most insurance companies have specific guidelines on the

WINDSHIELD

Use Procedure Explanation 13 with the following text.

R&I Windshield .... #3.0
#Includes Clean Up Old Adhesive as Necessary
R&I Reveal Mldg 2

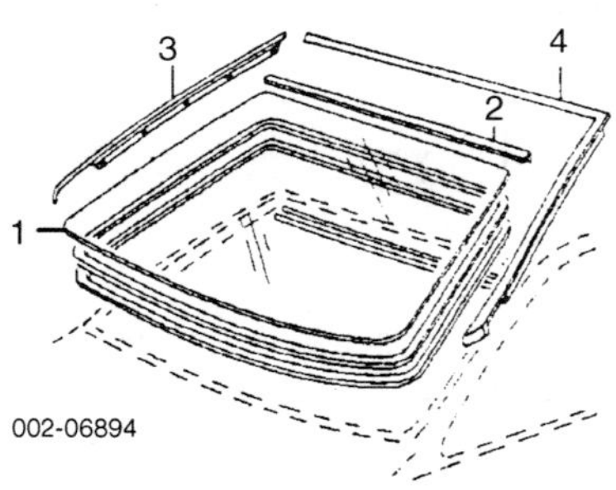

1 **Glass, Windshield**				
Ford				
Clear		E3SZ 6303100 A	#2.5	573.83
Tinted				
1989-93		FOSZ 6303100 A	#2.5	867.89
1994-97				
To 2-94		F4SZ 6303100 A	#2.5	868.61
From 4-94		F4SZ 6303100 B	#2.5	1027.06
NAGS				
Green/Blue Shade				
1989		DW01047GBN	#2.5	867.90
1990-93		DW01091GBN	#2.5	867.90
1994-97				
To 2-94		DW01171GBN	#2.5	868.65
From 2-94		DW01190GBN	#2.5	1027.10
#Includes R&R Roof Side Mouldings & Clean Up Old Adhesive as Necessary				
Urethane		EOAZ 19562 A		d63.61
2 **Moulding, Reveal Upr**	89-92	E9SZ 6303144 A9G	.2	26.55
Moulding, Reveal				

3 Right	93-97	F3WY 6603144 AAA	.2	49.83
4 Left	93-97	F3WY 6603145 AAA	.2	49.83
Mirror, Rear View¶				
Manual				
w/Sunroof		E9OY 17700 A	.2	144.15
w/o Sunroof	89	E5AZ 17700 B	.2	41.67
	90-91	E65Y 17700 A	.2	52.25
	92-93	FOTZ 17700 A	.2	14.20
	94	F4SZ 17700 A	.2	16.63
	95–97	F5SZ 17700 AA	.2	11.55
Power				
1989-91				
w/Headlamp Dimmer		E8OY 17700 B	.5	273.97
w/o Headlamp Dimmer		F19Z 17700 A	.5	264.75
1992-97		F29Z 17700 A	.5	269.88
¶Order by Application				
Visor, Sun¶				
1989				
w/Illuminated Mirror				
w/Sunroof	R	E9SZ 6304104 D9B	.3	d177.98
	L	E9SZ 6304105 D9B	.3	d177.98
w/o Sunroof	R	E9SZ 6304104 C9B	.3	241.47
	L	E9SZ 6304105 C9B	.3	241.47
w/o Illuminated Mirror	R	E9SZ 6304104 A9B	.2	d62.30
	L	E9SZ 6304105 A9B	.2	d62.30
1990-97				
w/Illuminated Mirror				
w/Sunroof				
w/Active Restraints	R	FOSZ 6304104 EBV	.3	199.40
	L	FOSZ 6304105 EBV	.3	199.40
w/Passive Restraints	R	FOSZ 6304104 DBV	.3	241.47
	L	FOSZ 6304105 DBV	.3	241.47
w/o Sunroof				
w/Active Restraints	R	FOSZ 6304104 CBV	.3	199.40
	L	FOSZ 6304105 CBV	.3	199.40
w/Passive Restraints	R	FOSZ 6304104 FBV	.3	241.47
	L	FOSZ 6304105 FBV	.3	241.47
w/o Illuminated Mirror				
w/Active Restraints	R	FOSZ 6304104 BBV	.2	44.32
	L	FOSZ 6304105 BBV	.2	44.32
w/Passive Restraints	R	FOSZ 6304104 ABV	.2	84.52
	L	FOSZ 6304105 ABV	.2	84.52

¶Blue listed, Order by Application

FIGURE 6–16 Glass prices in an estimating guide. Note OEM and NAGS prices. (Courtesy of Mitchell International.)

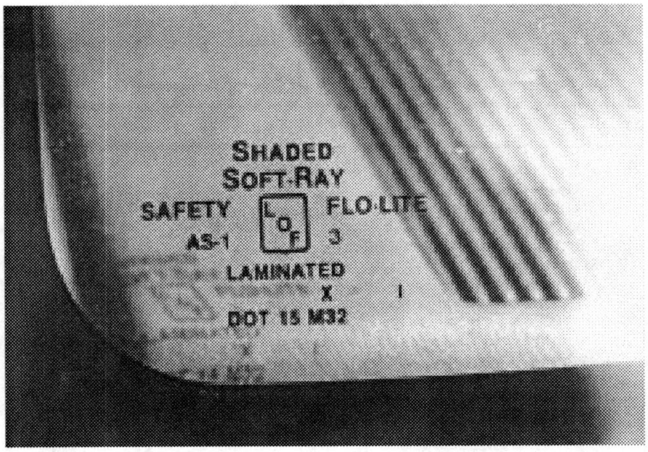

FIGURE 6–17 Typical glass label.

amount of profit margin they will allow the shops to make on glass parts.

SALVAGE PARTS

Replacing damaged panels with new, OEM parts quickly becomes expensive. To illustrate that fact, the Alliance of Auto Insurers conducts a study each year in which it compares the sticker price of a new car with the total cost of replacement parts that would be needed to rebuild the vehicle if it was demolished. In a recent study, research was done for a 1997 Pontiac Grand Am SE base model to determine estimated cost comparisons. The vehicle value was estimated at $14,874. However, the cost of rebuilding the car from the ground up using new replacement parts, without considering the cost of labor, was found to be estimated at $47,200, or 3.2 times the original retail value of the car (Figure 6–18). It should be evident that the estimator can quickly approach the total value of the damaged vehicle by replacing damaged panels with new OEM parts.

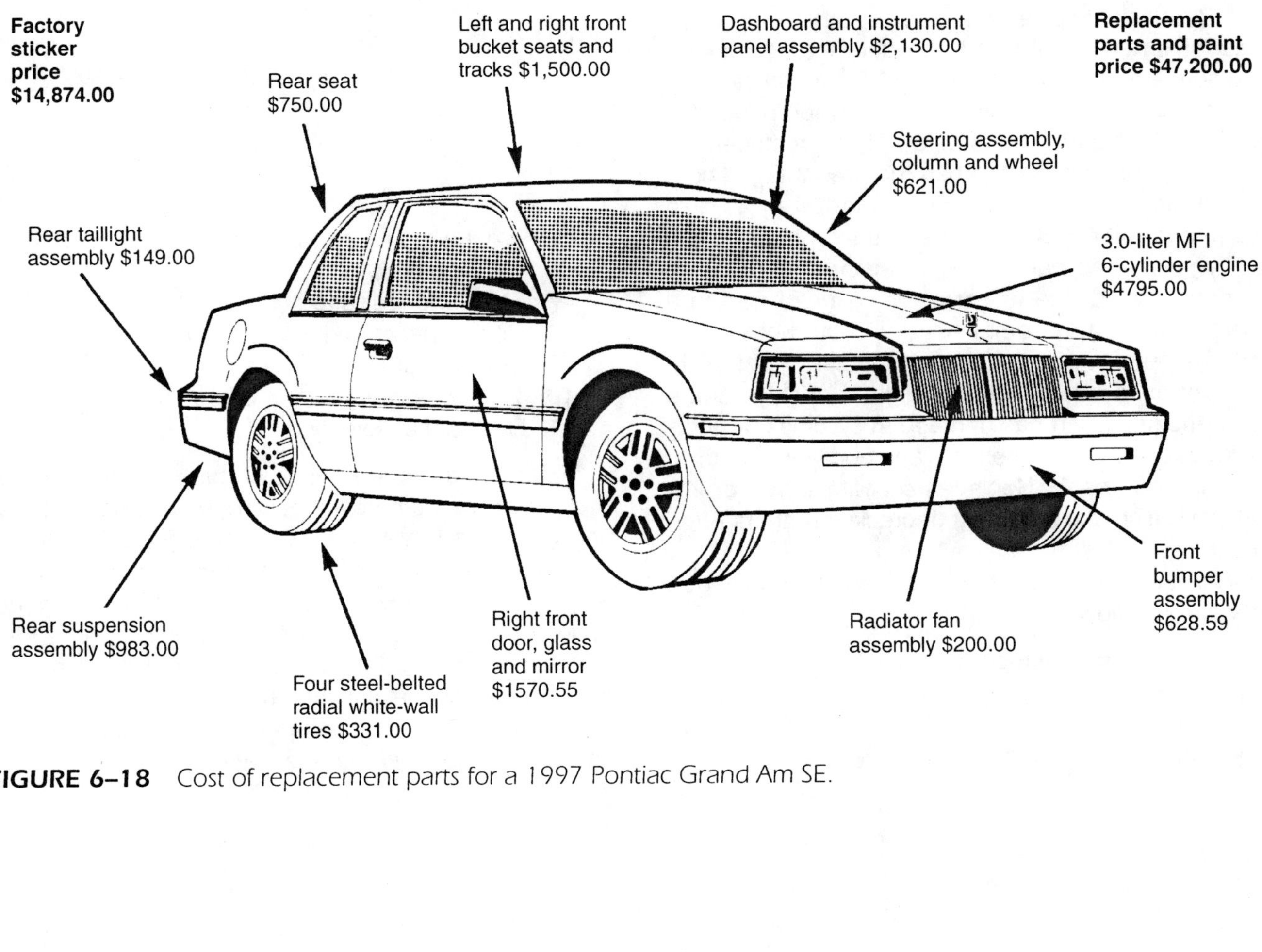

FIGURE 6–18 Cost of replacement parts for a 1997 Pontiac Grand Am SE.

FIGURE 6–19 This total loss vehicle has a usable front clip.

In order to keep repair costs down, salvage parts (Figure 6–19) are often used to replace damaged panels and mechanical components. Not only is the salvage part usually considerably less expensive than a new OEM part, but using salvage parts can shorten the downtime of the damaged car. During this period, the insurance company is sometimes obligated to provide a rental car for the insured. This courtesy to the car owner is referred to as "loss of use compensation." By eliminating the three to six days it normally takes for a new OEM part to be delivered, the insurance company can save the cost of the rental car.

When recommending the use of salvage parts, it is also important that the condition of the replacement panel be as good or better than the predamaged condition of the original part. This is referred to as "like kind and quality" (LKQ). It is not uncommon to find that the damaged part has incurred prior damage or has deteriorated due to corrosion. A used part in a salvage yard, on the other hand, is not subjected to road salts, parking lot dents, or stone dings and might be superior in quality to the original part. In some cases, the used part has damage, rust, dings or poor-quality paint. Use judgment for any additional labor to clean up used parts. Replacement with a part of like quality will also avoid figuring depreciation of the original part.

Parts that are most often replaced with salvaged parts are as follows:

1. Front or rear bumper assembly
2. Front fenders
3. Hood
4. Grilles and grille panels

FRONT DOOR

Use Procedure Explanations 16 and 28 with the following text.

SHEET METAL

Item	Part	Hours	Price
Refinish Door Outside			
Coupe/Convertible	R/L	**2.2**	
Sedan	R/L	**2.1**	
Add for Jambs	R/L	**.5**	
Add for Jambs & Interior	R/L	**1.0**	
R&I Door Assy			
w/Power Windows	R/L	**1.0**	
w/o Power Windows	R/L	**.5**	
Add to R&R Door Shell			
w/Power Windows	R/L	**.3**	
w/Power Locks	R/L	**.2**	
Add to R&R Door Repair Panel			
w/Power Option	R/L	**.5**	
Shell Assy, Door			
Coupe	95 R★ 22651366	**#4.8**	289.00
	L★ 22651367	**#4.8**	289.00
	96 R★ 22651366	**#4.8**	289.00
	L★ 22651367	**#4.8**	289.00
	97 R★ 22658924	**#4.8**	292.00
	L★ 22658925	**#4.8**	301.00
Convertible	95-96 R★ 22574406	**#4.8**	500.00
	L★ 22658929	**#4.8**	297.00
	97 R★ 22658928	**#4.8**	297.00
	L★ 22658929	**#4.8**	297.00
Sedan	95-96 ★ 22651362	**#4.8**	281.00
	L★ 22651363	**#4.8**	281.00
	97 R★ 22651362	**#4.8**	281.00
	L★ 22651363	**#4.8**	281.00
#Includes R&R Rear View Mirror			
Panel, Door Repair¶			
Coupe, Convertible (a)	R 22570424	**#5.5**	177.00
	L 22570425	**#5.5**	177.00
Sedan	R 22570446	**#5.5**	171.00
	L 22570447	**#5.5**	171.00

¶Included w/Door Shell Assy
(a) Rework Coupe Door Repair Panel for Use on Convertible
#Includes R&I Door Glass & Rear View Mirror

FIGURE 6–20 Parts prefixed with a star are interchangeable with a part from a similar body style.

5. Front end assemblies
6. Roof assembly
7. Front and rear doors
8. Rocker panel and center pillar
9. Quarter panels
10. Trunk lids, hatchbacks, and tailgates
11. Rear clip assembly

Finding a suitable salvage part is not always possible. To make the process easier, the collision estimating guides indicate interchangeable parts. Front and rear face bars and exterior sheet metal such as fenders, doors, and quarter panels are often interchangeable between models with similar body styles. In the Mitchell guides, interchangeability is indicated by a star preceding the part number (Figure 6–20). Motor collision estimating guides have an appendix in which interchangeable part numbers are listed.

FIGURE 6–21 Aftermarket parts catalog.

FIGURE 6–22 This aftermarket hood easily met the federal safety standard during a 30 mph front-into-barrier crash test performed by the Insurance Institute for Highway Safety. (Courtesy of the Insurance Institute for Highway Safety.)

AFTERMARKET CRASH PARTS

Aftermarket crash parts are another source of replacement panels (Figure 6–21). In the past, domestic car manufacturers redesigned their car models every other year or so; the cost of tooling dies for body styles that were only in production two or three years was prohibitive for aftermarket companies. However, when the domestic car manufacturers began to extend the production runs of many models, the market became favorable to the introduction of competition among body parts manufacturers. Aftermarket crash parts have suffered in the past from a reputation for poor quality. For years aftermarket crash parts manufactured for European and Asian markets simply did not meet American standards of quality for strength, corrosion resistance, and fit.

In recent years, crash parts manufacturers have responded to the trend of American car owners to keep cars longer. The increased demand for quality parts at an affordable price led to the upgrading of manufacturing processes to the point where many aftermarket crash parts today are approaching comparable quality with OEM parts. In fact, recent testing has shown that vehicles equipped with an aftermarket replacement hood easily met the Federal Motor Vehicle Safety Standards (FMVSS) (Figure 6–22).

In spite of the improved quality, there is an ongoing debate between OEM manufacturers, aftermarket distributors, and auto insurance companies over the acceptability of competition crash parts. Original equipment manufacturers have lobbied successfully in many states to legislate disclosure laws requiring body shops and insurance companies to inform the vehicle owner that aftermarket replacement panels are going to be installed on his or her vehicle.

In spite of the debate and unfavorable legislation, the use of aftermarket parts is increasing. Some insurance companies encourage the use of aftermarket crash parts to keep the cost of repairs to a minimum. Other companies only stipulate aftermarket parts when their use is necessary to avoid totaling the vehicle. Whenever aftermarket parts are used in a repair, it is important to buy parts from a reputable and established distributor. It might also be necessary to include time on the estimate for fitting the panel to the vehicle, due to the slight imperfections in size, shape, and fit often encountered with aftertmarket parts.

The advantages of using aftermarket crash parts are price and service. The lower prices result in lower estimates that may win more jobs (from the estimator's point of view) or save a vehicle from being totaled (from the appraiser's point of view). Aftermarket venders and jobbers tend to deliver parts more speedily, too. The shortened downtime results in savings for the insurance companies.

Most aftermarket crash parts are restricted to cosmetic panels such as fenders, grilles, and door panels. Bumpers are another common aftermarket sales item.

6.9 BETTERMENT

As has already been stated, the owner of a car is entitled to have his or her car restored to its precollision condition—no more and no less. If a new part is placed on a car several years old, the insurance company is not going to pay for the substantial improvement (betterment). The insurance claims adjuster and body shop estimator must depreciate the precollision value of the damaged part according to the years of prior use and service. The customer will then be responsible for paying for any betterment.

The areas of a vehicle that are commonly depreciated are as follows:

1. *Upholstery.* Upholstery is usually depreciated according to wear and age. For example, a convertible top is considered fully depreciated in three years and a vinyl top in five years. If the vinyl top on a vehicle two years old must be replaced, the estimator must depreciate the value of the top by 40 percent. The insurance company will pay only 60 percent of the price of replacing the vinyl top; the customer or car owner must make up the difference. Additional depreciation must be figured if the top has prior damage or is excessively worn or weathered.

   The plastic back window of a convertible top is depreciated 20 percent per year. The estimator and appraiser must also take into consideration the visual clarity of the window. Discoloration ranges from light amber to a speckled brown color.

   Interior trim and upholstery is depreciated over a ten-year span. If the interior is undamaged or stained and less than one year old, usually no betterment is charged to the owner. After the first year, however, interior upholstery is depreciated 10 percent per year, with additional betterment charged for discoloration and damage.

   Seat covers and upholstery are depreciated 25 percent per year.

   Radios are also depreciated 10 percent per year.
2. *Mechanical Parts.* Depreciation for mechanical parts is based on a 100,000-mile life span. If a vehicle with 25,000 miles sustains damage to a mechanical part, the insurance payment for the part is reduced by 25 percent. The customer is responsible for the remaining 25 percent cost of the part.
3. *Batteries.* Most batteries are stamped with the purchase date so that betterment is easily determined in terms of percentage of life expectancy left. For example, a 48-month battery that is 24 months old will be depreciated 50 percent.
4. *Paint.* Depreciation must also be figured on the age of the finish. Normal depreciation is 10 percent after the first year. The insurance company will pay to refinish only the damaged area. If the car is badly faded, it will be difficult to match the new paint with the old finish. Other faded areas of the car cannot be refinished by the insurance company settlement. However, refinishing the entire vehicle might be the only way to match overall gloss and color. The car owner will have to pay for the additional work and materials that are needed.

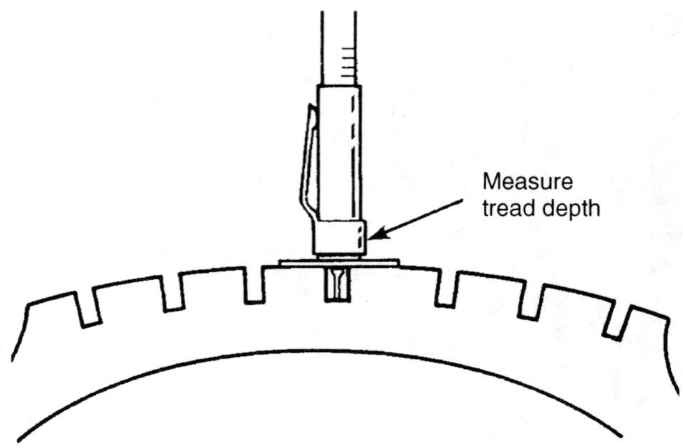

FIGURE 6–23 Checking tire tread depth with a tread depth gauge.

5. *Tires.* The full tread depth of a new tire is about 13/32 inch deep. To figure depreciation of a damaged tire, measure the depth of remaining tread as shown in Figure 6–23. Then calculate the percentage of tread left and discount the replacement price of a new tire accordingly. When estimating the price of the tire, be sure to record the amount of remaining tread, the tire manufacturer, and the size of the tire. If a damaged tire is completely bald, depreciation should be 100 percent. However, the local carcass price of a tire (about 3.00) should be discounted from the new tire price.
6. *Body Panels.* Body panels are not depreciated for age. If a new OEM panel is used to replace a damaged panel on a vehicle several years old, the car owner will not have to pay for betterment. However, body panels that show signs of prior damage are depreciated. Door dings, dents, gouges, and rust damage (Figure 6–24) are signs that a panel is inferior to a new OEM part. If a suitable salvage part cannot be obtained as a replacement for a panel with prior damage, betterment is charged to the vehicle owner for the substantial improvement the new panel will make to the vehicle. The amount of betterment charged to the owner is a subjective determination made by the insurance adjuster based on the amount of prior damage.

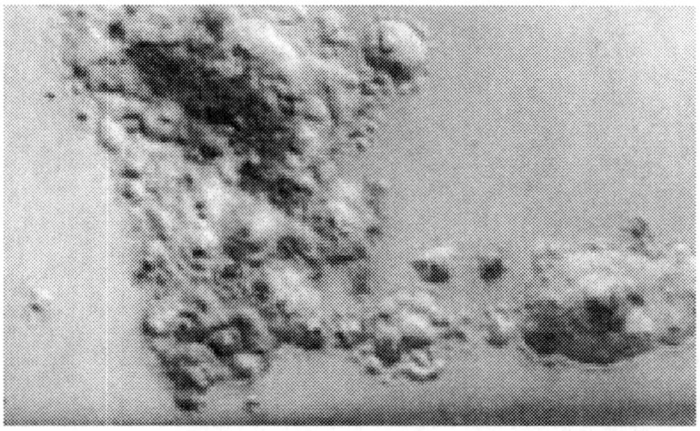

FIGURE 6–24 Surface rust caused by stone or door dings.

REVIEW QUESTIONS

1. Technician A says that time to aim lamps is included with headlight R&I. Technician B says that labor time to free rusted bolts must be added to the total labor time. Who is right?
   a. technician A
   b. technician B
   c. both A and B
   d. neither A nor B
2. The initials R&R mean to ________________.
3. Technician A says that overlap time must be deducted when a quarter panel and roof are replaced. Technician B says that overlap time must be deducted if a rear body panel and trunk floor are replaced. Who is right?
   a. technician A
   b. technician B
   c. both A and B
   d. neither A nor B
4. If a part needs a square hole, the labor time listed in the estimating guide will not apply.
   True False
5. Technician A allows 1.5 hours on each frame-damaged vehicle to mount and measure. Technician B figures a flat six hours on all frame-damaged vehicles. Who is right?
   a. technician A
   b. technician B
   c. both A and B
   d. neither A nor B
6. The initials LKQ mean ________________.
7. Technician A adds two hours of cleanup time on all used parts. Technician B says that used parts often have useable moldings on them. Who is right?
   a. technician A
   b. technician B
   c. both A and B
   d. neither A nor B
8. Which parts cost the most?
   a. OEM
   b. LKQ
   c. Aftermarket
   d. Used
9. Technician A says that betterment is deducted on damaged batteries. Technician B says that betterment is deducted from tires. Who is right?
   a. technician A
   b. technician B
   c. both A and B
   d. neither A nor B
10. If the repair time cost is 75 percent of the replacement cost, the part should be replaced.
    True False

Chapter 7

Assessing Damage to the Front Body

Objectives

After reading this chapter, you should be able to:

- *Assess damage to a bumper assembly and correctly appraise the labor involved in bumper R&I, R&R, and overhaul.*
- *Analyze damage to the grille, header panel, and headlights and enter the necessary repair operations on a damage report.*
- *Recognize various types of radiator supports and describe the options available for their repair or replacement.*
- *Explain how a hood is constructed, its component parts, and the factors involved in determining its flat rate replacement times.*
- *Determine the labor costs involved in replacing a fender.*
- *Explain the adjustments that must be made to moulding replacement times based on the type of moulding and the repair operation.*
- *Assess the labor involved in replacing an apron, a wheelhouse, and front body frame members on unibody and frame-body vehicles.*
- *Explain how to use the collision estimating guides to determine labor allowances for replacing damaged front-end body parts.*
- *Determine which operations are included and which are not included in the operation times listed in the collision estimating guides for front-end body parts.*

Key Term List

Energy absorption
LKQ
Radiator support
Header panel
Apron

Previous chapters have discussed how today's cars, trucks, and vans are constructed; how the vehicles react in a collision; how to analyze collision damage when writing a damage report; and the processes involved in restoring a vehicle to its precollision condition. Chapter 6 explained how to use the collision estimating guides to determine the labor and parts costs involved in repairing or replacing damaged panels. This chapter is going to show how this information can be used to determine labor allowances for repairing damage to the front end of a vehicle. Special emphasis will be placed on determining which repair operations are included in the listed operation times and which operations must be entered separately on the damage report. Attention will also be given to the headnotes and the footnotes in the collision estimating guides. These notes contain information that must be considered in determining the labor allowances for many repair operations. By following the guidelines contained in this chapter, the estimator and adjuster will be able to write a damage report that is both accurate and thorough.

7.1 BUMPERS

Bumpers are designed to protect the front and rear of a vehicle from damage during a low-speed collision. Prior to 1972, bumpers were rigidly bolted to the vehicle's frame. At best, the old bumpers only resisted the bending forces of an impact. They transferred the energy shock directly to the frame and thus to the occupants of the vehicle. Modern bumpers are designed to absorb the energy of a low-speed impact, minimizing the shock to both vehicle frame and the occupants of the vehicle. Between 1973 and 1983, car manufacturers were required by federal regulations to equip cars with bumpers that could withstand 5 mph collisions without incurring damage to the vehicle. (In 1983, the federal standards were relaxed to 2.5 mph.) In order to comply with the federal regulations, manufacturers fitted their bumpers with energy absorbers. The energy absorbers, mounted between the bumper face bar or reinforcement and the frame, provided a controlled amount of "give" in order to absorb and dampen some of the collision forces. Figure 7–1 shows a bumper and its energy absorbers and associated mounted hardware.

The traditional bumper is made of heavy-gauge spring steel and is plated with bright chromium metal. The chrome-plated steel bumper is still used on some larger passenger vehicles, vans, and trucks (Figure 7–2). However, most passenger cars have been equipped with plastic bumpers—urethane, polycarbonate or composite. The plastic covers may be painted to match the vehicle finish. When the bumper cover is integrated with the grille, header panel, and lower valance (Figure 7–3), the cover is called a fascia. Some compact cars are equipped with aluminum bumpers. Trucks are often fitted with painted steel bumpers.

When a chrome-plated face bar is damaged, it usually must be replaced. The chrome finish is easily cracked or chipped by the bending forces. Damaged bumpers often require rechroming, an operation that

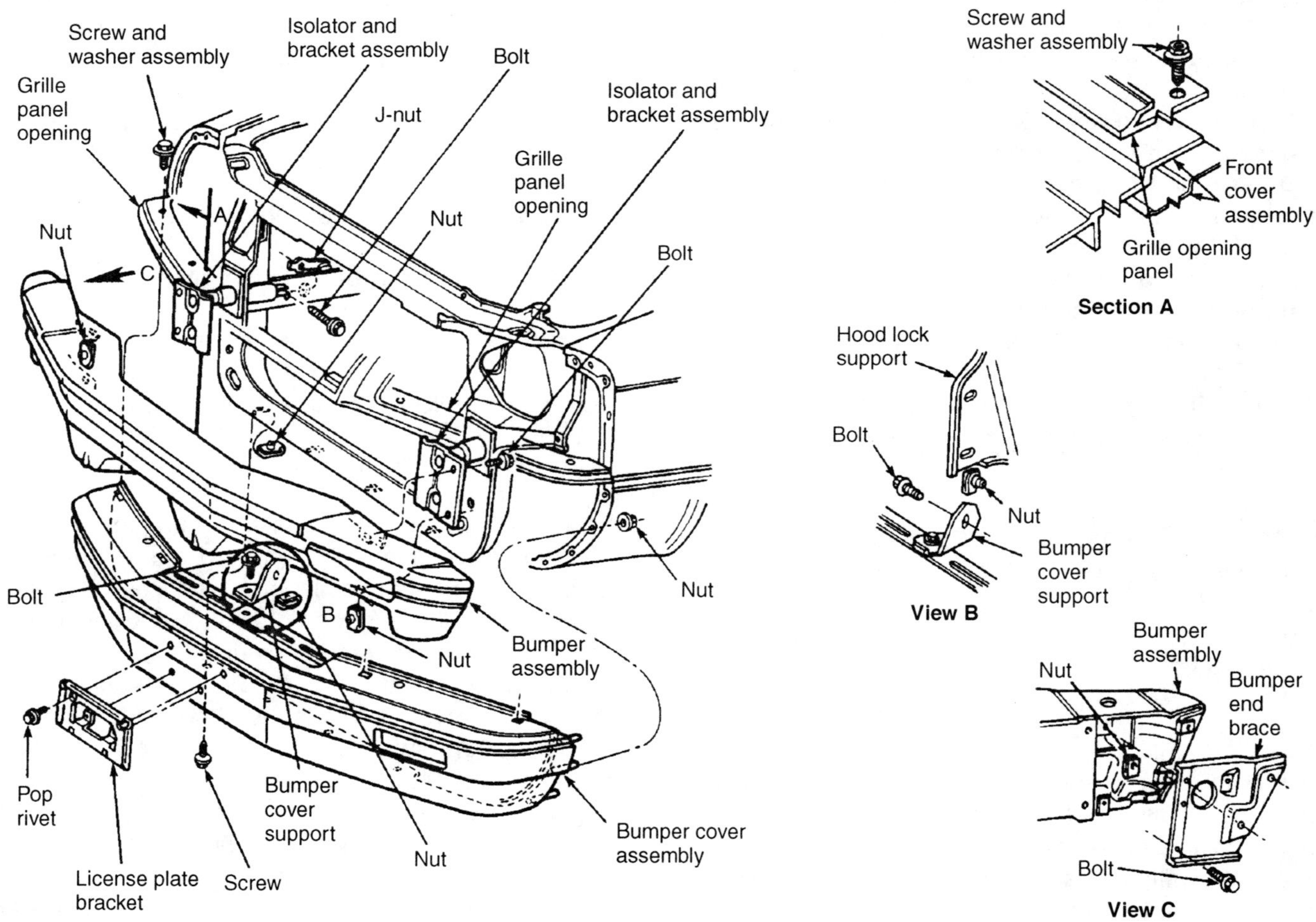

FIGURE 7–1 Energy absorbers are mounted between the bumper face bar and the vehicle frame.

only specialized bumper restoration shops perform. Also, most collision repair facilities are not equipped with the large hydraulic presses needed to bend the heavy-gauge steel face bars back into shape. Therefore, unless the damage is slight and can be worked out with conventional metalworking techniques, chrome-plated bumpers are replaced with new OEM, remanufactured, salvage, or aftermarket parts.

When a chrome-plated bumper needs to be replaced, insurance companies may require that a new OEM part be used only when the vehicle is less than two years old. If the vehicle is more than two years old, usually the damaged face bar will be exchanged for a rechromed bumper or a salvage part. Rechromed bumpers are available from specialty shops that specialize in bumper straightening and rechroming or from aftermarket parts distributors. Use of a reconditioned bumper can save from 25 to 40 percent of the price of an OEM part.

FIGURE 7–2 Chrome-plated steel bumpers are common on large cars and trucks.

Steel bumpers may be straightened by the collision repair facility and refinished. Light bends in aluminum bumpers can also be straightened. Scrapes and light scratches on an aluminum bumper can often be buffed out to restore the luster of the aluminum. However, when the cost to repair a painted steel or aluminum bumper exceeds 50 percent of the replacement cost, most insurance companies require that the bumper be replaced with reconditioned or LKQ (like kind and quality) parts.

FRONT BUMPER

Use Procedure Explanations 2 and 28 with the following text.

Refinish Bumper Cover	**2.6**
R&I Front Cover Assy	**1.4**
O/H Front Bumper Assy	**2.0**
Add to O/H Bumper Assy w/Fog Lamps	**.4**

NOTE 1: Bumper cover R&I/R&R or O/H requires the use of special cover retaining rivets.

NOTE 2: All Parts in this section are included in overhaul unless noted otherwise.

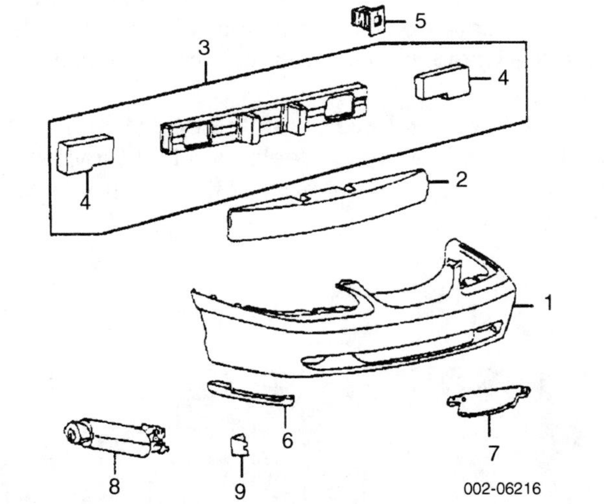

1 **Cover, Front**				
w/Cobra	94-95	F4ZZ 17D957 B	**1.6**	520.08
	96-97	F6ZZ 17D957 AA	**1.6**	700.00
w/o Cobra		F4ZZ 17D957 A	**1.6**	572.10
2 **Absorber, Reinf**		F4ZZ 17C947 A	**IOH**	75.58
3 **Reinf Assy, Bumper Ctr**		F4ZZ 17757 A		95.33
4 **Insert, Reinf¶**	R/L	F4ZZ 17C817 A-B	**IOH**	11.52
¶Included w/Bumper Center Reinf Assy				
5 **Absorber, Impact**				
See Front Inner Structure Section				
6 **Panel, Front Valance**	R/L	F4ZZ 17626 A-B	**.3**	53.70
7 **Cover, Fog Lamp Opng**	R/L	F4ZZ 17E810-1 A	**.1**	5.10
8 **Damper, Vibration**				
Convertible	R/L	F4ZZ 7611074 A	**#.3**	166.93
9 **Bracket, Damper**				
Convertible		F4ZZ 7611082 A	**#.3**	28.43
#R&R Complete, Not Included in O/H				

Figure 7–3 The bumper reinforcement on this '96 Ford Mustang is covered with a plastic reinforcement and plastic fascia. (Courtesy of Mitchell International.)

When replacing a bumper with a reconditioned face bar, the estimator and adjuster should allow extra time for fitting the part to the vehicle. It might also be necessary to include the cost of new mounting bolts. These bolts are often rusted because of their exposure to moisture and road salt.

Damage to plastic bumpers often involves damage to the cover. These plastic parts can be replaced with OEM, aftermarket, or LKQ parts. However, if the tear or hole is small, the damage can be repaired with either a plastic weld or an epoxy repair kit. If a polycarbonate bumper has damage to the reinforcement (boxed-in area) it must be replaced. Minor damage not in the reinforcement area may be repaired.

Plastics are used in other bumper parts, such as gravel deflectors, valances, and other filler panels. These parts can also be repaired with a plastic welder, but generally these parts are very inexpensive and should simply be replaced.

Some of the smaller, simpler bumpers have end caps (Figure 7–4). The caps are used to close the open ends of the face bar. If the caps are damaged, they must be included on the damage report as a replacement part. They can be made of rubber or steel. Large, chrome-plated bumpers often have bumper guards mounted on the front of the face bar. If damaged, the bumper guard and its rubber cushion must be replaced.

Some vehicles have a reinforcement behind the face bar. This is a heavy steel beam that provides additional protection for the front of the car. It should be replaced if it is damaged.

FIGURE 7–4 This Jeep bumper has end caps.

ENERGY ABSORBERS

The face bar and/or reinforcement is mounted to the energy absorbers. As stated earlier, the purpose of the energy absorber is to absorb the energy of a minor collision and allow the bumper to rebound to its original position without incurring damage to the vehicle.

There are several different kinds of energy absorbers (also called impact absorbers). Ford vehicles may be equipped with a rubber isolator (Figure 7–5). The isolator works in principle like a rubber motor mount. A rubber pad is sandwiched between the isolator and the vehicle's frame. Upon impact, the isolator moves with the force, stretching the rubber pad. The rubber "gives" and stretches, absorbing the energy of the impact. When the force is relieved, the rubber will return to its original shape (unless it is damaged by the impact) and will return the bumper to its original position.

The estimator and adjuster should inspect the isolators for bent mounting shafts and mounting plates and for torn rubber pads. Bend mounting plates can sometimes be straightened. Small tears between the mounting plates and the shaft can be welded (Figure 7–6A). However, if the shaft is bent (Figure 7–6B) or if the rubber pad is torn from its mounting, the isolator must be replaced. Some isolators have round inspection ports; if the rubber pad is visible in more than half of the inspection port, the isolator should be replaced. Isolators, like all energy absorbers, can be replaced individually; they do not have to be replaced as a pair.

Some GM vehicles have gas- or oil-filled impact absorbers (Figure 7–7) that operate much like a suspension shock absorber. Upon impact, a piston filled with inert gas is forced into a cylinder filled with hy-

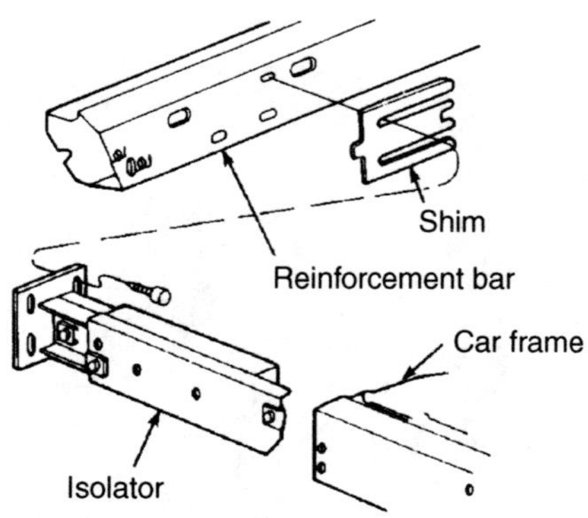

FIGURE 7–5 A rubber isolator found on Ford vehicles.

draulic fluid. Under pressure, the hydraulic fluid flows into the piston through a small opening. The controlled flow of fluid absorbs the energy of the impact. As the fluid flows into the cylinder, it bears against a floating piston and compresses the inert gas. When the impact forces are relieved, the compressed gases will force the hydraulic fluid out of the cylinder, causing the bumper to return to its original position. Another type of impact absorber, one often found on Chrysler vehicles, uses a spring rather than compressed gas to force the bumper back to its original position (Figure 7–8).

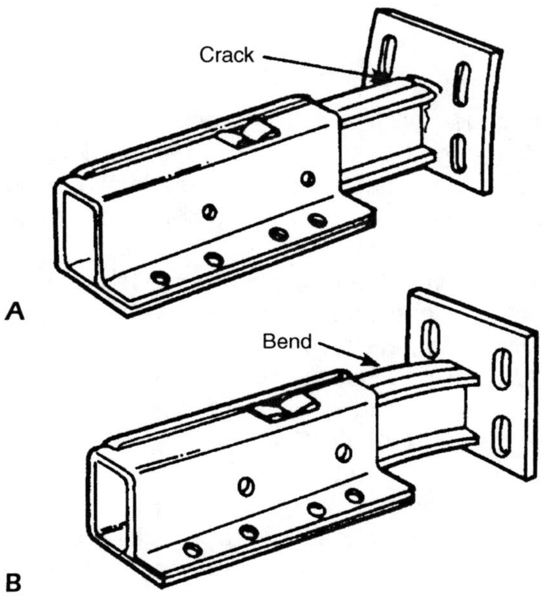

FIGURE 7–6 (A) A bent or cracked mounting plate can be repaired. (B) An isolator with a bent shaft must be replaced.

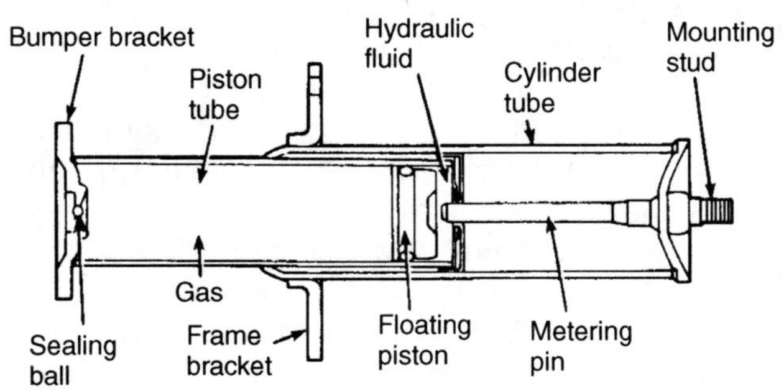

FIGURE 7–7 Cross section of a typical energy absorber found on GM vehicles.

When inspecting an impact absorber, look for signs of leaking fluid (Figure 7–9). Also check for bent cylinders, bent mounting plates, and elongated mounting holes. Also check the length of the absorbers; if the piston or cylinder is bent or otherwise damaged, the absorber will not return to its length. The shorter absorber must be replaced. Gas-filled absorbers cannot be straightened or welded.

Chrysler has an elastomeric energy absorber that was introduced on certain 1990 models (Figure 7–10). It consists of a hollow cylinder, an elastomeric core or "sleeve," a take-up spring, and a sliding member. Upon impact, the elastomeric sleeve compresses, then buckles, providing both an energy-absorbing and an energy-storing effect. After the impact, energy stored in the elastomeric sleeve, along with the clearance take-up spring, returns the bumper to its original position. The elastomeric bumper provides 5 mph protection but is much simpler in design than hydraulic energy absorbers.

Another type of impact absorber used on Chrysler products has a convoluted shaft designed to collapse in a controlled manner during an impact. Much like the crush zones built into structural body panels, the convoluted shafts absorb impact energy as they deform. These energy absorbers are designed for vehicles conforming to 2-1/2 mph bumper regulations.

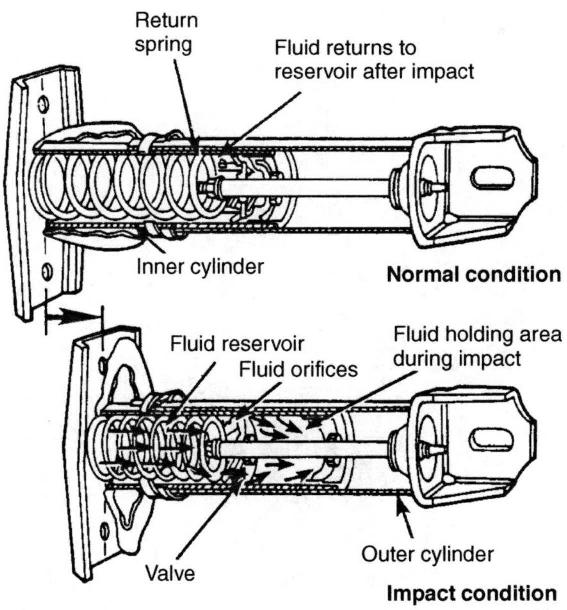

FIGURE 7–8 A spring-loaded energy absorber found on Chrysler vehicles.

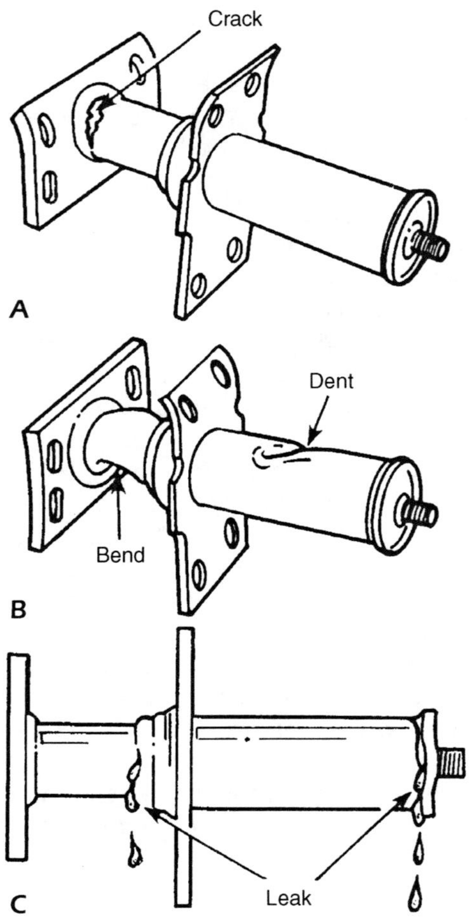

FIGURE 7–9 Gas and fluid-filled impact absorbers must be replaced if (A) cracked, (B) bent or dented, (C) leaking.

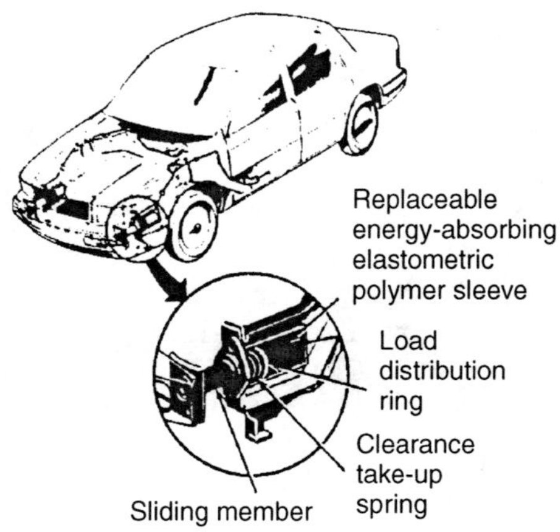

FIGURE 7–10 An elastomeric energy absorber used on Chrysler vehicles. (Courtesy of Chrysler Corporation.)

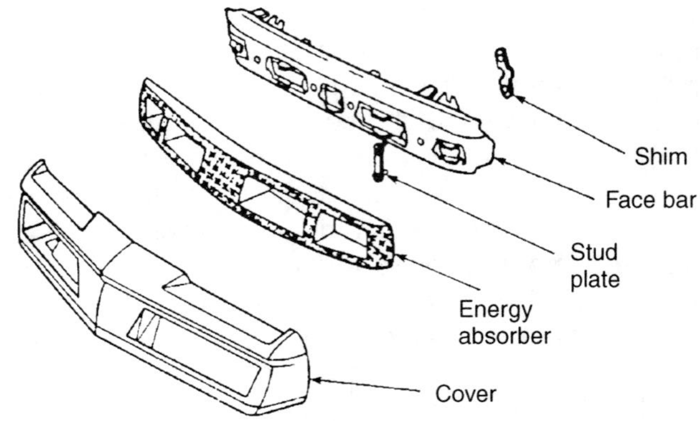

FIGURE 7–11 A urethane foam cushion provides impact absorption on some sports car models.

When inspecting this type of energy absorber, compare the lengths of the two absorbers to determine if any deformation has occurred. If the impact absorber is bent, cracked, or mashed by the impact forces, it should be replaced.

Another type of energy absorber is found on some light imports and on sports model vehicles. Instead of shock absorbers mounted between the frame and the face bar or reinforcement bar, a thick urethane foam pad is sandwiched between an impact bar and a plastic fascia or cover (Figure 7–11). The pad is designed to give the rebound to its original shape in a 2.5 mph collision. When inspecting this type of bumper, look for gouged and torn cushion and cover. If the cushion and cover cannot be repaired using plastic welding techniques or plastic repair materials, they should be replaced.

The collision estimating guides contain operation times for R&R, R&I, and overhaul of the bumper assembly (Figure 7–12). R&I times include time for removing and reinstalling the bumper assembly. This might be necessary to replace an energy absorber or to gain access to body panels. R&I times also include time for aligning the bumper assembly to the vehicle.

R&R times of the face bar include time for removal and reinstallation or replacement of the bumper guards and face bar cushions, unless noted otherwise in the text. R&R times also include time for aligning the new face bar to the vehicle. However, this assumes that the face bar is a new OEM part; if a reconditioned part is used, allow extra time for fitting the bumper to the vehicle, as stated earlier.

Overhaul times include time for removal of the bumper assembly from the vehicle, disassembly of the bumper, replacement of damaged parts, reassembly of the unit, and reinstallation of the bumper on the

BUMPER/GRILLE

Use Procedure Explanations 2 and 28 with the following text.

Refinish Front Cover **2.1**
R&I Bumper Assy **2.2**
O/H Front Bumper Assy (Includes R&I) .... **#3.2**
#Includes R&I/R&R Park/Side Marker Lamps
Add to R&R or O/H Front Bumper Assy
w/Fog Lamps .... **.3**
NOTE: All parts in this section are included in overhaul unless noted otherwise.

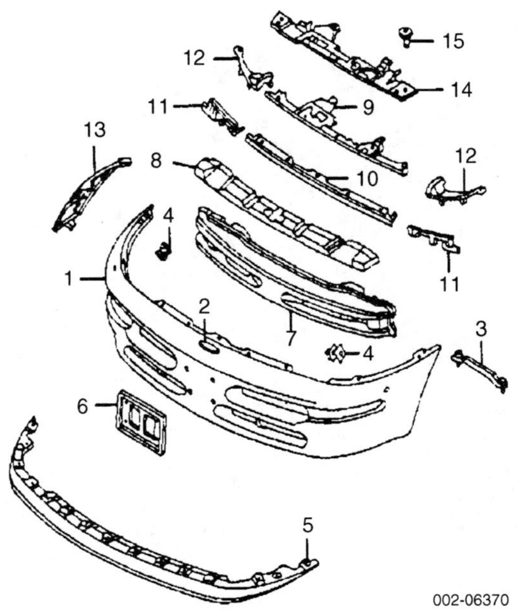

1 **Cover, Front (P)**				
Base Model		F32Z 17D957 B	**#3.2**	446.25
GT, SE Model		F32Z 17D957 A	**#3.5**	446.25
(P) Paint to Match				
#Includes R&I/R&R Park/Side Marker Lamps, Also Includes R&R Fog Lamps on GT				
2 **Emblem, Cover**		F32Z 6142528 A	**.3**	14.67
3 **Bracket, Valance**	R/L	F32Z 17C787 A	**IOH**	29.90
4 **Retainer, Frt Cover**	R/L	F32Z 17C756 T		8.72
5 **Panel, Valance**				
1993				
Base Model		F32Z 17626 A	**.2**	d63.57
GT, SE Model		F42Z 17626 A	**.3**	56.38
1994-97		F42Z 17626 A	**.2**	56.38
6 **Bracket, License**		F32Z 17A385 A	**.2**	20.85
7 **Face Bar**				
Base Model		F32Z 17757 A	**IOH**	215.95
GT, SE Model		F32Z 17757 B	**IOH**	214.97
8 **Support, Cover (Foam)**		F32Z 17C947 B	**IOH**	40.43
9 **Reinf, Cover Otr**		F42Z 17E814 A	**IOH**	32.07
10 **Reinf, Cover Inr Ctr**		F32Z 17E814 B	**IOH**	69.37
11 **Reinf, Cover Inr End**	R/L	F32Z 17E814 C-D	**IOH**	21.28
12 **Shield, Splash**	R/L	F42Z 13B159 C-D	**IOH**	6.42
13 **Retainer, Shield**				
Base, SE Model	R	F42Z 17C756 A		66.26
	L	F32Z 17C756 K		66.26
GT Model	R/L	F32Z 17C756 M-N		66.26
14 **Support, Bumper Upper**		F42Z 17C947 F	**.2**	22.95
15 **Fastener, Support (5)**		F32Z 17C756 C		1.54
Seal, Support	R/L	F42Z 17C947 C-D		11.05
Rubber, Support	R/L	F42Z 17756 A		16.67

FIGURE 7–12 Estimate guides contain labor rates for bumper R&I, O/H and refinish.

vehicle. Overhaul times also include time to align the bumper with the vehicle.

None of the operation times listed in the collision estimating guides allow for the following:

- Additional time for loosening frozen or broken fasteners
- R&R of the energy absorbers
- Refinishing the face bar or the plastic cover
- R&I of optional mouldings, nameplates, emblems, and ornamentation
- R&I of optional air bags
- R&I of optional accessories, such as lights, guards, and air dams
- Time to aim headlights
- Time required to straighten or align reconditioned or LKQ bumpers

7.2 GRILLE AND LIGHTS

The front exterior of a car contains several components in addition to the bumper assembly. The front of a typical car might have a grille, a header panel, headlight assemblies, and various mouldings and emblems. Depending on the vehicle, these parts are grouped in different ways in the collision estimating guides. For example, the grille might be part of the bumper fascia or it might be mounted to a header panel. Headlights might be included with the bumper assembly or with the grille, or be listed separately. For the sake of discussion, this section will treat the grille, header panel, and headlights together. The estimator should consult the collision estimating guides to identify component parts and their mounting positions.

GRILLE

The grille is located in the front center of the vehicle (Figure 7–13). Depending on the vehicle design, the grille might be mounted in the bumper fascia or it might be attached to the header panel (Figure 7–14) or radiator support or hood. Grilles have both an aesthetic and a functional purpose: they are used to conceal the radiator and to channel air into the radiator core. A grille can be made of one of several materials: aluminum, die cast metal, ABS, or urethane.

FIGURE 7–13 The grille hides the radiator and directs air through it.

There is a great variety in grille designs. Some grilles are constructed with several pieces; these pieces should be replaced individually rather than replacing the complete grille. Attaching emblems, ornaments, supports, brackets, mouldings, braces, and fillers can also be replaced separately. Slightly damaged plastic and urethane grilles can be repaired using plastic welding techniques or plastic repair methods. When repairs are not possible, new parts should be installed.

The estimator must be aware that on some models, parking lights and various mouldings are a part of the grille; these parts are included in the R&R times and should not be entered separately on the damage report. Also included in the operation times are standard equipment mouldings, nameplates, and ornamentation. Always consult the exploded views in the collision estimating guides to identify parts that must be ordered separately and to avoid duplicating parts on the damage report. The following operations are not included in the grille R&R times and must be addressed separately on the damage report:

- Replacing stripe tape and decals
- Replacing optional mouldings, nameplates, and ornamentation
- Replacing optional lights
- Aiming the headlights

GRILLE

Refinish Headlamp Mount Panel **1.8**
Add to Edge .... **.5**
R&I Header Panel **1.2**
R&I Grille .... **.2**
R&I Ornament .... **.2**

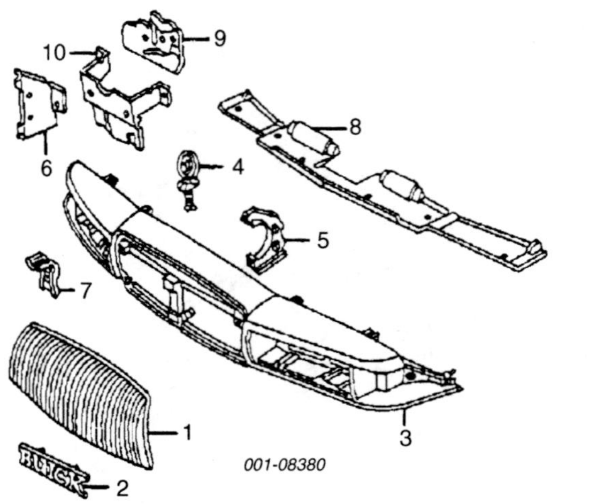

1 **Grille**				
1988-90				
Custom, Limited		14087960	**.3**	75.75
Gran Sport		14101732	**.3**	73.75
1991-94				
Custom, Limited		10200021	**.3**	103.00
Gran Sport¶		10200024	**.3**	103.00
1995-96¶		10244771	**.3**	103.00
¶White listed, Order by Application				
2 **Nameplate**				
"BUICK"	88-94	14101727	#.2	16.15
	95-96	10242357	.2	11.00
"BUICK GS"	88-94	10133606	#.2	7.55
#Included in R&R Grille				
3 **Panel, H/Lamp Mount**	88-94	10146028	**#2.0**	288.00
	95-96	10235405	**#2.0**	288.00
#Add .4 to Aim Lamps, Includes R&I Grille & Lamp Assemblies				
4 **Ornament**	88	14101730	**.2**	19.35
	89-96	12395304	**.2**	23.95
5 **Bracket, Front Panel**		14100477		5.70
6 **Brace, Bracket**		14102244		3.65
7 **Support, Panel**	88-89 R/L	14100478		7.85
Bracket, Panel	R/L	16508108-7		78.25
8 **Cover, Lamp Access**		10253824	**.2**	12.95
9 **Latch, Hood Primary**		10157899	**.3**	15.15
10 **Support, Hood Latch**		10130430	**#.3**	12.00
#w/Front Panel Removed				

FIGURE 7–14 Grille and header panel from a 1996 Buick Regal. (Courtesy of Mitchell International.)

HEADER PANEL

The header panel is usually mounted in front of the radiator support. Header panels are made of several materials: sheet metal, fiberglass, hard plastic, urethane, or aluminum. If the damage to the panel is not too extensive, plastic and fiberglass panels can be repaired using plastic welding techniques or an epoxy repair kit. (When repairing fiberglass panels, remember to add the cost of the repair material to the damage report.)

If the header panel must be replaced, the estimator must consult the footnotes to determine whether the R&R times include the grille and headlight assemblies. If not, additional time must be allowed to remove and install these parts. On some vehicles, the bumper must also be removed before the header panel can be removed. Header panels are available from the manufacturer, from aftermarket sources, and as salvage parts.

New OEM header panels normally come with predrilled holes for mouldings and emblems. If they do not, additional time must be allowed for fabricating the holes. The normal drill allowance for round holes is 0.2 hours. Deduct 0.1 hours for overlap for each additional moulding requiring drilling. If a square hole is required for installing ornamentation, add additional time.

Unless noted otherwise in the text or footnotes, the R&R times for header panels include R&I of the grille and headlight assemblies and R&I of fillers and extensions. The operation times also include realignment of the panel to the vehicle.

The header panel operation times do not include:

- Removal of the bumper assembly
- R&I of stripe tape, decals, and overlays
- R&I of adhesive-type mouldings, nameplates, and ornamentation
- Any necessary drill times
- Time for aiming the headlights

HEADLIGHT ASSEMBLIES

Headlight assemblies are mounted in various places: header panels, headlight mounting panels, side panels, radiator supports, and bumper fascia. Figure 7–15 shows an exploded headlight assembly and housing. The headlight door, or bezel as it is also called, might be made of aluminum, die-cast steel, or plastic. In some cases, the door is part of the grille. The retaining ring is used to hold the sealed beam or light in place. The mounting ring is used to support the light and to allow the light to be aligned (aimed). Also shown are a parking light, turn indicator light, and a side marker.

When the door or bezel must be replaced, it must be included as a separate item on the estimate. However, the retaining ring, light, and mounting ring are available as a complete assembly (also called a capsule): The estimator should also look for severed wires. Severed wires can be spliced and soldered or they can be rejoined using quick connectors that are crimped over the loose ends of the wires.

Composite headlights are made from plastic or glass (Figure 7–16). The bulb is replaceable. In some applications, the headlight mount is also replaceable. The aiming adjusters are usually built into the mount. Always check headlights for proper operation. Mounting tabs are often broken in a collision. If the headlight wiggles, a mounting tab is broken.

When one is estimating replacement of headlights, time must be included on the damage report for aiming the lights. However, headlight alignment is not necessary when only replacing the headlight door or retaining ring.

FRONT LAMPS

HEADLAMP

R&I Headlamp Capsule .... R/L #.3
#R&I Both One Side .5, R&I Both Sides Complete 1.0
R&R Does Not Include Aim Lamps 5

NOTE: All Parts in this section are included in R&R Headlamp Housing unless noted otherwise.

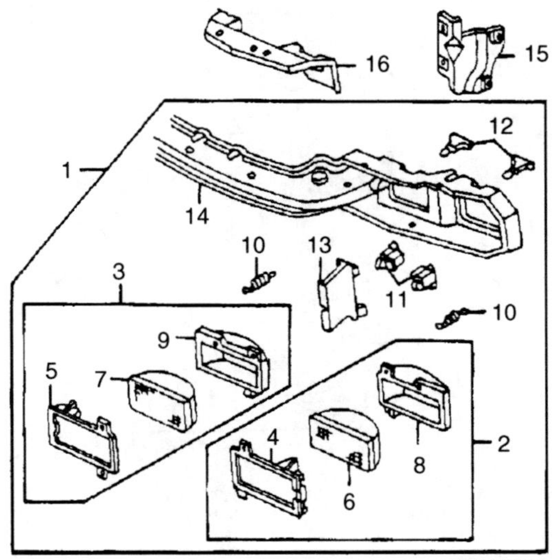

001-10099

1 **Headlamp & Panel Assy**		16521298	**#1.0**	399.00
#w/Front Bumper Cover Removed				
Capsule, Headlamp¶				
2 Low Beam	R	16513312	**#.3**	122.00
	L	16510744	**#.3**	122.00
3 High Beam	R	16510743	**#.3**	122.00
	L	16513310	**#.3**	122.00
#R&R Both One Side .5, R&R Both Sides Complete 1.0				
Ring, Retaining¶ (a)				
4 Low Beam	R/L	16513316-5	**.2**	34.50
5 High Beam	R/L	16513315-6	**.2**	34.50
Sealed Beam¶ (a)				
6 Low Beam	R/L	16510872	**#.2**	■40.92
7 High Beam	R/L	16510873	**#.2**	■60.75
#Includes R&R Retaining Ring				
Ring, Mounting¶ (a)				
8 Low Beam	R	16513313	**#.3**	43.00
	L	16513304	**#.3**	43.00
9 High Beam	R	16513308	**#.3**	43.00
	L	16513311	**#.3**	43.00
(a) Included w/Headlamp Capsule				
#Includes R&R Sealed Beam & Retaining Ring				
10 **Spring, Headlamp¶**				
2/Side	R/L	16514449		3.55
11 **Pivot, Horizontal¶**				
Outer	R/L	16514352		17.20
Inner	R/L	16516655		17.20
12 **Bolt w/Nut, Adjuster¶**				
2/Side	R/L	16516577		1.80
13 **Filler, Headlamp¶**	R/L	16516318-7		2.15
14 **Housing, Headlamp¶**		16519594	**#1.5**	174.00
¶Included w/Headlamp & Panel Assy				
#w/Bumper Assy Removed				
15 **Brkt, Housing Outer**	R/L	10280337-8	**#.1**	7.10
16 **Brkt, Housing Center**		10247777	**#.1**	7.00
#w/Headlamp Housing Removed, Not Included in R&R Headlamp Housing				
Module, Headlamp Control				
Canada		10098712	**.3**	114.00

FIGURE 7–15 Headlights parts on a 1996 Camaro. (Courtesy of Mitchell International.)

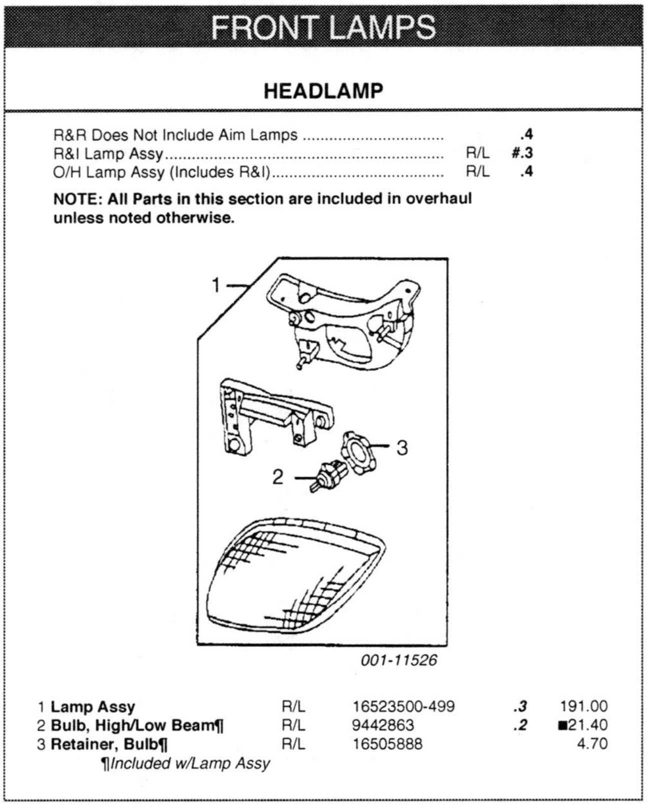

FRONT LAMPS

HEADLAMP

R&R Does Not Include Aim Lamps		.4
R&I Lamp Assy..	R/L	#.3
O/H Lamp Assy (Includes R&I).......................................	R/L	.4

NOTE: All Parts in this section are included in overhaul unless noted otherwise.

1 **Lamp Assy**	R/L	16523500-499	.3	191.00
2 **Bulb, High/Low Beam¶**	R/L	9442863	.2	■21.40
3 **Retainer, Bulb¶**	R/L	16505888		4.70

¶Included w/Lamp Assy

FIGURE 7–16 Composite headlight from a 1996 Sunfire. (Courtesy of Mitchell International.)

7.3 RADIATOR SUPPORT

The radiator support in today's unitized construction is welded to the apron and front cross member to form the front panel of the vehicle (Figure 7–17). On some body-over-frame construction vehicles, the radiator support is bolted to the fenders, wheel-houses, and frame assembly. Besides providing structural support to the front end sheet metal, it also provides support for the radiator and associated cooling system components. In the collision estimating guides, the radiator support is usually included with the cooling system components rather than with the fenders, frame, or grille.

The radiator support is a single panel on all body-over-frame vehicles and also on many unibody models; on other unibody vehicles, the radiator support is composed of several parts: an upper tie bar, a lower tie bar, and right and left side baffles (Figure 7–18).

Damaged radiator supports can be straightened and repaired using common frame-straightening equipment and techniques. When a support is damaged too severely to be repaired, however, it can be replaced with new or salvaged parts. If only parts of a

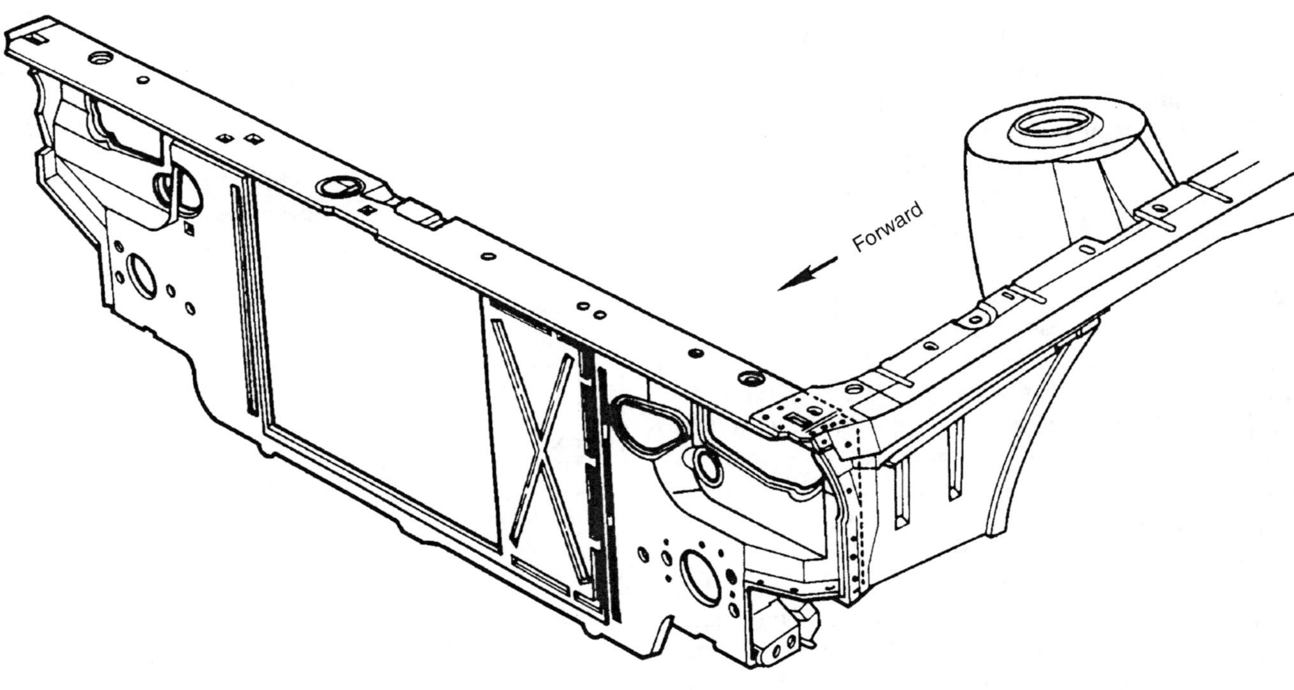

FIGURE 7–17 The radiator support forms the front of the unibody vehicle.

support, such as the upper tie bar or side baffle, are damaged, only the damaged part should be replaced. When the support is a whole panel, check the OEM specifications to see if sectioning is allowed. Sectioning saves considerable labor time by avoiding the necessity of drilling out all spot welds and rewelding the entire panel. Bolted-on radiator supports should not be sectioned; they should be replaced in their entirety.

When replacing a radiator support, be sure to check the headnotes and footnotes in the collision estimating guides to determine what operations are and are not included in the operation times. There is a great variety in radiator support design and replacement procedures. The operation times include the removal of the damaged panel, realignment of the new panel, and all necessary welding. The R&R times given in the Mitchell collision estimating guides allow for the removal of the radiator (except where noted otherwise in the text) but do not allow for the transfer of bolted-on parts, such as the washer or coolant reservoir, from the old panel to the new one. The Motor crash guides, on the other hand, do not allow for the removal of the radiator but do allow for the transfer of bolted-on parts, unless stated otherwise in the footnotes. The Motor guides also allow for the loosening of bolted-on fenders to provide access to the radiator support seams.

Unless stated otherwise in the text, the crash guides do not include:

- Evacuating and recharging the air-conditioning
- R&I of air-conditioning components
- R&I of the bumper, grille, headlights, the header panel, fenders, and extensions
- Refinishing (Some radiator supports require painting.)
- Applying undercoating or anticorrosion protection
- Aiming the headlights

7.4 HOOD PANEL

The hood, located over the engine compartment and between the two fenders, is designed to protect the engine from dirt and moisture. It also blocks much of the engine noise. Hoods are usually constructed of cold-rolled sheet metal, but hoods of aluminum, fiberglass, and plastic are also used on modern vehicles.

A typical hood (Figure 7–19) consists of an outer panel and an inner panel. These panels are spot welded together along the outside edges and bonded

FRONT INNER STRUCTURE

Use Procedure Explanations 8 and 28 with the following text.

FRONT

Operation		
Refinish Support Assy		1.5
Refinish Upper Tie Bar		.5
Refinish Front Panel		.5
Refinish Side Panel	R/L	.5
Refinish Lower Tie Bar		.5
Refinish Latch Support		.3
Add to R&R Support Assy w/Oil Cooler	m	#.3
#Include R&R Oil Cooler		
To Aim Lamps		.4
Evacuate & Recharge System	m	1.4
Refrigerant Recovery	m	.3

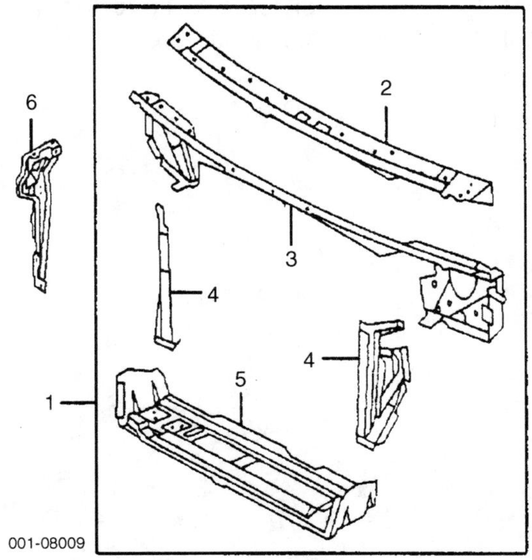

Part	Year	Part No.		Time	Price
1 **Support Assy**		25650047	s	**#8.9**	101.00
#Includes R&R Radiator, Shroud, Motors & Fan Assemblies, Impact Absorbers, H/Lamp Assemblies, H/Lamp Mount Panels, Latch Support, Hood Primary Latch, Hood Pop-Up Spring, Hood Bumpers, Front SIR Sensor, Hood Ajar Switch & Bracket, Battery, Air Intake Duct, A/C Condenser, Baffles, Brackets, Clips, Fasteners, Disconnect & Connect Hood Release Cable & Wiring Harness as Necessary					
2 **Tie Bar, Upper¶**	92-96	25650050	s	**2.0**	39.50
	97	25650050	s	**2.0**	39.50
3 **Panel, Front¶**					
1992-93		25615653	s	**#4.8**	34.00
1994-97-Service w/Support Assy					
#Includes R&R Upper Tie Bar, Disconnect & Connect Wiring Harness as Necessary, w/Bolted Parts & Assemblies Removed					
4 **Reinf, Panel¶**	92-93 R/L	25615656-7	s	**2.0**	17.50
5 **Tie Bar, Lower¶**		25630889	s	**2.5**	73.50
Included w/ Support Assy					
6 **Support, Latch**		25630768		**.3**	6.71
Sensor, Air Bag Front-See Air Bag System Section					
Switch, Hood Ajar¶		25620933	m	**#.4**	8.95
Brkt, Ajar Switch¶	92-94	25559981	m	**#.4**	1.95
¶Located on Radiator Support					
#R&R Complete, Switch & Bracket					

FIGURE 7–18 The radiator support for this 1996 Pontiac Bonneville is made of several parts. (Courtesy of Mitchell International.)

together with adhesive on the inner faces of each panel. A pilot bolt or striker, mounted on the underside of the front edge of the hood panel, engages a hood latch when the hood is closed. The latch is mounted on the radiator support on most cars. The pilot bolt or striker is released from the latch when a cable release is pulled from the inside of the vehicle. Hoods are also equipped with a safety latch; the safety latch prevents the hood from opening if the latch should suddenly fail to properly engage the striker. The hood is mounted to the cowl or inner fender skirt by two hinges. Some hinges use springs (Figure 7–20) or torsion bars to keep the hood up when it is opened. Other hoods are held open by a separate rod. Most hoods also have an insulator attached to the underside of the panel; the insulator is made of spun fiberglass and helps to reduce engine noises and also shields the hood panel and finish from high underhood temperatures. Hoods are also equipped with a variety of mouldings, ornaments, ram air scoops, decals, and so on.

A buckled hood is very difficult to straighten due to its double-panel construction. When a hood must be replaced, OEM, aftermarket, and LKQ parts are available. Besides the obvious cost advantage, salvage hoods are available as an assembly, complete with hinges, mouldings, and striker.

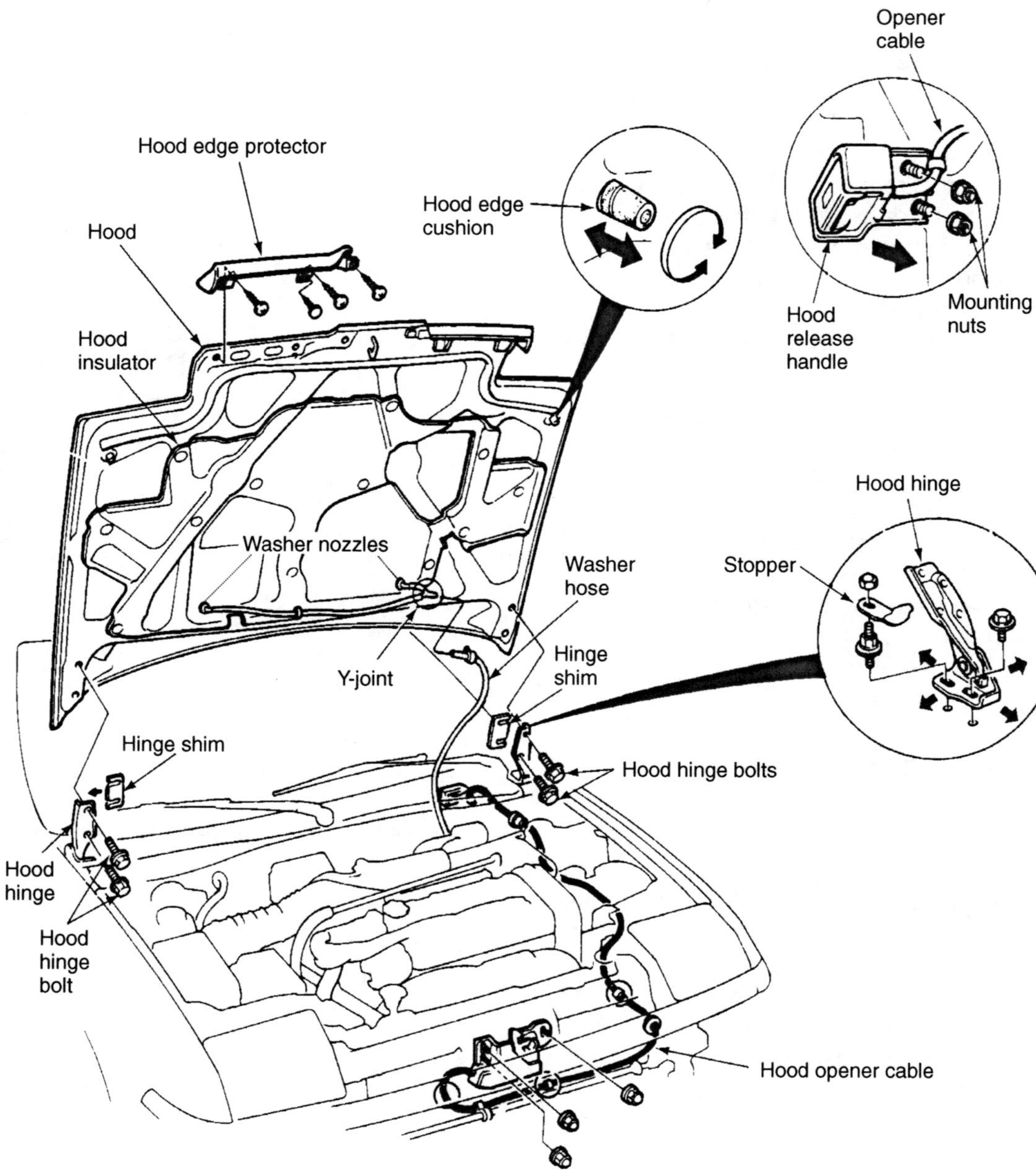

FIGURE 7–19 Components of a hood. (Courtesy of American Honda Motor Co., Inc.)

FIGURE 7–20 A spring-loaded hood hinge. (Courtesy of Mitchell International.)

Hood R&R times (Figure 7–21) allow for the removal and replacement of the hood, R&I of the hood insulation, and realignment of the hood on the hinges. (If the hood is not damaged but is only pushed out of alignment, 0.5 hours should be allowed for realignment.)

The striker and insulator panel are included in the hood R&R. However, the hood latch, safety catch, cable release, hinges, and mouldings are not included; these items must be written up separately. The striker, latch, and safety catch are safety-related items; if they are damaged, they must be replaced rather than repaired. The cable release should also be replaced if damaged.

If the hinges have incurred only slight damage, they can be repaired. However, if the hinges are sharply bent or twisted, they should be replaced. When replacing both the hood and the hinges, a deduction must be made in the flat rate time to account for the overlap, unless the footnote specifies that the hinge R&R assumes that the hood is already removed.

Hood mouldings (Figure 7–22) are constructed of various materials: chrome-plated die-cast steel, stainless steel, aluminum, plastic, or rubber. Hood mouldings are not as common on today's aerodynamically designed vehicles as they were prior to the 1980s. The most common application is a rear moulding that covers the back edge of the hood. A center moulding running the length of the hood is another popular application. Some hoods might also have a moulding running across the front edge of the hood. Front mouldings can be a single piece or three pieces (center, left, and right); when multipiece mouldings are damaged, only the damaged piece should be replaced. Because of the diversity in moulding design and application, mouldings are not included in the hood panel R&R times. They must be entered sepa-

HOOD

Use Procedure Explanations 4 and 28 with the following text.

Refinish Hood Outside		**3.0**
Add for Underside		**1.2**
Refinish Hinge	R/L	**.2**
R&I Hood Assy		**.6**

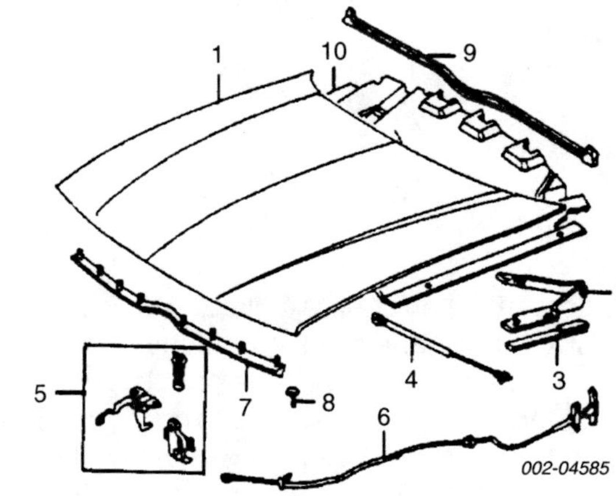

1 **Panel, Hood**				
L, GL, LX, SE		F2DZ 16612 A	**1.2**	400.12
SHO		F2DZ 16612 B	**1.2**	602.72
2 **Hinge**	R/L	E8DZ 16796-7 A	**#.2**	21.93
	#w/Hood Removed			
3 **Seal, Hinge Splash**	R/L	E6DZ 16C902 A		5.37
4 **Lift Hydraulic**	R/L	F2DZ 16C826 B	**.2**	33.40
5 **Latch**	92-93	F2DZ 16700 A	**.3**	16.55
	94-95	F4DZ 16700 A	**.3**	17.93
6 **Cable Hood Release**		F2DZ 16916 A	**.9**	28.33
7 **Seal, Radiator Supt**		E6DZ 16A238 A	**.2**	23.67
8 **Bumper, Hood**	R/L	F49Z 16758 A	**.1**	4.55
9 **Pad, Hood**		F4DZ 16740 A	**.2**	16.60
10 **Insulator, Hood**				
L, GL, LX, SE		F2DZ 16738 A	**.3**	36.70
SHO		F2DZ 16738 B	**.3**	76.08
Retainer, Insulator				
Clip		385595 S		.42
Pin		386553 S		.36

FIGURE 7–21 Labor times and parts on a 1996 Ford Taurus. (Courtesy of Mitchell International.)

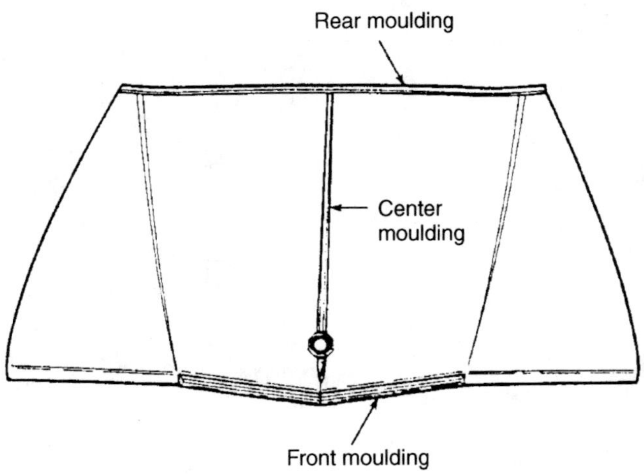

FIGURE 7–22 Hood mouldings.

rately on the damage report. Add drill time if the collision estimating guides allow for it.

The estimator or adjuster must be careful not to overlook hood ornaments, decals, stripe tape, emblems, emission control labels, and seals. Time to replace these items must be included on the damage report. Of course, time must also be included to refinish the panel.

7.5 FENDERS

The fenders of a vehicle are bolted to adjacent supporting structural panels. In a unibody vehicle, the fenders are bolted to the cowl side panels, aprons, radiator support, and splash shields. Together with the hood, header panel, and bumper assembly, the fenders form the outer, cosmetic contour of the front end of the vehicle. Figure 7–23 shows a fender and associated parts. Each part must be considered when writing a damage report.

FENDER PANELS

Although most fenders in the past were made of cold-rolled steel, fiberglass and plastics are being increasingly used in fender applications. Dented sheet metal fenders can be repaired with common metalworking techniques, unless excessive work hardening has made the metal no longer malleable. Gouged and shattered fiberglass panels can be patched or fabricated with a fiberglass repair kit (epoxy resin and fiberglass cloth). Plastic panels can be welded with a plastic welder or bonded.

When damage to the fender warrants replacement, OEM, aftermarket, or LKQ parts may be used. R&R times listed in the collision estimating guides (Figure 7–24) include the following:

- Removal and replacement of the fender
- Loosening the bumper and any filler panels (if necessary)
- R&I of all parts attached to the fender (except those noted in the following list as items not included in the operation times)
- R&I of standard equipment lights (side markers, etc.)

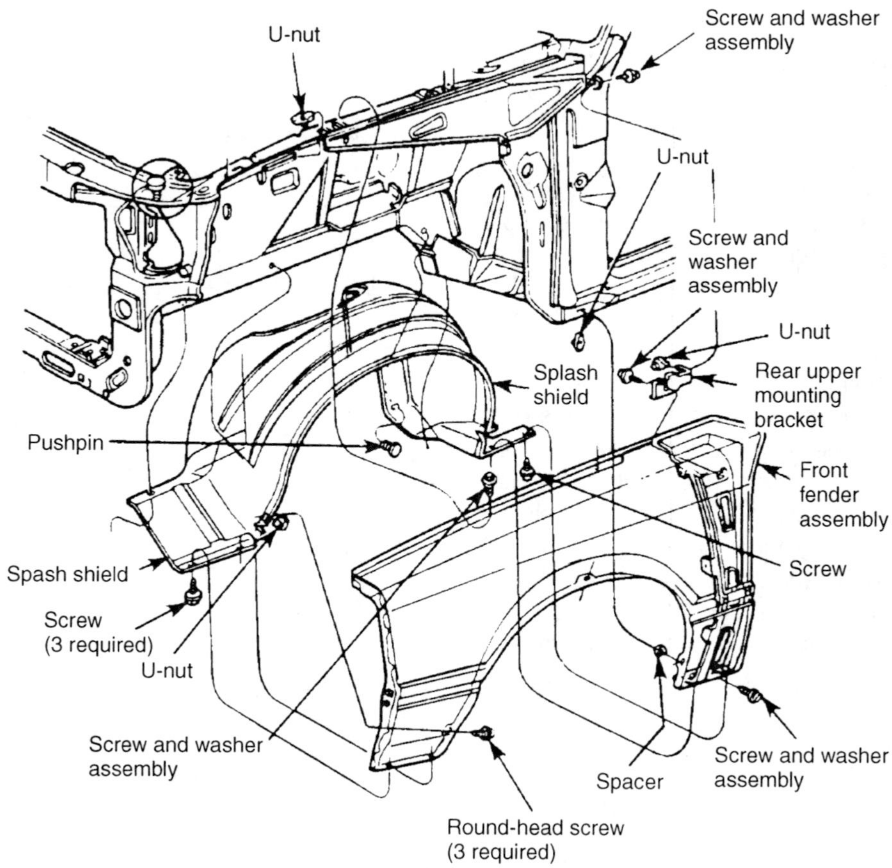

FIGURE 7–23 Typical front fender assembly.

FRONT FENDER

Use Procedure Explanations 7 and 28 with the following text.

Operation	Side	Time
Refinish Fender Outside	R/L	**2.2**
Add to Edge Fender	R/L	**.5**
R&I Fender Assy	R/L	**#1.0**
#w/Rocker Trim Panel & Front Bumper Assy Removed		
R&I Side Moulding	R/L	**#.5**
#Includeds R&I/R&R Spalsh Shield		
R&I Mudguard	R/L	**.2**
Add to R&R Fender		
To Apply Gravel Guard	R/L	**.3**
To Drill for Installation of Rocker Trim	R/L	**.3**

NOTE: When installing fender, apply silicone rubber sealer at fender splash shield.

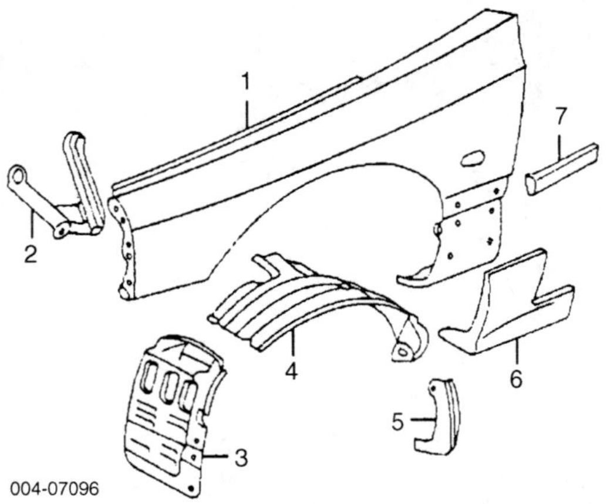

Part	Side	Part No.	Time	Price
1 **Fender**	R	MB637536	**#1.3**	295.00
	L	MB637535	**#1.3**	295.00
#w/Rocker Trim Panel & Front Bumper Assy Removed				
2 **Bracket, Fender**	R/L	MB637520-19		13.50
3 **Extension, Splash Shield**				
Base, ES Model				
1991-93	R/L	MB861146-5	**#.2**	23.75
1994-96	R/L	MR124476-5	**#.2**	23.00
R/T, R/T Turbo Model				
1991-93	R/L	MB861148-7	**#.2**	24.25
1994-96	R/L	MR124478-7	**#.2**	23.00
4 **Shield, Splash**	R/L	MB698864-3	**#.3**	25.50
R&R Both One Side .4				
5 **Mudguard**				
Base Model	R/L	MB641474-3	**.2**	18.75
6 **Panel, Rocker Trim**				
See Rocker/Pillars/Floor Section				
7 **Mldg Assy, Side¶**	91-94 R/L	MB698894-3	**#.2**	21.25
	95 R/L	MR124074-3	**#.2**	24.75
	96 R/L	MR273132-1	**#.2**	23.75

FIGURE 7–24 Labor times and part prices for a 1996 Dodge Stealth. (Courtesy of Mitchell International.)

The Mitchell collision estimating guides also allow for the transfer of clip-on type mouldings from the old panel to the new.

Except where noted in the text of the collision estimating guides, the following items are not included in R&R times for the fender:

- Refinishing
- Transfer of adhesive mouldings, nameplates, emblems, and ornamentation
- Application of stripe tape, decals, or overlays
- Installation of the antenna
- R&R of inner panels and wheelhouse
- Aiming the headlights
- Drilling holes for mouldings, nameplates, antenna, etc.
- R&I of mirrors
- Applying undercoating and anticorrosion material

SPLASH SHIELDS

Splash shields are located behind the fenders. They are designed to prevent moisture from splashing into the engine area and onto the backside of the fender. Splash shields are made of plastic and are usually replaced when damaged. Unless noted otherwise in the collision estimating guide text, the splash shields are not included in the fender R&R and must be listed separately on the damage report.

BRACES, BRACKETS, AND REINFORCEMENTS

Braces, brackets, and reinforcements (such as hood hinge reinforcements) are usually included in the fender panel R&R. However, these parts must be purchased separately, unless a salvage panel is obtained.

LIGHTS

As stated above, fender R&R time allows for the replacement or transfer of cornering lights, turn signal indicator lights, and side marker lights, unless otherwise in the collision estimating guide text. However, the replacement costs of these parts must be included on the damage report; fenders do not come equipped with lights.

ANTENNA

One of the items often overlooked on a damage report is the R&R time of the antenna. The antenna must be purchased separately from the fender panel, and time may need to be allowed for fabricating a mounting hole.

MOULDINGS, DECALS, AND OVERLAYS

One of the most commonly misunderstood and incorrectly listed items on damage reports is moulding replacement times and drill times. Fender R&R times, like other panel replacement times listed in the collision estimating guides, do not include time for installation of mouldings. Flat rate labor times for moulding installation is always given separately in the collision estimating guides. (If the mouldings are dealer installed rather than standard items, no times will be included in the collision estimating guides: in that case, the estimator and adjuster must negotiate a fair labor allowance.) However, there are different types of mouldings and repair situations that require making adjustments to the operation times. The following describes how to adjust the operation times for the type of moulding and/or the operation to be performed.

When an undamaged adhesive-type moulding must be removed and installed on a new fender, the full time listed in the collision estimating guides should be used. The listed times allow for the removal of the moulding and putting new tape on the back of the moulding prior to installing it on the new panel. However, when installing a new adhesive moulding on a new fender, only half of the listed time should be used. This reduction is justified because the old moulding does have to be removed from the old panel and because the new moulding already comes with an adhesive backing (Figure 7–25).

Figuring the installation time for bolt-on or clip-on mouldings is more complicated because of the drill time involved. When reinstalling an old moulding on a new panel, use the full time listed in the text and add time for drilling mounting holes in the new panel. If a moulding is to be removed from a fender so that the fender can be repaired and will be later reinstalled on the fender, the drill time does not apply.

In a new moulding is to be installed on a fender, use only half of the listed time, since time to remove the moulding from the old panel is not needed. When installing new moulding on a new panel, add drill time to the operation time.

In addition to the operation times listed in the collision estimating guides for moulding installation, the estimator and adjuster must take certain other adjustments into consideration. For example, when two or more mouldings must be installed, overlap must be figured into the total allowance, since the listed times include time for securing the necessary tools and equipment to perform the operation. This prep time is valid only for the first moulding. Time for each additional moulding application should be reduced by 0.1 hours to compensate for the overlap.

Another adjustment must be made to the listed times when square mounting holes must be fabricated. The drill times listed in the collision estimating guides are for round holes. Additional time must be allowed for fabricating square holes.

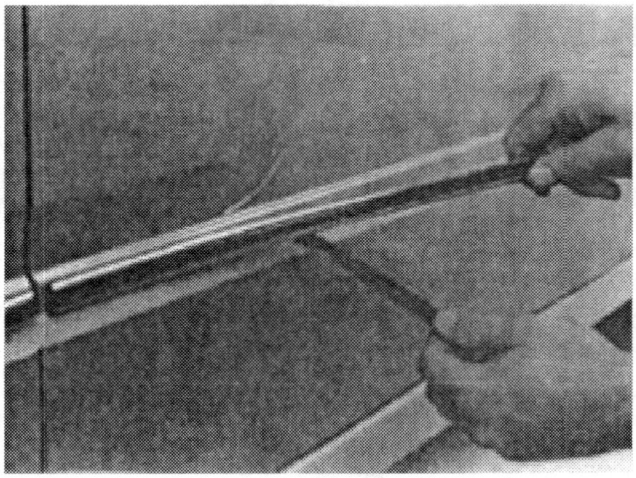

FIGURE 7–25 When installing a new adhesive moulding on a new panel, reduce the R&R time by one-half.

The operation times listed for decals, stripe tape, and overlays are for installation only. When these are being installed on repaired or salvage panels, the estimator and adjuster must allow for the time necessary for the removal of the old material. When figuring the labor involved in applying wood grain overlays, allowances must also be made for the removal and reinstallation of mouldings, nameplates, ornamentation, and door hardware.

7.6 APRONS AND WHEELHOUSES

Both unibody and body-over-frame vehicles have a panel called an apron or a wheelhouse. However, the construction and function of the apron or wheelhouse vary greatly between the two body designs.

The apron on the modern unibody vehicle performs several functions: it provides a mounting place for the suspension, it forms a wheel well to protect the engine from road splashes, it provides a mounting surface for various engine compartment components, and it supports the upper edge of the fender. The apron is also an integral part of the unitized support structure. Due to the various roles it plays, the apron might be grouped with either the fenders or with the frame in the collision estimating guides.

The apron is usually welded to the radiator support, the front rail, and the cowl side panel. The apron is usually comprised of several separate panels (Figure 7–26), some of which are made of cold-rolled steel and others of high-strength steel. For example, the strut tower is a load-bearing part and is often made of high strength, low-allow steel. Rails and reinforcements are also often made of high-strength steels. An apron might be available as a complete panel or it might be available in separate panels. It is best to separate these panels at factory seams. However, it is also permissible to section aprons, provided that the sectioning is performed in front of the strut tower and does not pass through any crush zones. Check with the OEM to see if sectioning is allowed.

It is very important that damage to the apron be carefully diagnosed. As stated in Chapter 5, the estimator and adjuster must look for split seams, popped welds, and other signs of distortion in the apron. Measurements should be taken of underhood control points to determine if misalignment of the suspension mounting panels has occurred. The Mitchell collision estimating guides contain strut tower dimensions that could be used when analyzing underhood damage.

Whenever possible, the apron should be repaired, using frame straightening equipment and repair methods. When repair of the panel is not the most economical means of restoring the panel to its precollision condition, aprons can be replaced with OEM or LKQ parts. Apron R&R times listed in the collision estimating guides allow for drilling out the OEM welds, removing the old panel, mounting and aligning the new panel to the car, and welding the apron in place. R&R times include time for removal and reinstallation of the interior carpet, insulation, and cowl trim. This is necessary due to the heat produced when welding the new panel to the cowl.

The listed R&R times do not include the time necessary to set up the vehicle on a frame straightening bench or time to measure and monitor the alignment of adjacent panels. Nor does it include time to remove and reinstall components bolted to the apron, such as fenders, the radiator support, struts, and other mechanical and electrical parts; R&I times for components attached to the apron must be included separately on the damage report. Whenever the apron is replaced or repaired, the damage report must also include time for aligning the suspension.

Most body-over-frame vehicles do not have an apron. The engine and suspension components are mounted to the heavy frame components. However, the engine compartment and the inner side of the fender are protected from road splashes by a wheelhouse (sometimes called a skirt). The wheelhouse also provides a mounting area for the upper edge of

SIDE

Operation		Time
Refinish Front Rail/Apron Assy	R/L	**2.0**
Refinish Front Side Rail Assy	R/L	**1.0**
Refinish Front Side Rear Rail	R/L	**.5**
Refinish Front Side Front Rail	R/L	**.5**
Refinish Front Apron Assy	R/L	**1.0**
Refinish Strut Tower	R/L	**.5**
Refinish Front Apron	R/L	**.5**
Section Frame Rail	R/L	**#4.0**
#Section 64mm (2.5 inches) Forward of the Leading Edge of the Steering Gear crossmember includes Section front Side Rail & R&R Bumper Mtg Reinforcement		

Note: For Sectioning Procedures Refer to Tech-COR Information Bulletin #95-5, See Disclaimer on Page P-3 concerning labor times.

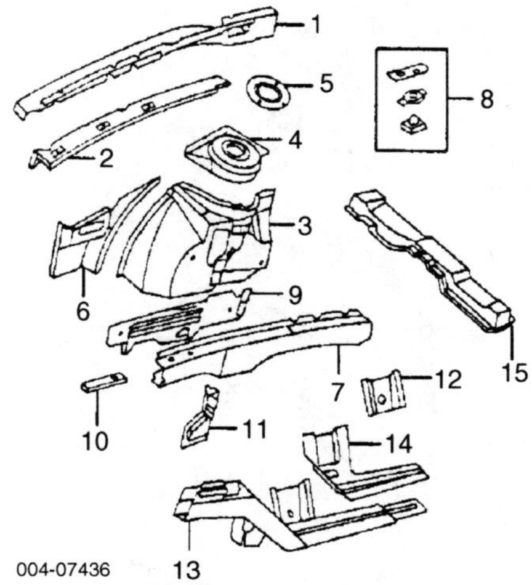

Item	Year/Side	Part No.	Labor	Price
Panel, Upper Beam (HSLA)				
1 Inner	94 R/L	4696718-9	**s #4.5**	42.50
	95-96 R/L	4696718-9	**s #4.5**	42.50
	97 R/L	4580718-9	**s #4.5**	42.50
2 Outer	94-96R	4624514	**s #4.5**	77.50
	L	4624515	**s #4.5**	87.50
	97 R/L	4871514-5	**s #4.5**	64.50
#R&R One Side Complete, w/H/Lamp Mtg Panel Removed Deduct 1.0				
3 **Panel, Strut Mtg**				
HSLA	93-96 R/L	4583100-1	**s #9.0**	85.00
	97 R	4583100	**s #9.0**	85.00
	L	4805155	**s #9.0**	75.50
#Includes R&R Upper Inner & Outer Beam Panels, Front Apron Panel, Strut Mtg Panel Cap & Strut Cap Plate, w/Radiator Support Removed Deduct 1.0, w/Front Side Rear Rail Removed Deduct 2.5				
4 **Cap, Strut Mtg Panel**				
HSLA	R/L	4583098-9	**s 2.5**	30.50
5 **Plate Strut Cap**	R/L	4646004	**#.5**	7.10
#Included in R&R Strut Mounting Cap				
6 **Panel, Front Apron**				
HSLA	93-96 R	4580664	**s #2.0**	16.00
	L	4696575	**s #2.0**	16.50
	97 R/L	4580664-5	**s #2.0**	16.00
#w/Strut Mtg Panel Remove Deduct 1.0, w/Headlamp Mtg Panel Remove Deduct .5				
7 **Rail, Front Side Rear**				
HSLA	R/L	4756706-7	**s #8.2**	33.50
#w/Strut Mtg Panel Remove Deduct 3.5, Includes R&R Front Side Front Rail & Bumper Mtg Reinf				
8 **Reinf, Susp Xmbr Mtg**	R/L	4624476-7		13.50
9 **Rail, Front Side (HSLA)**	R/L	4583944-5	**s**	47.25
10 **Reinf, Bumper Mounting**	R/L	4696540-1	**s**	2.85
11 **Support, Rail (HSLA)**				
Headlamp Mtg	R/L	4624310-1	**s #1.0**	9.60
12 **Reinf, Floor Pan**	R	4624526	**s #1.0**	78.00
	L	4624527	**s #1.0**	76.50
#w/Front Side Rear Rail Removed Deduct .5, w/Headlamp Mtg Panel Removed Deduct .5				
13 **Rail, Front Floor-See Rocker/Pillars/Floor Section**				
14 **Cap, Frt Floor Rail-See Rocker/Pillars/Floor Section**				
15 **Xmbr, Strg Gear (HSLA)**				
1994-96		4696946	**#2.0**	82.00
1997		4756926	**#2.0**	70.50
#w/Each Front Side Rail Removed Deduct 1.0				

FIGURE 7–26 Apron assembly labor times for a Chrysler LHS. Note section time. (Courtesy of Mitchell International.)

FRONT FENDER

Use Procedure Explanations 7 and 28 with the following text.

Operation	Side	Time
Refinish Fender Outside	R/L	**2.6**
Add for Edge & Inside	R/L	**1.0**
Refinish Wheelhouse Panel	R/L	**1.2**
Refinish Wheel Opening Flare	R/L	**.6**
R&I Fender Assy	R	**#1.8**
	L	**#1.3**
#Includes R&I Grille, Right Side Includes R&I/R&R Battery & Intermediate Air Duct		
R&I Wheel Opening Flare	R	**#1.9**
	L	**#1.4**
#Includes R&I Fender & Grille, Right Side Includes R&I/R&R Battery & Intermediate Air Duct		
Add to R&R Fender		
To Drill Holes for/R&R Flare	R/L	**.6**
w/Aux Battery	L	**.5**

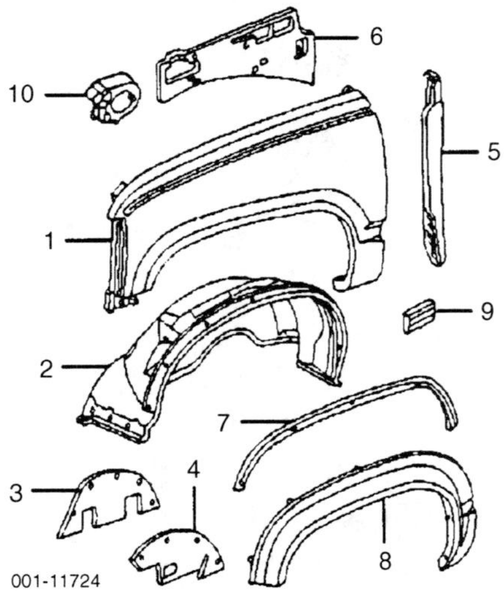

Ref	Part	Year	Side	Part No.	Time	Price
1	**Fender**					
	Right			★15998378	**#2.1**	124.00
	Left	88-95		★15998377	**#1.6**	124.00
		96-97		★15977659	**#1.6**	124.00
	#Includes R&I Grille, Right Side Includes R&I/R&R Battery & Intermediate Air Duct					
2	**Panel, Wheelhouse¶**					
	Right					
	w/15000 GVW			15995932	**#.7**	85.50
	w/o 15000 GVW			15990538	**#.7**	78.50
	#Includes R&R Battery Tray					
	Left					
	w/15000 GVW			15977655	**.4**	85.50
	w/o 15000 GVW			15977657	**.4**	76.50
	¶Order by Application					
	Shield, Splash Front					
	w/15000 GVW		R/L	15627726	**.2**	8.25
	w/o 15000 GVW		R/L	15553977	**.2**	6.95
	Shield, Splash Side					
3	2 WD					
	1500		R/L	15708731	**.2**	7.65
	2500					
	7200 GVW		R/L	15708731	**.2**	7.65
	8600 GVW		R/L	15569496	**.2**	7.65
	3500					
	w/15000 GVW		R/L	15967383	**.2**	9.00
	w/o 15000 GVW		R/L	15569496	**.2**	7.65
4	4 WD					
	1500		R/L	15569497	**.2**	8.25
	2500					
	7200 GVW		R/L	15569497	**.2**	8.25
	8600 GVW		R/L	15686091	**.2**	8.25
	3500		R/L	15686091	**.2**	8.25
	Retainer, Shield					
	7/Side		R/L	332364	**.2**	.18
5	**Seal, Fender**	88-96	R/L	15646326-5	**.2**	10.35
		97	R/L	N.A.	**.2**	N.A.
6	**Insulator**		R/L	15612898-7	**.2**	43.50
7	**Moulding, Wheel Opening**					
	Chrome		R/L	15546444-3	**.2**	16.05
	Black	88-93	R/L	15546448-7	**.2**	20.25
8	**Flare, Wheel Opening (P) ¶**					
	Right			15707094	**#2.0**	168.00
	Left			15707093	**#1.5**	168.00
	(P) Paint to Match					
	¶Order by Application					
	#Includes R&I Fender & Grille					
9	**Moulding, Fender Lower**					
	Adhesive Type					
	Chrome		R/L	15740120-19	**.2**	6.70
	Black		R/L	15607252-1	**.2**	8.25
	Black/Red		R/L	15607302-1	**.2**	6.70
	Protector, Fender Edge		R/L	15962534-3	**.1**	5.90
	Mast, Antenna			10289035	**.1**	9.30
	Cable w/Base, Antenna¶			15963441	**.7**	7.31
	¶Order by Application					
	Nut, Antenna			15591183		1.67
	Bezel, Antenna			15607056	**.2**	.21
	Relay, Fuel Pump (a)			14089936	**m .7**	8.87
	(a) Located on Right Side of Firewall Behind Battery Junction Box					
10	**Lamp, Underhood**					
	Reel type	92-97		15990293	**.2**	33.75
	Harness, Underhood Lamp					
	1992-94			12072864	**.3**	29.00
	1995			12137128	**.3**	29.75
	1996-97			N.A.	**.3**	N.A.

FIGURE 7–27 Fender and wheelhouse from a 1995 full-size Chevrolet pickup. (Courtesy of Mitchell International.)

the fender and for engine compartment components. The wheelhouse is bolted to the fender, the cowl, and the radiator support and, on some vehicles, to the frame. Some vehicles have an inner panel that separates the wheelhouse from the engine compartment (Figure 7–27).

R&R time for the wheelhouse does not include the removal of the fender. However, overlap is not figured when both the fender and the wheelhouse must be replaced, because the collision estimating guides specify that the listed time for wheelhouse R&R is "after the fender is removed."

7.7 FRAME MEMBERS

The remaining components of the front body are the frame members—the rails and cross members that support the major portion of the engine, transmission, and suspension components. In some unibody vehicles the front cross member is part of the radiator support and the side rails are part of the apron assembly. On the other vehicles, these parts, as well as other cross members, rail extensions, and reinforcements, are available separately (Figure 7–28). On a

FRONT INNER STRUCTURE

Use Procedure Explainations 8 and 28 with the following text.

FRONT

Operation		Hours
Refinish Radiator Support Complete		1.5
Refinish Upper Tie Bar		.5
Refinish Side Upper Support	R/L	.5
Refinish Side Lower Support	R/L	.5
Refinish Lower Tie Bar		.5
R&R Radiator Support Complete		#7.2

#Includes R&I Headlamp Assemlies, R&R Radiator & Shroud, Air Bag Sensors, Wiper Reservoir, Battery, Hood Latch & Disconnect/Connect Wiring Harness & Latch Cable

Add to R&R Radiator Support Complete		
W/A/C	m	.2
w/Auto Trans		.2
w/Fog Lamps		.2
To Aim Lamps		.4
Evacuate & Recharge System	m	1.4
Refrigerant Recovery	m	.3

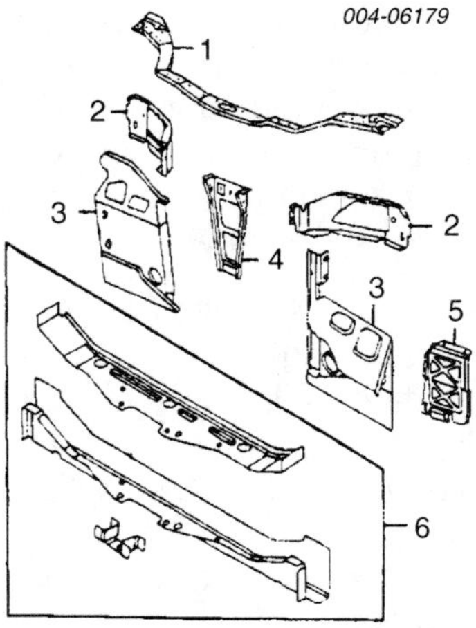

Part	Year / Side	Part No.	Labor	Price
1 **Tie Bar, Upper**		4655390	s 2.5	60.00
Shield, Tie Bar Snow		4495886	s	7.35
2 **Support, Side Upper**	R/L	4655520-1	s #1.0	23.50
#w/Upper Tie Bar Removed Deduct .3				
3 **Support, Side Lower**	R	4655214	s #2.0	20.00
	L	4655215	s #2.0	17.50
#w/Upper Tie Bar & Side Upper Support Removed				
Stud Plate, Air Bag Sensor				
Serviced w/Air Bag Sensor-See Air Bag System Section				
4 **Brace, Support**		4655474	s .5	33.75
5 **Baffle, Air**				
w/o A/C	95-96	4546761		9.85
	97	5278075		9.85
6 **Tie Bar, Lower**		4741415	s #3.0	55.00
#w/Lower Side Support Removed Deduct .3				

SIDE

Operation		Hours
Refinish Front Rail/Apron Assy	R/L	2.0
Refinish Side Rail Assy	R/L	1.0
Refinish Apron Assy	R/L	1.0
Refinish Upper Beam Panel	R/L	.5
Refinish Strut Mtg Panel	R/L	.5
Reinish Front Apron Panel	R/L	.5

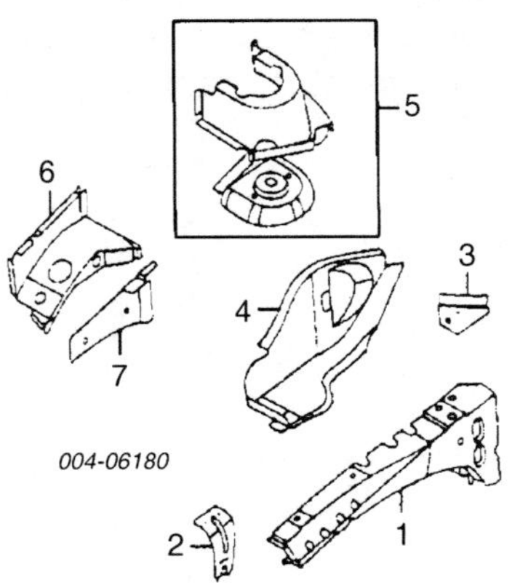

Part	Year / Side	Part No.	Labor	Price
1 **Panel, Upper Beam**	R/L	5256930-1	s #2.5	52.50
#w/Radiator Support Removed Deduct .5				
2 **Support, Panel**	R/L	5256918-9	s .5	20.50
3 **Plate, Hinge Pillar**	R	4655650		9.25
	L	4655651		16.50
4 **Shield, Splash**				
Right Only		4655092	.3	9.20
5 **Panel, Strut Mtg**	95-96 R/L	4741418-9	s #2.5	96.00
	97 R/L	4863854-5	s #2.5	97.00
#w/Upper Beam Panel Removed				
6 **Panel, Front Apron**	95-96 R	5256922	s #3.0	28.50
	L	4655099	s #3.0	28.50
	97 R/L	5256922-3	s #3.0	28.50
#w/Upper Beam Panel or Radiator Support Removed Deduct .5				
7 **Extension, Shield Panel**	R/L	4655104-5	s	10.25

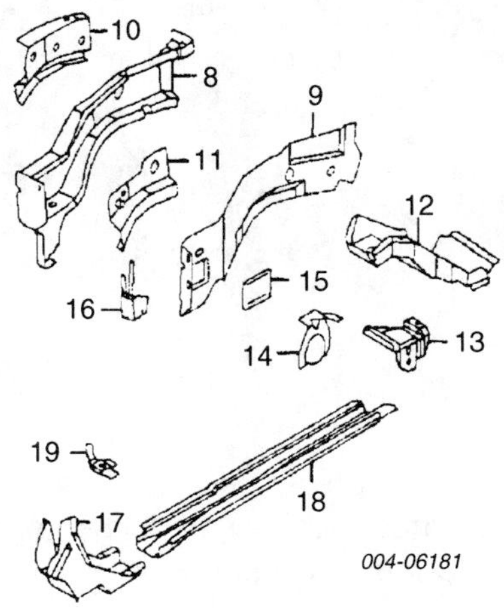

Part	Year / Side	Part No.	Labor	Price
8 **Rail Side**	95 R	4655064	s #6.5	37.25
	L	4655753	s #6.5	56.00
	96-97 R	5256926	s #6.5	39.25
	L	4783017	s #6.5	56.00
9 **Reinf, Rail**	95-96 R	5256924	s #6.5	31.00
	L	5256895	s #6.5	31.00
	97 R/L	5256924-5	s #6.5	31.00
#R&R One Side Complete, Includes R&R Eng Mtg Reinf or Front Inner Reinf, Left Rail also Includes Upr & Lwr Eng Mtg Brackets, w/Front Apron Panel, Strut Mtg Panel & Radiator Support Removed				
10 **Reinf, Eng Mtg**	R	4655514	s	30.50
11 **Reinf, Frt Inner**	L	4655565	s	30.50
12 **Brace, Rail Rear**	R	4655070	s	49.75
	L	4655071	s	39.75
13 **Bracket, Eng Mtg Upr**				
Left Only		4655515	s	26.00
14 **Bracket, Eng Mtg Lwr**				
Left Only		4655083	s	33.00
15 **Cover, Rail Access**		4655095		1.75
16 **Bracket, Susp Mtg**	R/L	4655072-3	s	11.00
17 **Rail, Front Side**	R/L	4655066-7	s	69.50
18 **Ext, Side Rail**	95-96 R/L	4783452-3	s	33.00
	97 R/L	4783452-3	s	33.00
19 **Retainer,Susp**	R/L	4655748-9		9.75

SUB-FRAME

Note: R&R Does not include align suspension.

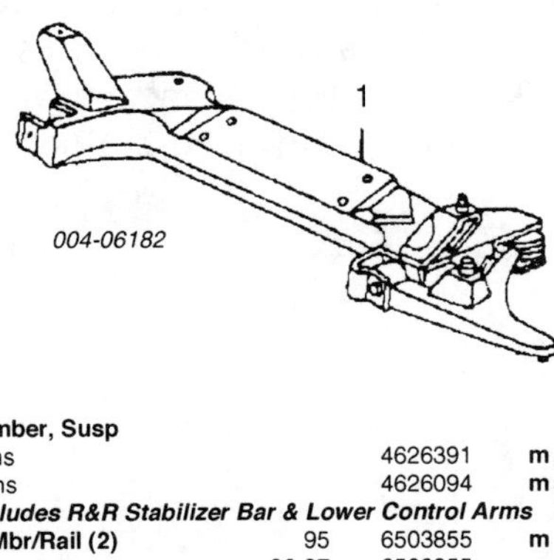

Part	Year / Side	Part No.	Labor	Price
1 **Crossmember, Susp**				
Man Trans		4626391	m #3.2	395.00
Auto Trans		4626094	m #3.2	420.00
#Includes R&R Stabilizer Bar & Lower Control Arms				
Screw, XMbr/Rail (2)	95	6503855	m	2.35
	96-97	6503855	m	2.35
Screw, Damper/XMbr (2)		6101760	m	1.45

FIGURE 7–28 Neon unibody frame members are available separately. (Courtesy of Mitchell International.)

FIGURE 7–29 When repairing a unibody frame, the damage report must include time for setup, measuring, and pulling. (Courtesy of DatalinerNicator.)

body-over-frame–type vehicles, some frame members are available separately or in subassemblies.

If a frame member on a unibody vehicle is damaged, it can often be straightened (Figure 7–29). When a repair is possible, the damage report must allow for setting the vehicle up on a bench, time for measuring the damage, and time for pulling the frame into alignment. It is also possible to section the rails and/or cross members; however, caution must be observed where the section cuts are made in the welding methods used. (Sectioning techniques are discussed in Chapter 4.)

When the frame members must be replaced, the damage report must allow for the frame to be rough pulled into position before the damaged parts are removed. The R&R times listed in the collision estimating guides allow for drilling out factory welds, removing the damaged panel, aligning the new panel with the frame, and rewelding the panel on the vehicle. These times also allow for the removal of all parts attached to the frame member. Always consult the pages and footnotes in the crash manual text to determine what operations are included and which ones must be listed separately on the damage report. The footnotes also provide other valuable information, such as manufacturers' suggested procedures, sectioning restrictions, and other repair operations (Figure 7–30). The estimator and adjuster must always consult the footnotes when listing frame R&R.

In addition to all necessary welding, the R&R times listed in the collision estimating guides include time for removing any electrical wiring that runs along the inside of the members and time to R&I floor mats, insulation, and interior trim that might be damaged by the heat of the welding process. The frame operation times do not include the following:

- Setup of the vehicle on the frame machine
- Diagnosis of damage
- Pulling out damage in adjacent panels
- R&I of all parts bolted to the frame members
- R&I of all body panels bolted to the frame members
- Removal of any adjacent welded panels
- Wheel alignment
- Applying undercoating, sound-deadening material, and/or anticorrosion material
- Refinishing the panels (if required)

A damaged full frame that cannot be repaired to preaccident condition should be replaced. OEM (Figure 7–31) or LKQ (Figure 7–32) frames can be utilized. In some cases, if allowed by the OEM, the frame can be sectioned, front or rear. Damaged cross members can be replaced. Welded cross members should be replaced with welding. Riveted or bolted members can be replaced with grade 8 bolts and nuts. Be sure to include the price of new fasteners in the estimate. Figure 7–33 shows what is included and not included in frame replacement.

Convolutions are found on some truck frames. Damaged convolutions cannot be repaired. The frame or frame section must be replaced in these cases (Figure 7–34).

SIDE

Refinish Upper Side Rail .... R/L .5
Refinish Apron Assy .... R/L 1.0
Refinish Front Apron Panel .... R/L .5
Refinish Side Lower Rail .... R/L 1.0
Section Frame Rail
1987–89 .... R/L #6.5
#Section 10 3/4' from Front Edge of Oval Hole on Engine Side of Frame Rail as Indicated by Cut Line in Illustration, w/Support Assy Removed, w/Front Apron Assy Removed Deduct 1.0
1990–91 .... R/L #7.0
#Section at Factory Seam located on Lower Firewall w/Support Assy Removed, w/Front Apron Assy Removed Deduct 1.0

Note: For sectioning methods, refer to TECH-COR Information Bulletin #01-88 for 1985-89 Models and #94-2 for 1990-91 Models. See disclaimer on page P3 concerning labor times.

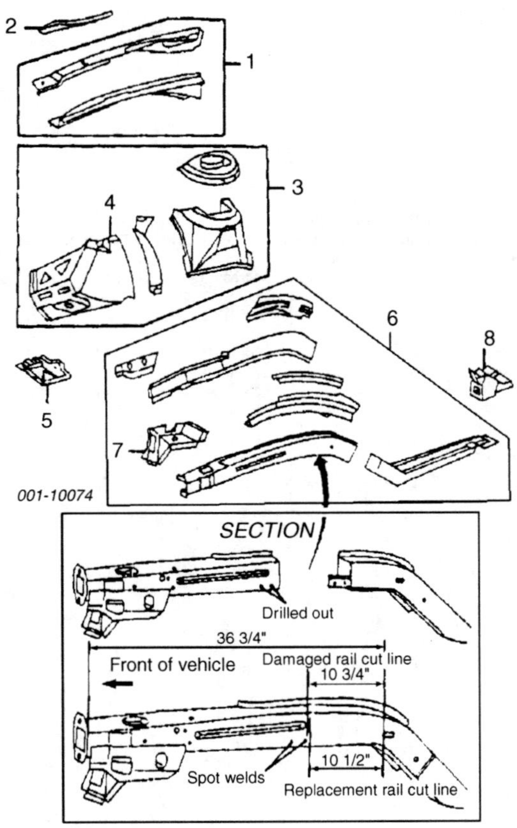

Part		Part No.			Price
1 **Rail, Side Upper**					
Right	87-89	20744316	s	**#3.5**	111.00
	90-91	25631317	s	**#3.5**	105.00
Left	87-90	20653541	s	**#3.5**	122.00
	90-91	25631318	s	**#3.5**	105.00
#Includes R&R Side Upper Reinforcement					
2 **Reinf, Side Upper Rail**	R	25601430	s		17.50
	L	25655082	s		14.90
3 **Apron Assy**	R	25655069	s	**#8.2**	236.00
	L	25647699	s	**#8.2**	236.00
#Includes R&R Upper Side Rail & R&I Wire Hareness, w/Bolted Parts & Assemblies Removed, w/Support Assy/Front Panel Removed Deduct 2.0, Right Side Add .5 to R&R Battery Tray					
4 **Panel, Apron Front**	R	25655071	s	**#6.2**	59.50
	L	25660103	s	**#6.2**	59.50
¶Included w/Apron Assy					
#w/Front Panel Removed Deduct 2.0, Right Side Add .5 to R&R Battery Tray, Includes R&I Wire Harness					
5 **Tray, Battery**	87-88	20566687			18.30
	89-91	25624017			18.30
Retainer, Battery		356668		**.2**	5.40
6 **Rail, side Lower**	87-89 R	20628938	s	**#10.0**	217.00
	L	20731219	s	**#10.0**	217.00
	90-91 R	25606477	s	**#10.0**	217.00
	L	25638595	s	**#10.0**	211.00
#Includes R&R Rail Inner Reinf, w/Support Assy Removed, w/Apron Assy Removed Deduct 3.5, Section & Cut line in Illustration Indicate Sectioning, Refer to Headnote					
7 **Reinf, Rail Front**	R	20416932	s		20.80
	L	25620949	s		20.60
¶Included w/Side Lower Rail					
Reinf, Rail Inner	87-89 R/L	20532301	s	**#**	17.50
	90-91 R/L	25617763	s	**#**	12.60
8 **Brace, Side Lower**	R	25541057	s	**#**	54.50
	L	20579965	s	**#**	57.00
#Included in R&R Side Lower Rail					
Servo, Cruise Control (a)		25074625	m	.5	216.00
(a) Located on Left Apron					
Solenoid					
Canister Purge¶ (b)		25526955	m		18.87
EGR		25532523	m		44.88
¶Order by Application					
(b) Located on Right Apron					
Relay, Fuel Pump	87	10038490	m	.5	10.46
	89-91	25530733	m	.5	6.94

FIGURE 7–30 The estimating guide contains lower side rail sectioning information for a 1991 Pontiac Bonneville. (Courtesy of Mitchell International.)

FRAME

Use Pocedure Explanation 9 with the folowing text.

Add to R&R Frame
- w/Auto Trans .... **.6**
- w/Air Cond .... **.5**
- w/4 WD **2.0**

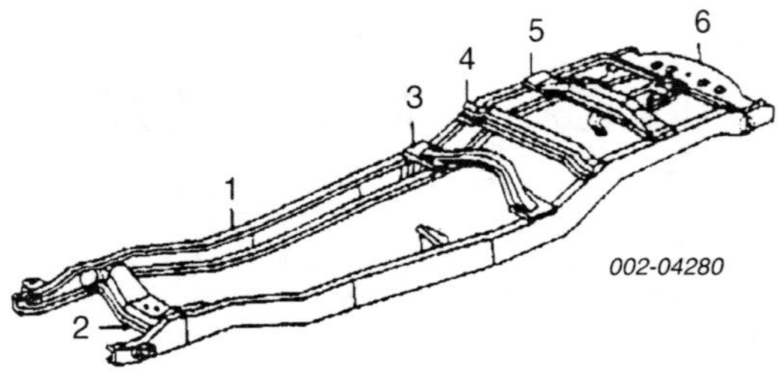

Item	Year	Side	Part No.	Time	Price
1 **Fram Assy**					
102" WB (2 Door)	91-92		F3TZ 5005 A	**25.6**	1204.12
	93-94		F3TZ 5005 A	**25.6**	1204.12
112" WB (4 Door)	91-92		F3TZ 5005 B	**25.6**	1278.62
	93-94		F3TZ 5005 B	**25.6**	1278.62
Ext Kit, Frame Horn					
1992		R	F2TZ 5B117 A	**#5.0**	459.63
		L	F2TZ 5B117 B	**#6.5**	459.63
1993-94		R	F3TZ 5B117 B	**#5.0**	459.63
		L	F3TZ 5B117 A	**#6.5**	459.63
#Includes D&C stabilizer Bar, Left Side Also Includes R&I Steering Gear					
2 **Crossmember, Font¶-Not Serviced**					
3 **Crossmember #2¶**			F1TZ 5030 A		d393.27
4 **Crossmember #3¶**					
102" WB			F1TZ 5A019 B		d185.20
112" WB			F1TZ 5A019 A		395.62
5 **Crossmember #4¶**			F1TZ 5037 A		389.25
6 **Crossmember, Rear¶**			F57Z 5K007 AA	**2.5**	94.57

¶Included w/Frame Assy

FIGURE 7–31 The frame for a body-over frame vehicle is usually available as a complete assembly. (Courtesy of Mitchell International.)

FRAME ASSEMBLY

Use Pocedure Explantion 6 with the following text.

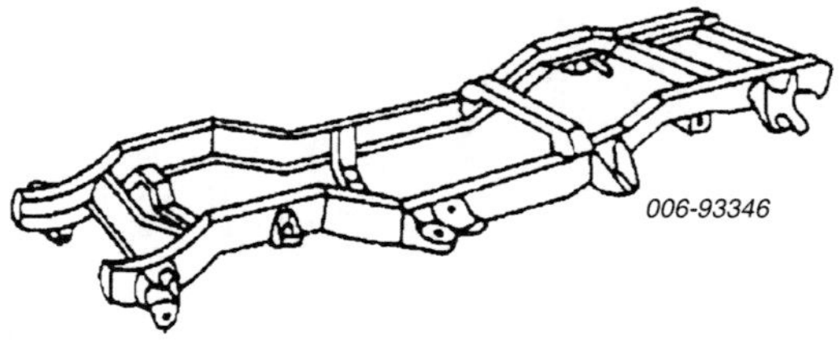

Replace .... f **28.0**
- Add to Replace
  - w/A/C **2.1**
- Add to
  - Clean, Visually Inspect & Paint Frame **2.5**

Adjust Font Alignment .... m **1.0**

FIGURE 7–32 Used or LKQ frame replacement time from the replacement assemblies estimating guide.

FRAME

NOTE: It may be necessary to transfer parts from the damaged frame to the used frame section. This cannot be determined until you see the used frame section. The additional time needed for this operation must be estimated.

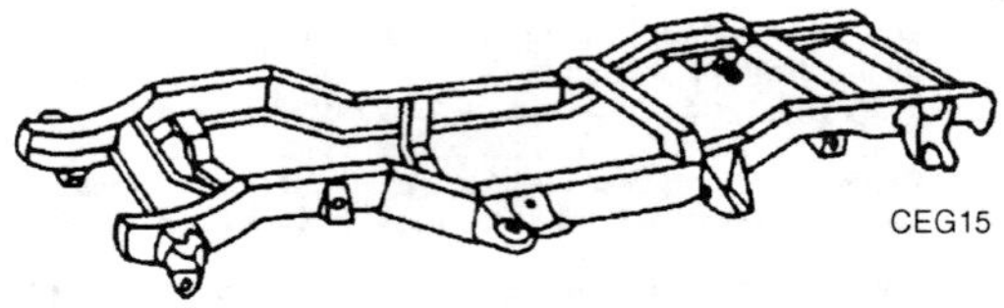

FRAME ASSEMBLY R&R

INCLUDED OPERATIONS
- Remove and install:
  - Front and rear bumpers
  - Front sheet metal as an assembly
  - Body assembly
  - Pickup bed
  - Mechanical assemblies (example: engine, transmission, suspension, fuel tank, differential and components)
- Remove and install or replace:
  - Wiring and/or wiring harness if necessary

NOT INCLUDED OPERATIONS
- Overhaul assemblies or replace parts of assemblies that are transferred
- Remove and install or replace:
  - Welded or riveted assemblies of the frame (see Note above)
  - Optional accessories (example: auxiliary lamps and fuel tanks, brush guard, triler hich, trailer connector)
- Wheel alignment
- Adjust control linkage
- Bleed brakes if necessary
- Evacuate and recharge Air Conditioning if so equipped and necessary
- Refigerant recovery if necessary

FIGURE 7–33 Included and not included operations from P pages of replacement assemblies guide.

FRAME

Use Procedure Explanation 9 with the following text.

Add to R&R Frame		
w/A/C	m	.5
w/Auto Trans	m	.6
w/Pwr Strg (2 WD Only)	m	.8

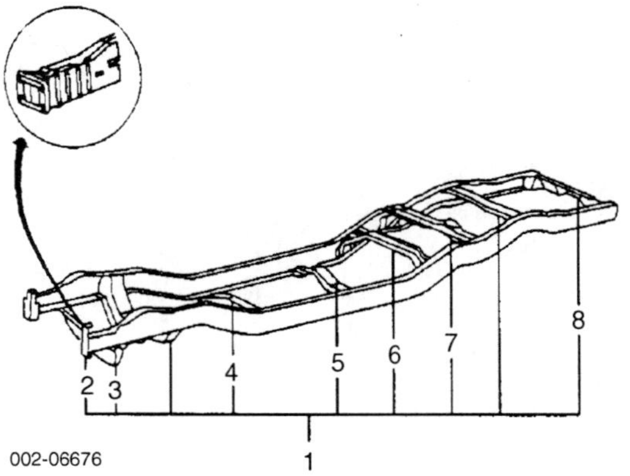

1 **Frame Assy**			
2WD			
108" WB	F67Z 5005 RA	**22.0**	747.92
114" WB	F57Z 5005 B	**22.0**	1326.68
125" WB	F57Z 5005 E	**22.0**	1326.68
4 WD			
108" WB	F77Z 5005 RC	**24.8**	808.25
114" WB	F77Z 5005 RD	**24.8**	671.83
125" WB	F57Z 5005 F	**24.8**	1326.68
2 **Ext Kit, Frame Horn¶**			
Regular Cab			
Right	F37Z 5B117 B	**#5.0**	599.90
#Includes D&C Stabilizer Bar			
Left	F37Z 5B117 A	**#6.0**	678.15
#Includes D&C Stabilizer Bar & R & I Steering Gear, w/Pwr Stg Add .5			
Super Cab			
Right	F37Z 5B117 D	**#5.0**	678.15
#Includes D&C Stabilizer Bar			
Left	F37Z 5B117 C	**#6.0**	678.15
#Includes D&C Stabilizer Bar & R&I Steering Gear, w/Pwr Strg Add .5			
3 **Crossmember at Bumper¶**	F2TZ 5C128 C	**#.5**	76.58
#Time is for Bolted Part, If Part is Welded, Add Time as Necessary			
4 **Crossmember, Trans Mount¶**			
4 Cyl	E9TZ 6A023 E	**#.4**	154.22
V6			
3.0L Eng			
2 WD	E9TZ 6A023 E	**#.4**	154.22
4 WD	F0TZ 6A023 A	**#.4**	66.57
4.0L Eng	F0TZ 6A023 A	**#.4**	66.57
#Included in R&R Trans Mount			
5 **Crossmember #3¶**			
108"/114" WB	F2TZ 5025 C	**2.0**	124.82
125" WB			
2 WD Only	F2TZ 5A051 A	**2.0**	125.97
6 **Crossmember #4¶**			
108" WB	F2TZ 5026 B	**#2.0**	130.62
114"/125" WB	F2TZ 5026 D	**#2.0**	130.62
7 **Crossmember #5¶**	F0TZ 5030 B	**#2.0**	129.35
8 **Crossmember, Rear¶**	F0TZ 5K007 A	**#2.0**	96.55
¶Included w/Frame Assy			
#w/Pickup Bed Removed			

FIGURE 7–34 Replacement times on a Ford pickup truck frame.

REVIEW QUESTIONS

1. Technician A says that leaking impact absorbers should be replaced. Technician B says that in some cases, a bent impact absorber can be repaired. Who is right?
   a. technician A
   b. technician B
   c. both A and B
   d. neither A nor B
2. Bumpers on pickup trucks can withstand a 5 mph collision with no damage.
   True False
3. Technician A says that composite headlights should be checked for operation after a front collision. Technician B says that composite headlights have a replaceable bulb. Who is right?
   a. technician A
   b. technician B
   c. both A and B
   d. neither A nor B
4. Which of the following is not included in the overhaul time for most bumper assemblies?
   a. removal of face bar
   b. replacement of torn fascia
   c. replacement of impact absorbers
   d. installation of replacement face bar
5. Technician A X-checks the hood opening before installing a hood and fenders. Technician B says that an X-check is not necessary in most cases. Who is right?
   a. technician A
   b. technician B
   c. both A and B
   d. neither A nor B
6. Which of the following is included in the R&R time for a full frame?
   a. front bumper
   b. wheel alignment
   c. trailer hitch
   d. bleed brakes
7. Technician A says that the hood R&R time includes hood hinge replacement. Technician B says that hood R&R time does not include refinishing the hood. Who is right?
   a. technician A
   b. technician B
   c. both A and B
   d. neither A nor B
8. Fender R&R time includes the time to replace the fender liner.
   True False
9. Technician A says that the time to pull damaged frame rails is not included in R&R on an apron. Technician B says that R&R time for a frame rail includes the time to remove the wiring that may be damaged by welding. Who is right?
   a. technician A
   b. technician B
   c. both A and B
   d. neither A nor B
10. The upper tie bar, lower tie bar and side baffles make up the ________________ .

Chapter 8

Assessing Damage to the Body Shell

Objectives

After reading this chapter, you should be able to:

- **Identify the panels that make up the cowl and dash assembly and windshield frame.**
- **Name the methods and materials involved in replacing a windshield and other body glass.**
- **Identify the components that make up the body side panels.**
- **List the parts that make up the vehicle roof.**
- **Recognize the various body parts that make up the rear body.**
- **Explain when sectioning is a valid repair option on body shell panels.**
- **Name the parts of a rear bumper and know how to determine operation times for bumper R&R, R&I, and overhaul.**
- **Describe which operations are included and not included in the operation times listed in the collision estimating guides for replacement of body shell parts.**

Key Terms

Rocker panel
Center pillar
Quarter panel
Body section
Rear clip
Supplemental restraint system

Chapter 7 discussed how to analyze and appraise the cost of repairing damage to front-end body parts on both body-over-frame and unibody vehicles. This chapter will cover the rest of the vehicle body—the portion usually referred to as the body shell. The body shell is made up of all-sheet-metal parts that are welded together to form a complete unit (Figure 8–1). On a unibody vehicle, the body shell actually contains all welded body parts, including the front end sheet metal and frame members; only the fenders, header panels, bumpers, and other bolt-on parts are not included in the body shell. However, on a conventional body-over-frame vehicle, the body shell begins with the cowl and dash panels and extends rearward to the rear body panel; it does not include any of the frame members or the front-end sheet metal. In either case, this chapter will focus on appraising damage to body parts not dealt with in Chapter 7, starting with the cowl and dash panel and working back to the rear body panel.

8.1 COWL AND DASH ASSEMBLY

The cowl and dash assembly are similar for most vehicles. As shown in Figure 8–2, a typical cowl consists of a top panel, a dash panel, and side panels. The top panel is usually a finished panel on modern vehicles. On older models, the top cowl panel might be vented. The lower cowl panel, usually called the dash panel or firewall, is welded to the floor pan and on some body-over-frame vehicles is a part of the floor pan. Often

TYPICAL BODY PANELS ON UNITIZED BODY

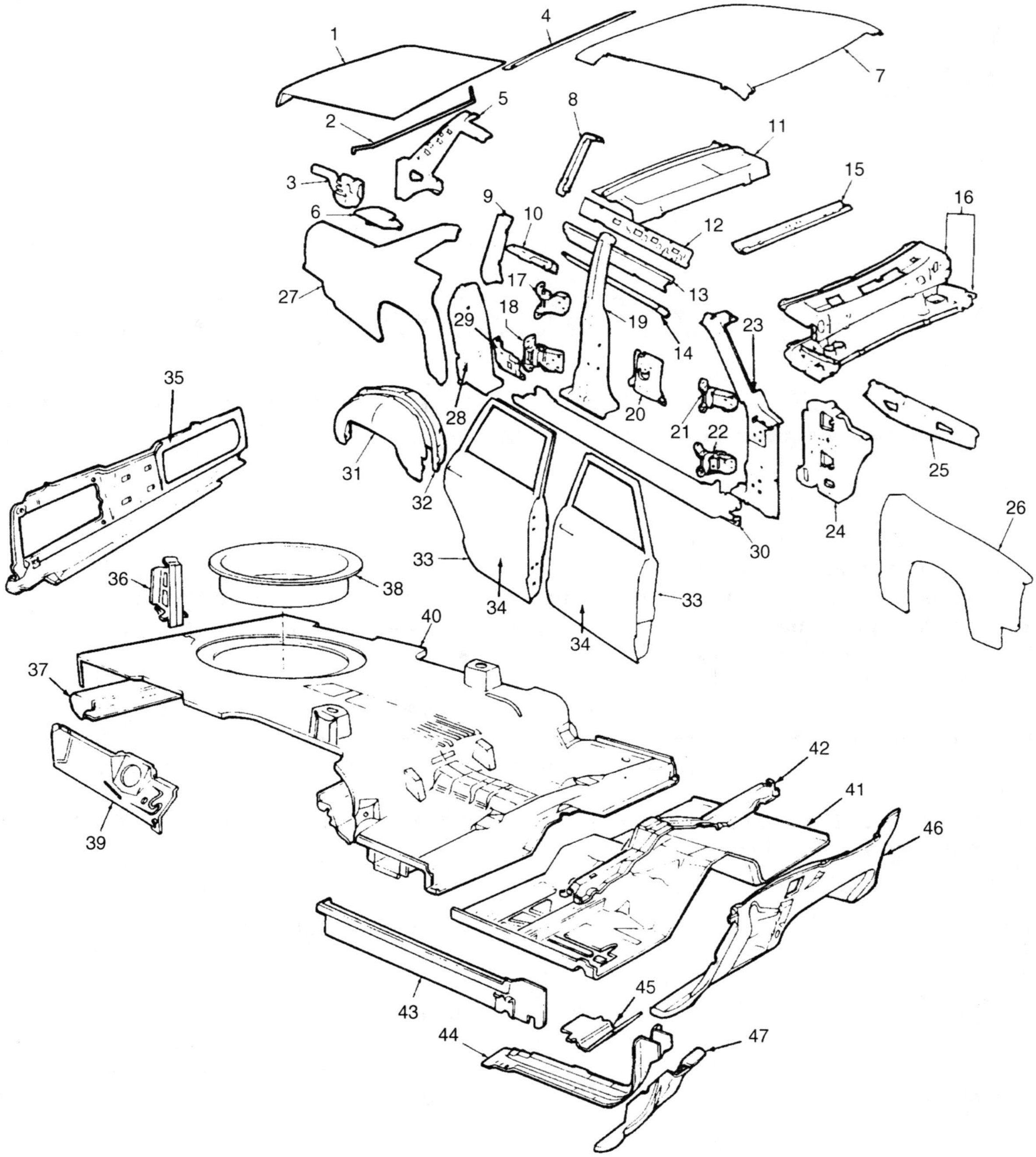

1 Trunk lid
2 Torque rod
3 Hinge
4 Rail above rear window
5 Rear roof rail
6 Quarter window reinforcement
7 Roof panel
8 Roof rail extension
9 Rear quarter upper lock pillar
10 Quarter panel reinforcement at belt line
11 Upper rear body panel
12 Inner roof side panel
13 Outer roof side rail
14 Center roof side rail
15 Windshield header panel
16 Cowl top panel
17 Rear door upper hinge
18 Rear door lower hinge
19 Center body pillar
20 Center body pillar extension
21 Front door upper hinge
22 Front door lower hinge
23 Hinge and windshield pillar
24 Cowl inner side panel
25 Front fender reinforcement
26 Front fender
27 Quarter panel
28 Rear quarter panel lower lock pillar
29 Lock pillar reinforcement
30 Rocker panel
31 Outer wheelhouse
32 Inner wheelhouse
33 Door assembly
34 Door outer panel
35 Rear body panel
36 Rear body panel support
37 Rear floor cross member
38 Spare tire well
39 Extension from quarter panel to floor
40 Rear floor panel
41 Front floor panel
42 Front floor panel support
43 Inner rocker panel
44 Front side member
45 Front side member reinforcement
46 Dash panel
47 Gusset to dash panel

FIGURE 8–1 *An exploded view of a body shell. (Courtesy of Motor Information Systems.)*

COWL & DASH

Use Procedure Explanations 14 and 28 with the following text.

Refinish Upper Dash Panel **1.0**

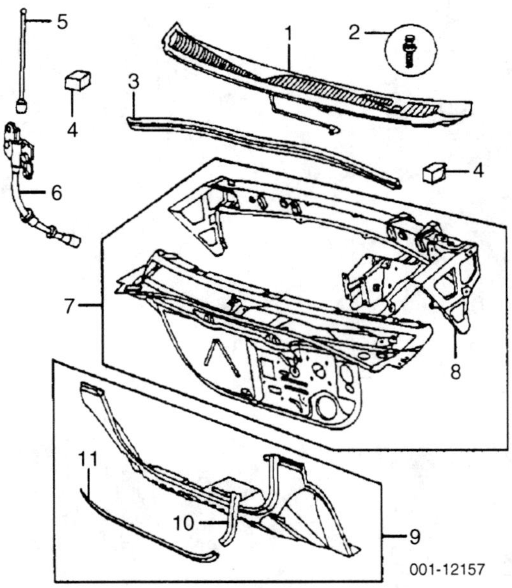

Item		Part No.	Time	Price
1 **Grille, Cowl Vent**		21099025	**#.5**	55.63
#Includes R&R Wiper Arms				
2 **Retainer, Grille (2/Side)**	R/L	21077123		.18
3 **Seal, Hood Rear**		21098515	**.1**	17.15
4 **Filler, Hood Cowl Seal**	R/L	21096137		1.51
5 **Mast, Antenna**		21022765	**.1**	17.43
6 **Base, Antenna**		21023273	**#.7**	22.34
#Includes R&R Mast				
7 **Panel Assy, Upper Dash**				
1996				
To VIN TZ358844		21123480	**#17.5**	382.75
From VIN TZ358845		21123480	**#17.5**	382.75
1997		21123266	**#17.5**	369.80
#R&R Lower Dash Panel Assy Deduct 2.0				
8 **Beam, Reinf¶**	96	21122955	**3.0**	121.68
	97	21123267	**3.0**	117.57
¶Included w/Upper Dash Panel Assy				
9 **Panel Assy, Lower Dash**		21126098	**7.5**	105.51
Seal, Dash Panel¶				
10 Outer	R/L	21080920-1	**#.5**	4.61
11 Center		21080629	**#.5**	5.52
¶Included w/Lower Dash Panel Assy				
#R&R Complete				

FIGURE 8–2 An exploded view and collision estimating guide text for a cowl. (Courtesy of Mitchell International.)

the side panels are incorporated with the hinge and windshield pillars. Other parts that will be encountered as part of a cowl and dash assembly include various supports, extensions, braces, brackets, and reinforcements. Be sure to consult the exploded views in the collision estimating guides so that all attached parts are included on the damage report.

The cowl and dash assembly is usually welded to the front floor pan, the left- and right-side rocker panels, and the hinge pillars. On a unitized vehicle, the aprons and front rails are welded to the side cowl

Item	Price	Time	Part No.
Moulding	4.48	**0.4**	4335138
Wiper arm	17.25	**0.2**	4360021
blade	5.85		3799828
motor	109.88	**0.8**	4339450
washer reservoir			
w/sensor	16.22	**0.3**	4360003
w/o sensor	16.22	**0.3**	4360002
pump	166.93		4334469
nozzle	3.24		4360016

 # After fenders and glass are removed. Time is for cutting in windshield opening.
§ After fender and door are removed.
† After glass and outer panel is removed.

FIGURE 8–3 These operations must be listed separately on the damage report. They are not included in the cowl or dash panel R&R times. (Courtesy of Motor Information Systems.)

panels, as well; therefore, the side cowl panels are the strongest part of the cowl and dash assembly.

It is very important that the estimator and adjuster consult the footnotes to determine what operations are included in the listed labor times and which operations must be performed prior to the operation in question. For example, when replacing the top cowl panel on a 1996 Dodge Dakota truck, the fenders and windshield must be removed prior to removing the cowl (Figure 8–3). Time for these operations must be entered separately on the damage report.

Although the cowl and dash panels can be removed and replaced at factory seams, the more cost-effective method is often sectioning the damaged area and replacing the damaged panels with LKQ parts.

The cowl and dash assembly R&R times given in the collision estimating guides include the following operations:

- R&I of the cowl side trim so that the trim is not damaged by the welding operation
- R&I of the door weatherstripping and sill plates
- R&I of the jamb switch
- R&I of any insulation material between the dash panel and the passenger compartment
- Drilling or cutting out factory welds
- Aligning the new panel and welding the new panel in place
- Grinding, filling, and smoothing welded seams

Operations that are not usually included are as follows:

- R&I of the instrument panel and crash pad (often necessary to avoid heat damage)
- R&I of the windshield reveal and garnish mouldings
- R&I of the windshield
- R&I of the fenders

- R&I of the doors
- Loosening the front edge of the headliner
- R&I of air conditioner and heater parts
- R&I of the roof rail moulding
- Refinishing any of the panels
- Applying anticorrosion material to inner panel surfaces

8.2 WINDSHIELD

Windshields have been going through a time of change, just like the rest of the modern vehicle body. At one time, most windshields were flat and were held in place with a rubber gasket (Figure 8–4A). A locking strip in the gasket secured the glass in place. Many windshields today, however, are curved to match the aerodynamic shape of the vehicle and are held in place with adhesives (Figure 8–4B). Some, in fact, are an integral part of the unibody structure and play an important part in the rigidity of the body shell.

Labor times and parts prices for windshield R&R are included with the cowl and dash assembly text in the Motor Crash Estimating Guides. In the Mitchell Collision Estimating Guides, the windshield is listed separately. In either case, the guides provide R&R times that can also be used for R&I times. The collision estimating guides also contain times for the body parts that make up the windshield frame, such as windshield pillars and a header panel. An example is shown in Figure 8–5.

As discussed in Chapter 6, the collision estimating guides list both the manufacturer's suggested list price and the National Auto Glass Specifications (NAGS) suggested list price for vehicle glass. The estimator and adjuster should include the lower of the two list prices on the damage report, minus a 20 to 25 percent discount.

It is important that the correct type of glass is listed on the damage report. The collision estimating guides contain information on clear, tinted, shaded, and heated glass and glass with and without an antenna printed or embedded in it (Figure 8–6). The type of glass can be identified by looking at the decal on the glass.

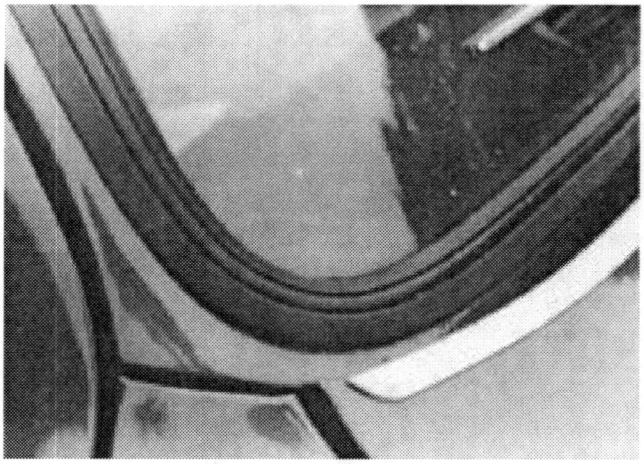

A

B

FIGURE 8–4 Some windshields are installed with (A) gaskets or (B) adhesive.

The estimator and adjuster must be aware that on many new unibody vehicles, the windshield is considered part of the structural unit. It stiffens the body shell. These windshields are installed with a urethane adhesive. When replacing a windshield with urethane adhesive, all the old adhesive must be removed and

WINDSHIELD

Use Procedure Explanation 13 with the following text.

R&I Windshield Glass .... **#2.4**

#Includes Clean Up Old Adhesive as Necessary

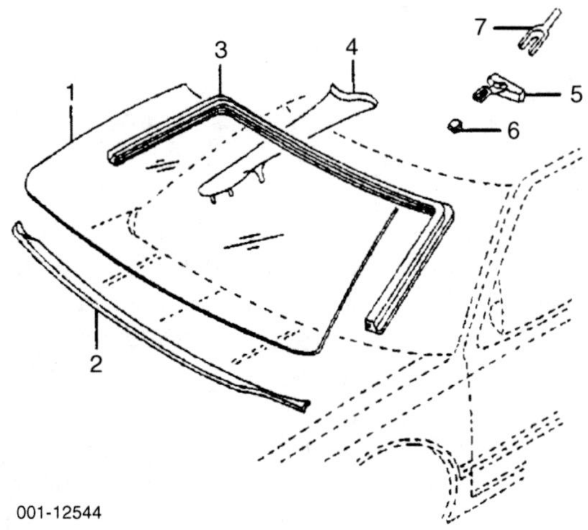

Item		Part No.	Time	Price
1 **Windshield Assy**				
Pontiac				
w/Head-Up Display		10293669	**#2.0**	725.00
w/o Head-Up Display		10293667	**#2.0**	744.00
NAGS		N.A.	**#2.0**	N.A.
#Includes Clean Up Old Adhesive as Necessary				
2 **Strip, Windshield Sealing**		10241733		14.05
3 **Moulding, Reveal**		10293831	**#2.0**	12.50
#Includes R&R Windshield & Clean Up Old Adhesive as Necessary				
4 **Mldg, W/S Garnish¶**	R/L	10294523-4	**.3**	10.80
¶Graphite listed, Order by Application				
5 **Mirror, Rear View**				
Standard		10240815	**.2**	46.25
Auto Day/Night		10277770	**.2**	135.00
6 **Supt, Rear View Mirror¶**		20626174	**#.2**	7.70
¶Included w/Windshield Assy				
#w/Mirror Removed				
7 **Cover, Mirror Wiring**				
w/Auto Day/Night		10277693	**.1**	3.10

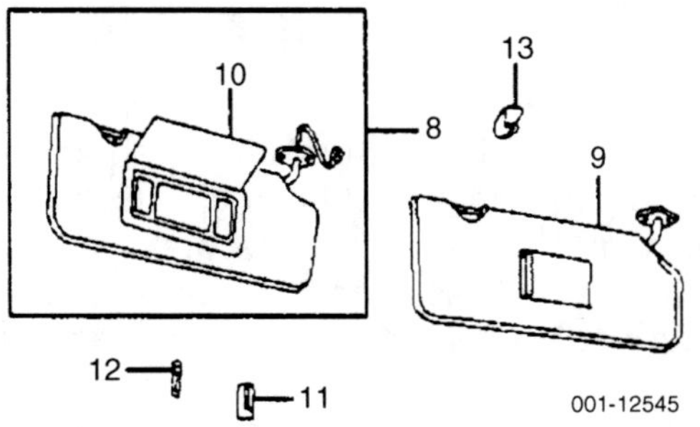

Item		Part No.	Time	Price
Visor, Sun¶				
8 w/Illum Mirror	R/L	10293569-70	**.2**	53.50
9 w/o Illum Mirror	R/L	10248668-9	**.2**	33.50
10 **Mirror, Illum-Serviced w/Sun Visor**				
11 **Lens, Mirror (2)**	R/L	12330075	**#.1**	4.55
12 **Bulb, Mirror (2)**	R/L	9442395	**#.1**	■1.50
#R&R Complete				
13 **Retainer, Sun Visor¶**	R/L	10272016		1.45
¶Graphite listed, Order by Application				

FIGURE 8–5 Windshield replacement times. (Courtesy of Mitchell International.)

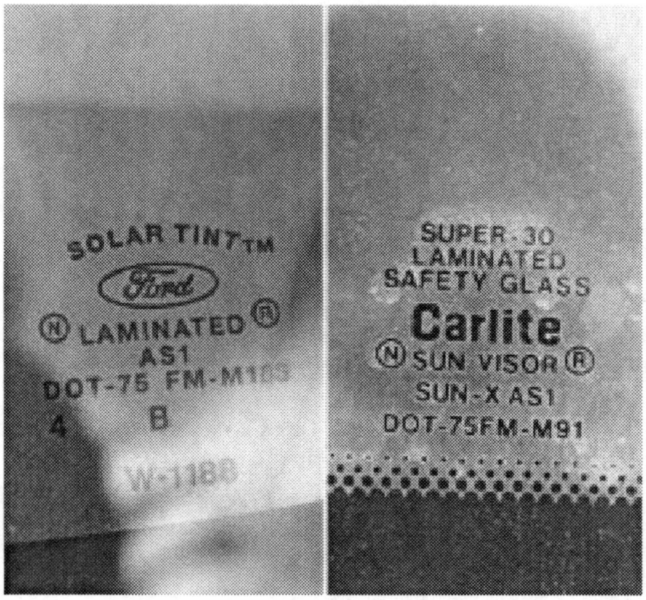

FIGURE 8–6 Check the windshield decal to determine the type of glass.

fresh urethane applied to the pinch weld (Figure 8–7). The labor time for replacing a windshield with urethane adhesive is normally quite a bit more than the time allowed for replacing a conventional windshield using the short method.

The damage report must also include the cost of the adhesive and other installation materials. If the windshield is urethane mounted, the collision estimating guides might give the suggested list price of a urethane kit. This price must be included on the damage report.

The windshield R&R times given in the collision estimating guides include removal of the reveal mouldings (if present) and garnish mouldings from the windshield. They also included removal of the adhesive from the windshield glass. Removal of the adhesive from the windshield frame pinch welds is not included, nor is removal of broken glass. If the rearview mirror is mounted on the glass, removal of the mount is included in the labor allowance. If the windshield wipers interfere with the removal of the windshield, R&I of the wiper arms is also included in the windshield R&R.

The collision estimating guides group various other components with the cowl and windshield subassembly. Usually the windshield wipers and washer components are listed (Figure 8–8), as well as garnish mouldings, visors, and rearview mirrors.

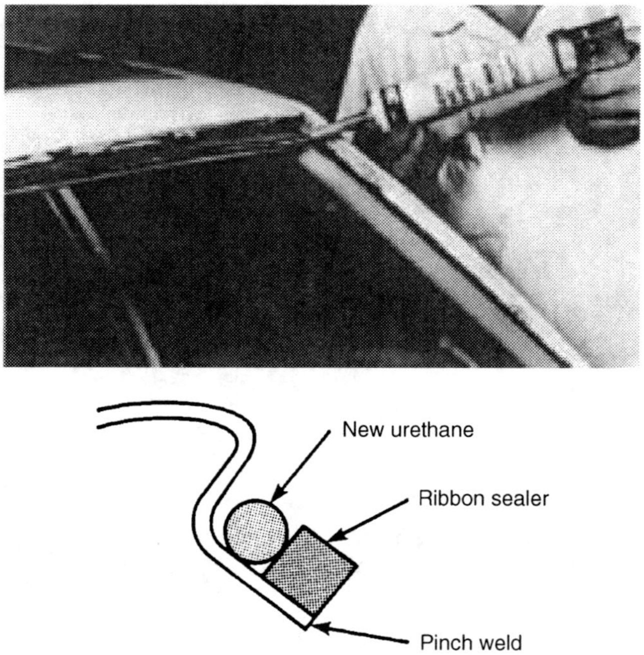

FIGURE 8–7 Applying urethane adhesive to the windshield frame pinch weld. (Courtesy of I-CAR.)

Item		Price	Time	Part No.
Windshield glass				
GM		715.00	**4.5**	1644586
NAGS			**4.5**	W1030
sealing strip		42.50		1635173
Reveal mouldings				
upper		20.75	**0.4**	1635180
lower		75.50	**0.4**	1635179
side	RT	32.50	**0.4**	1635171
	LT	32.50	**0.4**	1635172
Interior mouldings				
upper		30.75†	**0.3**	1643089
side **(p.3)**	RT	180.00†	**0.3**	1641032
(p.3)	LT	180.00†	**0.3**	1639884
Wiper arm	RT	32.75	**0.3**	22063274
	LT	32.75	**0.3**	22063275
blade		2.70	**0.2**	22063276
motor		117.85	**1.0**	22054198
Washer reservoir		30.25	**0.5**	22063266
pump		9.65	**0.4**	22049373
Mirror			**0.3**	NO LISTING
Visor	RT	129.00†	**0.5**	1642349
	LT	†	**0.5**	1642350

 # After fender, door, windshield, and trim is removed.
§ After outer pillar is removed.
† Order by color.
‡ Time is for inner and outer after roof panel, windshield and necessary trim is removed.

FIGURE 8–8 This collision estimating guide entry includes moulding and wiper information in the cowl and windshield assembly. (Courtesy of Motor Information Systems.)

8.3 BODY SIDE PANELS

The side panels of a body shell form the door openings and provide structural support for the passenger compartment. The body side panels consist of the hinge and windshield pillars, rocker panels, roof rails, and quarter panels. If the vehicle is a four-door model, a center pillar will also be part of the side panels. On some of the newer unibody vehicles, the side panels (not including the quarter panel) are available as a subassembly (Figure 8–9). This reduces the amount of time needed to separate individual parts at factory seams and to weld the part together and to the vehicle.

However, the individual parts of the side body are usually listed separately in the collision estimating guide or as part of other subassemblies. For example, this chapter has already mentioned that hinge and windshield pillars are usually listed with the cowl and dash assembly, since the pillars are often part of the cowl side panel. Rocker panels and center pillars are also often listed together. These parts are often available as combined repair panels. Always consult the footnotes for information relating to repair panels and/or ask the parts dealer if combined repair panels are available.

The following paragraphs will discuss the repair of each body side panel individually.

HINGE AND WINDSHIELD PILLARS

The hinge and windshield pillars (also called A-pillars) often consist of inner and outer panels that are welded together to form a strong, closed structure. These panels can be replaced separately, or they can be sectioned and replaced with partial LKQ panels. For example, if a vehicle rolls in a collision, the windshield pillars might be damaged and require replacement. Rather than replace the entire hinge pillar and windshield pillar assembly, the windshield pillar can usually be sectioned and an LKQ part welded in place.

Before making the decision to section, make sure that the section will not compromise the structural integrity of the vehicle. A section is only allowable when both sides of a closed member are accessible. A section cut cannot be made through reinforcements, either. Windshield pillars are usually sectioned through the middle, since this area is usually free of internal reinforcements.

ROCKER/PILLARS/FLOOR

Use Procedure Explanations 14,15 and 28 with the following text.

ROCKER/PILLARS
SHEET METAL
COUPE

Refinish Body Side Panel Assy Complete........................R/L **2.8**
Refinish Hinge Pillar W/S Post..................................R/L **1.0**
Refinish Rocker Panel..R/L **1.4**
Section Door Frame Opening.......................................R/L **#6.0**
#Section in Center of the Front Hinge Pillar & Rear Lock Pillar, w/Fender, Door & Interior Trim Removed, w/Quarter Panel Removed Deduct 2.0
Section Hinge Pillar & W/S Post..................................R/L **#6.5**
#Section at Top of Hinge Pillar & in Rocker Opening w/Fender, Door, Windshield, Instrument Panel & Interior Trim Removed
Section Rocker Panel...R/L **#6.0**
#Section in Door Opening, w/Door & Interor Trim Removed, w/Quarter Panel Removed Deduct 2.0

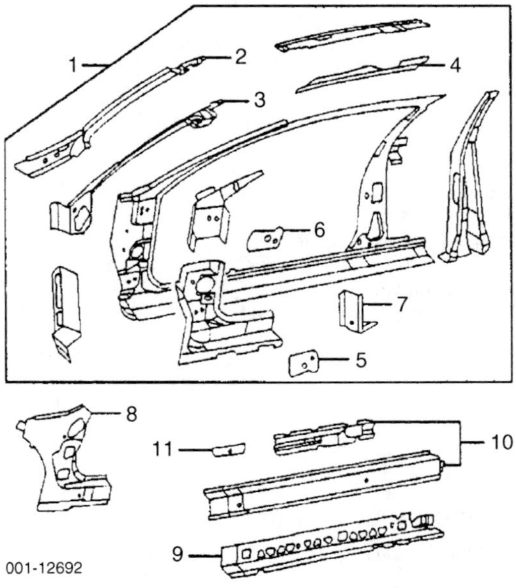

Item	Year/Side	Part No.		Time	Price
1 **Panel Assy, Body Side**	95 R	12363064	s	**#15.0**	498.00
	L	12363065	s	**#15.0**	514.00
	96 R	12363064	s	**#15.0**	498.00
	L	12363065	s	**#15.0**	N.A.
	97 R	N.A.	s	**#15.0**	N.A.
	L	N.A.	s	**#15.0**	N.A.
#w/Windshield, Fender, Door, Seat, Instrument Panel & Trim Removed, w/Quarter Panel Removed Deduct 2.0					
2 **Frame, W/S Side Outer ¶**	R/L	22644732-3	s	**3.0**	50.00
3 **Frame, W/S Side Inner ¶**	R/L	22568000-1	s	**1.5**	50.00
4 **Rail, Roof Side Inner ¶**	95 R	22656066	s	**.5**	22.15
	L	22567997	s	**.5**	22.15
	96-97 R	22656066	s	**.5**	22.15
	L	22567997	s	**.5**	22.15
5 **Plate, Upr Hinge Anchor¶**	95 R/L	22645204-3	s	**.8**	1.85
	96-97 R/L	22645204-3	s	**.8**	1.85
6 **Plate, Lwr Hinge Anchor¶**	95 R/L	22645203-4	s	**.8**	1.85
	96-97 R/L	22645203-4	s	**.8**	1.85
7 **Reinf, Pillar Otr Panel¶**	R/L	22578633-4	s	**.5**	2.85
¶Included w/Body side Panel Assy					
8 **Pillar, Hinge Inner**	95 R/L	22570785-6	s	**#2.0**	40.50
	96 R	22570785	s	**#2.0**	40.50
	L	22639921	s	**#2.0**	40.50
	97 R/L	22639920-1	s	**#2.0**	40.50
#w/Body Side Panel or Hinge Pillar Removed					
9 **Reinf Rocker Otr Panel**	R/L	22651000-1	s	**#2.5**	38.50
10 **Panel, Rocker Inner**	95-96 R/L	12367162-3	s	**#3.0**	N.A.
	97 R/L	12367162-3	s	**#3.0**	N.A.
#w/Body Side Panel Assy or Rocker Panel Removed, R&R Both Deduct 1.0					
11 **Plate, Seat Belt Anchor**	R/L	20320556	s		2.25
Pan, Floor-Not Serviced					

FIGURE 8–9 The body side panel for this 1997 Pontiac Sunfire can be purchased as a unit. Individual parts are also available. (Courtesy of Mitchell International.)

FIGURE 8–10 This SUV is undergoing A-pillar, center pillar, and rocker panel sectioning. The inner rocker panel remains in place as well as the inner A-pillar and inner center pillar.

The repair operations included in the hinge and windshield pillar R&R times are identical to the included/not included operations given earlier in this chapter for the cowl and dash assembly. Usually the operation times assume that some prior operations have been completed (such as removal of the door, fender, and windshield) before R&R of the pillars. The estimator and adjuster must always be careful to consult the footnotes in order to determine included/not included operations.

ROCKER PANELS

The rocker panel (Figure 8–10) is a very important structural component on unibody vehicles, providing support for the floor of the passenger compartment. (On a conventional body-over-frame vehicle, the floor is supported by the vehicle frame.) Rocker panels on unibody vehicles are made of high-strength steels and are galvanized on both sides to resist corrosion. A rocker panel is usually composed of an inner panel and an outer panel. On some vehicles, however, the outer panel is welded directly to the floor pan. On four-door vehicles, a reinforcement panel is usually present where the center pillar joins to the rocker panel.

Rocker panels are often grouped with the floor pan in the collision estimating guides (Figure 8–11). Rocker panels are welded to the floor pan, cowl side panels, center pillars (on four-door vehicles), and to the quarter panels. On some unibody vehicles, the front rails are welded to the rocker panels as well.

ROCKER/PILLARS/FLOOR

Use Procedure Explainations 14,15 and 28 with the following text.

Operation	Side	Hours
Refinish Body Center Pillar Assy	R/L	**2.8**
Refinish Inner Lower Pillar	R/L	**1.0**
Refinish Outer Pillar	R/L	**.5**
Refinish Rocker Panel Complete	R/L	**1.6**
Refinish Rocker Panel Front	R/L	**1.2**
Refinish Rocker Panel Rear	R.L	**1.2**
R&I Rocker Moulding Front & Rear	R/L	**.3**
R&R Front Rocker Panel	R/L	**6.0**
R&R Rear Rocker Panel	R/L	**#6.0**

#Deduct 1.5 w/Quarter Panel Removed, Deduction Includes Common Surfaces & Included Operations

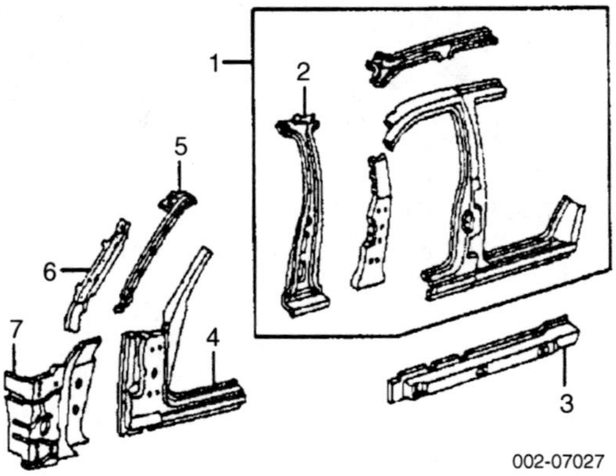

Item	Year	Side	Part No.	Labor	Price
1 **Pillar Assy, Ctr Body**		R	F2MY 54243A00 A	**#16.0**	969.70
		L	F2MY 54243A01	**#16.0**	969.83
#Deduct 1.5 w/Quarter Panel Removed, Deduction Includes Common Surfaces & Included Operations					
2 **Pillar, Center Body¶**	92-94	R/L	F2AZ 5424300-1 A		36.37
	95-97	R	F5MY 5424300 A		814.18
		L	F5MY 5424301 A		N.A.
¶Included w/Center body Pillar Assy					
3 **Reinf, Ctr Body Pillar**					
Lower		R/L	F5AZ 5420402-3 A	**3.5**	54.48
Mldg Assy, Rocker		R	F6AZ 5410176 AAA	**.4**	61.43
		L	F6AZ 5410177 AAA	**.4**	72.38
Cap, Rocker Mldg¶		R/L	F6AZ 54101C48-9 AAA	**.1**	7.57
¶Included w/Rocker Mldy Assy					
4 **Pillar, Inner Lower**	92-94	R/L	F2AZ 54025A00-1 A	**7.5**	340.25
	95-97	R/L	F5AZ54025A00-1	**7.5**	617.22
5 **Pillar, Upper Outer**		R	F2AZ 5402508 A	**3.5**	33.28
		L	F2AZ 5402509 A	**3.5**	38.28
6 **Pillar, Upper Inner**		R/L	F2AZ 5402504-5 A	**1.0**	180.58
7 **Reinf, Hinge Upper**	92-94	R/L	F2AZ5422842-3 A	**1.5**	d46.30
	95-97	R/L	F5AZ 5422842-3 A	**1.5**	50.65
Panel, Pillar Trim¶					
Upper		R/L	F2AZ 5424346-7 AAT	**#.6**	31.93
#Includes R&I/R&R Lower Trim Panel & Scuff Plates					
Lower		R	F2AZ 5424347 BAT	**#.4**	105.08
		L	F2AZ 5424347 BAT	**#.4**	33.53
¶Blue listed, Order By Application					
#Includes R&R Scuff Plates					

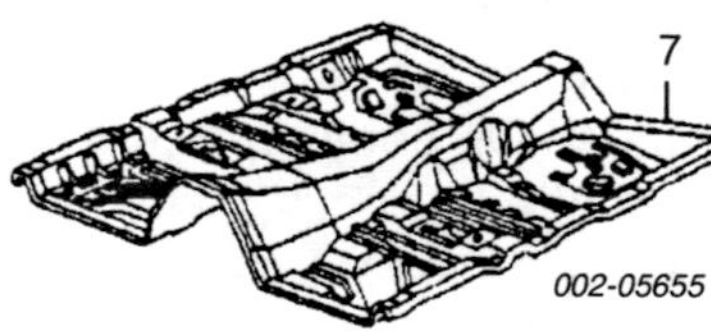

Item	Year	Side	Part No.	Labor	Price
7 **Pan, Front Floor**	92-94		F2AZ 5411135 A	**#20.5**	481.35
	95-97		F5AZ 5411135 A	**#20.5**	527.20
#Deduct 4.5 w/Each Ctr Body Pillar Assy Removed					
Plate, Frt Scuff¶		R/L	F2AZ 5413208-9 A	**.2**	18.80
Plate, Rear Scuff¶		R/L	F2AZ 5413228-9 A	**.2**	9.33
¶Blue listed, Order by Application					

FIGURE 8–11 Rocker panel and floor for a 1996 Grand Marquis. Note that the rocker panel is in two sections. (Courtesy of Mitchell International.)

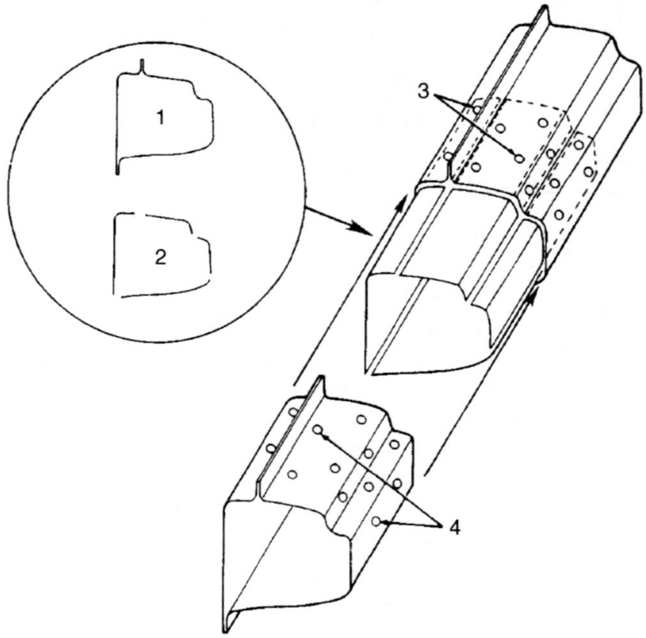

1. Cross section of rocker panel insert material before cutting lengthwise
2. Insert cut lengthwise into sections
3. Insert inside rocker panel, secured with plug welds or sheet metal screws
4. 5/16" holes for plug welds

FIGURE 8–12 When a rocker panel is sectioned, metal inserts are used to reinforce the joint.

The inner and outer rocker panels can be purchased separately or they can be replaced as a unit. When purchased separately, both of the panels must be recorded on the damage report as separate operations. As mentioned earlier, a rocker panel is sometimes available as a subassembly with either the center pillar or windshield pillar attached.

Sectioning is often an alternative to replacing a complete rocker panel (Figure 8–12). Rocker panels are sectioned between the pillars to avoid cutting through inner reinforcements. When only the damaged area of a rocker panel is cut out and a section of new panel is welded in its place, a reduction in the panel R&R time should be made on the damage report.

The inner rocker panel is not an appearance item; metal finishing and painting are not necessary. However, an allowance should be made on the damage report for applying anticorrosion material to the panel after all welding is complete.

Most rocker panels have a sill or scuff plate. The sill plate is made of plastic or aluminum and is screwed to the rocker panel. If more than slightly damaged, the sill plate should be replaced.

FIGURE 8–13 Center pillar on a four-door car.

The rocker panel R&R times include R&I of the cowl trim, center pillar trim, and the quarter panel trim. Removal of the floor mats, sill plates, and rear seat cushions are also included in the operation times. If the front fender must be loosened to gain access to the front edge of the rocker panel, this is also included in the operation times.

The operation times for rocker panel R&R do not include the following:

- R&I of the door
- R&I of fuel lines, control cables, and electrical wiring that might be connected to or near the rocker panel
- R&I or replacement of mouldings, nameplates, emblems, or ornamentation
- Installation of decals or stripe tape
- Applying anticorrosion or undercoating
- Refinishing the outer rocker panel

When replacing adjacent panels, such as a center pillar or quarter panel, 1.0 hour should be deducted from the operation time for each overlapping (common) seam. if the rocker panel shows signs of previous damage or rusting, a deduction for betterment must also be made on the damage report.

CENTER PILLAR

The center pillar (Figure 8–13) on a four-door vehicle has several functions: it supports the roof panel, provides a locking surface for the front door and serves as a hinge post for the rear door. A center pillar (also called a B-pillar) is welded to the rocker panel, floor pan, and roof rail (except on convertible and hardtop vehicles in which the center pillar only extends to the car's belt line). Usually it has both an inner and outer panel, forming a closed structure.

Both OEM and LKQ parts are available for replacing a damaged center pillar. However, it is common practice to section the center pillar below the roof. Labor time for sectioning the center pillar is often given in the collision estimating guides (Figure 8–14). When specifying a section repair, the estimator and adjuster should be aware that the section must be made below the seat belt D-ring to avoid cutting into the D-ring reinforcement that is often a part of the center pillar.

When a center pillar is damaged in a side collision, the rocker panel may be damaged as well. When that is the case, it is best to obtain a salvaged center pillar and rocker panel subassembly to use as a replacement.

The cutting and welding operations involved in replacing a center pillar necessitate removing the rear door, the front seats, loosening the headliner, and rolling back the floor mats and carpet. These operations are included in the center pillar R&R times given in the collision estimating guides. Replacing the center pillar will also require R&I of the pillar trim, sill plate, door weatherstripping, and door striker. These operations are also included in the listed R&R times. The R&R times also include time to perform all necessary cutting, welding, grinding, filling, and smoothing of the welded seams.

Operations that are not covered by the R&R times given in the collision estimating guides include the following:

- R&I of mouldings, emblems, nameplates, and other ornamentation
- R&I of any optional lights
- Installation of stripe tape, decals, and overlays
- Time for drilling mounting holes
- R&I of any electrical wiring running through the roof rails, center pillar, or rocker panel
- Applying anticorrosion material
- Refinishing the inner and outer panels

8.4 ROOF AND BACK GLASS

The roof encloses the top of the body shell and is made up of several components. The roof panel is usually a single metal sheet; however, on some vehicles, an inner panel is joined to the outer panel and both are serviced as a complete assembly. The structural components of the roof are the front and rear headers and the side rails (Figure 8–15). Some roofs, particularly those on large sedans and station wagons, also have roof bows or reinforcements running

ROCKER/PILLARS/FLOOR

Use Procedure Explanation 14,15, and 28 with the following text.

SHEET METAL

Refinish Hinge Pillar	R/L	**1.0**
Refinish Center Pillar	R/L	**1.0**
Refinish Outer Rocker Panel	R/L	**1.6**

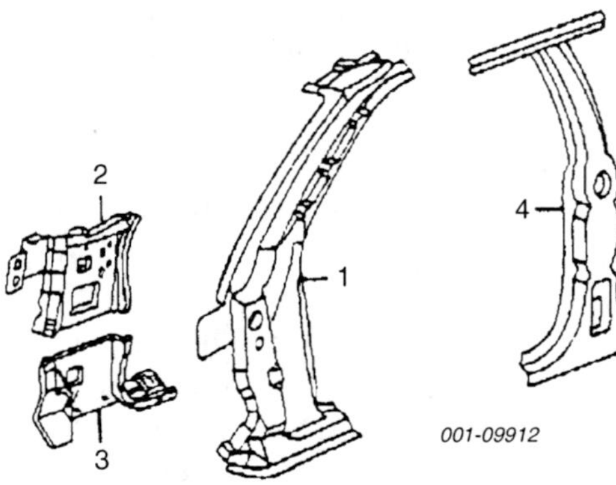

Item	Year/Side	Part No.	Labor	Price
Panel, Hinge Pillar				
1 Outer	91-95 R	10278410	**#7.0**	391.00
	L	10274343	**#7.0**	381.00
	96 R	10278410	**#7.0**	391.00
	L	10274343	**#7.0**	381.00
2 Inner	91-93 R	12510344	**#7.0**	206.00
	L	12514514	**#7.0**	209.00
	94-96 R	12514513	**#7.0**	243.00
	L	12514514	**#7.0**	209.00
#R&R One Side Complete, Time is to Cut in Windshield Opening				
3 **Extension, Hinge Pillar¶**				
Inner	R/L	10253366-7		23.85
¶Included w/Inner Hinge Pillar Panel				
Baffle, Hinge Pillar				
1991-93	R/L	10203604		8.10
1994-96-Use Foam Kit Sealer				
4 **Pillar, Center**	R/L★	10258095	**#6.0**	393.00
#Time is to Cut Below Roof, w/Power Windows Add .3				
Anchor, Lock Striker¶	R/L	10131164		2.75
¶Included w/Center Pillar				
Striker, Door Lock	R/L	21090040	**.2**	10.20
Applique, Pillar	91-92 R/L	10114578-9	**#.2**	21.90
#Time is for Installation Only				

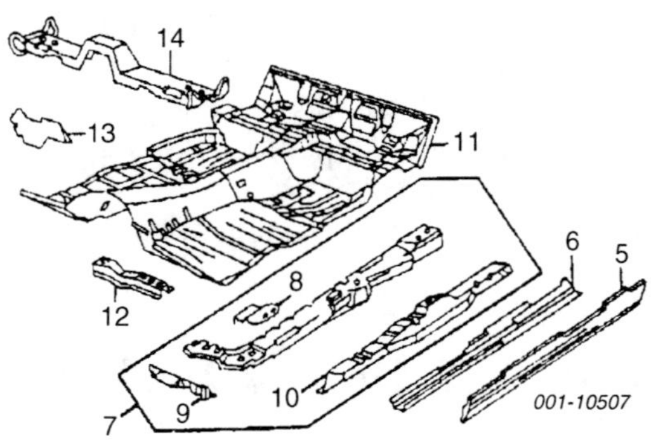

Item	Side	Part No.	Labor	Price
5 **Panel, Outer Rocker**	R★	10283082	**7.5**	79.25
	L★	10283083	**7.5**	87.50
6 **Panel, Upper Rocker**	R/L★	10219984-5		47.00
7 **Panel, Inner Rocker**	R/L	10200772-3		222.00
8 **Bracket, Heat Shield¶**		20165658		3.75
9 **Brace, Dash/Frame¶**	R/L	10142956-7		79.00
10 **Reinf, Inr Rocker¶- Serviced w/Inr Rocker Panel**				
¶Included w/Inner Rocker Panel				
11 **Pan, Floor-Not Serviced**				
12 **Reinf, Floor Pan**		14015974		26.00
13 **Filler, Floor Pan**	R	20004482		32.00
	L	20047424		32.00
14 **Bar, Floor Pan-Not Serviced**				

FIGURE 8–14 Center pillar (4) for a 1996 Caprice. (Courtesy of Mitchell International.)

ROOF

Use Procedure Explanations 17 and 28 with the following text.

Refinish Roof Outside	
2 Door	**4.5**
4 Door	**4.7**
Add to Edge Roof	**1.2**
R&I Roof Console	**.3**
Add to R&R Roof Panel	
w/Luggage Rack	**#3.0**
#includes Time to Drill for & Install	

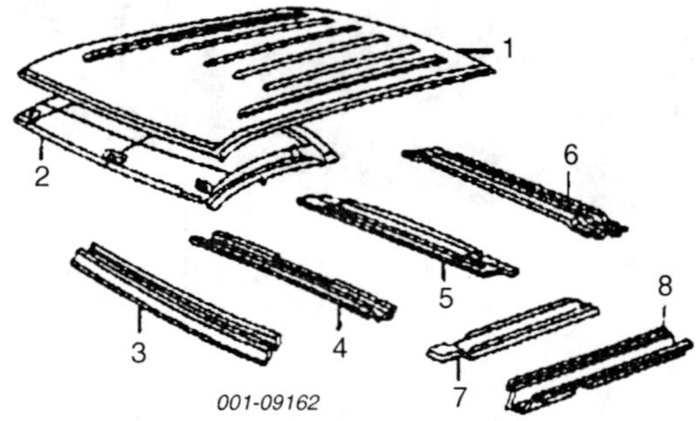

Item	Year/Side	Part No.	Labor	Price
1 **Panel, Roof Outer**				
2 Door		15645970	**21.5**	199.00
4 Door				
w/Luggage Carrier		15740023	**23.0**	307.00
w/o Luggage Carrier		★15739865	**23.0**	177.00
2 **Panel, Roof Inner**		15724154	**#7.5**	89.75
#w/Outer Roof Panel Removed				
3 **Panel, Header**		15646573	**1.5**	47.50
Bow, Roof Reinf				
4 Front		15690200	**1.0**	38.75
5 Center		15995667	**1.0**	36.75
6 Rear-See Gate Opening Frame in Rear Body Section				
Rail, Roof Side				
7 Upper				
2 Door	R	15645972		47.75
	L	15645971		46.00
4 Door	R/L	15970846-5		42.75
8 Lower				
2 Door	92-95 R/L	15645974-3		75.25
	96-97 R/L	15976608-7		75.25
4 Door	92-94 R	15981780		64.00
	L	15976547		64.00
	95-97 R/L	15970844-3		37.00
Retainer, Roof Rail W/Strip				
See Rocker/Pillars/Floor Section				
W/Strip, Roof Rail				
2 Door	R/L	15720260-59	**.2**	9.05
4 Door	92-95 R/L	15720264-3		30.75
	96-97 R/L	15741028		30.75
Headliner¶		15697194	**1.5**	320.00
¶Order by Application				

FIGURE 8–15 Exploded view of parts and labor times for a 1996 Blazer. (Courtesy of Mitchell International.)

horizontally across the underside of the roof panel; these reinforcements are generally the strongest members of the roof assembly. The center reinforcement usually supports the dome light. Other parts that are part of a roof assembly include smaller reinforcements, rain gutters, and drip rails. Some roofs also come equipped with a sun roof. Care must be taken that all these are accounted for on the damage report.

The estimator and adjuster should be aware that some members of the roof assembly might be listed with other subassemblies. For example, the front header is often included with the cowl and windshield assembly. The outer panels of the roof rails are often listed with the body side panels. The estimator and adjuster must be familiar with the possible location of these parts in the collision estimating guides so that no part is overlooked when the damage report is written.

The R&R times listed in the collision estimating guides for replacing the roof panel do not include replacement of the headers and rails or the time needed to straighten them. If the headers or rails are damaged beyond repair, R&R time for replacing them must be entered separately on the damage report. R&R times for these structural panels assumes that the roof panel will be removed before the headers or rails are removed and replaced. Always consult the footnotes for information concerning prior and included operations.

Roof panel R&R times do not include optional vinyl roof covers, either. The replacement of a vinyl-covered roof usually requires increasing the panel R&R times by the number of hours specified in the footnotes (Figure 8–16). The cost of the vinyl cover must also be included on the damage report. Vinyl tops usually have mouldings that are not present on nonvinyl-top vehicles; these must not be overlooked when writing the damage report.

Additional consideration must also be given to sunroofs, T-roofs, and convertible tops. Again, the footnotes must be consulted for instructions concerning adjustment to the listed times.

R&R of the roof panel requires that many parts of the vehicle be temporarily removed while the repair is being made. Following is a list of components for which R&I time is included in the panel R&R times:

- Windshield glass
- Back window and reveal mouldings
- Glass runs and channels
- Quarter glass
- OEM sunroofs or T-top
- Lift gate
- Reveal mouldings and other interior mouldings
- Visors
- Coat hooks
- Lights
- Front and rear seats
- Weatherstripping and retainers
- Weathercords
- Quarter trim panels (except trucks)
- Headliner (except trucks)
- Shoulder harness
- Rear package tray

ROOF

Use Procedure Explanations 17 and 28 with the following text.

SHEET METAL

Refinish Roof Outside		**3.7**
Refinish Garnish Moulding	R/L	**.3**
Add to R&R Roof		
w/Landau Roof		**4.0**
w/Simulated Convertible		**4.3**
w/Full Vinyl Cover		**5.0**
R&I Headliner		
w/Sunroof		
2 Door		**1.0**
4 Door		**1.5**
w/o Sunroof		**.8**

1-6453

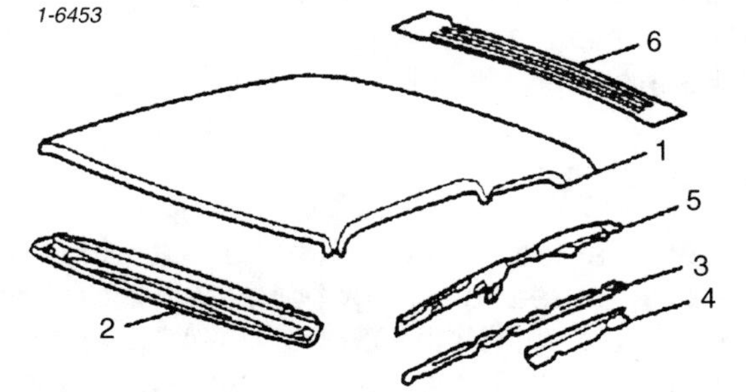

Item	Year / Side	Part No.	Time	Price
1 **Panel, Roof**				
w/ Sunroof				
2 Door		25616307	**25.0**	807.00
4 Door				
w/ Full Vinyl Trim		25551505	**#32.5**	807.00
#Includes R&R Vinyl Cover				
w/o Full Vinyl Trim	89	20744323	**25.0**	807.00
	90-93	25551506	**25.0**	807.00
w/o Sunroof				
2 Door		25609501	**22.0**	408.00
4 Door		25544852	**22.0**	331.00
2 **Frame, Upper (HSLA)**		20551533	**1.5**	45.75
Rail, Side Outer				
3 Upper				
2 Door	R/L	20509078-9		85.25
4 Door	R/L	20730760-1		44.25
4 Lower				
2 Door	R/L	20482240-1		25.25
4 Door	R	25619207		51.25
	L	25619210		51.25
5 **Rail, Side Inner**				
2 Door	R/L	20509122-3		95.50
4 Door	R/L	25542418-9		65.00
6 **Bow, Roof**		3525672	**1.5**	15.05
Headliner				
w/Sunroof				
2 Door		12535851	**1.2**	302.00
4 Door		25615881	**1.7**	302.00
w/o Sunroof				
2 Door		12535845	**1.0**	302.00
4 Door		12535862	**1.0**	302.00
¶Order by Application				
Moulding, Inside Garnish				
Fleetwood (a)				
2 Door	89-90 R/L	20746118-9	**.3**	64.50
	91-93 R/L	16653546-7	**.3**	64.00
4 Door				
Front	89-90 R/L	20746118-9	**#.3**	64.50
	91-93 R/L	16652904-5	**#.3**	51.75
Rear	89-90 R/L	20746092-3	**#.3**	64.50
	91-93 R/L	16652920-1	**#.3**	36.00
#R&R One Side Complete				
DeVille (P)				
2 Door	R/L	20574604-5	**.3**	64.50
4 Door				
Front	R/L	20574656-7	**#.3**	64.75
Rear	R/L	20735266-7	**#.3**	53.50

(a) Blue listed, Order by Application
(P) Paint to Match
#R&R One Side Complete

FIGURE 8–16 Roof R&R times for a vinyl roof. (Courtesy of Mitchell International.)

R&R of the roof panel does not include the following:

- R&R of the header panels
- R&R of the roof rails
- R&R of the roof reinforcements
- R&R of a new roof cover
- R&R of luggage racks
- R&I of mouldings, nameplates, emblems, or ornamentation
- R&I of electrical wiring and components
- R&I of optional lights
- Removal of shattered glass
- Installation of stripe tape and decals
- Application of sound-deadening material]
- Refinishing

REAR GLASS

Much of what was written earlier about windshields applies to the rear glass (or back lite as it is sometimes referred to). Both OEM and NAGS prices are listed in the collision estimating guides for quarter glass and rear glass. The methods of installation are also the same for both windshields and rear glass. The estimator and adjuster must determine whether the rear glass is heated or not when listing the glass price. If urethane was used to install the glass, urethane must again be used in order to maintain the structural integrity of the vehicle.

R&R of the rear glass includes the following:

- R&R of the rubber channel (where applicable)
- R&I of reveal and garnish mouldings
- R&I of wiper arms (where applicable)
- Testing for leaks
- Disconnecting and reconnecting the heated glass connectors

R&R of the rear glass does not include:

- Removal of shattered glass
- R&I of high-mounted stoplights
- Cost of installation material
- Checking the window opening for proper size

A

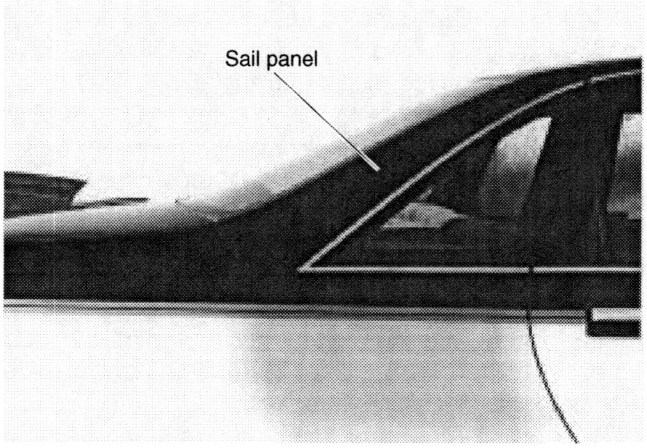

B

FIGURE 8–17 *(A) Damaged quarter panel and (B) sail panel.*

8.5 QUARTER PANEL

A quarter panel (Figure 8–17) extends from the rocker panel and roof to the rear body panel, forming one side of the rear body shell. A typical quarter panel assembly (Figure 8–18) is composed of the following parts: the outer panel; sail panel or roof extension; inner panel; lock pillar; outer wheelhouse; inner wheelhouse; and various fillers, extensions, gussets, braces, and reinforcements. In today's unibody vehicle, many of these parts are made of high-strength steels and are galvanized on both sides.

OUTER PANEL

The outer panel is a comestic panel that might have some or all of the following components: body and window mouldings, stripe tape, emblems, door lock reinforcement, quarter window, and fuel door. The outer quarter panel is welded to the rocker panel, roof rail, lock pillar, rear floor pan extension, parcel shelf,

QUARTER PANEL

Use Procedure Explanations 20 and 28 with the following text.

SHEET METAL

Operation		Hours
Refinish Quarter Outside	R/L	**2.2**
Add to Edge	R/L	**.5**
Add for Pillar	R/L	**.5**
R&R Both Quarter Panels		**27.6**
R&R Roof & Both Quarter Panels		**44.4**
R&R Rear Body Panel & Both Quarter Panels		**30.6**
R&R Roof, Both Quarter Panels & Rear Body Panel		**47.4**
Section Quarter Panel	R	**#12.5**
	L	**#13.0**
Section Both Quarter Panels		**#23.6**
R&R Rear Body Panel		**#26.6**

#Horizontal Cut Below Back Glass & Vertically into Trunk Opening, Refer to Cut Line in Illustration

Add to R&R Quarter Panel		
To R&I Back Glass		**.3**

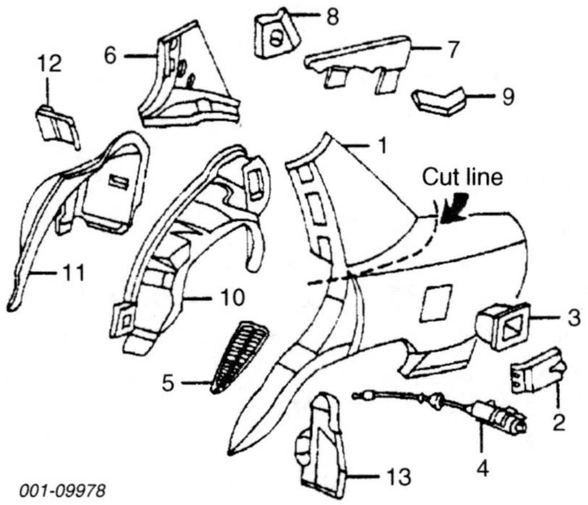

No.	Part	Year/Side	Part No.	Labor	Price
1	**Panel, Qtr Outer**	92-95 R	12539822	**#15.5**	427.00
		L	12539825	**#16.0**	424.00
		96-97 R	12539816	**#15.5**	427.00
		L	12539817	**#16.0**	427.00
	#Deduct 4.4 w/Roof Removed, Deduction Includes Common Surfaces & Included Operations				
2	**Door, Fuel**		25611820	**.3**	34.50
	Latch, Fuel Door		25636110		12.55
3	**Pocket, Fuel Filler**		25554268		20.30
4	**Actuator, Power Lock**		25606839		31.25
5	**Vent, Pressure Release**	R/L	3549154		5.15
6	**Panel, Inner Upper**	R/L	25628601-2	**1.5**	25.50
7	**Baffle, Panel Inner**				
	1992-94	R/L	25606036-7	**#.5**	6.70
	1995-97-Not Serviced				
8	**Retainer, Inner Baffle**				
	1992-94	R/L	8732474	**#.5**	1.25
	1995-97-Not Serviced				
	#R&R Complete				
9	**Baffle, Outer Panel**				
	1992-94	R/L	25606024-5	**.5**	6.70
	1995-97-Not Serviced				
10	**Panel, Wheelhouse Outer**	R/L	25630789-90	**#3.0**	124.00
11	**Panel, Wheelhouse Inner**	R	25630779	**s #2.5**	141.00
		L	20740807	**s #2.5**	141.00
	#w/Quarter Panel Removed				
12	**Filler, Wheelhouse**	R	25558663		11.25
13	**Shield, Filler Pipe-See Fuel Tank Section**				
	Antenna Assy				
	Manual		10289035	**#1.0**	9.30
	Power		22138200	**#1.0**	91.11
	Mast, Antenna				
	Power¶		22154379	**#1.5**	18.88
	¶Included w/Power Antenna Assy				
	#w/Necessary Trim Removed				

FIGURE 8–18 Estimate guide quarter panel replacement. Note section line. (Courtesy of Mitchell International.)

FIGURE 8–19 Quarter panels are often sectioned at the belt line to avoid removing the quarter and rear glass.

rear body panel(s), and outer wheelhouse. When damaged, the outer panel can be removed at the factory seams and replaced with OEM, aftermarket, or salvage parts, R&R times for this operation are given in the collision estimating guides.

However, it is a common procedure to section the quarter panel below the window or belt line if allowed by the OEM. This eliminates some of the labor required to R&I certain parts that are necessary to replace the entire quarter panel. For example, when replacing the outer quarter panel at factory seams, the rear glass and quarter window must be removed and reinstalled after the repair is complete. Not only is this a labor-intensive operation, but there is always the chance that the glass will chip or break, resulting in a costly replacement procedure. By cutting through the damaged panel below the glass area (Figure 8–19) and replacing the damaged sheet metal with a section of a new or salvaged panel, these and other operations are avoided.

For many models, the collision estimating guides contain labor times for sectioning the quarter panel at the belt line. However, smaller areas of the quarter panel can also be replaced with a partial panel when the damage is not extensive. For example, if only the rear corner of the quarter panel is damaged, a section

FIGURE 8–20 A plastic rear extension panel.

cut can be made behind the wheel well. A reasonable labor allowance must be agreed on by the estimator and adjuster for this operation.

The quarter panels on some vehicles have rear extension panels. Rear extension panels are very common on body-over-frame vehicles (Figure 8–20) but are not often found on unibody vehicles. These small panels are usually bolted to the outer panel, filling the space between the outer quarter panel and the rear body panel and sometimes forming the mounting surface for the rear lights. The rear extensions are usually made of die-cast white metal, fiberglass, or plastic and are normally replaced when damaged. The rear extensions and all other bolted-on extensions and filler panels are included in the outer quarter panel R&R. The collision estimating guides also list separate labor allowances for replacing the rear extensions.

UNISIDE

A uniside consists of all welded outer panels on one side from the cowl back. This includes the A-pillar, rocker panel, center pillar (on four-door vehicles), quarter panel and the side of the roof. Collision estimating guides list the times for the replacement of the entire uniside as well as the recommended sectioning locations.

SAIL PANEL

The sail panel is that part of the quarter panel that extends above the belt line to the roof panel. It is usually part of the outer quarter panel, but on some vehicles, the sail panel is called an extension panel and is part of the roof panel assembly. Sail panels are included in the R&R of the outer quarter panel. However, R&R time for the roof extension must be entered separately on the damage report if the panel must be replaced. If the sail panel is covered with a vinyl roof covering, time must be allowed for the covering to be loosened and pulled away from the repair area.

INNER PANEL

Some vehicles have both an outer quarter panel and one or more inner quarter panels. On those vehicles, the inner quarter panel is welded to the outer panel, lock pillar, wheelhouse assembly, and floor pan. The inner panel forms the regulator panel for the passenger compartment and the sides of the luggage compartment. If the vehicle has a movable quarter window, the regulator, run channel, and handle will be attached to the inner panel. The inner panel is usually covered with interior trim; thus it is a nonappearance item.

The collision estimating guides list R&R times for the inner panels. Inner panels are not included in the outer panel operation times; they must be listed separately on the damage report. However, since the inner panels are nonappearance panels, often they can be straightened by pulling and metalworking techniques.

LOCK PILLAR

On some vehicles, the outer quarter panels or the wheelhouses are formed in such a way that their leading edges form the door jambs. On other vehicles, separate panels called lock pillars (Figure 8–21) provide the surfaces against which the doors are shut. Generally, the lock pillar is welded to the rocker panel and to the inner and outer quarter panels. On some vehicles, the lock pillar might be composed of an inner and outer panel; other vehicles only have a single panel. On some vehicles, the lock pillar only extends up to the belt line. On other models, the pillar extends to the roof and might consist of both a lower and an upper pillar. It is very important that the estimator and adjuster consult the exploded view of the quarter panel in the appropriate collision estimating guide to determine the actual construction of the lock pillar.

The lock pillar panels are not included in the outer quarter panel R&R times, except where they are a part of the quarter panel. If the lock pillar panels are damaged, their replacement labor allowances must be entered separately on the damage report.

QUARTER PANEL

Use Procedure Explanations 20 and 28 with the following text.

2 DOOR HATCHBACK
SHEET METAL

Operation		Time
Refinish Quarter Outside	R/L	**2.8**
Add for Pillar	R/L	**.5**
Add to Edge	R/L	**.5**
Refinish Fuel Door		**.3**
R&R Both Quarter Panels		**24.3**
R&R Roof & Both Quarter Panels		
w/Sunroof		**40.3**
w/o Sunroof		**38.3**
Add to R&R Quarter Panel		
To Drill for & Install Spoiler Extension	R/L	**2.**

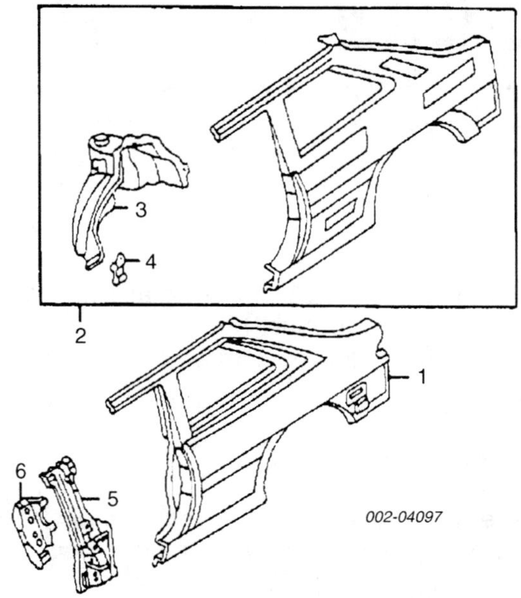

Part		Part Number	Time	Price
1 **Panel, Quarter Outer**				
1991-92				
w/GT	R	FOCZ 6127840 A	**#12.3**	624.88
	L	FOCZ 6127841 A	**#13.3**	732.55
w/o GT	R	F4CZ 6127840 A	**#12.3**	621.67
	L	F4CZ 6127841 A	**#13.3**	621.67
1993-96	R	F4CZ 6127840 A	**#12.3**	621.67
	L	F4CZ 6127841 A	**#13.3**	621.67
#Deduct 2.5 w/Roof Removed, Deduction Includes Common Surfaces & Included Operations				
Door, Fuel				
w/Remote Release	91-93	F1CZ 61405A26 B	**.2**	28.92
	94-96	F4CZ 61405A26 B	**.2**	38.02
w/o Remote Release	91-93	F1CZ 61405A26 A	**.2**	28.92
	94-96	F4CZ61405A26 A	**.2**	17.22
Cable, Release		F1CZ 5828610 A	**1.0**	19.58
Spring, Fuel Door		F3CZ 58282A42 A		7.75
Actuator, Fuel Door		F32Z 6128612 A	**.4**	19.13
2 **Panel Assy, Inner**	R	F4CZ 6127858 A**s**	**6.5**	400.85
	L	F4CZ 6127859 A**s**	**6.5**	354.60
3 **Pnl, Whlhse Inr¶**	91-93 R	FOCZ 5827886 A**s**	**2.5**	d136.50
	L	F4CZ 5427887 B**s**	**2.5**	211.35
	94 R	F4CZ 5827886 BB**s**	**2.5**	192.97
	L	F4CZ 5827887 B**s**	**2.5**	N.A.
	95-96 R/L	F5CZ 5427886-7 A**s**	**2.5**	N.A.
4 **Brace, Qtr/Whlhse¶**	R	FOCZ 5827800 A	**.3**	8.70
	L	FOCZ 5827801 A	**.3**	24.50
¶Included w/Inner Panel Assy				
5 **Reinf, Center Pillar**	R	F3CZ 6124382 AA	**#1.5**	129.58
	L	F3CZ 6124383 AA	**#1.5**	109.52
6 **Supt, Seat Belt Retainer**	R/L	F3CZ 6160292-3 AA	**#1.5**	45.18
#R&R Complete, w/Outer Panel Removed				

FIGURE 8–21 This quarter panel on a 1995 Ford Escort includes the lock pillar; the inner wheelhouse is separate. (Courtesy of Mitchell International.)

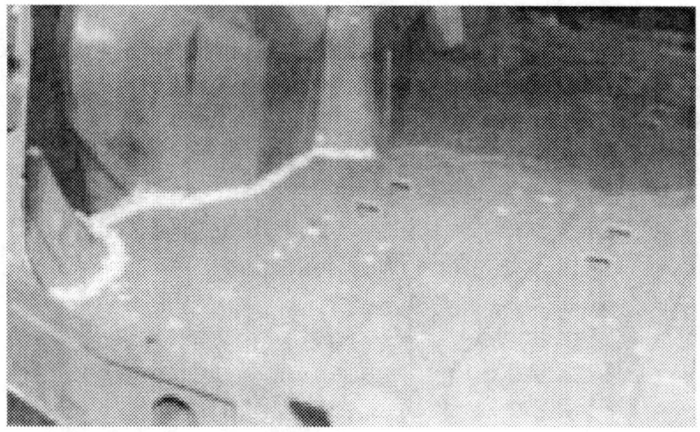

FIGURE 8–22 The wheelhouse forms the side of the luggage compartment on some vehicles. (Courtesy of Tech-Cor.)

WHEELHOUSE

The wheelhouse (Figure 8–22) forms the splash shield for the rear wheel, protecting the backside of the quarter panel and the trunk area from road splashes. It also helps to form the side of the luggage compartment. The wheelhouse is usually made up of two panels: an inner panel and an outer panel. The inner wheelhouse is welded to the outer wheelhouse, parcel shelf, rear floor pan, and roof side rail. The outer wheelhouse is welded to the inner wheelhouse and outer quarter panel and is a nonappearance panel.

The wheelhouse panels are not included in the quarter panel R&R times. Labor allowances and parts prices for replacing the wheelhouse must be listed separately on the damage report.

INCLUDED/NOT INCLUDED OPERATIONS

The quarter panel assembly is a very complex subassembly on the body shell, and it is important that the estimator and adjuster understand which operations are included in the operation times listed in the crash guides. In addition to the drilling, welding, grinding, filling, and smoothing involved in replacing the damaged outer quarter panel, quarter panel R&R times include the following:

- Loosening the headliner
- R&I of the garnish moulding
- R&I of the rear window and the reveal mouldings (if necessary)
- R&I of the rear seat
- R&I of the rear package tray

- R&I of the quarter trim panel
- R&I of the quarter glass and regulator assembly
- R&I of stationary quarter glass (unless otherwise noted in the text)
- Detaching the convertible top
- R&I of the sill plates
- R&I of bolt-on quarter extensions and fillers
- R&I of seal strips, weatherstrips
- R&I of the door striker
- R&I of lights
- R&I of the bumper assembly (if necessary)
- R&I of the luggage compartment trim

If the panel is sectioned below the window openings, only the following are included in the sectioning time:

- R&I of the rear seat
- R&I of the quarter trim panel
- R&I of the sill plates
- R&I of bolt-on quarter extensions and fillers
- R&I of seal strips, weatherstrips
- R&I of the door striker
- R&I of lights
- R&I of the bumper assembly (if necessary)

Quarter panel R&R and sectioning below the belt line do not include the following operations:

- Refinishing
- Applying undercoating, sound-deadening, and anticorrosion protection
- R&I of the wheels
- R&I of the antenna
- R&R of panels inward of the quarter panel
- R&I or rollover of the roof cover
- R&I of optional lamps
- R&I of optional mouldings, nameplates, emblems, and ornamentation
- R&I of stripe tape, decals, and overlays
- Removal of shattered glass

CAB CORNERS

Cab corners or cab side panels on extended-cab pickup trucks can be damaged in a side impact. Also, the rear of the cab is often damaged by the bed when the frame flexes in a front or rear impact (Figure 8–23). The panels can be replaced or sectioned in the same manner as quarter panels (Figure 8–24). OEM, aftermarket, or LKQ panels may be used. Labor operations included in an estimate guide are: to remove, install, or replace seats, sill plates, quarter window, striker caulking, carpet, and mouldings. The following may be required and must be figured separately, if needed: R&R or R&I, back glass, headliner, bed, and sound-deadening trim. If broken glass needs to be cleaned up, the time must be added as well as the time to drill holes for mouldings, install stripes, and refinish.

FIGURE 8–23 *This Ranger hit a tree head-on. When the frame flexed, the front of the bed hit the rear of the cab.*

VAN SIDE PANELS

Van side panel replacement is similar to quarter panels and pickup truck cab corner replacement (Figure 8–25). Internal braces, floor, and wheelhouse are attached to the side panel. Check these for hidden damage. Some OEMs sell side panels in front or rear sections. Labor time includes the following: R&I side door, rear door, rear bumper, rear lamp, filler panel, R&R side windows, sill plates and striker, caulk and move headliner. Operations that are not included in labor time and must be added: repaint, R&I interior trim, gas tank, wheel, seat and sound deadening, install stripes, clean up broken glass and cut holes for trim.

Conversion vans may have a considerable number of extra interior finish panels. In some cases, these will need to be removed for access. Figure the labor time on an individual basis. Also, these vans usually have many stripes—some painted and some tape.

BED SIDE

If pickup truck bed side damage repair labor time exceeds the cost of a bed side, consider bed side replacement (Figure 8–26). Collision estimating guide labor times include the removal and installation of rear lamps, gas door, rear bumper and taillight. Removal

CAB SHEET METAL

Use Procedure Explanations 14, 15, 17, 18 and 28 with the following text.

REGULAR CAB

Operation		Hours
Refinish Roof Outside		2.8
Refinish Inner Panel		1.8
Refinish Back Panel Complete		2.3
Refinish Corner Panel	R/L	2.0
Refinish Lock Pillar	R/L	1.0

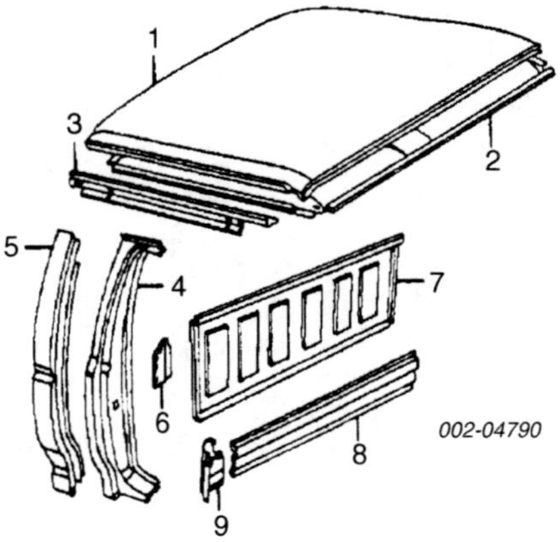

Part		Part No.	Labor	Price
1 **Panel, Roof¶**				
1992-93 (a)		F4TZ 1550202 A	**11.0**	213.12
1994-97				
w/Running Lamps		F4TZ 1550202 B	**12.4**	213.12
w/o Running Lamps		F4TZ 1550202 A	**11.0**	213.12
¶Cut Holes for Running Lamps if Required				
(a) Replacement Panel Requires High Mount Stop/Cargo Lamp				
Headliner, Roof¶		E9TZ 1551944 A3A	**1.2**	245.73
¶Blue listed, Order by Application				
2 **Panel, Inner**	92-93	E7TZ 15502B84 A	**#4.5**	105.50
	94-97	F6TZ 15502B84 A	**#4.5**	89.05
3 **Rail, Drip Side**	R	F2TZ 1551700 A	**#1.0**	29.88
	L	F2TZ 1551701 A	**#1.0**	32.37
#w/Roof Panel Removed				
4 **Pillar, Lock**	92-93 R	F7TZ 1528160 AA	**#2.5**	89.08
	L	F2TZ 1528161 A	**#2.5**	93.12
	94-97 R	F7TZ 1528160 AA	**#2.5**	89.08
	L	F4TZ 1528161 A	**#2.5**	97.77
#w/Corner Panel Removed				
5 **Panel, Corner**	R/L	F3TZ 15280A58-9 A	**#7.5**	161.82
#w/Bed Removed, Includes R&I/R&R Back Glass, Trim Panel, Back Trim Panel, Upper & Lower Garnish Mldgs				
6 **Filler, Corner**	R/L	F3TZ 83277A48-9 A	**.3**	8.80
7 **Panel, Back**		E9HZ 6040300 A	**#9.8**	122.27
#w/Bed Removed, Incudes R&I/R&R Back Glass, Right & Left Trim Panels, Back Trim Panel & Lower Garnish Mldg, w/Corner Panel Removed Deduct 4.2				
Panel, Back Inner				
8 Center		E4TZ 1540492 A	**2.0**	44.17
9 Outer	R/L	F3TZ 1540476-7 A	**1.0**	42.68
Lamp Assy, High Mount-See Rear Lamps Section				

FIGURE 8–24 Cab components. (Courtesy of Mitchell International.)

ECONOLINE VAN 1992-97

VAN SIDE PANEL

SHEET METAL RIGHT SIDE

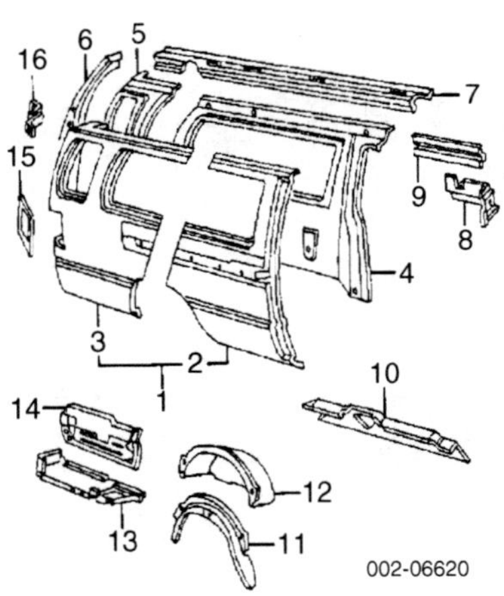

Part		Part No.	Labor	Price
1 **Panel Assy, Body Side**				
Standard Van/Wagon				
Sliding Door				
w/Window		F2UZ 15278A96 E	**26.5**	1132.83
w/o Window		F2UZ 15278A96 F	**24.0**	1125.87
Hinged Door				
w/Window		F2UZ 15278A96 G	**25.3**	893.05
w/o Window		F2UZ 15278A96 H	**22.8**	887.42
Extended Van/Wagon				
Hinged Door				
w/Window		F2UZ 28278A96 D	**30.5**	1121.53
w/o Window		F2UZ 28278A96 E	**26.5**	1200.33
Sliding Door				
w/Window		F5UZ 28278A96 B	**29.3**	810.20
w/o Window		F5UZ 28278A96 A	**25.3**	777.42
Panel, Body Side Outer¶				
2 Front				
Standard Van/Wagon				
Sliding Door				
w/Window		F2UZ 15279D44 E	**26.0**	639.13
w/o Window		F2UZ 15279D44 F	**27.5**	6054.87
Hinged Door				
w/Window		F2UZ 15279D44 C	**22.8**	465.27
w/o Window		F2UZ 15279D44 D	**26.3**	498.42
Extended Van/Wagon				
Sliding Door				
w/Window		F2UZ 15279D44 E	**26.8**	639.13
w/o Window		F2UZ 15279D44 F	**28.8**	604.87
Hinged Door				
w/Window		F2UZ 15279D44 C	**28.0**	465.27
w/o Window		F2UZ 15279D44 D	**30.0**	498.42
3 Rear				
Standard Van/Wagon				
Sliding Door				
w/Window		F2UZ 15279D46 A	**27.2**	674.33
w/o Window		F2UZ 15279D46 B	**30.2**	680.92
Hinged Door				
w/Window		F2UZ 15279D46 C	**26.0**	563.35
w/o Window		F2UZ 15279D46 D	**29.0**	538.73
Extended Van/Wagon				
Hinged Door				
w/Window		F2UZ 28279D46 A	**32.0**	711.35
w/o Window		F2UZ 28279D46 B	**31.0**	698.13
Sliding Door				
w/Window		F5UZ 28279D46 B	**33.2**	711.35
w/o Window		F5UZ 28279D46 A	**34.2**	618.57
¶Included w/Body Side Panel Assy				
4 **Panel, Inner**				
w/Window		F2UZ 1527864 C	**8.0**	312.62
w/o Window		F2UZ 1527864 A	**6.0**	312.62
5 **Extension, Inner Rear**				
Standard Van/Wagon		F2UZ 1527998 A	**2.0**	64.25
Extended Van/Wagon				
W/Window	92-94	F2UZ 2827998 A	**3.0**	63.10
	95-97	F5UZ 2827998 A	**3.0**	67.93
w/o Window		F5UZ 2827998 B	**2.5**	71.33
6 **Pillar, Hinge**				
Rear Door		F4UZ 15429A20	**4.5**	N.A.
7 **Reinf, Inner Panel**				
Standard Van/Wagon				
w/XLT Trim		F2UZ 1527947 A	**4.0**	114.22
w/o XLT trim		F2UZ 1527947 B	**4.0**	99.32
Extended Van/Wagon		F2UZ 2827946 A	**4.5**	76.32
8 **Filler, Side Panel Opening**				
Sliding Door		F2UZ 15278D72 A	**.5**	24.95
9 **Support, Inner Panel**				
Hinged Door		F2UZ 15280C06 A	**.5**	51.60
10 **Ext. Pan Side Front**		F2UZ 15113A44 A	**4.0**	20.83
11 **Wheelhouse, Outer**		F2UZ 1527886 B	**1.5**	175.18
12 **Wheelhouse, Inner**		F2UZ 1527886 A	**2.0**	175.18
13 **Ext, Pan Side Rear**				
Standard Van/Wagon		F2UZ 1511252 A	**1.5**	27.75
Extended Van/Wagon		F2UZ 1511252 A	**2.0**	27.75

FIGURE 8–25 Van side replacement parts and labor times. (Courtesy of Mitchell International.)

CHEVROLET/GMC TRUCK C/K

PICKUP BED

SHEET METAL STEPSIDE

Operation		Time
Refinish Side Panel		
Outside	R/L	**1.8**
Inside	R/L	**1.8**
Refinish Front End Panel (Complete)		**2.0**
Refinish Inside		**1.0**
Refinish Floor (Complete)		**#3.0**
#Includes Floor Sill		
Refinish Fender	R/L	**2.8**
Refinish Floor Sill		**.8**
Refinish Tailgate		
Complete		**2.3**
Outside		**1.9**
Inside		**1.0**
Refinish Fuel Door		**.3**
R&I Bed Assy		**2.5**
R&I Tailgate Assy		**.2**
Add to R&R Tailgate		
w/Welded Hinge (Each)	R/L	**.5**

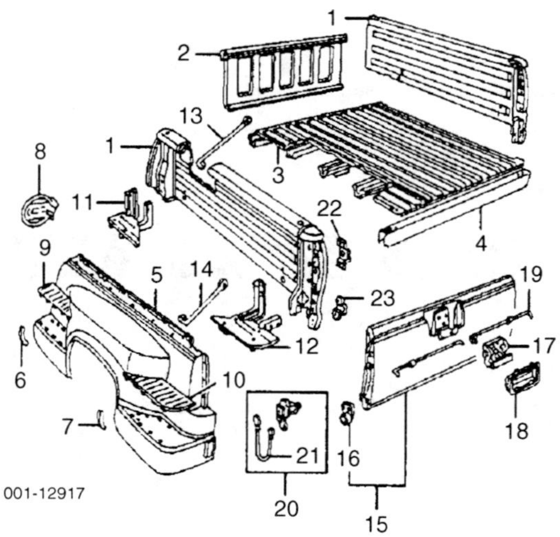

Part		Part No.	Time	Price
Box Assy		15663478	**#9.5**	2058.00
#Includes Transfer Right & Left Fenders, Brackets & Tailgate				
1 **Panel, Side¶**	R	15678726	**#7.5**	606.00
	L	15678725	**#7.5**	606.00
#Includes R&I Bed Assy, w/Rear Fender Removed				
2 **Panel, Front End¶**		15593140	**#5.5**	114.00
#Includes R&I Bed Assy, w/Side Panel Assy Removed Deduct 1.0				
3 **Panel, Floor¶**		15611232	**#18.5**	520.00
#Deduct 4.0 for Each Side Panel Removed				
4 **Sill, Floor¶**		15611240	**#4.5**	79.00
¶Included w/Bed Assy				
#Included in R&R Floor Pan, Deduct 1.0 for Each Side Panel Assy Removed				
5 **Fender**	R/L	15613092-1	**#3.5**	537.00
#Includes R&I Pads				
Protector, Fender				
6 Front Edge				
1988-91	R/L	15962536-5	**.1**	37.00
1992	R/L	15962540-39	**.1**	4.20
1993-94	R/L	15962530-29	**.1**	4.20
1995-97	R/L	15706730-29	**.1**	4.20
7 Wheel Opening	R/L	15962540-39	**.1**	4.20
8 **Door, Fuel Tank**		15630142	**.3**	20.35
Pad, Step				
9 Front	R/L	15552208-7	**.3**	14.55
10 Rear	R/L	15552210-09	**.3**	12.05
Bracket, Fender				
11 Front	R	15552222	**#.3**	60.50
	L	15552221	**#.3**	65.00
12 Rear	R/L	15552224-3	**#.3**	55.50
#w/Fender Removed				
Brace, Fender Lower				
13 Front	R/L	15552212-1	**.2**	8.65
14 Rear	R/L	15606924-3	**.2**	9.55
15 **Tailgate Assy**		12540227	**1.1**	209.00
16 **Hinge, Inner¶**				
1998-92				
2 Bolt Mtg	R	15603560	**#.2**	7.45
	L	15545203	**#.2**	7.45
1 Bolt Mtg	R/L	15691088-7	**#.2**	8.45
1993-97	R/L	15691088-7	**#.2**	8.45
¶Included w/Tailgate Assy				
#w/Tailgate Removed				
17 **Handle, Tailgate**		159917685	**#.3**	30.25
18 **Bezel, Handle**		15991786	**#.3**	10.05
#R&R Complete				
19 **Rod, Latch**	R/L	15562722		5.35
20 **Latch & Cable Assy (a)**				
1988-91¶	R/L	15724158-7	**.4**	19.75
1992-95	R/L	15724158-7	**.4**	19.75
1996-97	R/L	15724158-7	**.4**	19.75
¶Replaces Original Latch/Link Assy				
(a) Manufacturer Recommends Replacement in Pairs				
21 **Cable, Latch¶**	R/L	15673251	**.2**	7.45
¶Included w/Latch & Cable Assy				
22 **Striker**	R/L	15724162-1	**.2**	3.15
23 **Hinge, Outer**	R	15603558	**#.2**	5.00
	L	15521579	**#.2**	5.20

FIGURE 8–26 Bed side replacement times. (Courtesy of Mitchell International.)

and replacement of rear fender, running board, gas neck, and caulking are also included. The following are not included operations and must be figured separately if needed:

- R&I wheel
- R&I options: topper, bed liner, fender flares
- Install tape stripe and decals
- Drill or cut holes for trim or lamps
- R&I or R&R mouldings or trim
- Refinish bed assembly

In some cases, a good used or take-off box can be installed. This straightforward operation may be more cost-effective than bed side replacement (Figure 8–27).

8.6 FRONT AND REAR DOORS

A door is one of the most complex and expensive body panels. A typical door has an inner shell and an outer panel called a skin (Figure 8–28). The shell is made of sheet metal, but the door skin may be metal, fiberglass, or plastic. The skin is either welded or crimped and bonded to the inner shell. A reinforcement bar is welded to the door shell just inside the door skin. The reinforcement, also called an intrusion bar, is made of high-strength steel and is designed to

PICK-UP BED

Refinish Pick-Up Bed Assy	8.4
Add to Refinish if Required	
R&I Wheel Opening Moulding	.2
R&I Mudguard	.2

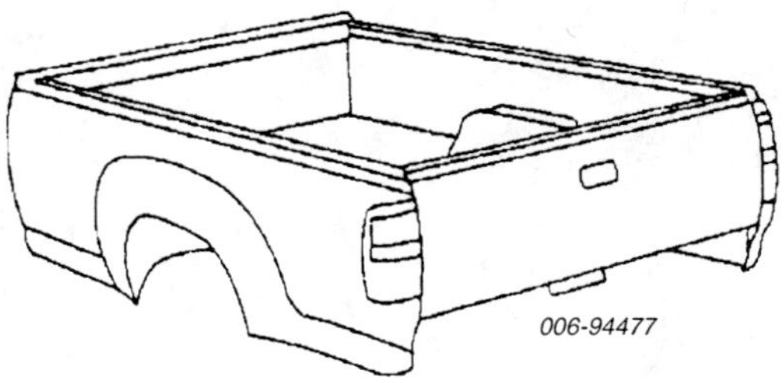

Replace	2.5
Add to R&R	
Tail Lamp (Each)	.3
Wheel Opening Moulding	.2
Mudguard	.2
Fuel Tank m	1.8

FIGURE 8–27 Replacement assembly estimating guide listing for bed replacement in a Toyota Tacoma pickup truck.

prevent collision forces from bending the door in on the vehicle occupant. The door is connected to the hinge pillar (front or enter) by hinges. Most doors are bolted to the hinges and the hinges either bolted or welded to the pillar.

In addition to the shell, skin, and hinges, a typical door assembly has many attachments and internal components (Figure 8–29). On the exterior of the door, the following might be attached: door handles, lock cylinders, latch, side mirror, mouldings, decals, emblems, stripe tape, or overlays. Inside the door frame is the window glass, glass channels and runs, regulator (either manual or electric), wiring harnesses, locking mechanisms, and exterior mirror controls. Most if not all of the interior surface of the door shell is covered with trim panels. Door pulls, arm rests, control panels, and window regulator handle are also mounted on the interior side of the door shell.

The outer door panel can be repaired using suction cups, pry bars and picks, or stud welders and a pulling ram. If the damage is severe, however, the door skin can be replaced separately from the door shell. This procedure should be specified whenever possible rather than replacing the door shell. Door skins are available from the OEM manufacturer and from aftermarket distributors, but they are not available in LKQ parts.

The replacement procedure usually requires removing the door from the vehicle, and removing the glass from the frame to avoid breaking it. Then the

FRONT DOOR

Use Procedure Explanations 16 and 28 with the following text.

SHEET METAL

Refinish Door Outside	R/L	**2.1**
Add for Jambs & Interior	R/L	**1.0**
Add for Jambs	R/L	**.5**
R&I Door Assy	R/L	**.4**
Section Door Repair Panel	R/L	**#5.3**

#Time is to Cut in Window Opening, Includes R&R Mirror & Door Glass

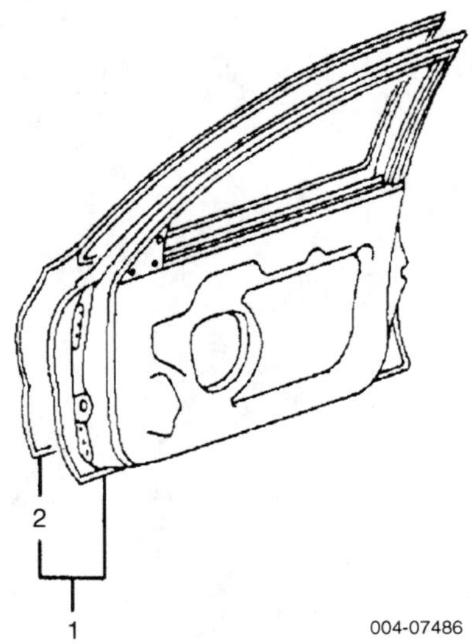

1 **Shell Assy, Door**	R	4646466	**#4.5**	400.00
	L	4814851	**#4.5**	400.00
#Includes R&R Mirror				
2 **Panel, Door Repair¶**	R	4646468	**#6.0**	108.00
	L	4646469	**#6.0**	107.00
¶Included w/Door Shell Assy				
#Includes R&R Mirror & Door Glass				

EXTERIOR TRIM

R&I Rear View Mirror	R/L		**#.6**	
#Includes R&I Trim Panel				
Mirror, Rear View				
1 Manual Remote	R	4646802	**#.2**	92.50
	L	4646803	**#.2**	91.00
2 Power Remote	R/L	4646308-9	**#.2**	108.00
#w/Trim Panel Removed				
Bezel, Inside				
3 Man Remote Mirror	R/L	4630404-5		9.10
4 Pwr Remote Mirror	R/L	4630402-3		7.35
Moulding, Belt-Serviced w/Belt Weatherstrip				
See Glass & Parts Section				
5 **Moulding, Side¶**				
Breeze				
Black	95-96 R	LD44SX8	**.2**	37.25
	L	LD45SX8	**.2**	37.25

FIGURE 8–28 Front door shell, door skin, and exterior trim. (Courtesy of Mitchell International.)

spot welds are drilled out and any brazed metal is ground off. The door skin is removed from the door shell by grinding through the hem flange along the edge of the door (Figure 8–30). After the removal of any leftover flange or weld material, the door shell flange is cleaned and straightened if necessary. A

GLASS AND PARTS

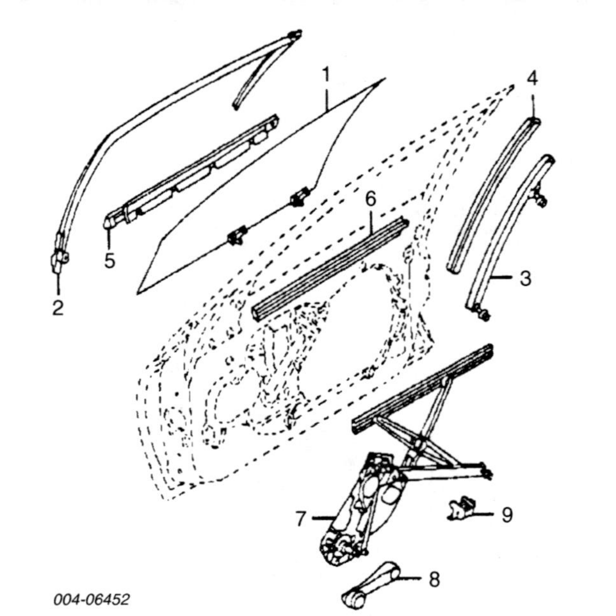

1 **Glass, Front Door**				
Chrysler-Tinted	R/L	4646398-9	**.9**	189.00
NAGS-Green	R/L	DD08738-9GTY	**.9**	405.00
2 **Channel, Run**	R/L	4646826-7	**#1.2**	153.00
#w/Door Glass Removed Deduct .9				
3 **Guide, Rear Lower**	R/L	4646406-7	**#.6**	7.00
#w/Door Glass Removed Deduct .5				
4 **Channel, Rear Lower**	R/L	4646964		9.85
5 **W/Strip, Belt Outer**	R/L	4630190-1	**.3**	27.50
¶*Includes Belt Moulding*				
6 **W/Strip, Belt Inner**	R/L	4646504-5		26.00
7 **Regulator, Glass**				
Manual	95 R	4646272	**#.7**	45.50
	L	4646273	**#.7**	44.25
	96-97 R/L	4696998-9	**#.7**	49.75
Power	95-96 R/L	4778574-5	**#.7**	148.00
	97 R/L	4778576-7	**#.7**	117.00
#w/Trim Panel Removed Deduct .4				
8 **Handle, Man Regulator**	R/L	FE86RX9		6.60
Motor, Pwr Regulator	R/L	4696579-8	**#.2**	74.00
#w/Regulator Removed				
9 **Brkt, Door Pull-See Interior Trim Section**				

FIGURE 8–29 Door glass and parts. (Courtesy of Mitchell International.)

new door skin is positioned over the door shell and the hem flange crimped over the shell flange. The skin is secured to the shell either with new spot weld or with adhesives. The upper edge of the skin is usually MIG welded to the top of the door shell to seal out moisture and dust and to provide a closed, finished seam.

Operation times for R&R of the door skin are given in the collision estimating guides and include the following operations:

- R&I of the door from the vehicle
- R&I of the inner trim panel
- R&R of caulking
- Either R&R or R&I of the outside handle, the lock cylinder and the weatherstrip on the door edge
- Replacing clip-on type moulding

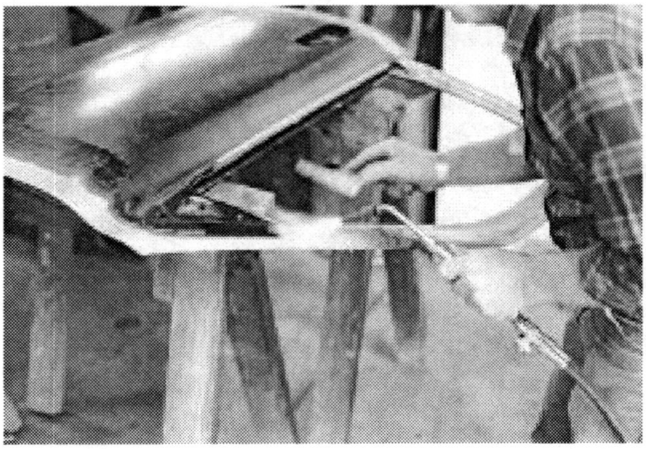

FIGURE 8–30 Door skin replacement. The skin has been installed. A splice joint has been made in the window frame. This shop uses only lead as filler.

The times listed for R&R of the door skin do not include:

- Either R&R or R&I of the door glass, hardware, mirror, and channels
- Replacing sound-deadening material
- Recoding the lock cylinder
- R&I of adhesive-type exterior trim or installing new adhesive trim
- Installing stripe tape, decals, transfers, or overlays
- Drilling holes for installing exterior trim
- Cleaning up broken glass
- Refinishing the panel and shell

If the collision forces have damaged the door shell to the point where it is not possible or economically feasible to repair the shell, a new door shell can be purchased from the manufacturer or from an aftermarket vendor. The new door shell comes with the door skin attached. The new shell also comes equipped with the inner door reinforcement bar and other reinforcements. However, the window glass, runs, and regulator and other parts attached to the door shell are not included in the OEM or aftermarket shell. These parts must be removed from the old vehicle and installed on the new.

It is very important to note that the R&R times for a door shell include transferring parts from the old door shell to the new. These items should not be listed separately on the damage report. For example, the time required to remove a window regulator from the old door shell and install it on the new one is included in the door shell R&R time. The following is a list of operations that are included in the shell operation times:

- Replacing or transferring all parts attached to the door (door handles, glass, run channels, regulator vent assembly, and weatherstripping)
- Removing and installing the trim panel
- R&I of the lock cylinder
- Applying new caulking to the factory seams
- Replacing standard equipment, clip-on type mouldings

The listed operation times do not include time for removing window glass from the regulator channel. When this is necessary, add 0.2 hour to the door shell R&R time for removing the lower channel from the glass and 0.3 hour for removing both the upper and lower channels. Additional time must be added to the damage report for the cleanup of any broken glass.

Some other operations that are not included in the listed R&R times are the following:

- Replacing the sound-deadening material
- Removing and installing or replacing the rearview mirror
- R&R or R&I of weld-on–type hinges
- R&R or R&I of hinge anchor plates
- R&R or R&I of decals (such as EPA or body identification)
- Recoding the lock cylinder
- R&I or replacing adhesive trim
- Installing stripes, decals, transfers, or overlays
- Drilling holes for moulding installation
- Refinishing the door

When any of these operations are necessary, time must be allowed and entered separately on the damage report. The collision estimating guides list flat rate labor times for some of the operations; others must be negotiated between the estimator and adjuster.

Salvage doors should be considered when not only the door skin but the frame and interior components are damaged. A salvage door comes as a complete assembly, with glass, regulators, and other components in place. When replacing a damaged door with a LKQ unit, list the R&I time for the door and the R&R time for any part such as inner trim and lock cylinder that must be replaced or transferred to the replacement door.

It is very important to refer to the head notes and footnotes in the collision estimating guide when determining flat rate times (Figure 8–31). Additions must be made to the operation times when the vehicles contain options not found on the base model. For example, additional time must be added if the door is equipped with a theft-deterrent system, a remotely controlled mirror, illuminated locks, speakers, electronic locks, or electric regulators. The footnotes often indicate prior operations, as well. For example, when replacing the reinforcement bar, it is assumed in the operation times that the door skin has been removed.

The collision repair guides list R&R times for individual parts. As stated earlier in this chapter, these individual times are included in the door shell R&R. If, however, a part or parts must be replaced individually (for example, broken glass or a nonworking window regulator), list the part separately and include the trim panel R&I time listed in the manual.

8.7 REAR BODY AND LIGHTS

The rear of the body shell has several components. These components vary from body style to body style. On today's unibody vehicle, the rear body (Figure 8–32) consists of a rear body panel, the rear floor pan, rear rails and various cross members, reinforcements, filler panels, and extensions. On some vehicles, an upper rear panel is situated between the lower edge of the rear window and the luggage lid and is welded to the quarter panels (Figure 8–33). On other vehicles, the upper rear panel is a part of the package tray. Some vehicles also have a finish panel that is attached to the outside of the rear body panel. When determining the actual repairs necessary to correct rear body damage, the estimator and adjuster must keep in mind that many rear body panels are not appearance panels. For example, the parcel shelf is normally covered with a plastic trim, and rough metalworking is sufficient to restore the metal panel to its basic shape and dimensions. Floor pans and rails can be straightened using hydraulic pulls and rams (Figure 8–34).

However, when the damaged panels cannot be economically repaired, the collision estimating guides give operation times and parts prices for their replacement. R&R of the rear body panel includes removal and reinstallation of the rear bumper; filler panels, stone defector and/or the lower valance; any weatherstripping attached to the panel; and the rear light assemblies. Also included is time to either R&R or R&I the impact absorbers or mounting arms (when this operation is necessary before the rear panel can be removed). R&I or R&R of the lock cylinder, the latch, and the lock striker are included when these parts are attached to the panel. R&R of the factory caulking and clip-on type moulding is also included in the listed times.

FRONT DOOR

Use Procedure Explanations 16 and 28 with the following text.

SHEET METAL

Operation			
Refinish Door Outside			
Coupe, Convertible	R/L	**2.2**	
Sedan	R/L	**2.1**	
Add for Jambs	R/L	**.5**	
Add for Jambs & Interior	R/L	**1.0**	
R&I Door Assy			
w/Power Windows	R/L	**1.0**	
w/o Power Windows	R/L	**.5**	
Add to R&R Door Shell			
w/Power Windows	R/L	**.3**	
w/Power Locks	R/L	**.2**	
To Drill Holes for Lower Side Moulding	R/L	**.4**	
Add to R&R Door Repair Panel			
w/Power Option	R/L	**.5**	
To Drill Holes for Lower Side Moulding	R/L	**.4**	

Part	Part No.	Labor	Price
Shell Assy, Door			
Coupe	95 R★22651366	**#4.8**	289.00
	L★22651367	**#4.8**	289.00
	96 R★22651366	**#4.8**	289.00
	L★22651367	**#4.8**	289.00
	97 R★22658924	**#4.8**	292.00
	L★22658925	**#4.8**	301.00
Convertible	95-96 R★22574406	**#4.8**	500.00
	L★22658929	**#4.8**	297.00
	97 R★22658928	**#4.8**	297.00
	L★22658929	**#4.8**	297.00
Sedan	95-96 R★22651362	**#4.8**	281.00
	L★22651363	**#4.8**	281.00
	97 R★22651362	**#4.8**	281.00
	l★22651363	**#4.8**	281.00
#Includes R&R Rear View Mirror			
Panel, Door Repair¶			
Coupe, Convertible (a)	R 22570424	**#5.5**	177.00
	L 22570425	**#5.5**	177.00
Sedan	R 22570446	**#5.5**	171.00
	L 22570447	**#5.5**	171.00

¶Included w/Door Shell Assy
(a) Rework Coupe Door Repair Panel for Use on Covertible
#Includes R&I Door Glass & Rear View Mirror

EXTERIOR TRIM

Operation		
Refinish Side Lower Moulding	R/L	**.6**
Refinish Rear View Mirror	R/L	**.5**
R&I Rear Frame Applique	R/L	**.2**
R&I Belt Moulding	R/L	**#.9**
#Includes R&I/R&R Trim Panel & Rear View Mirror		
R&I Rear View Mirror	R/L	**#.7**
#Includes R&I Trim Panel		

001-11539

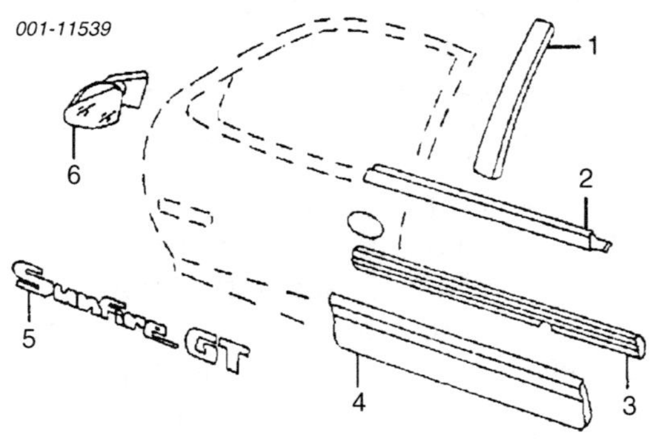

Part		Part No.	Labor	Price
1 **Applique, Rear Frame**				
Coupe	R/L	22639454-5	**.3**	14.70
Sedan	R/L	22639456-7	**.3**	14.70
2 **Moulding, Belt**				
Coupe	R/L	22649232-3	**#.9**	30.75
Sedan	R/L	22657630-1	**#.9**	14.35
#Includes R&I/R&R Trim Panel & Rear View Mirror				
3 **Moulding, Side¶**				
Coupe, Convertible				
Aqua	R/L	22588620-1	**.2**	43.25
Black	R/L	22588618-9	**.2**	43.25
Blue (26U)	R/L	22588622-3	**.2**	43.25
Red (81U)	R/L	22588624	**.2**	43.25
White	R/L	22588614-5	**.2**	43.25
Sedan				
Aqua	R/L	22588642-3	**.2**	34.25
Black	R/L	22593670-1	**.2**	34.25
Blue (26U)	R/L	22588644-5	**.2**	34.25
Red (81U)	R/L	22588646-7	**.2**	34.25
White	R/L	22588638-9	**.2**	34.25
¶Order by Application				
4 **Mldg, Side Lower (P)**	95-96 R/L	12356862-3	**.3**	146.00
	97 R/L	N.A.	**.3**	N.A.
(P) Paint to Match				
5 **Nameplate¶**				
"Sunfire"				
Aqua	R/L	22596472	**.2**	5.40
Black	R/L	22596471	**.2**	5.40
Blue (26U)	R/L	22596474	**.2**	5.40
Red (81U)	R/L	22596473	**.2**	5.40
White	R/L	22596470	**.2**	5.40
"GT"				
Aqua	R/L	22596481	**.2**	9.80
Black	R/L	22596480	**.2**	9.80
Blue (26U)	R/L	22596483	**.2**	9.80
Red (81U)	R/L	22596482	**.2**	9.80
White	R/L	22596479	**.2**	9.80
¶Order by Application				
6 **Mirror Assy, Rear View (P)**				
Manual Remote				
Coupe	95 R	22648704	**#.7**	62.25
	L	22648781	**#.7**	62.25
	96 R/L	22648780-1	**#.7**	62.25
	97 R/L	22652876-7	**#.7**	62.25
Convertible	95 R/L	22643976-7	**#.7**	62.25
	96-97 R/L	22648712-3	**#.7**	62.25
Sedan	95 R	22642278	**#.7**	62.25
	L	22648785	**#.7**	62.25
	96-97 R/L	22648784-5	**#.7**	62.25
Power Remote				
Coupe	95 R/L	22642284-5	**#.7**	107.00
	96 R	22648782	**#.7**	107.00
	L	22652879	**#.7**	107.00
	97 R/L	22652878-9	**#.7**	107.00
Convertible	95 R	22648714	**#.7**	107.00
	L	22643979	**#.7**	107.00
	96-97 R/L	22648714-5	**#.7**	107.00
Sedan	95 R/L	22642280-1	**#.7**	107.00
	96-97 R/L	22648786-7	**#.7**	107.00
(P) Paint to Match				
#Includes R&I Trim Panel				
Glass, Mirror¶				
Manual	R	22636132	**.3**	33.00
	L	22638227	**.3**	28.25
Power	R	22636132	**.3**	33.00
	L	22636187	**.3**	34.00
¶Included w/Rear View Mirror Assy				

FIGURE 8–31 Additional labor and parts for a door. (Courtesy of Mitchell International.)

REAR BODY

Use Procedure Explanations 26 and 28 with the following text.

SHEET METAL

Operation		Time
Refinish Rear Body Panel		**1.2**
Add for Inside		**.7**
Refinish Rear Panel Extension	R/L	**.4**

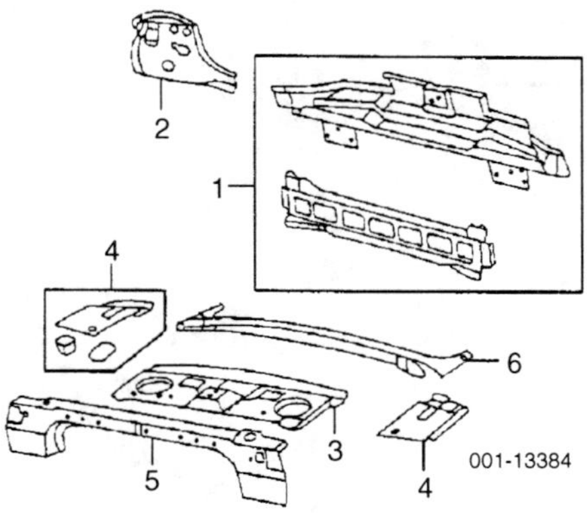

Part	R/L	Part No.	Time	Price
1 **Panel, Rear Body**		22652760	**#6.0**	N.A.
#Deduct .5 w/Each Rear Panel Extension Removed, w/Quarter Panel Removed Deduct 1.6, Deduction Includes Common Surfaces & Included Operations				
2 **Ext, Rear Panel**	R/L	22652746-7	**1.0**	14.25
3 **Panel, Package Shelf**		N.A.	**5.0**	N.A.
4 **Ext, Package Shelf**	R/L	22650898-9	**1.0**	33.50
5 **Reinf, Rear Pkg Shelf**	N.A.		**3.0**	N.A.
6 **Panel, Rear Window Lwr**		22593129	**3.5**	18.50

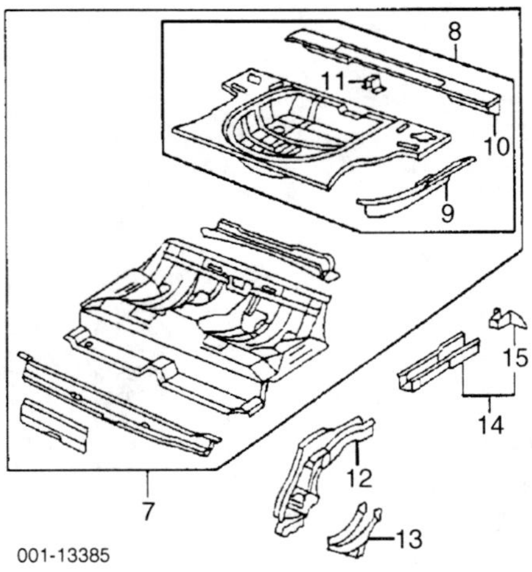

Part	R/L	Part No.	Time	Price
7 **Panel Assy, Rear Compt**	N.A.		**#14.5**	N.A.
#w/Rear Body Panel Removed, Deduct 1.5 for Each Inner Wheelhouse Panel Removed				
8 **Panel Assy, Rear Floor¶**		22602235	**#10.0**	625.00
#w/Rear Body Panel Removed, Deduct 10.0 for Each Inner Wheelhouse Panel Removed				
9 **Shield, Heat¶ (a)**		22648469	**1.5**	11.10
10 **Panel, Rear¶ (a)**		22593149	**2.5**	30.00
11 **Plate, Anchor¶ (a)**		22637799	**.4**	9.20
¶Included w/Rear Compt Panel Assy				
(a) Included w/Rear Floor Panel Assy				
12 **Rail, Front Floor**	R/L N.A.		**s #6.5**	N.A.
#w/Rear Floor Rail Removed Deduct .5				
13 **Bracket, Rear Susp**	R/L N.A.		**s 1.5**	N.A.
14 **Rail Assy, Rear Floor**	R/L	22654442-3	**s #5.0**	N.A.
#w/Rear Body Panel Removed				
15 **Reinf, Rear Rail¶**				
Outer	R/L	22592858-9	**s .5**	3.90
Inner	R/L	22592859-8	**s**	3.90
¶Included w/Rear Floor Rail Assy				

FIGURE 8–32 Rear body parts and labor times. (Courtesy of Mitchell International.)

FIGURE 8–33 An upper rear body panel.

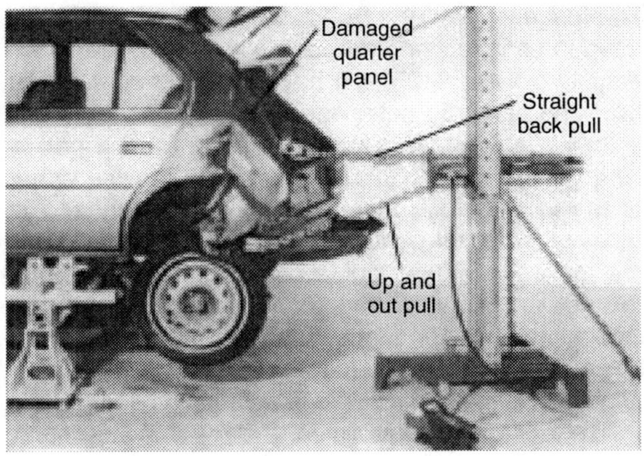

FIGURE 8–34 Rear rails and pans can often be pulled into alignment. (Courtesy of Wedge Clamp Intl., Inc.)

The following are not included in the rear panel R&R times:

- Removing and installing the fuel tank assembly
- Removing and either installing or replacing the wiring and/or wiring harness
- Recoding the lock
- Removing the installing adhesive moulding
- Installing stripe tape, decals, transfers, or overlays
- Drilling holes for installing exterior trim
- Cutting holes for installing lights

The operation times given for removing and replacing the rear side rails and the floor pan only include the time necessary to drill out the factory welds and to align and reweld the new panels in place. The flat rate times listed in the collision estimating guides do not include time for removing and installing parts attached to the damaged panels (such as suspension assemblies, axle assembly, fuel tank, brake and fuel lines) or R&I of parts that must be removed for access. These operations must be entered separately on the damage report. The damage report must also include labor allowance for setting the vehicle up on a bench, pulling damaged adjacent panels into alignment, and measuring the damage. Correct realignment of rear panels is especially important on four-wheel drive vehicles and vehicles with independent rear suspensions.

8.8 LUGGAGE LID

The luggage lid, or rear deck lid, is very similar to the hood in construction and operation. A luggage lid assembly (Figure 8–35) will have most if not all of the following parts: an outer panel, an inner panel, a lock cylinder, a latch assembly, a striker, and a pair of hinges. The outer panel is spot welded to the edges of the inner panel (Figure 8–36) and the inner surfaces of the inner panel are glued to the underside of the top panel. This double panel construction limits the amount of metal straightening and finishing that can be done on the luggage lid. When the lid is damaged beyond a reasonable cost of repair, the upper and lower panels are replaced as a unit.

The operation times listed in the collision estimating guides for removing and replacing the luggage lid include time for removing and either reinstalling or replacing the lock cylinder, latch, lock striker, weatherstripping, light assemblies, and wiring harnesses, if these are attached to the lid. Otherwise, these items are not included in the R&R time. Time is also allowed for replacing clip-on type mouldings and seam sealer.

The luggage lid R&R times given in the collision estimating guides do not include time for:

- R&I or R&R of the hinges
- R&I or R&R of luggage racks or spoilers
- R&I of decals, such as jacking instructions and cautions
- R&I or adhesive type mouldings or installation of new adhesive moulding
- Installing stripe tape, decals, transfers, or overlays
- Drilling holes for mounting exterior trim
- Recoding the lock cylinder
- Refinishing the luggage lid

LUGGAGE LID

Use Procedure Explanations 24 and 28 with the following text.

Operation		Time
Refinish Luggage Lid Outside		**2.5**
Add for Underside		**1.1**
Refinish Hinge	R/L	**.4**
R&I Luggage Lid		**.5**
R&I Lower Applique		**.4**

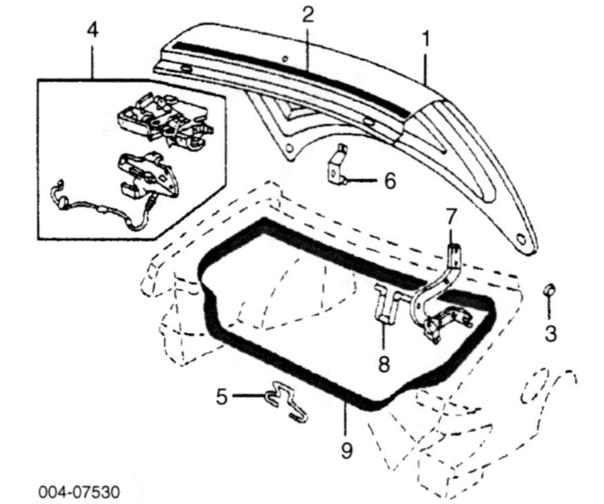

Part	Year/Side	Part No.	Time	Price
1 **Lid, Luggage (HSLA)**	94	4723823	**1.5**	520.00
	95-96	4778189	**1.5**	445.00
	97	4874152	**1.5**	495.00
2 **Applique, Lid Lower**				
New Yorker		4584274	**.2**	94.50
LHS		4630479	**.2**	97.00
Nut, Attaching (a)		6502589		1.00
(a) Use as Required				
3 **Bumper, Hinge Stop**	R/L	4265379		1.45
4 **Latch**		4624762	**.4**	38.75
Bracket, Latch Retainer		4646853		5.00
5 **Striker, Latch**	94-96	4646079	**.2**	11.00
	97	4756483	**.2**	13.25
Cylinder & Key, Lock	94	4746672	**.2**	12.50
6 **Bracket, Lock Cyl Mtg**	95-97	4778145	**.2**	11.50
To 6-16-93		4624788		2.70
From 6-16-93		4756140		2.60
7 **Hinge, Lid**				
1994-95				
To 3-26-93	R	4756522		75.00
	L	4624393		72.50
From 3-26-93	R	4756522		75.00
	L	4575553		67.00
1996	R/L	4769164-5		73.00
1997	R/L	4575552-3		67.00
8 **Bar, Torsion**	R	4624556	**#.5**	13.25
	L	4624557	**#.5**	11.00
#R&R One or Both				
9 **Weatherstrip, Lid**		4583769	**.2**	61.50
Panel, Lid Trim		4758235	**.2**	98.50
Retainer, Trim Panel¶		6030441		1.10
Push-Pin, Panel¶		6035507		.20
¶Use as Required				
Lamp, Luggage		4601012		10.75
Bulb, Lamp		154549		1.25

FIGURE 8–35 Luggage lid and related parts. Note that the lid is HSLA steel. (Courtesy of Mitchell International.)

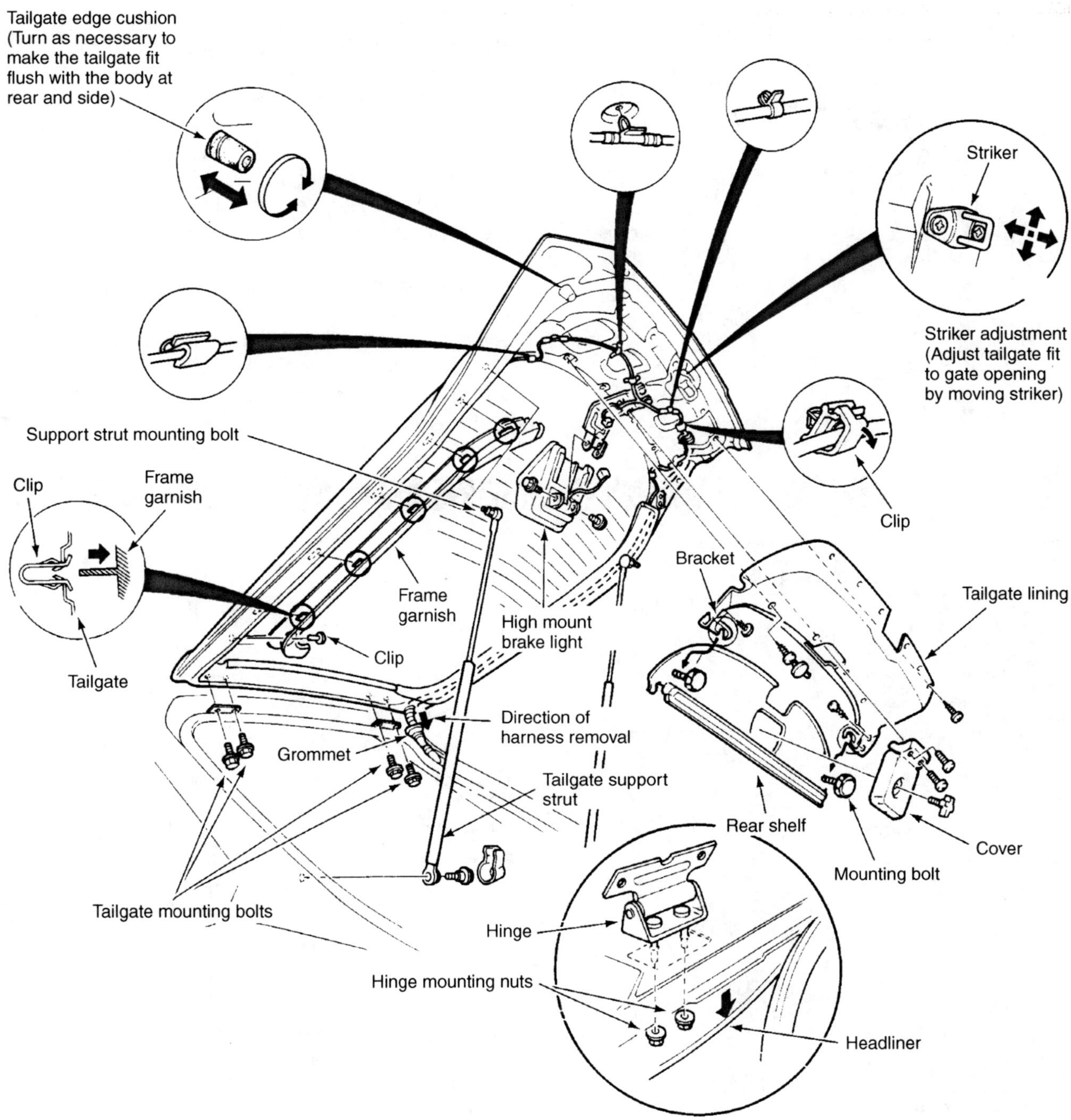

FIGURE 8–36 The luggage lid is made of two panels welded and glued together. (Courtesy of American Honda Motor Co., Inc.)

8.9 TAILGATES AND LIFTGATES

The labor rates given in the collision estimating guides for replacing a tailgate on a station wagon and a liftgate on a hatchback or van (Figure 8–37) include the same items as described for a luggage lid. However, some additional operations are included in the R&R times because the tailgates and liftgates contain the rear glass. The R&R times for a tailgate or liftgate (Figure 8–38) include time for R&R of the rear glass and time to replace or transfer parts attached to the gate(such as glass run channels, regulator, outside handles, weatherstripping, and standard light assemblies).

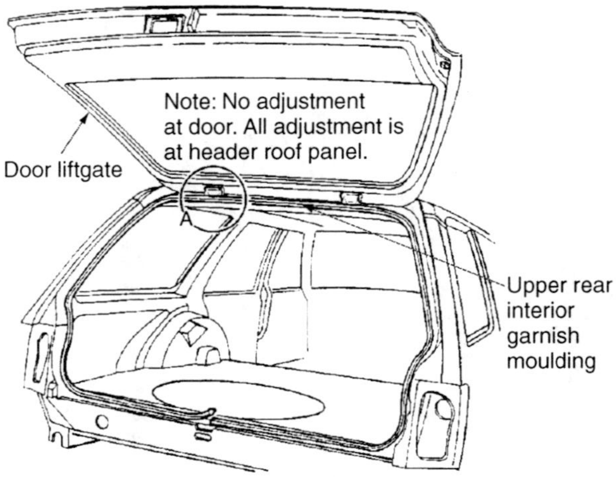

FIGURE 8–37 Rear liftgate.

However, like the operation times for a door, the R&R times for a tailgate or liftgate do not include time for removing the channels from the glass; 0.2 hour should be added to the listed time for the removal of the lower channel and 0.3 hour for the removal of both the upper and lower channels. Nor is time included for transferring or replacing wiper and washer assemblies or hinge anchor plates. Additional time for these operations must be included on the damage report.

The outer panel of a tailgate can be replaced separately from the gate shell. The procedure used to replace the outer panel is similar to the procedure described earlier for replacing door skins. When this procedure is the chosen repair option, the repair panel R&R times given in the collision estimating

LIFTGATE

Use Procedure Explanations 25 and 28 with the following text.

SHEET METAL

Operation		
Refinish Liftgate Outside		**2.5**
Add for Jambs & Interior		**1.2**
Refinish Hinge	R/L	**.2**
R&I Lock Cylinder		**#.6**
#Includes R&I Trim Panel		
R&R Lock Cylinder		**#.2**
#w/Trim Panel Removed		
Add to R&R Liftgate		
To Transfer Glass		**.8**
w/Power Lock		**.2**

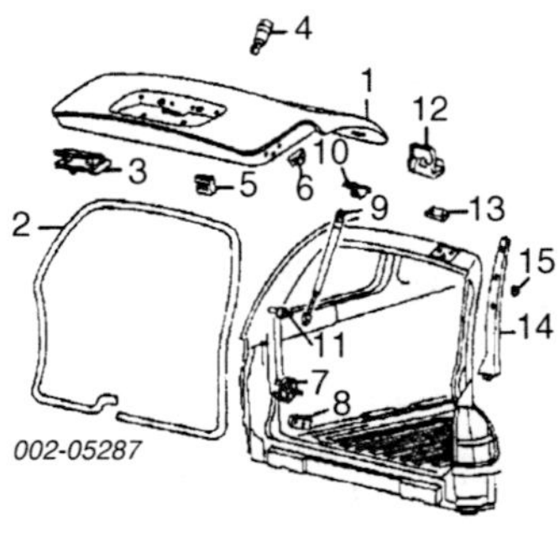

Part	R/L	Part No.	Labor	Price
1 **Shell, Liftgate**				
To 12-19-95		F68Z 1640010 AA	**#4.0**	453.33
From 12-19-95		F68Z 1640010 AA	**#4.0**	453.33
#Includes R&R Liftgate Glass, Wiper & Motor				
2 **Weatherstrip, Shell**				
Van		F58Z 16404A06 B	**.4**	74.53
Wagon		F58Z 16404A06 A	**.4**	74.53
3 **Handle, Liftgate**		F58Z 1643400 A	**#.3**	16.95
#w/Trim Panel Removed				
4 **Cylinder & Key-See Steering Wheel/Column Section**				
5 **Retainer, Cyl**		F58Z 1622023 AA		2.75
6 **Wedge & Plate, Liftgate**		F58Z 16404C04 A	**.1**	3.43
7 **Striker, Latch**		F58Z 16404A42 A	**.2**	7.18
8 **Wedge, Body Side**		F58Z 16404C28 A	**.2**	2.90
9 **Lift, Hydraulic**	R/L	F58Z 16406A10 A	**.1**	37.83
10 **Bracket, Lift Upper**	R/L	F1TZ 78427A48 B	**#.1**	8.18
11 **Retainer, Lower (2)**	R/L	E89Z 1141836 B	**#.1**	1.74
#w/Lift Removed				
12 **Hinge, Door**	R/L	F58Z 1642900 A	**#.2**	19.72
#w/Liftgate Removed				
13 **Pad, Hinge**	R/L	F58Z 1642840 A		3.04
14 **Moulding, Side**	R/L	F58Z 16410A06-7 A	**.2**	16.67
15 **Clip, Moulding (3/Side)**	R/L	D7VY 6050074 A		.29

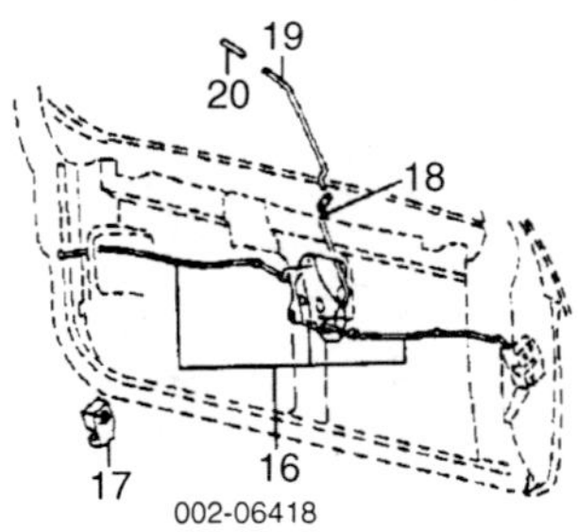

Part	R/L	Part No.	Labor	Price
16 **Control, Remote**		F58Z 1643170 A	**#.5**	80.93
17 **Latch, Gate**	R/L	F58Z 1643150-1 B	**#.2**	33.42
#w/Trim Panel Removed				
18 **Rod, Cylinder/Release**		F58Z 1643876 A		5.57
19 **Link, Latch Release**		F58Z 16431A78 A		6.07
20 **Knob, Rod**		D9ZZ 6621850 B		3.15

GLASS & PARTS

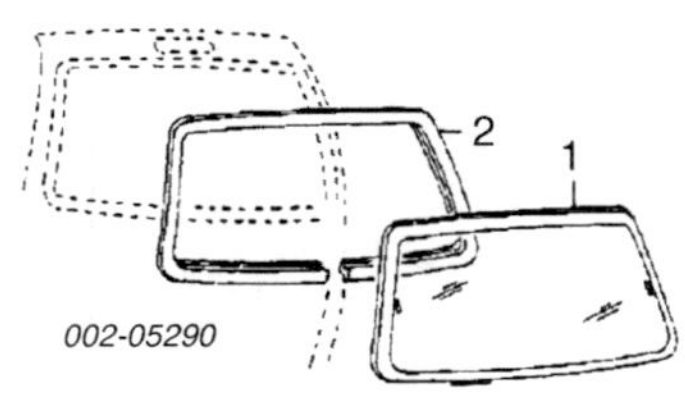

Part	Part No.	Labor	Price
1 **Glass, Liftgate**			
Ford			
Heated			
Green	F58Z 1642006 B	**#1.5**	1035.41
Gray-Privacy	F58Z 1642006 A	**#1.5**	1375.27
Non-Heated			
Green	F58Z 1642006 D	**#1.5**	776.77
Gray-Privacy	F58Z 1642006 C	**#1.5**	1031.77
NAGS			
Heated			
Green	DB08482GTY	**#1.5**	1035.45
Gray-Privacy	DB08483YPY	**#1.5**	1375.25
Non-Heated			
Green	DB08484GTY	**#1.5**	776.80
Gray-Privacy	DB08585YPY	**#1.5**	1031.80
#Includes Clean Up Old Adhesive as Necessary			
2 **Weatherstrip, Glass**	F58Z 1642084 A		40.52
Blade, Wiper	F58Z 17528 AC	**.1**	15.89
Arm, Wiper	F58Z 17526 C	**#.2**	45.23
#Includes R&I/R&R Blade			
Motor, Wiper	F58Z 17508 A	**#.3**	142.36
#w/Trim Panel Removed			
Nozzle, Washer	F58Z 17603 B		6.80

FIGURE 8–38 Estimate guide entry for liftgate and related parts. (Courtesy of Mitchell International.)

guides include time for removing the gate assembly from the vehicle and the glass and trim panel from the gate shell.

When the glass in the tailgate or liftgate must be replaced, the estimator and adjuster must determine whether the glass is heated or unheated when listing the glass price on the damage report. Also, the cost of the installation material must be included on the damage report if the liftgate glass is installed with an adhesive or urethane rather than the older rubber gasket method of installation.

8.10 BODY SECTION

In some cases, severe unibody rear or side damage is best repaired with a used assembly. A rear cliptop consists of the rear section of a car, from the windshield pillars and the floor under the dash to the rear bumper. This assembly includes the roof, floor and rocker panel sections, center pillar (on four-door cars), both quarter panels and wheelhouses, trunk floor, deck lid, and rear body panel (Figure 8–39, Figure 8–40, Figure 8–41). A rear clip does not include the roof or anything ahead of the rear seat (Figure 8–42). This type of repair involves less welding and disrupts fewer factory seams than individually replacing the parts. The savings in labor and parts may prevent a vehicle from being totalled. In a rear cliptop, the floor, rocker panels, and windshield pillars are sectioned (Figure 8–43, Figure 8–44). This repair requires the highest skill in measuring, welding, and corrosion protection. Pickup cabs can also be sectioned as cliptops.

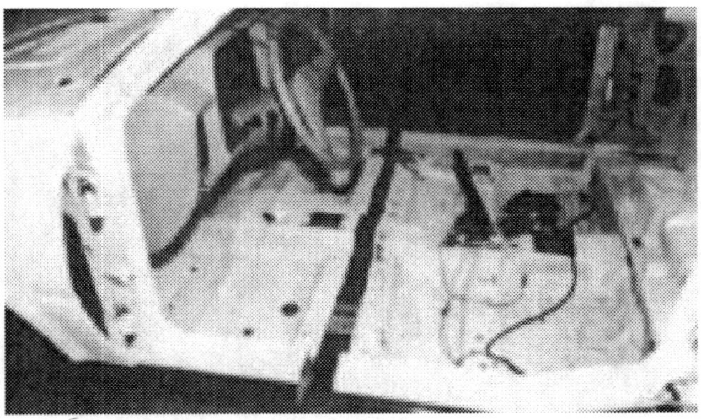

FIGURE 8–39 This is how the floor of the damaged vehicle is separated.

8.11 INTERIOR TRIM AND UPHOLSTERY

Chapter 5 discussed how to inspect for damage in the passenger compartment and how damage to the interior trim and upholstery (Figure 8–45) can be caused by both the collision forces and by the impact of passengers and objects. The damage to plastic parts can

FIGURE 8–40 This shows how the rear cliptop will be joined to the damaged vehicle, with splices in the windshield pillars, rocker panels, and floor.

REAR SECTION & ROOF

Use Procedure Explanations 25 and 32 with the following text.

Refinish Rear Section & Roof 18.7

NOTE: See "Adds" Under Quarter Panel & Luggage Lid for Additional Operations.

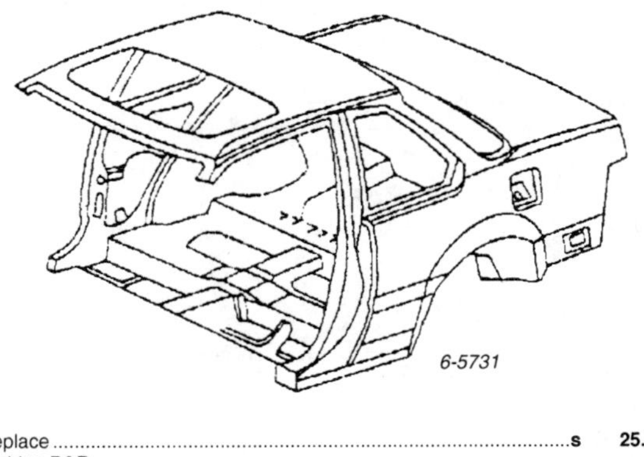

Replace .... s 25.0
Add to R&R
Each Moulding 2
One Tail Lamp 5
One Quarter Glass 1.5
Back Glass 2.0
Headliner 2.5

FIGURE 8–41 Rear cliptop for a Honda Prelude. (Courtesy of Mitchell International.)

REAR SECTION

Use Procedure Explanations 24 and 32 with the following text.

Refinish Rear Section **16.1**

NOTE: See "Adds" Under Quarter Panel & Luggage Lid for Additional Operations.

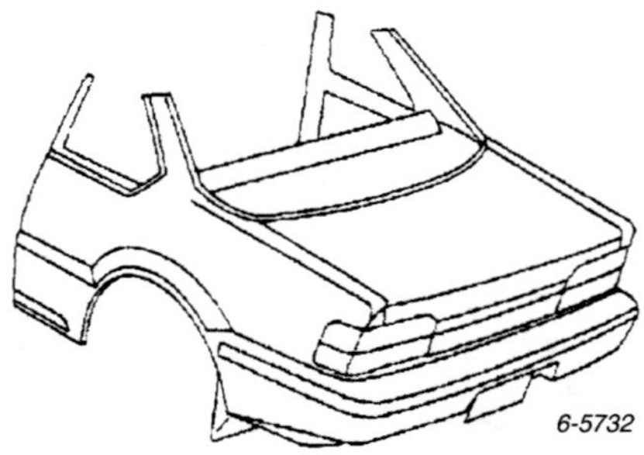

Replaced .... **s 25.0**
Add to R&R
Each Moulding .... **.2**
One Tail Lamp Assembly .... **.5**

FIGURE 8–42 Rear clip for a Honda Prelude. (Courtesy of Mitchell International.)

25 REAR SECTION AND ROOF

IMPORTANT REMINDER: No windshield, back or quarter glass (if applicable), or flush mounted reveal mouldings can be removed without the possibility of breakage. **Agree beforehand** who will incur the charge for damage occurring during normal R&R or R&I and whether additional labor time will be required to perform R&I operations on glass secured by urethane.

REAR SECTION R&R

INCLUDED OPERATIONS

- Cut in windshield openings and at base of lock pillar or dog legs
- Remove and install or replace:
  - Quarter trim
  - Windshield, seals or mouldings
  - Rear seat
  - Center pillar trim
  - Sill plates
  - Rear bumper
  - Luggage compartment trim
  - Lock cylinder
- Loosen or pull back carpet and insulation as required
- Remove and replace caulking

NOT INCLUDED OPERATIONS

- Refinish rear section
- Remove and install or replace:
  - Preformed or molded headliner
  - Restraint systems
  - Suspension assembly
  - Antenna
  - Fuel tank
  - Control cables (latch release, etc.)
  - Fluid lines
- Recode lock cylinder
- Remove from salvaged assembly
  - Trim panels
  - Lock cylinder

FIGURE 8–43 Rear cliptop included/not included operations. (Courtesy of Mitchell International.)

REAR SECTION

IMPORTANT REMINDER: No back glass or flush mounted reveal mouldings can be removed without the possibility of breakage. Unintentional damage to glass that is considered a structural member by the vehicle manufacturer may occur when attempting alignment pulls to correct unibody collision damage. **Agree beforehand** who will incur the charge for damage occurring during normal R&R or R&I, and whether additional labor time will be required to perform R&I operations on glass secured by urethane.

REAR SECTION R&R

INCLUDED OPERATIONS

- Cut at base of root line and at base of lock pillar or dog legs
- Remove and install or replace:
  - Quarter trim
  - Side and back glass, seals or mouldings
  - Rear seat
  - Sill plates
  - Rear bumper
  - Luggage compartment trim
  - Lock cylinder
- Loosen and pull back cloth type headliner, carpet or insulation as required
- Remove and replace caulking

NO INCLUDED OPERATIONS

- Refinish rear section
- Remove and install or replace:
  - Preformed or molded headliner
  - Restraint systems
  - Suspension assembly
  - Antenna
  - Fuel tank
  - Control cables (latch release, etc.)
  - Fluid lines
- Recode lock cylinder
- Remove from salvaged assembly
  - Trim panels
  - Lock cylinder

Estimating Information Pages P1-P3 Must Be Used With The Procedure Pages For An Accurate Damage Report

FIGURE 8–44 Rear clip included/not included operations. (Courtesy of Mitchell International.)

FIGURE 8–45 Interior trim and upholstery.

sometimes be repaired with a plastic welder and refinished to both match the color and texture of the trim. Soiled and stained carpets and upholstery can be cleaned and deodorized. Vinyl and textile upholstery can be repaired or replaced by a qualified auto upholstery shop. The advisability of repairing damaged trim and upholstery depends on the extent of the damage, the cost and availability of replacement parts, and skill and expertise of the collision repair facility personnel in upholstery repair.

When replacement of trim or upholstery is necessary, the collision estimating guides provide flat rate labor allowance and parts prices for cowl trim, door trim, center pillar trim, quarter trim, headliner, package tray, and luggage compartment. The estimator and adjuster must consult the exploded views of the assembly affected by the damage to determine the actual components involved. Figure 8–46 is an excerpt from a collision estimating guide showing the parts of an instrument panel and console assembly. Individual labor times and part prices are shown for each component. The footnotes indicate that some parts must be ordered by color and description. Other parts must be painted to match the interior.

As shown in the head note in Figure 8–47, part prices and numbers might vary from model year to model year due to changes in style, design, and color. The estimator and adjuster should consult a local dealer to obtain actual part numbers and prices of interior trim panel and upholstery. Time and materials for painting unfinished panels must also be included on the damage report.

The R&R times listed in the collision estimating guide include all operations necessary to remove and replace damaged trim components, unless otherwise stated in the footnotes. The exception to this is the headliner. The headliner R&R includes R&I of the visors, lights, coat hooks, garnish mouldings, rear seats, and seat belts (optional lights and other equipment are not included).

8.12 RESTRAINT SYSTEMS

Most new vehicles are equipped with a supplemental restraint system or air bag. Air bags can be driver side only, driver and passenger side only, and side impact. The air bag is intended to instantly deploy in a moderate to severe impact to protect the occupants.

Air bag systems consist of the following components (Figures 8–48 and 8–49):

- *Impact Sensor.* Detects rapid deceleration, sends an electrical impulse
- *Safing Sensor.* Also detects rapid deceleration, sends an electrical impulse
- *Computer.* Detects impulse from the sensors, tells module to deploy, also checks the system for defects each time the vehicle is turned on
- *Module.* Driver's side or passenger's side, air bag inflates when it receives an impulse from the computer
- *Clock Spring.* Maintains an electrical connection to the computer and module and allows the steering wheel to be turned

After deployment (Figure 8–50), the system computer will signal with a series of dashboard air bag lamp flashes. These signals can be interpreted using a chart or a scanner can also be used to retrieve trouble codes from an air bag system. The interpretation will explain what type of problem exists. When replacing an air bag module, always check the signals and follow the manufacturer's recommendations on what components should be checked or replaced (Figure 8–51). The component replacement is not difficult if you follow the procedures (Figure 8–52). Once the parts are in place, the system will self-check. If the dashboard air bag lamp goes out after the module replacement, the system is functional.

Some manufacturers install force-sensitive labels at the seat belt attachment point. If the seat belt is in use during a collision, the force of the occupant may be strong enough to overextend the seat belt. The label will tear. A seat belt with a torn label should be replaced. Check the label by pulling the seat belt completely out of its retractor. If the seat belt doesn't have a label, check for defects that are shown in Figure 8–53. Replacement labor times are shown in Figure 8–54.

THUNDERBIRD 1989-97

INSTRUMENT PANEL

1994-97

R&I Instrument Panel Assy .... m #7.0

#Includes R&I Steering Column & Console

NOTE: All Parts in this section are included in R&R Instrument Panel unless noted otherwise.

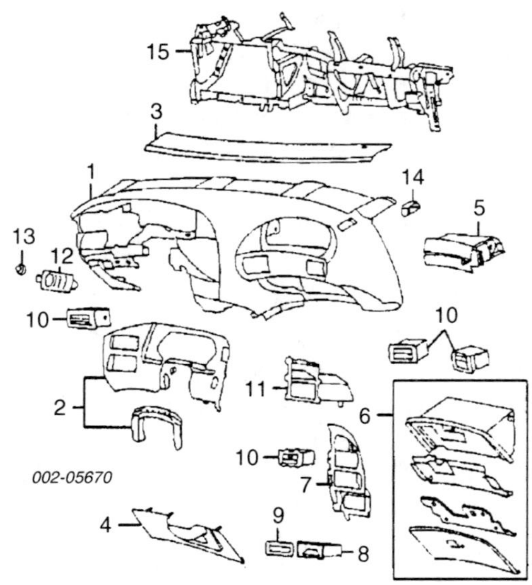

Part		Part No.	Time	Price
1 **Panel, Instrument¶**	94-96	F4SZ 6304320 C	m #8.3	1210.00
	97	F7SZ 6304320 AAB	m #8.3	624.17
¶Blue listed, Order by Application				
#Includes R&I Steering Column & Console				
2 **Panel, Cluster Finish**				
w/Telephone		F4SZ 63044D70 C	.8	118.77
w/o Telephone		F4SZ 63044D70 D	.8	120.23
3 **Grille, Defroster Opng¶**				
w/Auto Air		F4SZ 63044E82 A	.2	79.28
w/o Auto Air		F4SZ 63044E82 C	.2	87.42
4 **Cover, Strg Column¶**		F4SZ 6304459 D	.2	78.33
¶Blue listed, Order by Application				
5 **Module, Air Bag-See Air Bag System Section**				
6 **Compartment, Glove¶**		F6SZ 6306010 AAC	.3	126.60
7 **Panel, Cluster Center¶**		F4WY 66044D70 C	#1.0	68.77
¶Blue listed, Order by Application				
#Includes R&R Cluster Finish Panel				
8 **Pocket, Utility**		F4SZ 6313570 A	#.2	10.18
9 **Cover, Utility Pocket**		F4SZ 6313598 A	#.2	10.77
#R&R Complete				
10 **Vent, A/C¶**				
Passenger	R	F4SZ 19893 E	1.0	24.85
	L	F4SZ 19893 G	1.0	26.63
Driver	R	F4SZ 19893 F	1.0	d26.45
	L	F4SZ 19893 H	1.0	23.32
¶Blue listed, Order by Application				
11 **Housing, Inst Pnl Cont**		F 4SZ 63047A26 A		70.08
12 **Panel, Inst Finish Left**				
w/Auto Headlamp Cont		F4SZ 6304338 B		14.82
w/o Auto Headlamp Cont		F4SZ 6304338 A		14.82
13 **Knob, Light Switch**		F4SZ 11666 A		6.45
14 **Vent, Defrost¶**	R/L	F4SZ 63046A76-7 D		6.90
¶Blue listed, Order by Application				
15 **Reinf, Inst Pnl-Serviced w/Inst Panel**				
Radio				
AM/FM		F4SZ 18806 C	.9	N.A.
AM/FM Std Cassette		F4SZ 18806 A	.9	671.52
AM/FM Premium Cassette		F4SZ 18806 B	.9	671.52
AM/FM CD Player		F4SZ 18D806 A	.9	869.03

CENTER CONSOLE

Use Procedure Explanation 14 with the following text.

1994-97

R&I Console Assy 6

NOTE: All Parts in this section are included in R&R Console Panel Unless noted otherwise.

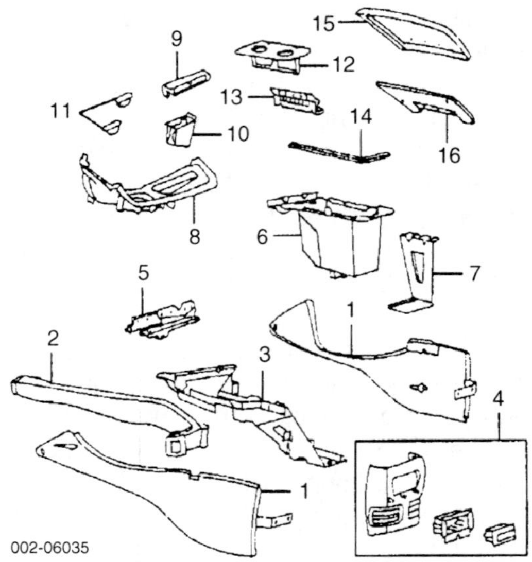

Part	Part No.	Time	Price
1 **Panel Assy, Console¶**			
1994-95			
LX	F4SZ 63045A36 D	1.0	463.55
Super coupe	F4SZ 63045A36 J	1.0	463.55
1996	F7SZ 63045A36 AAC	1.0	463.55
1997	F7SZ 63045A36 AAC	1.0	463.55
¶Blue listed, Order by Application			
2 **Duct, Rear Seat A/C**	F4SZ 18C420 A		22.42
3 **Brace, Console**	F4SZ 6304684 A		69.28
4 **Panel, Finish End¶**			
LX	F4SZ 63045D72 D		34.38
Super Coupe	F4SZ 63045D72 U		34.38
¶Blue listed, Order by Application			
5 **Bracket, Support**	F4SZ 63043B92 A	#.1	17.22
#w/Console Assy Removed, Not Included in R&R Console Panel			
6 **Compartment, Console¶**	F6SZ 6306010 AAC		126.60
¶Blue listed, Order by Application			
7 **Support, Console**	F4SZ 63045B30 A	#.1	26.10
#w/Console Assy Removed, Not Included in R&R Console Panel			
8 **Panel, Console Top¶**	F4SZ 63044D90 L	.4	138.35
9 **Door, Ashtray¶**	F4SZ 6304786 E		11.67
¶Order by Application			
10 **Ashtray**	F4SZ 6304788 A		8.07
11 **Trays, Console (2)**	F4SZ 6306202 A		11.47
12 **Cupholder**	F5SZ 6313530 AAA		25.65
13 **Brkt, Cupholder**	F4SZ 6306050 A		20.88
14 **Bezel, Console**	F4SZ 63061A34 A		6.27
15 **Door, Compartment**	F4SZ 6306024 A	.2	67.32
16 **Latch, Door¶**	F4ZZ 6106072 D		11.20
¶Blue listed, Order by Application			

FIGURE 8–46 Instrument panel and console assembly from a 1996 Ford Thunderbird. (Courtesy of Mitchell International.)

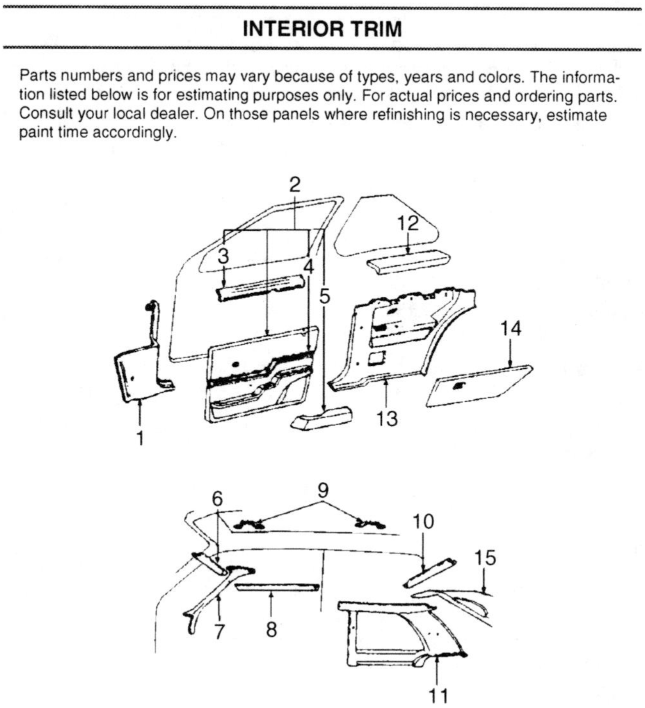

FIGURE 8–47 Important ordering information contained in the head notes. (Courtesy of Motor Information Systems.)

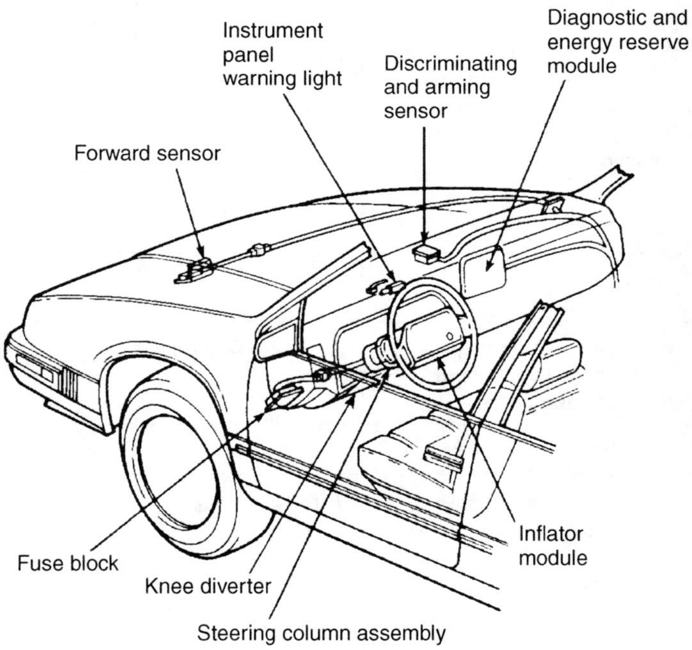

FIGURE 8–48 Components of a typical supplemental restraint system, driver's side air bag only.

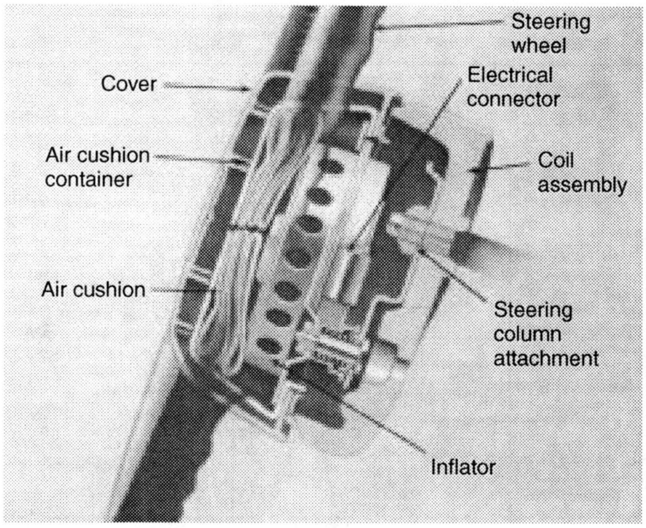

FIGURE 8–49 Cross section of a driver's side air bag module.

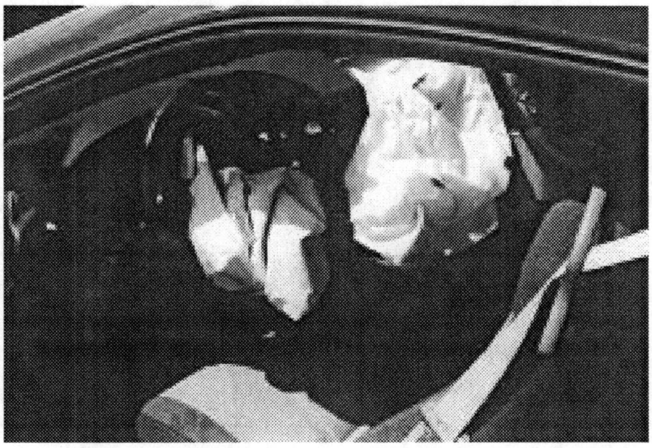

FIGURE 8–50 Deployed driver and passenger side air bags.

GENERAL MOTORS LIGHT TRUCK & VAN APPLICATION TABLE (1993–96)

Make/Model	Year	Location	Table
CHEVORLET			
Astro	1996	D/P	GM-1
	1993-95	DS	GM-2
Blazer	1995-96	DS	GM-2
Chevy Van	1996	D/P	GM-1
	1994-96	DS	GM-2
C/K Pickup	1995-96	DS	GM-2
Express Van	1996	D/P	GM-1
Lumina APV	1994-96	DS	GM-2
Sportvan	1994-96	DS	GM-2
Suburban	1195-96	DS	GM-2
S 10 Pickup	1995-96	DS	GM-2
Tahoe	1995-96	DS	GM-2
GMC			
Jimmy	1995-96	DS	GM-2
Rally/Vandura	1994-96	DS	GM-2
Safari	1996	D/P	GM-1
	1993-95	DS	GM-2
Savana	1996	D/P	GM-1
Sierra	1995-96	DS	GM-2
Sonoma	1995-96	DS	GM-2
Suburban	1995-96	DS	GM-2
Yukon	1995-96	DS	GM-2
Oldsmobile			
Bravada	1996	DS	GM-2
Silhouette	1994-96	DS	GM-2
Pontiac			
Trans Sport	1994-96	DS	GM-2

Location Definitions: D/P=Driver's & Passenger's Side, DS=Driver's Side Only

TABLE GM-1

Air Bag Application	Replace After Deployment	Inspect & If Damaged, Replace Component (Even if Air Bag Did Not Deploy)	Comments
See Application Table	• Air Bag Module(s) • Forward Discriminating Sensor[1] • Sensing & Diagnostic Module (SDM) • Sensors In Area Of Accident Damage[2]	• Forward Discriminating Sensor[1] • Instrument Panel Braces • Instrument Panel Steering Column Reinforcement Plate • Knee Bolsters & Mounting Points • Seat Belts & Mounting Points • Sensors In Area Of Accident Damage (Without Deployment)[2] • SIR Coil Assembly • Steering Column	• Any sensor which the Sensing & Diagnostic Module (SDM) indicates as bad must be replaced. • If any components are damaged or bent, they must be replaced. • SRS wiring can be repaired following manufacturer's instructions [1] Except Aurora, Camaro, Firebird & Riviera [2] Aurora & Riviera Only

TABLE GM-2

Air Bag Application	Replace After Deployment	Inspect & If Damaged, Replace Component (Even if Air Bag Did Not Deploy)	Comments
See Application Table	• Air Bag Module(s) • Sensors In Area Of Accident Damage	• Instrument Panel Braces • Instrument Panel Steering Column Reinforcement Plate • Knee Bolsters & Mounting Points • Seat Belts & Mounting Points • Sensors In Area Of Accident Damage (Without Deployment) • SIR Coil Assembly • Steering Column • Steering Wheel[1]	• Any sensor which the Diagnostic Energy Reserve Module (DERM) or Sensing & Diagnostic Module (SDM) indicates as bad must be replaced. • If any components are damaged or bent, they must be replaced. • SRS wiring can be repaired following manufacturer's instructions. [1] 1995 Aurora, Bonnevill, Concours, DeVille, Eighty Eight, Eldorado, LeSabre, Ninety Eight, Park Avenue, Riviera, Seville & 1995-96 Fleetwood

FIGURE 8–51 This chart lists what air bag system components should be replaced and inspected after deployment. (Courtesy of Mitchell International.)

AIR BAG SYSTEM

Disable & Enable Air Bag System.......................................m .3
Diagnose Air Bag System (If Required)...................................m .5
NOTE 1: R&R Does Not include disable and enable or diagnose Air Bag System.

NOTE 2: Refer to Procedure Explanation 30 for Supplemental Restraint/Air Bag System Component Inspection and Replacement information.

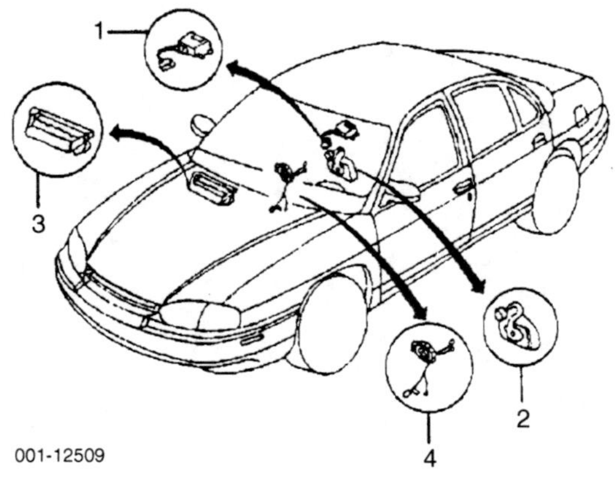

	Part No.			Price
1 **Module, Sensing & Diagnostic Module, Air Bag**	16239683	m	.7	■306.00
2 Driver				
Blue	16757515	m	.4	550.80
Gray	16757085	m	.4	550.80
Neutral	16757517	m	.4	550.80
Red	16757516	m	.4	550.80
3 Passenger	10298114	m	#5.5	■780.00
#Includes R&I Instrument Panel				
4 **Coil, SIR¶**				
w/Electronic Control	26049191	m	.9	■100.00
w/o Electronic Control	26049192	m	.9	■104.00
¶Included w/Steering Column Assy listed in Steering Wheel/Column Section				

FIGURE 8–52 Air bag system components in a Buick Regal. (Courtesy of Mitchell International.)

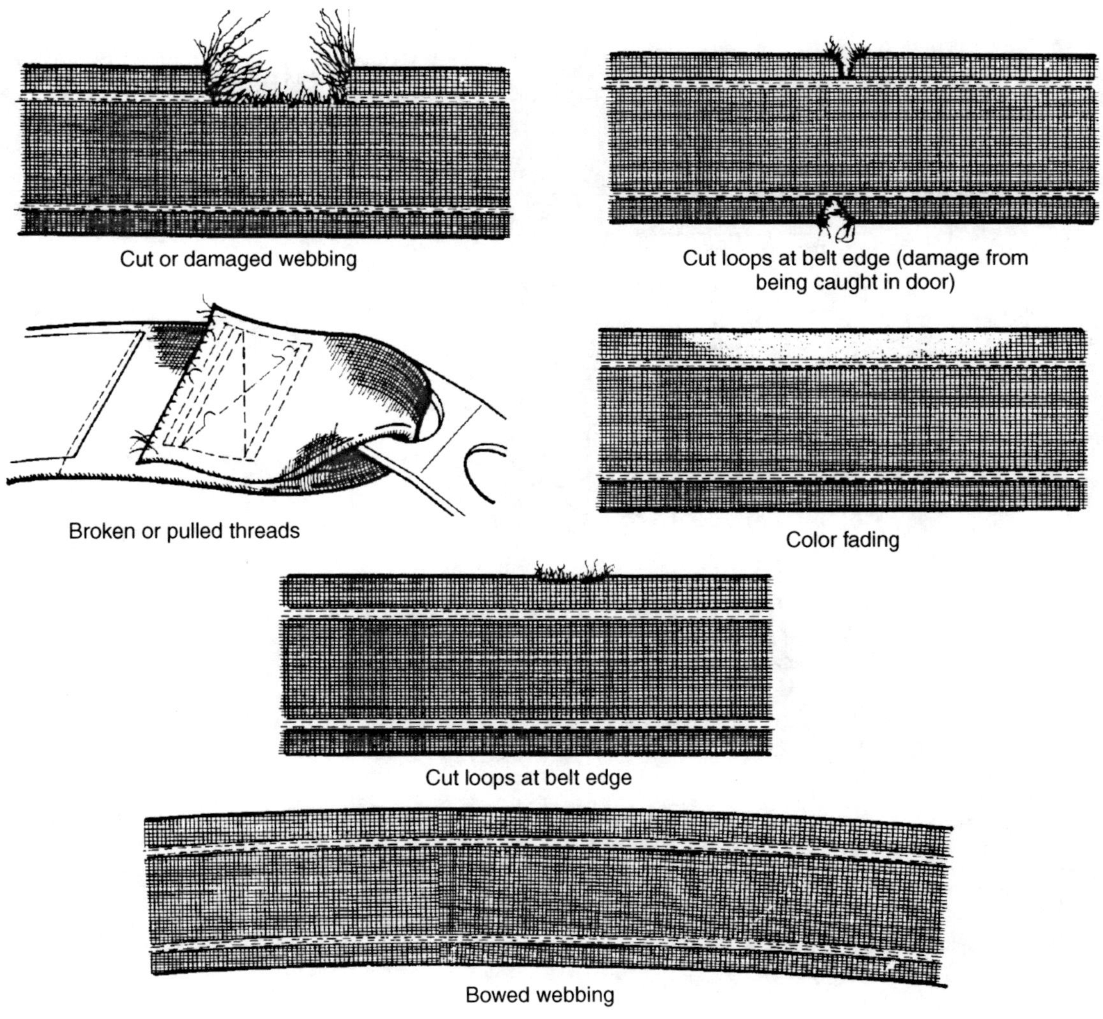

FIGURE 8–53 Seat belt defects.

SEAT BELTS

FRONT BUCKET SEAT

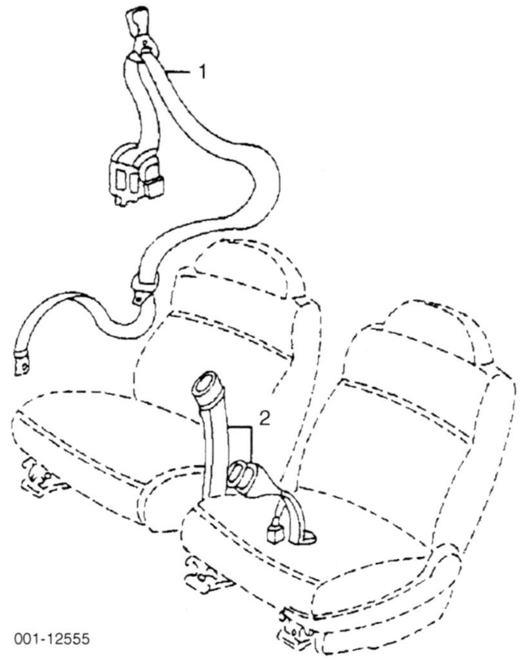

Belt, Front Seat¶

1 Retractor Side	R/L	12531106	**#.6**	56.00
#Includes R&I/R&R Lower Center Pillar Moulding				
2 Buckle Side				
Driver		12533407	**.3**	51.50
Passenger		12533404	**.3**	51.50

¶Gray listed, Order by Application

FRONT SPLIT BENCH SEAT

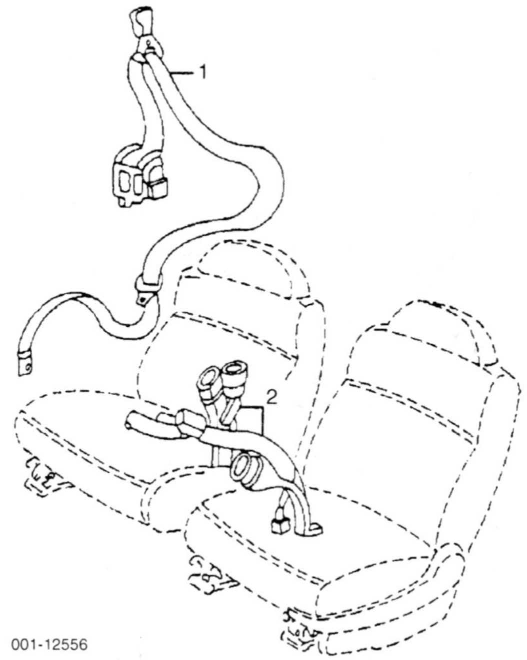

Belt, Front Seat¶

1 Retractor Side	R/L	12531106	**#.12**	56.00
#Incudes R&I/R&R Lower Center Pillar Moulding				
2 Buckle Side (Set)		12530579	**.3**	42.25

¶Gray listed, Order by Application

REAR

Belt, Rear Seat¶

1 Retractor Side	R/L	12531114	**#1.2**	48.75
#Includes R&I/R&R Rear Seat & Package Shelf				
2 Buckle Side (Set)		12534817	**#.6**	N.A.

¶Gray listed, Order by Application
#Includes R&I Rear Seat Cushion

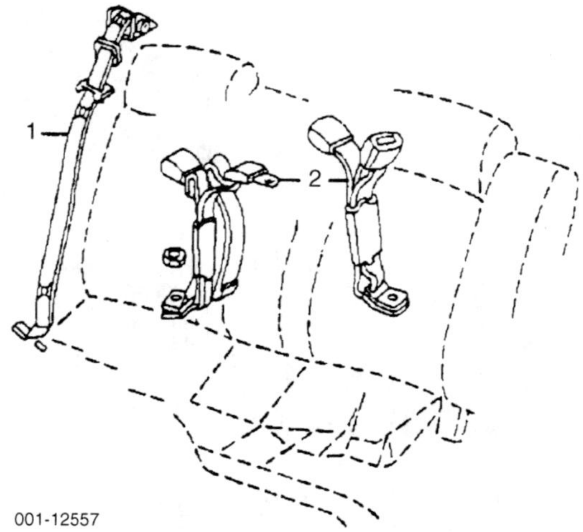

FIGURE 8–54 Seat belt labor time and part prices. (Courtesy of Mitchell International.)

REVIEW QUESTIONS

1. Technician A says the center pillars are usually sectioned rather than totally replaced. Technician B says that a fender is considered part of the body shell. Who is right?
   a. technician A b. technician B
   c. both A and B d. neither A nor B
2. A damaged A-pillar is usually sectioned.
   True False
3. Technician A says that on some unibody vehicles, the windshield is considered a part of the structural unit. Technician B says that the cost of windshield installation materials must be added to an estimate. Who is right?
   a. technician A b. technician B
   c. both A and B d. neither A nor B
4. Which of the following is not usually included in windshield R&R times?
   a. removing old adhesive
   b. removing broken glass
   c. straightening the windshield pillar
   d. all of the above
5. Technician A says that the back glass R&I time is included in roof R&R. Technician B says that the application of sound-deadening material is included in roof R&R. Who is right?
   a. technician A b. technician B
   c. both A and B d. neither A nor B
6. The ________________ connects the quarter panel to the roof.
7. Technician A says that the rear bumper R&I is included in the quarter panel R&R. Technician B says that the R&I of quarter glass is included in quarter panel R&R. Who is right?
   a. technician A b. technician B
   c. both A and B d. neither A nor B
8. Door skins can be replaced by
   a. bonding.
   b. welding.
   c. crimping only.
   d. A and B only.
9. Technician A replaces all sensors on a deployed air bag system. Technician B says that you should read the manufacturer's recommendations first before replacing air bag system components. Who is right?
   a. technician A b. technician B
   c. both A and B d. neither A nor B
10. The ________________ is the assembly from the cowl to the rear bumper.

Chapter 9

Assessing Mechanical Component Damage

Objectives

After reading this chapter, you should be able to:

- **Identify the basic components of the cooling system and explain damage diagnosis of the system.**
- **Identify the basic components of the A/C system and explain damage diagnosis of the system.**
- **Describe the car's basic electrical systems.**
- **Describe what type of damage the engine and transmission are likely to sustain in a collision.**
- **Identify the basic components of the exhaust system and explain damage diagnosis of the system.**
- **Describe the different front suspension systems.**
- **Explain damage diagnosis of the wheels.**
- **Describe damage diagnosis of the steering system.**
- **Describe damage diagnosis to brake-system.**

Key Terms

Drivetrain
Front wheel drive
Rear wheel drive

In many accidents, damage is not confined to exterior sheet metal and plastic panels, trim components and lights, paint and surface finishes, windshields and glass, and other areas normally associated with collision repair work. Mechanical damage to engines and transmissions, steering and suspension systems, brakes, exhaust systems, air-conditioning (A/C) systems and cooling systems often occurs. Electric wiring and an ever-changing array of electrical sensors and relays might be damaged. Oftentimes, the damaged vehicle will not run, or it will operate poorly.

Mechanical and electrical damage must be accurately assessed to write a complete damage report or appraisal. Collision repair facilities that consistently leave the mechanical repair sections of their bids open or incomplete will not elicit confidence from their customers or the insurance adjusters that deal with the shop. Likewise, insurance adjusters with a limited understanding of mechanical systems and their repair will face a never-ending cycle of damage supplements along with the strain they place on shop/insurance company relations.

All collision estimating guides list replacement part costs and labor times for many mechanical components commonly damaged during collisions, so working up detailed, accurate reports on these systems is possible provided the estimator has good troubleshooting ability.

In the past, many collision repair facilities relied on outside garages or specialty shops to assess and repair nearly all types of mechanical damage. And while this is still the case with many types of engine, transmission, and electrical repairs, more and more collision repair facilities are performing certain mechanical repairs in-house. Suspension, steering, cooling, brake, and exhaust system work are areas in which modern facilities often perform repairs to cut outside spending and increase profits.

9.1 COOLING SYSTEM

On most gasoline engines, the method of cooling is to circulate coolant through the cylinder block and cylinder head (Figure 9–1). Liquid cooling is preferable to air cooling because it is less noisy and better able to maintain a constant temperature at the cylinders. It also lets the engine operate more efficiently and makes a ready supply of hot coolant available to operate a heater for the passenger compartment.

The cooling system is made up of the following:

- *Pump.* Circulates the cooling liquid through the system. The liquid is a mixture of water and antifreeze; referred to as *coolant.*
- *Water Jackets.* Cored passages in the cylinder block and cylinder head that carry the coolant around the cylinders and combustion chambers.
- *Radiator.* Transfers heat in the coolant to the outside air as coolant flows through its tubes.
- *Fan.* A device that pulls cool outside air through the fins of the radiator to pick up heat.
- *Pressure Cap.* Maintains a pressure in the system to raise the boiling point of the coolant to a higher temperature. Also provides relief from excess pressure or vacuum.
- *Hoses.* Connect the components of the system to one another.
- *Thermostat.* Blocks off circulation in the system until a preset temperature is reached to speed engine warm-up. Also controls engine temperature at a predetermined level.
- *Temperature Indicator.* Warns the driver in case of overheating.
- *Shroud.* Plastic case that directs air flow.

The water pump is usually belt driven by the engine crankshaft. The engine cooling fan can be mounted on the water pump or operated electrically. Coolant is pumped or circulated by the water pump from the bottom or side of the radiator through internal passages in the engine block and heads. The internal passages are often called the *water jacket.* The water pump then pumps the coolant through the thermostat to the top tank of the radiator. From the top tank the coolant flows slowly through a system of tiny tubes surrounded by cooling fans. These tubes and fins are called the radiator core. When the coolant reaches the opposite radiator tank, the process begins again. When the thermostat is open, the coolant constantly circulates through the engine and radiator. As the coolant travels through the engine block and heads, heat caused by combustion is transferred to the coolant. As the coolant flows through the radiator core, this heat is transferred to the air as the air

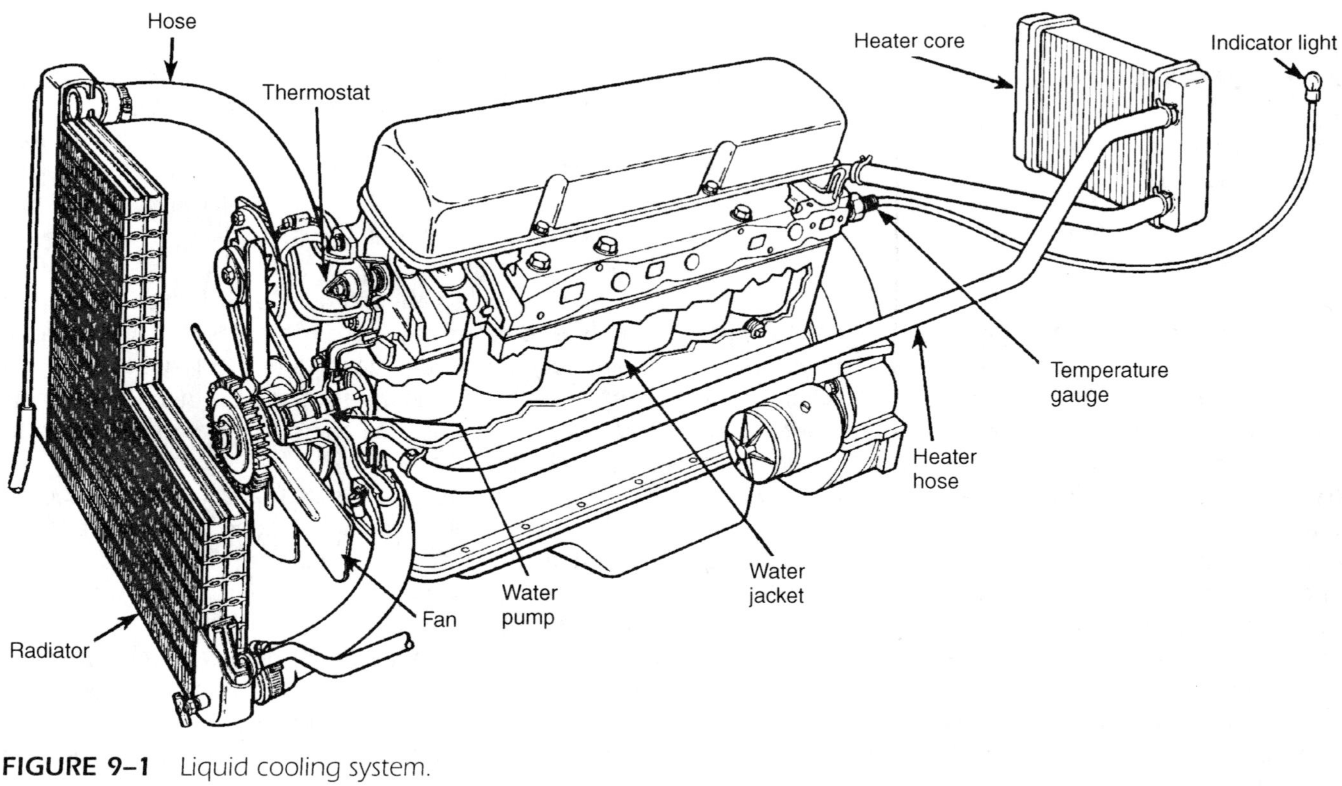

FIGURE 9–1 Liquid cooling system.

passes through the core around the fins. The fan is used to move air through the radiator when the engine is idling or operating slowly, as in traffic. This insures sufficient air movement and good heat transfer at slow speeds. The cooling system's job is twofold:

1. It carries away unwanted heat from the engine.
2. It must be able to maintain the engine temperature at the right point for efficient operation.

Front wheel drive vehicles have at least one thermostatically controlled electric fan assembly mounted to the rear of the radiator. This assembly consists of an electric motor, a fan, and a plastic shroud (Figure 9–2). The fan's on-and-off operation is controlled by engine temperature. In front impacts, the condenser, radiator, and fan assembly can be driven back into the engine (Figure 9–3).

CAUTION: As long as the cooling system is pressurized, the temperature can be considerably higher than the boiling temperature of the solution in the radiator without causing the solution to boil. Removing the radiator cap while the engine is hot and the system is pressurized will cause the solution to boil instantaneously and possibly with explosive force, spewing the solution over engine, fenders, and the person removing the cap.

COOLING

Use Procedure Explanations 5 and 28 with the following text.

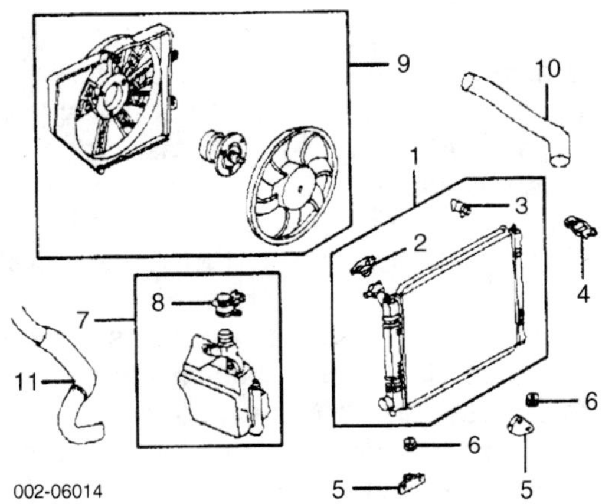

1 **Radiator Assy**				
1.8L Eng.				
Man Trans		FOCZ 8005 D	**1.1**	350.32
Auto Trans		FOCZ 8005A	**1.3**	347.43
1.9L Eng				
Man Trans				
w/ A/C		FOCZ 8005 D	**1.1**	350.32
w/o A/C		FOCZ 8005 C	**1.1**	330.58
Auto Trans		FOCZ 8005 A	**1.3**	350.32
2 **Cap, Radiator¶**		E92Z 8100 B		6.52
3 **Drain, Radiator¶**		E8BZ 8115 A		2.67
¶Included w/Radiator Assy				
4 **Bracket, Upper**				
1.8L Eng	R	F1CZ 8A193 C	.1	41.85
	L	F5CZ 8A193 BA	.1	19.50
1.9L Eng	R/L	F5CZ 8A193 AA	.1	45.83
5 **Bracket, Lower**	91-92 R	FOCZ 8B068 A	.1	13.03
	L	F1CZ 8B068 A	.1	13.03
	93-96 R/L	F3CZ 8B068 A-B	.1	10.30
6 **Insulator, Mount**				
To 6-94	R/L	FOCZ 8125 A		7.85
From 7-94	R/L	F5CZ 8125A		N.A.
7 **Tank, Coolant**				
1.8L Eng		F1CZ 8A080 A	.3	39.22
1.9L Eng		F1CZ 8A080 B	.3	36.25
8 **Sensor, Low Level**				
1.8L Eng		FOFZ 10D968 C	.3	32.66
1.9L Eng		F5CZ 10D968 AA	.3	30.21
Bracket, Coolant Tank				
#w/Coolant Tank Removed				
9 **Fan Assy**				
1.8L Eng				
1991-92				
Man Trans				
To 12-90		F5CZ 8C607 D	**#.3**	242.89
From 12-90		F5CZ 8C607 F	**#.3**	242.89
Auto Trans				
To 12-90		F5CZ 8C607 C	**#.3**	266.66
From 12-90		F5CZ 8C607 E	**#.3**	253.39
1993				
Man Trans		F5CZ 8C607 B	**#.3**	257.50
Auto Trans		F5CZ 8C607 A	**#.3**	275.48
1994				
Man Trans		F5CZ 8C607 B	**#.3**	257.50
Auto Trans		F5CZ 8C607 A	**#.3**	275.48
1995-96				
Man Trans		F5CZ 8C607 B	**#.3**	257.50
Auto Trans		F5CZ 8C607 A	**#.3**	275.48
1.9L Eng				
1991-92				
w/A/C				
To 12-90		F5CZ 8C607 C	**#.3**	266.66
From 12-90		F5CZ 8C607 E	**#.3**	253.39
w/o A/C				
To 12-90		F5CZ 8C607 D	**#.3**	242.89
From 12-90		F1CZ 8C607 N	**#.3**	212.11
1993				
w/ A/C		F5CZ 8C607 A	**#.3**	275.48
w/o A/C		F5CZ8C607 B	**#.3**	257.50
1994				
w/ A/C		F5CZ 8C607 A	**#.3**	275.48
w/o A/C		F5CZ 8C607 B	**#.3**	257.50
1995-96				
w/ A/C		F5CZ 8C607 A	**#.3**	275.48
w/o A/C		F5CZ 8C607 B	**#.3**	257.50
#Included in R&R Radiator				
10 **Hose, Rad Upper**				
1.8L Eng		F1CZ 8260 A	**#.5**	14.19
1.9l Eng		FOCZ 8260 A	**#.5**	10.15
11 **Hose, Rad Lower**				
1.8L Eng		FOCZ 8286 B	**#.6**	12.00
1.9L Eng				
1991				
To 5-91		FOCZ 8286 A	**#.6**	12.00
From 5-91		F1CZ 8286 A	**#.6**	26.64
1992-96				
To 5-95		F5CZ 8286 CA	**#.6**	25.53
From 6-95		F5CZ 8286 CA	**#.6**	25.53
#R&R Both Upper & Lower .9				
Pump, Water				
1.8 L Eng		E9JY 8501 B	**m 3.0**	113.55
1.9L Eng				
To 4-91		FOCZ 8501 A	**m 2.3**	94.06
From 5-91		F4CZ 8501 A	**m 2.3**	100.25
Belt, Drive¶				
1.8L Eng	91	F1CZ 8620 F	**.5**	31.72
	92-96	F1CZ 8620 H	**.5**	32.62
1.9L Eng	91-92	F1CZ 8620 C	.5	24.91
	93	F2CZ 8620 A	**.5**	20.40
	94-96	F4CZ 8620 C	**.5**	37.02

FIGURE 9–2 Cooling system components on a 1995 Ford Escort. (Courtesy of Mitchell International.)

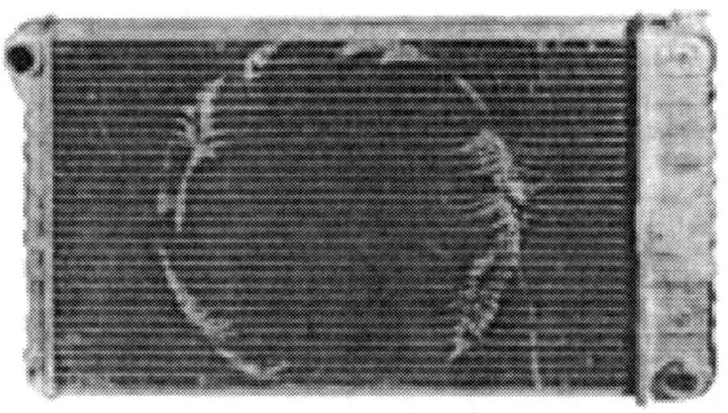

FIGURE 9–3 This radiator has been driven back into the fan.

COOLING SYSTEM DAMAGE DIAGNOSIS

The most vulnerable component in the cooling system with respect to collision repair is the radiator, due to its position between the grille and the motor. There are several kinds of damage the radiator can suffer in a collision, but the one encountered most often is core damage. Depending on the severity of the crash, the fan might only do cosmetic damage to the core (which can be easily repaired) or it might destroy it. Crushed fins can be straightened with a special tool designed for that purpose and tubes that are not too badly mangled can be soldered. But if large sections of cooling fins have been pulled loose or if multiple tubes have been crushed or ruptured, a new core is recommended. The price of recoring a radiator is usually 50 percent less than the cost of a new radiator. Recore prices are not listed in the collision estimating guides. If, in addition to the core damage, the end tanks have been damaged or if the radiator is badly corroded, complete replacement might be the better choice.

CAUTION: Sometimes a radiator might appear to have survived a collision without any visible damage. Yet the impact has created hairline cracks around the hose connections, along the core header or tube seams. If there is any reason to suspect hidden damage (low coolant in a radiator with no obvious damage), the radiator should be pressure tested for a leak.

Note that the operational times given for radiator replacement in the collision estimating guides do not include the time for removal and reinstallation of the fan, pulley, and belts. However, these times do include the following:

- Draining, checking, and refilling the fluids (Motor estimating guide only)
- Disconnecting and reconnecting the hoses
- Removing and reinstalling an electric fan assembly
- Disconnecting and reconnecting the transmission lines
- Disconnecting and reconnecting the fan shroud

Remember also that the cost of the fluid is not included in the cost given.

No attempt should be made to repair and reuse a fan blade that is bent or damaged. A bent or damaged fan assembly should always be replaced with a new fan assembly.

A damaged or bent fan clutch must also be replaced. It is not a repairable item. If the water pump hub or shaft is damaged, it should be replaced. Since they are often made of plastic, damaged fan shrouds can usually be repaired by plastic welding if the damage is not severe.

Belts and hoses often survive crashes without damage because they are flexible. However, sometimes a good belt must be cut from a pulley to free a damaged part. Also, any belt that has been nicked, cut, scraped, or frayed should be replaced. If a hose is ripped, punctured, split, age cracked, burned, chaffed, or soft, replace it. Particular attention should be paid to the lower radiator hose. It has wire reinforcing inside to prevent collapse when the water pump creates suction vacuum at high speed. If the lower hose has been crushed, so has the reinforcement spring inside. If it is not replaced, the vehicle can suffer overheating at high speed. OEM hose clamps should be replaced with the hoses. Cracks on polypropylene fan shrouds can be plastic welded.

9.2 AIR CONDITIONING

An air-conditioning unit works on the simple principle that when a liquid is converted to a gas, it absorbs heat from its surroundings. When it is reconverted from this gaseous state back to a liquid, it gives up this heat. The air-conditioning system consists of the following components (Figure 9–4):

- Compressor
- Condenser
- Receiver/dryer
- Refrigerant control
- Evaporator

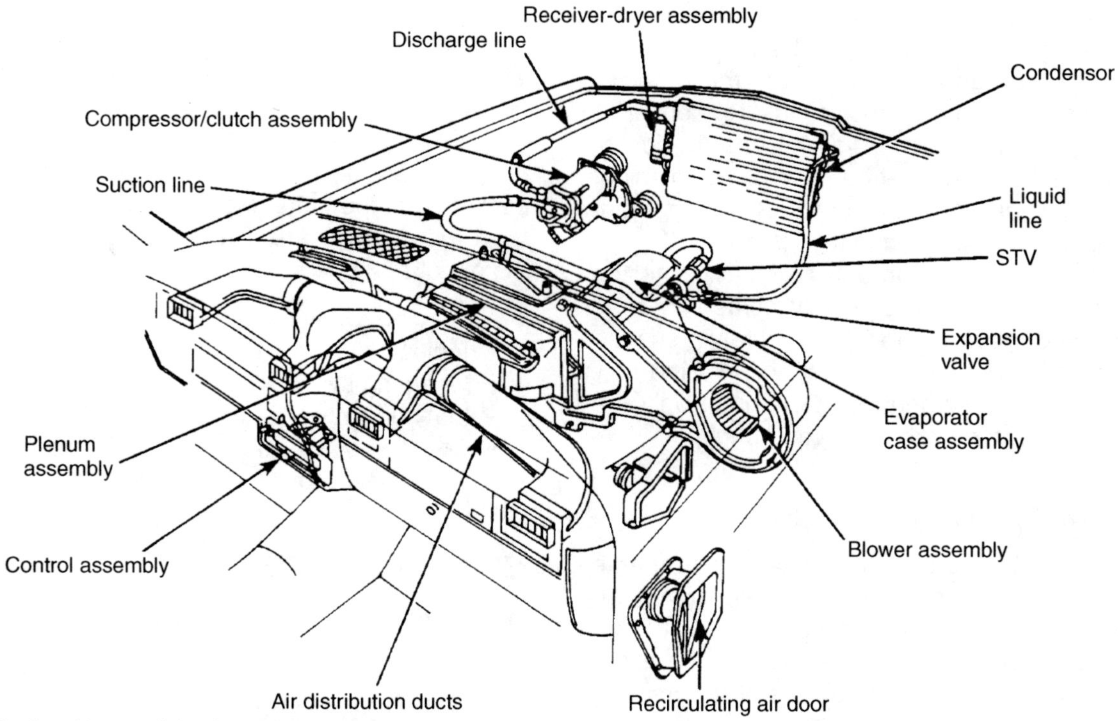

FIGURE 9–4 *Air-conditioning system components.*

Refrigerants are the chemicals that transfer heat by absorbing it during evaporation and releasing it during condensation. There are two types of refrigerants used in vehicles today:

- *R12.* Used in vehicles manufactured before 1994, not produced in the U.S. anymore because the chlorofluorocarbons it contains are harmful to the ozone layer.
- *R134a.* Used in vehicles manufactured in 1994 and after, not as efficient as compare to R12. It must operate at a higher system pressure, but it is not as harmful to the environment if it were to be accidently released.

These two refrigerants are not compatible. An air-conditioning system designed for R12 cannot operate with R134a. By law, all refrigerants must be recovered and recycled. A collision repair facility that does its own A/C work must have two separate systems to recover and store the two types of refrigerants.

The compressor is nothing more than a pump driven by the crankshaft with a belt. It picks up a gaseous refrigerant from the evaporator inside the car and compresses it. The compressor uses an electromagnetic clutch to permit it to be turned off when not needed.

The refrigerant is metered into a cooling coil (evaporator) normally located inside the car on the fire wall, at low pressure. The refrigerant is in liquid form at this point. Its boiling point at low pressure is just above the freezing point of water. The refrigerant therefore tends to boil, absorbing heat from the coil.

A blower forces either inside or outside air, depending on the setting, through the evaporator. The air then passes into the passenger compartment. As the air passes through the evaporator, heat and moisture are removed.

The refrigerant boils inside the evaporator then passes into the compressor as a gas, where its pressure is increased. Refrigerant pressure as it leaves the compressor is high. The refrigerant then enters the condenser, a heat-exchanging coil resembling a radiator and usually located in front of the car's radiator. The high pressure caused by the compressor is put to work at this point and raises the boiling point of the refrigerant. When outside air is passed over the thin tubes and fins of the condenser, it cools and changes the refrigerant back to a liquid, losing the heat it picked up from the interior of the car.

The liquified refrigerant then enters the receiver (dryer), a small tank located next to the condenser or on one of the fender wells. This unit has the job of separating liquid refrigerant from any gas that might have left the condenser, and also filters the refrigerant and absorbs any moisture it might contain.

Refrigerant then flows through a liquid line to the expansion valve. This valve is usually located near the evaporator, on or near the firewall. The valve, shaped like a mushroom on some systems, controls the flow of refrigerant to the evaporator. It provides only the amount of flow that the evaporator can handle.

A/C DAMAGE DIAGNOSIS

There are a number of air-conditioning components susceptible to damage in a collision. While some can be repaired, others must be replaced.

When the compressor is damaged from collision, the clutch and pulley assembly are usually damaged first. These can be disassembled and repaired or replaced separately from the compressor. When damaged, the compressor itself can be disassembled and repaired. There is a seal at the front of the compressor. It prevents refrigerant and refrigerant oil from leaking out around the compressor shaft. When this seal is defective, it should be replaced.

Because of its location, the condenser is susceptible to damage in front-end collisions. Its air directional fins, like those on a radiator, can be cleaned and straightened, and leaks can be repaired by silver soldering.

When the condenser is damaged, the receiver/dryer should also be checked for damage. If the unit is damaged, it must be replaced. This unit might also have to be replaced if the system has been open to the atmosphere for a period of time due to collision. This repair is not included in the operational time or cost given for the original malfunction.

The evaporator, thermostatic expansion valve, and suction throttling valve are seldom damaged due to collision. If the evaporator is damaged, the case and the core can be replaced separately. A damaged thermostatic expansion valve should be replaced. If a suction throttling valve is damaged, replacement parts and repair kits are available.

Whenever any operation requires a refrigerant line disconnect, additional operational time must be included for evacuating and recharging the system. Time must also be added for the removal of any parts for accessibility and their reinstallation. The cost of refrigerant and oil should be added to the estimate (Figure 9–5).

9.3 ELECTRICAL SYSTEM

Automobiles have many circuits that carry electrical current from the battery to individual components. The total electrical system (Figure 9–6) includes such major subsystems as:

- Starting system
- Ignition system
- Charging system
- Lighting circuit
- Other electrical circuits, including horn, windshield wipers, and flashers

CHARGING SYSTEM

The charging system performs two basic tasks:

1. Maintains the battery's state of charge
2. Provides electrical power for the ignition system, air conditioner, heater, lights, radio, and all electrical accessories

In addition to the battery, the charging system includes

- Alternator
- Voltage regulator
- Indicator light
- Necessary wiring

Turned by a drive belt driven by the engine crankshaft, the alternator (Figure 9–7) converts mechanical energy into electrical energy. When the electrical current flows into the battery, the battery is said to be charging. When the current flows out of the battery, the battery is said to be discharging.

The battery must be charged in order to provide the source of current for starting the engine. The battery also supplies electrical power when the demands of the system exceed the output of the alternator.

An alternator has a self-limiting feature that controls current flow in the charging circuit. It does not, however, control voltage output. The function of the voltage regulator is to limit the output voltage to a present value by controlling the strength of the alternator field current. Without the regulator, the voltage generated at high engine speeds would overcharge the battery.

LIGHTING SYSTEMS

Automotive lighting systems include headlights, taillights, side marker and park lamps, turn signals, license plate lights, and indicator lights on the dash.

A/C REFRIGERANT CAPACITIES

A/C R-134a Refrigerant Capacity
Standard A/C: 37 Ozs (2.3 Lbs)
Standard & Aux A/C: 58 Ozs (3.7 Lbs)

AIR CONDITIONING/HEATER

Use Procedure Explanation 6 with the following text.

Evacuate & Recharge System .... **m 1.4**
Refrigerant Recovery .... **m .3**

NOTE: R&R Does Not include recover, evacuate and recharge system.

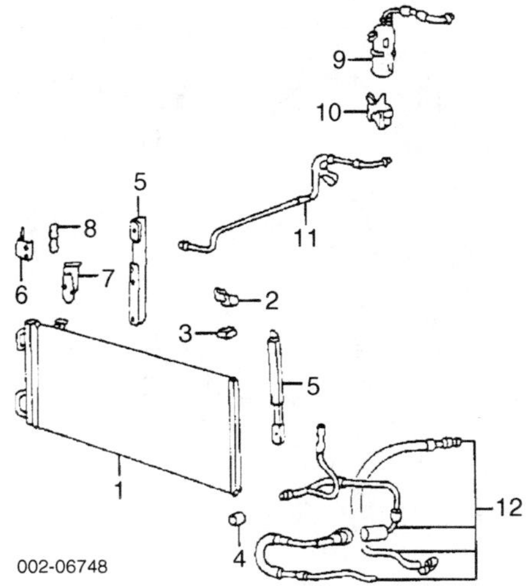

Item	Side	Part No.	Labor	Price
1 **Condenser**		F65Z 19712 AB	**m #1.4**	189.52
#w/Radiator Removed Deduct .5				
2 **Brkt, Cond Mtg Upr**	R/L	F65Z 19702 BB		7.78
3 **Insulator, Cond Upr**	R/L	N.A.		N.A.
4 **Insulator, Cond Lwr**	R/L	F65Z 19C566 AA		5.33
5 **Seal, Cond**	R	F65Z 19E572 DA		4.57
	L	F65Z 19E572 CB		4.38
6 **Brkt, Cond Inlet Tube Otr**		F65Z 19D720 BB		3.15
7 **Brkt, Cond Inlet Tube Inr**		F65Z 19D720 AA		3.03
8 **Clamp, Cond Inlet Tube**		N.A.		N.A.
9 **Drier/Accumulator**		F65Z 19C836 AA	**m #.7**	84.81
#Included in R&R Condenser				
10 **Brkt, Drier/Accumulator**		F65Z 19D607 AB		5.88
11 **Hose, Suction**				
w/Aux Air		F75Z 19837 BD	**m .5**	45.13
w/o Aux Air		F75Z 19837 AD	**m .5**	36.42
12 **Hose, Discharge**				
w/Aux Air		F75Z 19D850 BE	**m .6**	211.06
w/o Aux Air		F75Z 19D850 AD	**m .6**	194.71

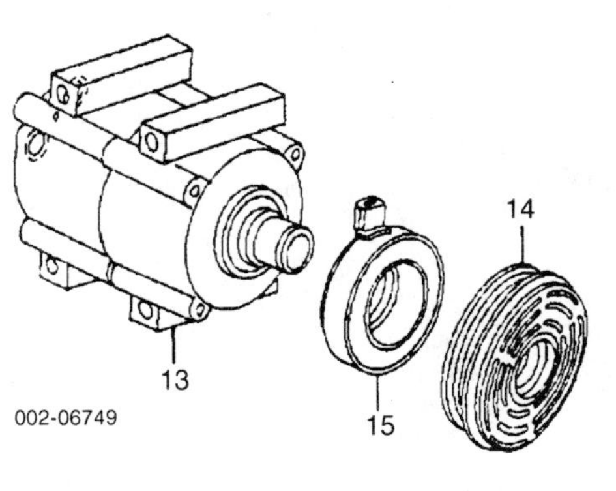

Item	Part No.	Labor	Price
13 **Compressor**	F6LZ 19703 AA	**m 1.1**	337.33
14 **Pulley, Compressor**	F7UZ 19D784 A	**m #1.0**	N.A.
15 **Coil, Compressor**	F5VY 19D798 A	**m #1.0**	39.13
#R&R Complete, included in R&R Compressor			
Belt, Drive-See Cooling System Section			

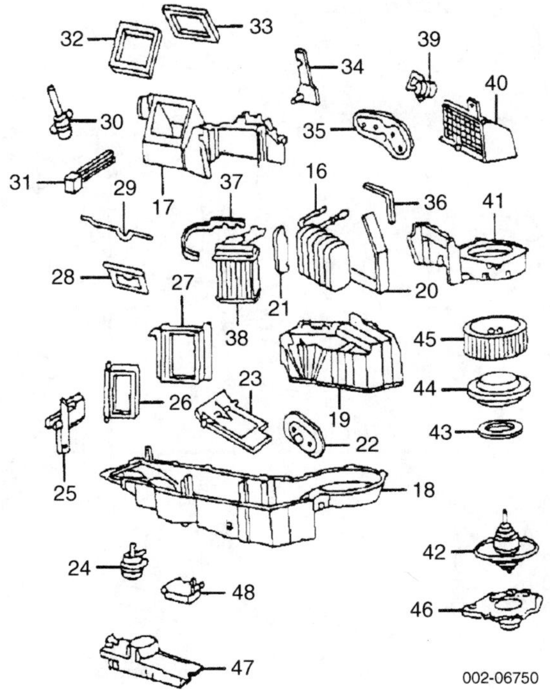

Item	Part No.	Labor	Price
16 **Core, Evaporator**			
Main	F65Z 19860 AA	**m #7.0**	184.27
#Includes R&I Instrument Panel			
Aux	F75Z 19860 AA	**m .9**	184.27
Valve, Evaporator	N.A.	**m .5**	N.A.
17 **Hsg, Evap Upr**	F65Z 19A626 AE	**m #8.0**	9.17
18 **Hsg. Evap Lwr**			
w/Floor Console	F75Z 19C831 BH	**m #8.0**	67.92
w/o Floor Console	F75Z 19C831 AA	**m #8.0**	63.75
#R&R Complete, includes R&I Instrument Panel & R&R Evaporator Core, Heater Core & Blower Motor Hsg			
19 **Cover, Evap Hsg**	F75Z 19C946 AA	**m #6.5**	11.70
#Includes R&I Instrument Panel, Included in R&R Evaporator Core			
20 **Seal, Housing**	F50Y 19A672A		6.27
12 **Seal, Evap**			
Main	F65Z 19D578 AA		2.65
Aux	F75Z 19D578 AB		6.05
22 **Seal, Drain Tube**	F65Z 19B739 AB		6.67
23 **Door, Outlet**			
Main	F65Z 19B749 AC	**m**	15.10
Aux	F75Z 19B749 AB	**m**	13.37
24 **Motor, Vacuum Cont Upr**	F65Z 18A318 AB	**m**	10.56
25 **Baffle, A/C**	F65Z 19W555 AB	**m**	2.70
26 **Door, Control Valve**	F65Z 18B545 AE	**m**	15.03
27 **Frame, A/C Cont**	F65Z 19D963 AA	**m**	18.07
28 **Door, Defrost**	F65Z 18A478 AB	**m**	9.63
29 **Shaft, Defrost Door**	F65Z 19D640 AB	**m**	2.70
30 **Motor, Heat Door Vacuum**	F65Z 18A318 BB	**m**	13.15
31 **Harness, Vacuum**	F75Z 19C827 BC	**m**	d10.45
32 **Duct, Evap Outlet**	F65Z 19643 AA	**m**	5.83
33 **Duct, Defrost**	F65Z 18C367 AA	**m**	5.27
34 **Brkt, Evap Case**	F65Z 19A804 AA		5.68
35 **Seal, Evap Case**	F65Z 19C593 AD		7.90
36 **Seal, Evap Core**	F65Z 9B738 BA		3.15
37 **Seal, Heater Core**	F65Z 18658 BA		5.45
38 **Core, Heater**			
Main	F65Z 18476 AA	**m #7.0**	27.33
#Includes R&I Instrument Panel			
Aux	F75Z 18476 AA	**m 1.5**	80.60
39 **Motor, Duct Vacuum**	F65Z 18A318 AB	**m**	10.56
40 **Duct, A/C**	F65Z 19A618 AB	**m**	29.27
41 **Hsg, Blower Motor**	F65Z 19D818 AC	**m**	4.77
42 **Motor, Blower**	F75Z 19805 FA	**m #.9**	71.45
43 **Seal, Blower Motor**	F65Z 18N260 BB	**m #.9**	3.08
44 **Spacer, Blower Motor**	F65Z 19C671 AB	**m #.9**	7.83
45 **Fan, Blower Motor**	F58Z 9834 AA	**m #.9**	11.72
46 **Cover, Blower Motor**	F65Z 18515 AA	**m #.9**	7.40
#R&R Complete			
47 **Duct, Heater Floor**	F65Z 18C433 AC		14.42
48 **Motor, A/C Door**	F75Z 19E616 BA		114.61

FIGURE 9–5 Air-conditioning components for a Ford Expedition. Note the refrigerant capacity.

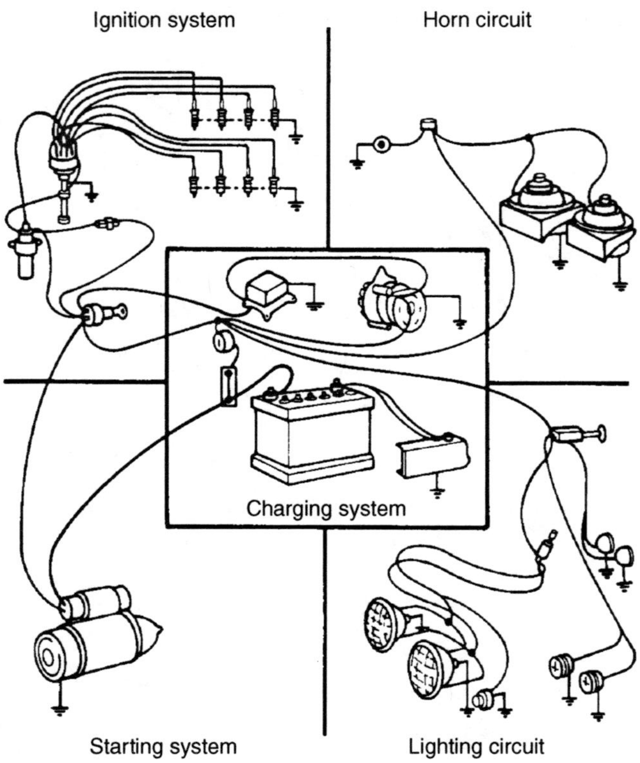

FIGURE 9–6 Diagram of electrical system.

The wires for these lighting systems are grouped together in harnesses with wires for other electrical circuits in the same local area (Figure 9–8).

Headlights are either replaceable bulbs or sealed beams (Figure 9–9). The replaceable bulb assembly consists of a lens, bulb, adjusting screws, and a

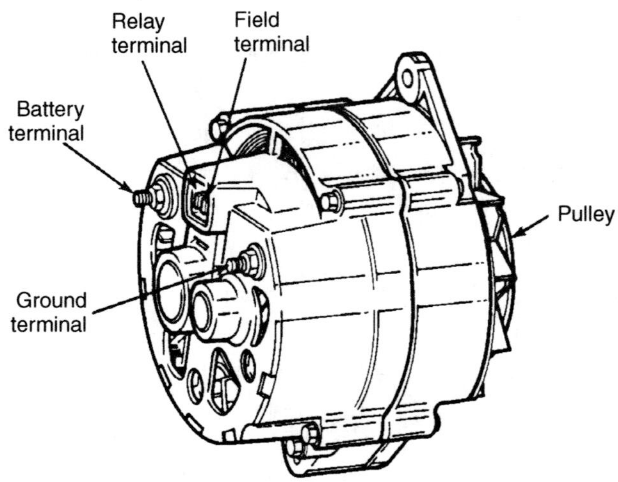

FIGURE 9–7 Alternator.

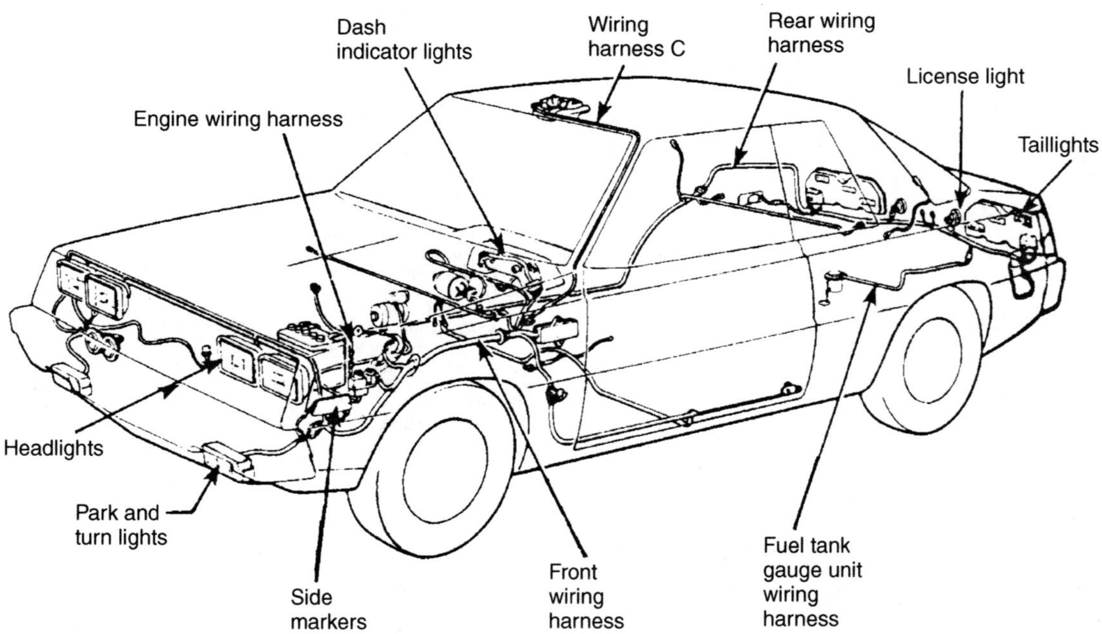

FIGURE 9–8 Wire harness groupings and lighting systems.

mounting bracket (Figure 9–10). Always check the headlights for proper operation and looseness. If the headlight is loose, often the tab that fastens the lens to the adjusting screw is broken. In this case, replace the lens. If the lens and mounting bracket are broken, replace the assembly. A sealed-beam headlight consists of a sealed-beam unit, adjusting screws, and mount. Check it for proper operation when writing an estimate. Include time to aim the headlights whenever a headlamp is moved or replaced.

Taillights, side marker, park lamps, and turn signals are each available as an assembly and in some cases, as separate parts (Figure 9–11). In cases of minor damage, the lens is often replaced instead of the entire assembly.

A

B

FIGURE 9–9 (A) Replaceable bulb or composite headlights; (B) sealed-beam headlights.

FRONT LAMPS

HEADLAMP

R&I Headlamp	R/L	.2
Aim Lamps		.4

NOTE: R&R Does Not include aim lamps.

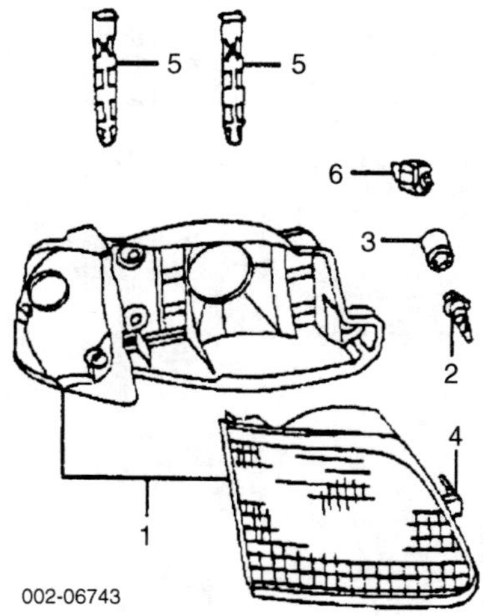

1 **Headlamp**				
To 7/96	R/L	F65Z 13C192-3 AA	#.3	119.60
From 7/96	R/L	F75Z 13008 A-B	#.3	N.A.
#Includes R&R Park/Signal Lamp Assy				
2 **Bulb/Retainer**	R/L	F5TZ 13N021 AA	.2	24.42
3 **Hsg, Bulb/Retainer**	R/L	E9FZ 13N019 A		1.88
4 **Screw, Adjusting**	R/L	F75Z 13032 BA		2.50
5 **Clip, Mounting**	R/L	F65Z 13N129 AD		8.08
6 **Module, Running Lamp**				
Canada	N.A.		m .5	N.A.

PARK/SIGNAL LAMP

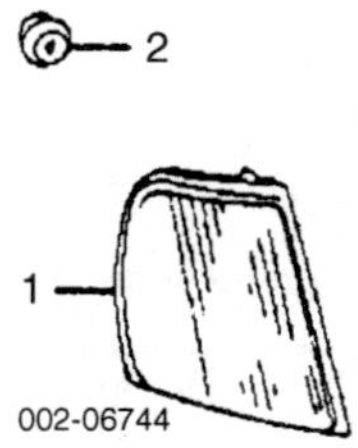

1 **Lamp, Park/Signal**				
To 6/96	R/L	F65Z 13200-1 AD	#.2	70.83
From 6/96	R/L	F75Z 13200-1 AC	#.2	42.42
2 **Bulb, Park/Signal**	R/L	F5AZ 13466 A	#.2	4.76
#R&R One Side Complete				

FOG LAMP

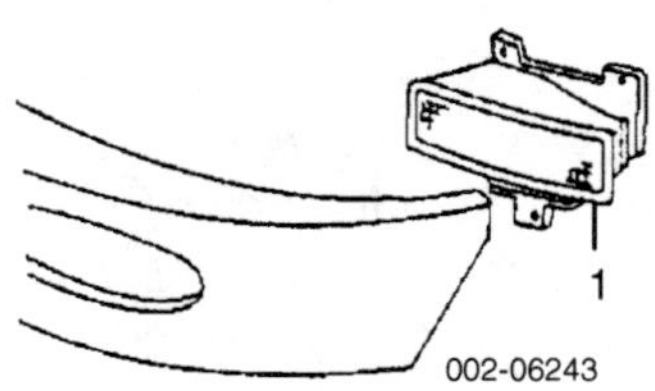

FIGURE 9–10 Headlamp components on a Grand Prix. (Courtesy of Mitchell International.)

REAR LAMPS

COMBINATION LAMP

R&I Lamp Assy	R/L		.3	

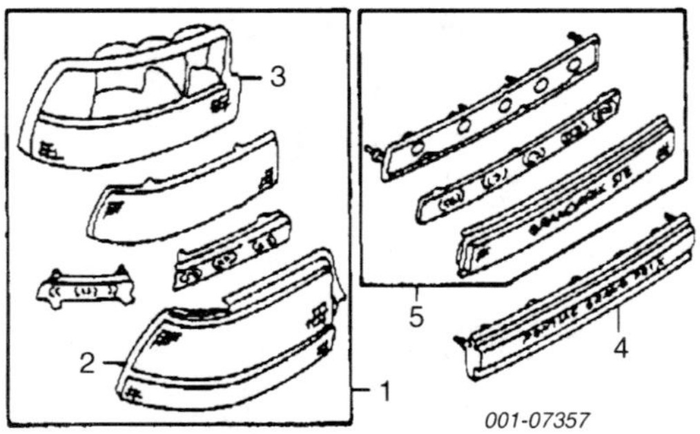

Item		Part No.		Price
1 **Lamp Assy**				
1990-91				
LE,SE	R/L	5976208-7	.3	258.00
STE	R/L	5977916-5	.3	275.00
1992-93	R/L	5976714-3	.3	275.00
1994-96	R/L	5976716-5	.3	275.00
Lens, Tail Lamp				
2 Outer¶				
1990-01				
LE,SE	R/L	16508910-09	#.7	96.25
STE	R/L	5977916-5	#.7	275.00
1992-93	R/L	16508928-7	#.7	93.50
1994-96	R/L	5977916-5	#.7	275.00
#R&R One Side Complete				
3 **Housing, Tail Lamp¶-Not Serviced**				
¶Included w/Tail Lamp Assy				
Filler, Tail Lamp				
1990-93				
4 LE		16509728	.3	152.00
SE	91	16509728	.3	152.00
	92-93	16518783	.3	48.25
5 STE		16508569	.3	182.00
1994-96		5977242	.3	161.00
Housing, Tail Lamp Filler				
SE	92-93	16508564		80.50
Socket w/Wire				
Tail/Stop¶	R/L	12083005		33.75
Back-Up/Signal¶	R/L	12083005		33.75
Side Marker	R/L	6298886		7.45
¶Order by Application				
Bulb				
Tail/Stop/Back-up	R/L	9441839	#.3	■2.95
Side Marker	R/L	9421330	#.3	■.47
#R&R One Side Complete, Included in R&R Lamp				

FIGURE 9–11 Rear lamp components on a Grand Prix. (Courtesy of Mitchell International.)

OTHER ELECTRICAL CIRCUITS

There are two other electric circuits in every automobile: windshield wipers and washers, and horn. Other electrical devices include radios, cassette players, speaker systems, chimes, buzzers, graphic displays, analog instruments, computer commands, power seat positioners, power windows and locks, power-positioned outside mirrors, automatic headlight dimmers, air bag systems, and cruise control.

9.4 A FRONT WHEEL DRIVETRAIN

ENGINE

All engines, both gasoline and diesel, are classified as *internal combustion engines* because the combustion or burning that drives the pistons takes place inside the engine. Front wheel drive vehicles have a transverse-mounted inline four-cylinder, V-6, or V-8 engine (Figure 9–12).

Collision damage to internal engine parts is possible. The cylinder head and overhead camshaft can be damaged if a transverse-mounted engine vehicle sustains a hard impact above the bumper. Some external engine parts that may be damaged in a collision are pulleys, belts, engine mounts, timing covers, oil pans, and air cleaners (Figure 9–13).

The crankshaft pulley turns belts to power accessories, such as the air-conditioning compressor, the power steering pump, and the water pump. Engine mounts hold the engine in place and isolate engine vibration. The mounts are usually found at the right, left, and front sides of the engine. Some engines may also have a rear mount (Figure 9–14). The timing cover protects the timing gears or timing belt. The oil pan acts as a sump for engine oil. The air cleaner filters engine air.

In a side impact, the lower frame rail can be moved enough to bend a pulley, then bounce back. When you inspect the damage, be aware that even though there may be a gap between the rail and the pulley, the pulley may be damaged (Figure 9–15). It is best to start the engine and observe for wobble in the pulleys. Damaged pulleys cannot be repaired; they must be replaced. Also, if the pulley is damaged, there is a chance that the pump or compressor or whatever the pulley is attached to may be damaged. Check the part for proper operation and leaking fluid. Note the possibility of damage on the estimate. The drive belts should be checked for tears.

Engine mounts may be damaged in a severe front or side impact. As a lower rail or radiator support is moved in a crash, whatever is attached to it will move also. Engine mounts are often bent in this manner. Observe the position of the mount, the engine, and the frame rail. Usually the mount is at a right angle to the engine and rail (Figure 9–16). Any angle other than a right angle indicates movement of the engine or rail. Often repair to the rail will restore the proper angle to the mount. If the mount is bent, it should be

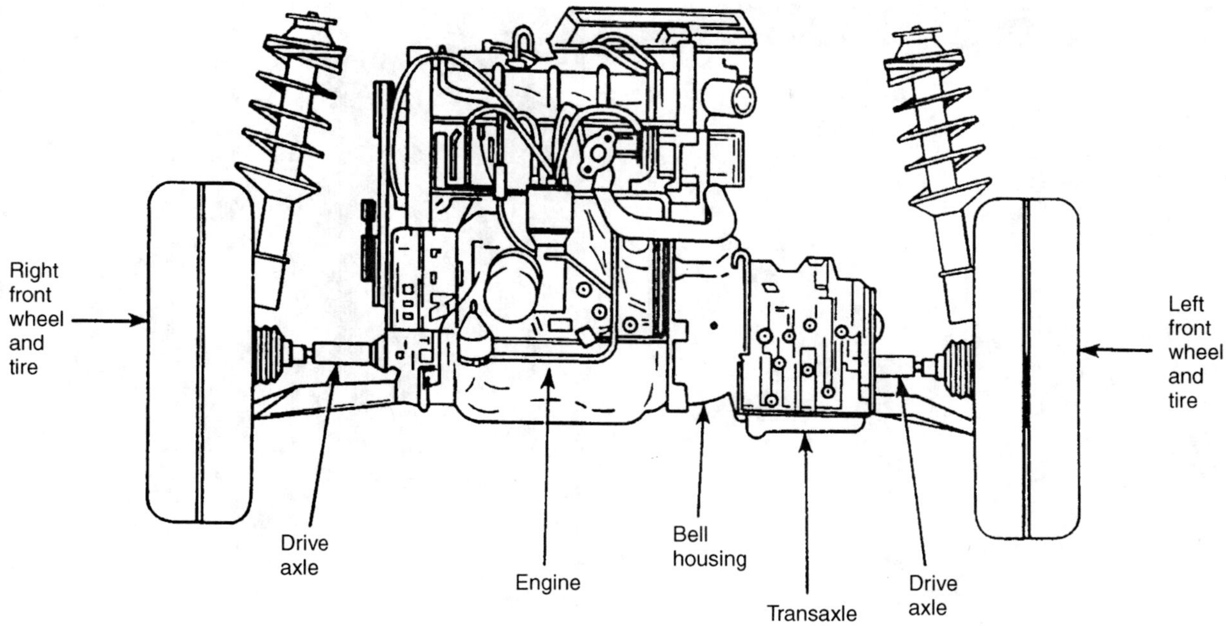

FIGURE 9–12 Front wheel drive power train components.

ENGINE

Use Procedure Explanation 12 with the following text.

Item	Year	Part No.		Labor	Price
R&I Engine Assy					
Man Trans			m	**6.5**	
Auto Trans					
1.8L Eng.			m	**5.6**	
1.9L Eng			m	**4.9**	
Add to R&I Engine Assy					
W/Air Cond			m	**.9**	
w/Pwr Strg			m	**.3**	
Pulley, Crankshaft					
1.8L Eng					
To 2-91		FOCZ 6312 A	m	**.8**	96.00
From 2-91		F1CZ 6312 A	m	**.8**	96.00
1.9L Eng		FOCZ 6316 A	m	**.7**	95.77
Cover, Timing					
1.8L Eng					
Upper					
Outer		FOCZ 6019 B	m	**.7**	75.67
Inner		FOCZ 6019 C	m		21.33
Lower					
Outer		F3CZ 6019 A	m	**1.1**	30.62
Center		F1CZ 6019 B	m		33.50
1.9L Eng		FOCZ 6019 A	m	**.7**	14.15
Pan, Eng Oil					
1.8L Eng					
Man Trans	91	FOCZ 6675 B	m	**#2.2**	213.75
	92-96	F3CZ 6675 D	m	**#2.2**	204.75
Auto Trans	91	F3CZ 6675 B	m	**#2.2**	219.08
	92-96	F3CZ 6675 B	m	**#2.2**	219.08
1.9L Eng		F3CZ 6675 C	m	**#2.2**	139.97
Gasket Eng Oil Pan					
1.8L Eng¶		F2HZ 6020 BA	m	**#2.2**	10.11
1.9L Eng		FOCZ 6710 A	m	**#2.2**	24.78

¶*Sealant*
#R&R Complete

FIGURE 9–13 Engine parts that may be damaged in a collision. Note the **M** for mechanical. The operations may be charged at a higher labor rate than body or painting repairs. (Courtesy of Mitchell International.)

replaced. Mounts can be broken in a severe impact. To check for a broken mount, jack up the engine. If the engine moves up, the mount may be broken. To check for a broken engine mount in a car with automatic transmission, start the engine, push the brake pedal down and put the car into "drive." Do not release the brake, but press lightly on the gas pedal. If the engine jumps up, the mount may be broken. Replace a broken or bent engine mount.

If the timing cover or oil pan are made of stamped sheet metal and have been slightly dented, they may be repaired by removal and metal finishing. Damaged covers and pans made from cast iron or aluminum and sheet metal, if badly damaged, should be replaced. Air cleaners are often mounted just behind the radiator support. A front or side impact can easily damage them. Check carefully, because damage can be difficult to see. A plastic air cleaner case or mount tab can be repaired by plastic welding adhesives.

9.5 TRANSAXLE

The transmission enables the automobile to move forward and backward. It also provides different gear ratios for various driving speeds in the forward direction.

ENGINE/TRANS MOUNTS

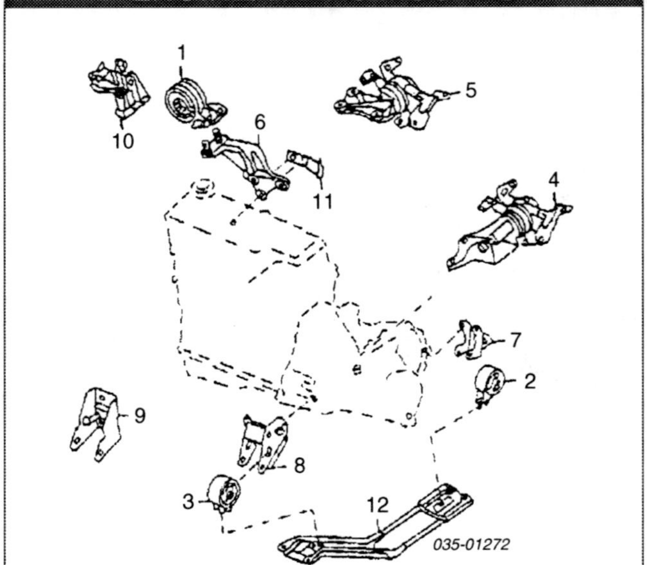

1 **Mount, Eng Right**	91-94	F4CZ 6038 C	.6	58.40
	95-96	F5CZ 6038 B	.6	58.40
2 **Mount, Eng Left**				
Man Trans		F4CZ 6068 B	.6	18.97
Auto Trans		F4CZ 6068 A	.6	35.65
3 **Mount, Eng Front**				
1991-93		F5CZ 6038 CA	.6	24.30
1994-96				
Man Trans		F5CZ 6038 CA	.6	24.30
Auto Trans		F4CZ 6038 D	.6	59.81
Mount, Eng Rear				
4 Man Trans				
1.8L Eng				
To 4-91		F1CZ 6038 D	.8	61.58
From 4-91		F1CZ 6038 J	.8	78.67
1.9L Eng		F4CZ 6038 B	.8	34.78
5 Auto Trans		F1CZ 6038 H	.8	78.67
6 **Bracket, Mount Rear**				
1.8L Eng		F1CZ 6030 A	#.2	32.83
1.9L Eng		FOCZ 6030 A	#.2	34.68
7 **Bracket, Mount Left**				
Man Trans				
1.8L Eng		F1CZ 6E042 D	#.2	15.92
1.9L Eng		F1CZ 6E042 B	#.2	14.62
Auto Trans				
To 2-28-94		F4CZ 6E042B	#.2	20.83
From 2-28-94		F4CZ 6E042B	#.2	20.83
Bracket, Mount Front				
8 Man Trans				
1.8L Eng		FOCZ 6031 A	#.2	34.38
1.9L Eng		FOCZ 6031 B	#.2	16.67
9 Auto Trans	91-93	FOCZ 6031 C	#.2	14.50
	94-96	F4CZ 6031 A	#.2	14.78
10 **Bracket, Eng Mount**		FOCZ 6F028 A	#.2	13.85

FIGURE 9–14 Engine mounts on a 1996 Escort. (Courtesy of Mitchell International.)

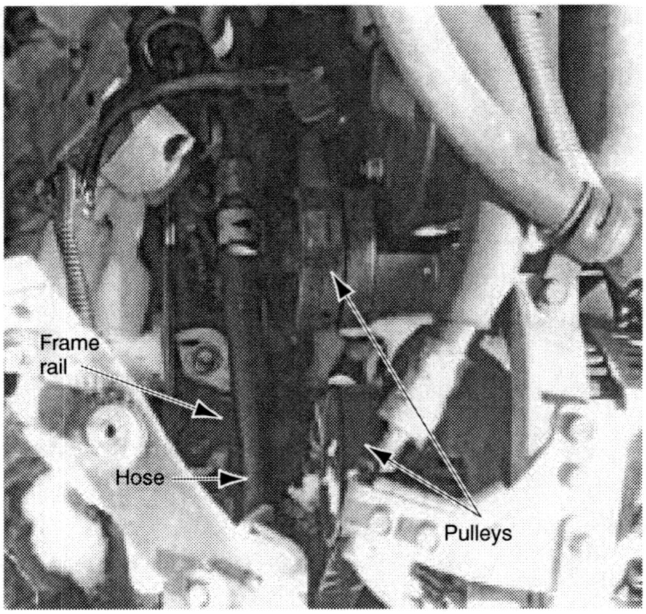

FIGURE 9–15 This Escort was hit on the right side. The frame rail moved enough to damage the pulleys.

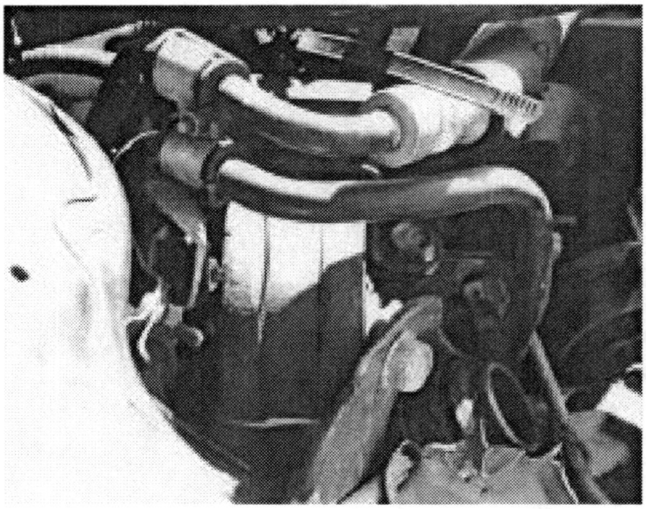

FIGURE 9–16 This is the same right side hit Escort as Figure 9–15. Notice how badly distorted the engine mount is.

The many different types of manual and automatic transmissions have made it possible for auto repair shops and technicians to specialize in the diagnosis and repair of specific units, and body shops and insurance adjusters normally rely on their professional assistance.

TRANSAXLE

In a front wheel drive car, the transaxle is in a position to be damaged in a front impact.

MANUAL TRANSAXLE

Most manual transaxles are five-speeds—that is, there are five forward speeds, including one overdrive, and one reverse. The transmission gears are housed in a cast aluminum case. The internal parts of the transaxle are the gears, clutch assembly, and shift forks. The external transaxle parts are linkages, shift lever, clutch master cylinder, and clutch slave cylinder. For proper operation, the transaxle and hydraulic clutch must have fluid, and linkages must be aligned. A collision can

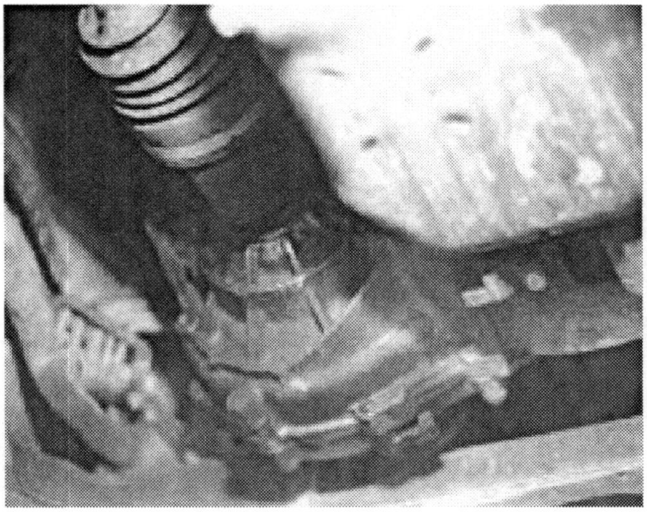

FIGURE 9–17 Manual transmission from a Cavalier.

FIGURE 9–18 Damaged automatic transmission. This drivetrain was hit hard in the right wheel; the transmission case is broken at the left front.

damage these by cracking the transmission case, breaking the hydraulic system, or misaligning the linkage (Figure 9–17). Raise the vehicle on a lift and check for fluid leakage around the transaxle. Look for bent linkage rods. Any damaged parts must be replaced.

AUTOMATIC TRANSAXLE

An automatic transmission (Figure 9–18) consists of one or more planetary gear sets, bands, servos, clutches, side gears, and oil pumps. These parts are enclosed in a transmission case and cover. Again, if the case is broken or cracked, it should be replaced. The automatic transmission has a stamped steel oil pan attached to the bottom of the transmission case for storage of fluid. If it is damaged in the seal area, it should be removed, straightened, and reinstalled using a new gasket. Whenever this pan is removed after damage, the internal parts should also be inspected.

The transmission may need to be disassembled and inspected if the external parts are damaged or if internal damage is suspected. Because wear can cause transmission malfunction, the customer and the insurance company should agree beforehand on who will pay for the disassembly. Typically, if the problem is caused by collision, the insurance company pays. If wear is the cause of the problem, the customer pays. However, the transmission should not be removed from the vehicle for this troubleshooting operation. In fact, many repairs can be made without removal.

If the transmission is in Park when the vehicle is struck, it could break the parking pawl, which is designed to break before any other parts would be damaged.

DRIVE AXLES

The power produced by a front wheel drive engine reaches the drive wheels by way of two drive axles or half shafts (Figure 9–19). To permit the drive wheels to be steered, each half shaft has two CV (constant velocity) joints, one at each end of the half shaft. Each CV joint consists of a cage, bearings, driver or tripod, housing, and boot. Proper operation depend's on grease retained by the boot. A severe impact on a drive wheel can pull the half shaft from the transaxle or even break a CV joint (Figure 9–20). Whenever a drive wheel is damaged, the half shaft should be inspected. Look for a torn boot. Pull the half shaft and check for looseness. Damaged boots and CV joints can be replaced. In some cases, the entire axle shaft will need to be replaced.

ENGINE CRADLE

An engine cradle—sometimes referred to as the subframe—is found in some front wheel drive cars (Figure 9–21). The engine, transaxle, and occasionally the rack and pinion are mounted to this steel framework. This framework resembles the frame of a full-frame vehicle (Figure 9–22). The cradle is mounted to the unibody by four bolts, two at the radiator support and two at the cowl. By loosening the four bolts and raising the vehicle, one can easily remove the disconnected drive train as a unit. Due to the cradle's intricate structure, damage to two or more mounting areas call for cradle replacement, although repairs involving only one mounting area should be attempted.

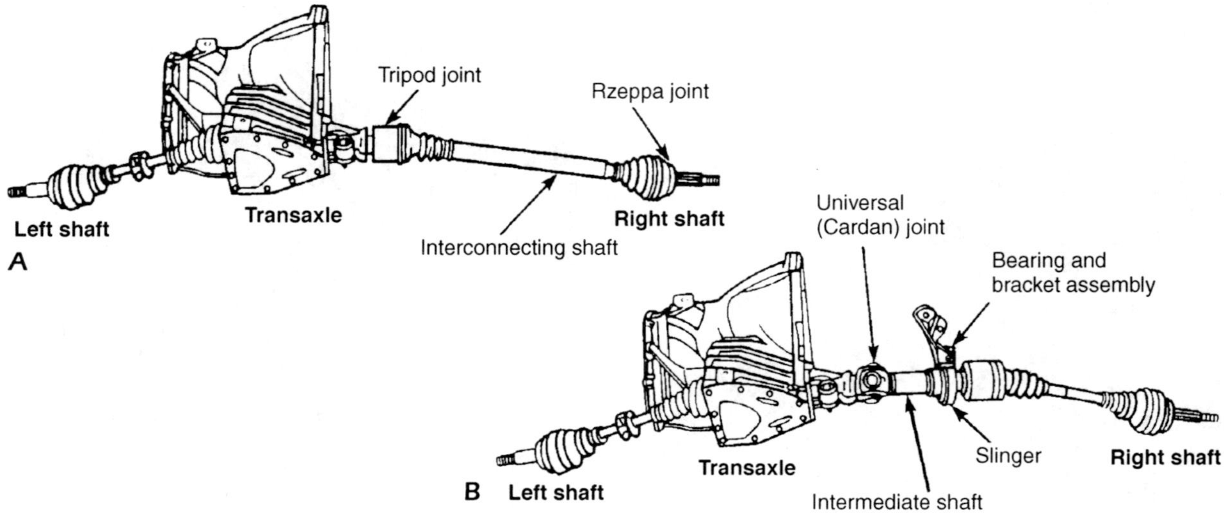

FIGURE 9–19 Two types of half shafts: (A) unequal length, (B) equal length.

FIGURE 9–20 An impact on the wheel pulled this half shaft out of the transaxle.

FIGURE 9–22 This is the engine cradle from an Astro van. Notice the damage near the right front mount.

SUB-FRAME

NOTE 1: R&R Does Not include alignment or bleed system.

NOTE 2: Manufacturer advises whenever frame mount bolts are loosened or removed, they must be replaced with new parts and torqued to avoid damage to sub-frame, drive train or suspension.

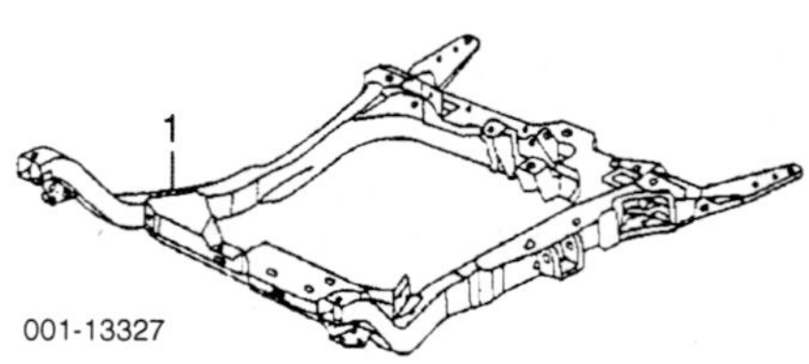

1 **Sub-Frame**	22638045	**m #5.3**	255.00

#Includes R&I/R&R Lower Control Arms, Stabilizer Bar, Front Exhaust Pipe, Steering Gear & Support Engine

FIGURE 9–21 Sub-frame from a 1997 Chevy Malibu. Notice the warning about the mount bolts and that the labor operation is listed as **m.** (Courtesy of Mitchell International.)

9.6 REAR WHEEL DRIVE DRIVETRAIN

ENGINE

The engine of a rear wheel drive vehicle is mounted longitudinally and consists of four, six, eight, or ten cylinders (Figure 9–23). As with the front wheel drive engine, collision damage to the block, heads, and internal components is unlikely. External parts that may be damaged include the pulleys, belts, oil pan, engine mounts, and air cleaner. These are checked and repaired in the same way as with front wheel drive vehicles.

FIGURE 9–23 Rear drive engine.

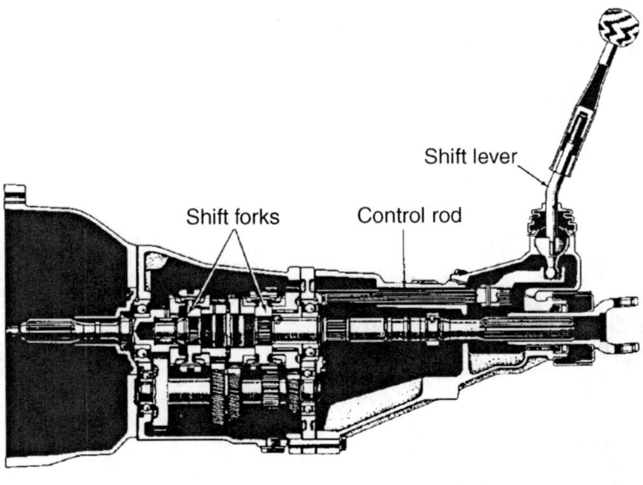

FIGURE 9–24 Rear wheel drive manual transmission.

MANUAL TRANSMISSION

A manual transmission consists of gears, five forward and one reverse, and a clutch assembly enclosed in an aluminum case (Figure 9–24). External components are the master and slave cylinders and the shift linkage. The clutch master cylinder is mounted in the cowl and can be damaged in a crash. When inspecting a manual transmission for damage, look for leaking fluid first. A cracked transmission case or master cylinder must be replaced. Check the area where the transmission is mounted to the cross member for proper alignment. Note any suspected damage on the estimate.

AUTOMATIC TRANSMISSION

An automatic transmission consists of many of the same parts as an automatic transaxle: Planetary gear sets, bands, servos, clutches, and oil pumps inside an aluminum case. A slightly damaged oil pan can be repaired. Check the transmission for leaks and the mounting points for distortion. Always replace a cracked case (Figure 9–25).

DRIVESHAFT

The driveshaft transfers the engine and transmission power to the differential (Figure 9–26). The shaft is a hollow tube with two universal joints, one at each end. Some vehicles may have a two-piece driveshaft. In that case, they have three U joints. Driveshafts can be pulled out from the transmission by a severe impact on the rear axle. No attempt should be made to repair a damaged driveshaft. Replace the entire assembly if damaged.

REAR AXLE ASSEMBLY

The drivewheels are powered by the driveshaft and axles. The differential gears allow the rear wheels to rotate at different rates on turns. The rear axle assembly consists of an axle housing, two drive axles, and a set of differential gears (Figure 9–27). An impact on a rear wheel can bend the axle or even break the axle housing. To check for a bent axle, first jack up the rear end and support the axle housing with jack stands. If the wheel is bent, install a good wheel. Spin the wheel. Look for wobble by standing at the rear. It may help to place a pointer as a reference point (Figure 9–28). If the wheel wobbles, the axle is bent. To check for a bent housing, measure side to side from a reference point. Bent axles or housings should be replaced.

FOUR WHEEL DRIVE

A four wheel drive vehicle, commonly a pickup truck or sport utility, has a transfer case, front driveshaft and a front axle assembly. The transfer case is mounted behind the transmission and powers the front driveshaft and front axles (Figure 9–29). In some cases, there are CV joints at the front wheels. Damage to these components can occur if there is an impact on the front wheel or if there is severe frame damage. Check the transfer case for leaks or mounting point damage. The front driveshaft and front axle are inspected and repaired in the same manner as the rear driveshaft and rear axle.

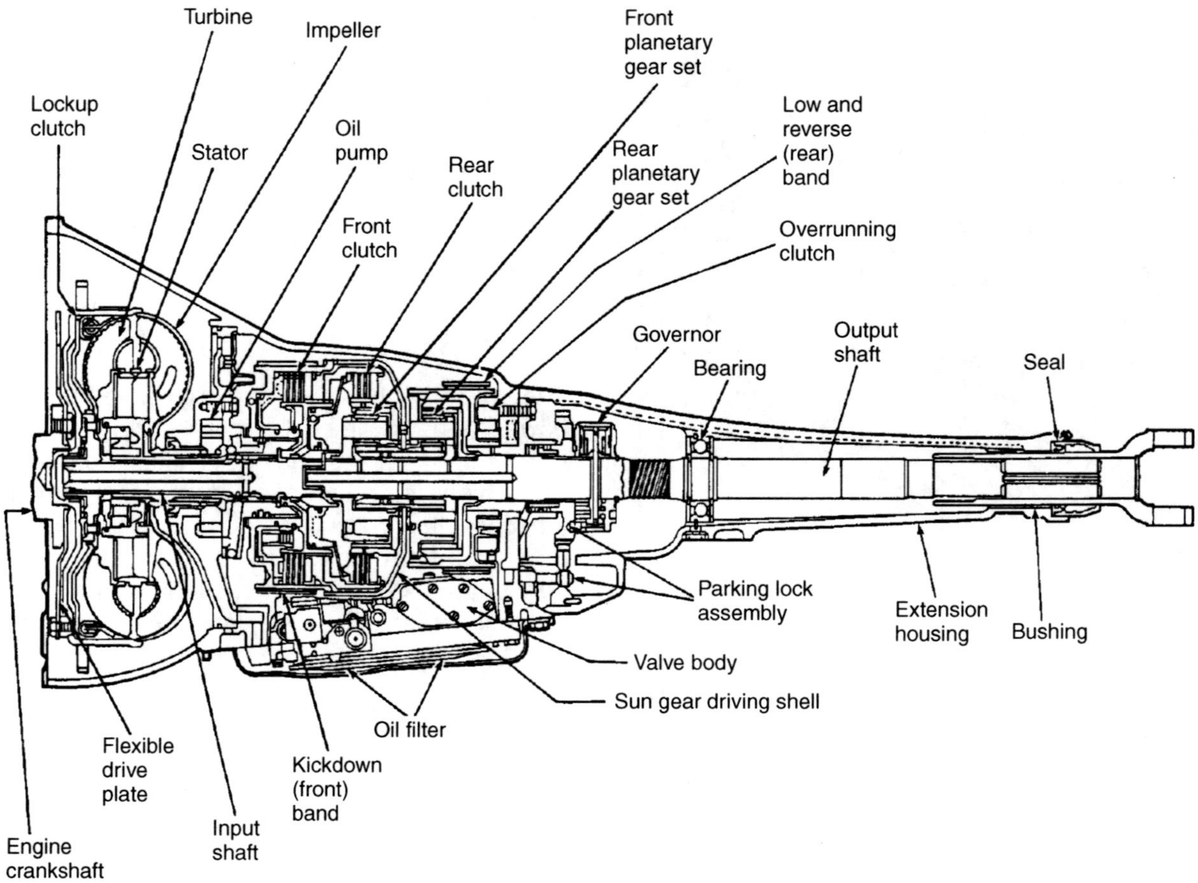

FIGURE 9–25 Rear wheel drive automatic transmission.

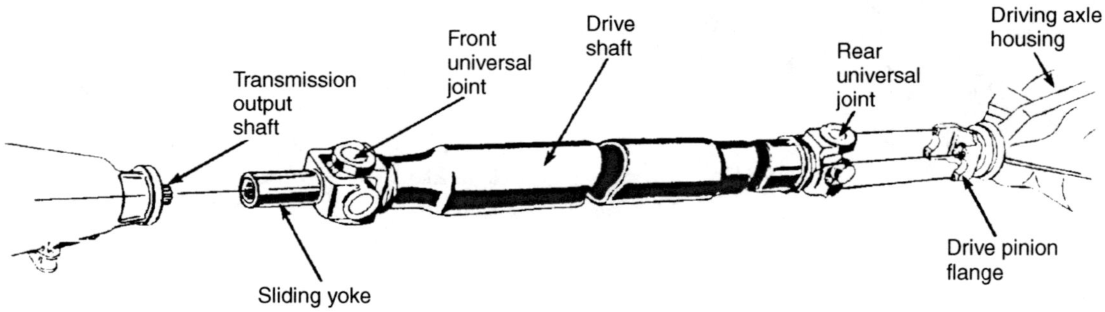

FIGURE 9–26 Drive shaft on a rear wheel drive vehicle.

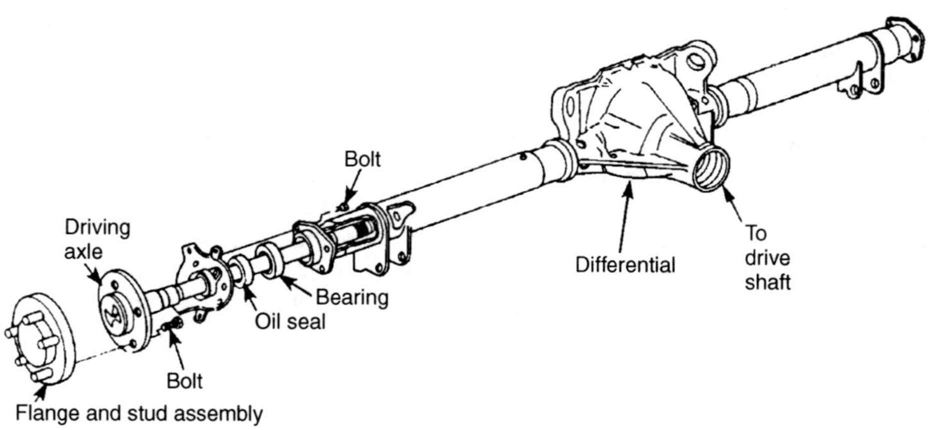

FIGURE 9–27 Rear axle housing on a rear wheel drive vehicle.

FIGURE 9–28 This simple setup—jack stand and pencil—can be used as a reference point to check for a bent wheel. Spin the wheel and watch to see if the reference point to the wheel distance changes.

9.7 EXHAUST SYSTEM

The exhaust system is used to collect and discharge the exhaust gases caused by the combustion of the air/fuel mixture within the engine. It also reduces the noise of the engine and contributes significantly to the cooling of the engine. The exhaust system found on today's automobile must also aid in the control of exhaust emissions. The components of the exhaust system (Figure 9–30) include the following:

- Exhaust manifold and gasket
- Exhaust pipe, seal, and connector pipe
- Intermediate pipes
- Catalytic converter
- Muffler
- Resonator
- Tailpipe
- Heat shields
- Clamps, gaskets, and hangers

The collection of exhaust gas begins in the exhaust manifold or header. This is the first part of the exhaust system and bolts directly to the cylinder head. The exhaust manifold is designed so that it will mate correctly with the exhaust ports of the cylinder head. Most in-line engines are equipped with a single

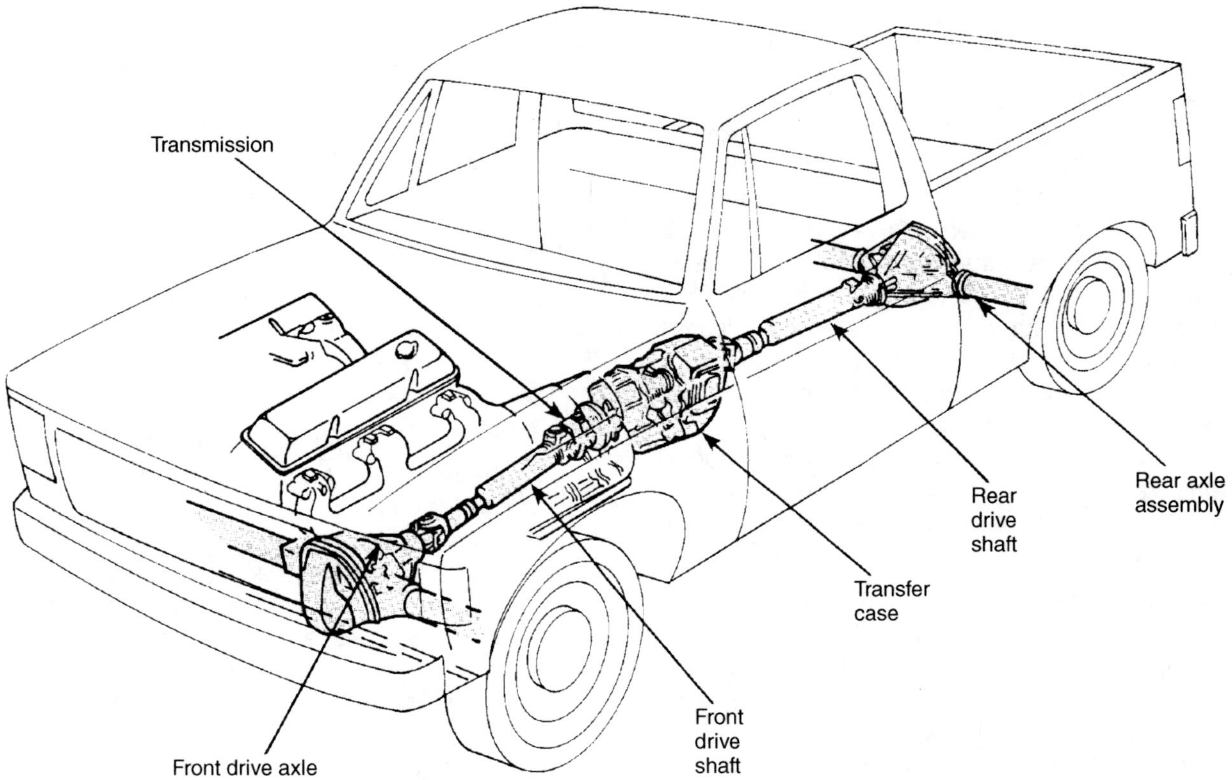

FIGURE 9–29 Four wheel drive consists of front and rear driveshafts and axles.

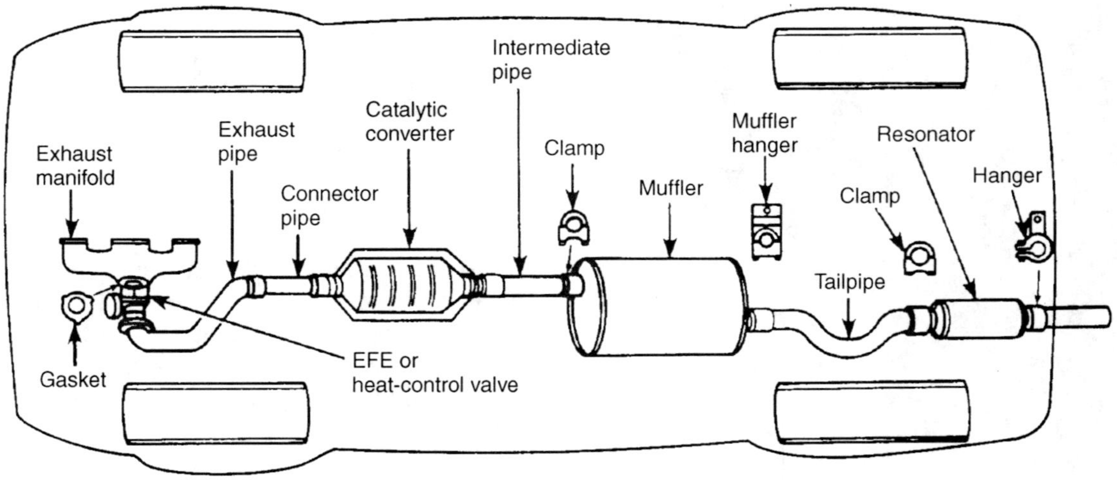

FIGURE 9–30 Exhaust system components.

exhaust manifold. While V-design engines have one manifold for each cylinder head, in older automobiles it was common for a V-design engine to be equipped with a complete dual exhaust system. Each side of the engine had its own system to collect and discharge exhaust. On most late-model automobiles the two exhaust manifolds are connected by the exhaust or inlet pipe, which is manufactured in the form of a Y. All exhaust gas is then discharged through a single muffler and tailpipe arrangement.

In all types of engines the exhaust gas flows from the manifold to the exhaust or inlet pipe. The inlet pipe then connects to a series of exhaust pipes made up of a catalytic converter, muffler, and interconnecting pipes. Some large engine exhaust systems also have a second muffler or resonator to further muffle the noise of the exhaust.

The muffler is simply a can or container with an inlet and outlet pipe attached. Inside the container is a series of baffles designed to deflect the exhaust within the container. The baffling or deflecting of the exhaust quiets the sound of combustion as it leaves the tailpipe.

The catalytic converter is an emission control device that changes harmful carbon monoxide gas to carbon dioxide and ignites and burns the hydrocarbons that were not ignited during the engine's combustion process. There are two basic designs of catalytic converters:

1. *Monolithic.* This design, used on Chrysler and Ford vehicles (Figure 9–31A), contains a ceramic honeycomb-shaped substrate, or carrier, on which the precious metal catalyst is deposited. The catalyst is usually platinum and/or palladium. This substrate is housed in a stainless steel casing that is insulated to produce a skin temperature of between 250 and 300 °F.
2. *Pelletized.* This design (Figure 9–31B) is wide and flat (looks like a hot water bottle) and contains a substrate made up of thousands of tiny porous beads or pellets that are coated with the same precious metals as those found on the honeycomb or the monolithic design. This pellet design is housed in a stainless steel casing, which is also insulated.

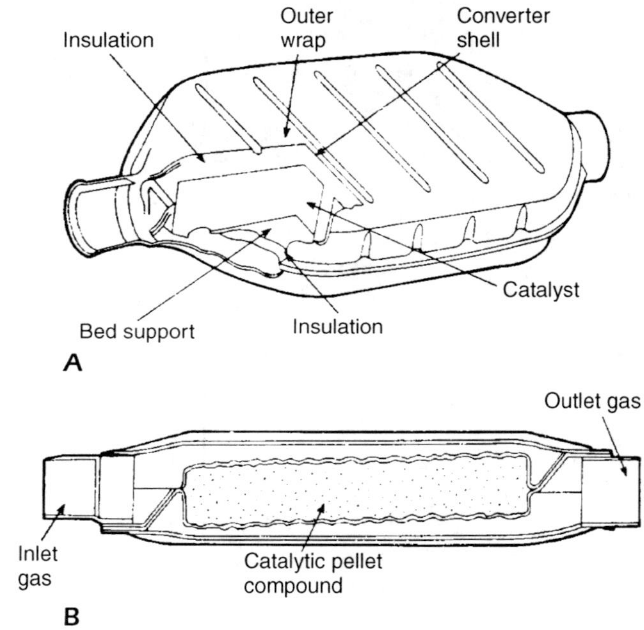

FIGURE 9–31 Catalytic converters (A) monolithic type, (B) pelletized type.

EXHAUST SYSTEM ANALYSIS

Damage to the exhaust system might be detected by a sharp increase in engine noise. Exhaust parts should also be visually inspected for cracks or bends. In any case where the damage is more extensive than minor scrapes or very slight bends, replacement of the damaged part is required. Short areas of damage at the end of the tailpipe can be removed and replaced with a tailpipe extension. Catalytic converters must be replaced if they are damaged or not functioning properly. Catalytic converters must be replaced with new original equipment, aftermarket, or rebuilt parts that have been certified by the Environmental Protection Agency. A salvage catalytic converter may not be installed unless it first passes a test that proves it can still clean up 50 percent of the unburned hydrocarbon (HC) and carbon monoxide (CO) emissions within two minutes and 75 percent of the HC and CO emissions within 200 seconds. Aftermarket replacement converters likewise must be certified to show they will continue to work for at least 25,000 miles and not fail due to construction defects for five years or 50,000 miles.

Many exhaust parts have a short life expectancy. These parts often fail due to rusting. Therefore betterment should be taken when applicable.

9.8 FRONT SUSPENSION

Front suspension systems are fairly complex because they have somewhat contradictory jobs. They must keep the wheels rigidly positioned and at the same time let them steer right and left. In addition, because of weight transfer during braking, the front suspension system absorbs most of the braking torque. While accomplishing this, it must provide good ride and stability characteristics.

There are three basic types of independent front suspensions (IFS) found on passenger cars. They are as follows:

- Coil spring suspension
- Torsion bar suspension
- Single control arm (strut type)

Coil springs and torsion bars were the traditional suspensions used on American cars until the recent popularity of strut systems. MacPherson strut suspensions are a lighter weight, which helps save gas; and they take up less space in the engine compartment, which is very important with the transverse-mounted engines used in most front wheel drive cars. Twin I-beam, another type of front suspension, is used on vans and trucks.

It should be noted that when individual suspension parts are replaced, the estimated time in the collision estimating guides does not include the time needed for wheel alignment. A suspension overhaul includes disassembly, inspection, cleaning, replacement of old parts, and reassembly. It also includes disconnection of the tie-rod end at the steering knuckle or arm. However, a suspension overhaul does not include:

- Wheel alignment (front or rear)
- Removal and installation of the steering gear, steering linkage, or stabilizer bar
- Bleeding and adjusting the brakes

SHORT-LONG ARM SUSPENSIONS

The unequal control arm or short-long arm (SLA) suspension system has been common on domestic-made vehicles for many years. Since each wheel is independently connected to the frame by a steering knuckle, ball joint assemblies, and upper and lower control arms, the system is often described as an independent suspension. The two SLA suspensions in common use are the coil spring suspension and torsion bar suspension.

Coil Spring Suspension

This system is used in older vehicles, new heavier full-size cars, and pickups. The essential components of this system are discussed below.

Knuckle and Wheel Spindle Assembly. In most cases, a steering knuckle and wheel spindle are forged to form a single piece. The wheel spindle is connected to the wheel through wheel bearings. The steering knuckle is attached to the upper and lower control arms by ball joints (Figure 9–32). A straightedge or dial indicator can be used to determine if the assembly is bent. The tolerance is 0.007 to 0.012. If the assembly does not meet these specifications, it must be replaced.

Ball Joints. The ball joints connect the steering knuckle to the control arms, allowing the steering knuckle to pivot between the control arms when the car is steered. They also permit up and down movement of the knuckle assembly. These ball joints, which consist of a ball and a socket, can be visually inspected for distortion, tears, or binds. If they are damaged, they must be replaced. They are serviced separately and are subject to betterment if there are any signs of wear.

Control Arms. The upper and lower control arms on the traditional independent front suspension function primarily as locators, fixing the position of the system and its components relative to the car. They are made of heavy-gauge sheet metal with grooves and channels stamped into them for additional strength

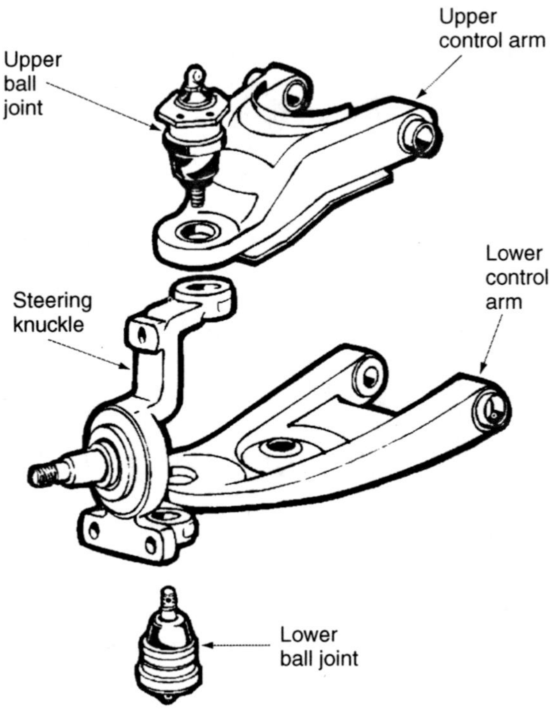

FIGURE 9–32 Steering knuckle with upper and lower control arms and upper and lower ball joints. (Courtesy of Moog Automotive.)

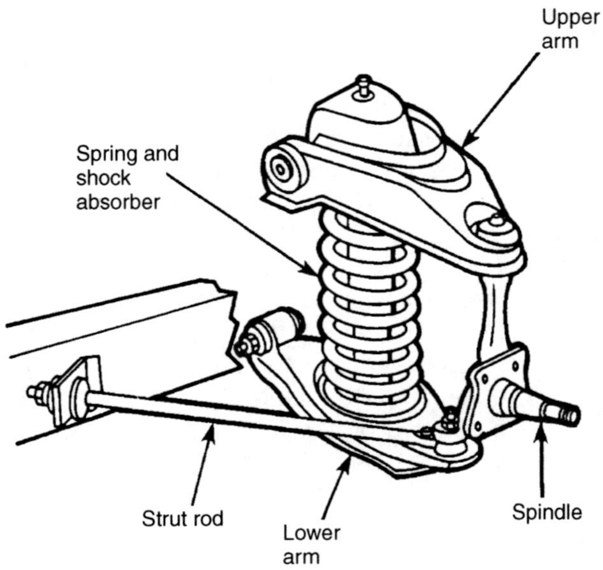

FIGURE 9–33 Short-long arm suspension.

and rigidity. If the control arms are merely mispositioned with no sharp creases or buckles, a major front end alignment is possible. However, if the control arms are buckled with sharp creases, they must be replaced.

Cross Shafts. The cross shafts attach the control arms to the frame. They are located at the inner end of the control arms. A front end alignment is possible as long as there is no visible damage, but if the cross shafts are bent, they must be replaced.

Shock Absorbers. Shock absorbers control the deflection and rebound rate of the coil springs and also help to prevent the vehicle from swaying. On the conventional frame, the bottoms of the shock absorbers are connected to the lower control arms, and their tops are connected to the vehicle frame. Another design places the shock absorber between the upper control arm and the frame. On the unibody, the tops are connected to the apron, whether the bottoms are connected to the upper or lower control arms. In all of these designs, the shock absorbers are usually located inside the coil springs.

The shocks should be visually inspected for bends or leaks. Slight dents caused by stones kicked up by the front wheels will not interfere with the normal operation of the vehicle. A mounting loop that is loose on the open end can be welded. Some repairs are possible on the bolt-type connection.

Shock absorbers should be replaced in pairs. Replacements must be of the like kind and quality. If a right front shock absorber is replaced, the left front shock absorber should be replaced also.

Coil Springs. The coil springs control the riding height of the vehicle. They also provide support and smoothness of ride. These heat-treated spring steel coils should be visually checked for cracks or permanent bends. As a further check for permanent bends, the springs can be rolled on a flat surface. If the springs are damaged, they must be replaced in pairs. Salvage coil springs can be considered for use as replacements.

Strut Rods. Strut rods (Figure 9–33) are used in some systems to restrict the forward and backward movement of the lower control arm. They are attached between the lower control arm and the frame. Often, both ends of the strut rod are threaded to allow for the caster angle adjustment when performing a wheel alignment. If a strut rod is not severely damaged, it can be repaired by straightening or adjusting it.

Stabilizer. The stabilizer or sway bar is a spring steel bar shaped like a wide, shallow U that extends from one lower control arm to the other. Its purpose is to reduce the sway of the vehicle as it goes around curves or travels over uneven road surfaces. The upward movement of one of the control arms causes a twisting force to be exerted on the stabilizer bar. This twisting force is transmitted to the opposite control arm, causing it to move upward to help keep the vehicle level and reduce sway. If the stabilizer is bent, it should be replaced.

FIGURE 9–34 The front suspension on this 4WD Dodge pickup is bent.

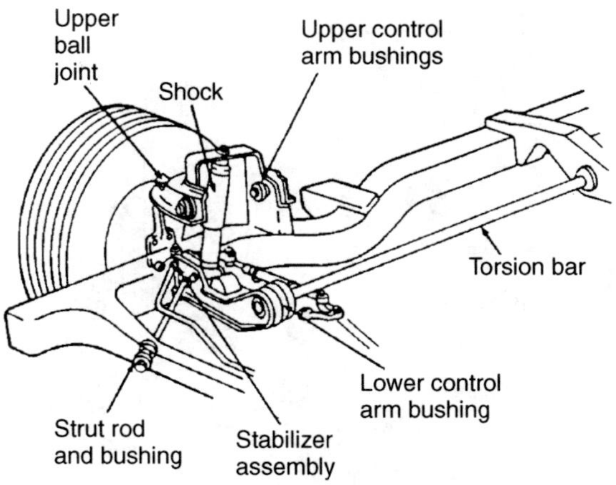

FIGURE 9–35 Conventional torsion bar system.

The stabilizer is connected to the left and right lower control arms with stabilizer links. If these links are damaged, they must be replaced. These links can be purchased separately.

The stabilizer is also supported by mounting brackets at the left and right frame side rails. If damaged, these brackets can be repaired or, in most cases, replaced. On those models that do not have the brackets available separately, extensive repairs can be made to the brackets.

Rubber Bumping Blocks. Rubber bumping blocks are attached to the upper and lower control arms. When the vehicle "bottoms," that is, when a spring completely compresses, such as it does when the vehicle passes over a pothole, these rubber blocks prevent metal-to-metal contact that could damage suspension parts. They are usually cone- or wedge-shaped, and they come in contact with a striking pad secured to the frame side rail when "bottoming" occurs. They should be visually checked for splits or breaks. Be alert to prior damage. If the rubber blocks are damaged, they might be replaced separately.

These rubber bumpers also provide the means for giving all suspensions a quick visual check. By observing whether or not these blocks strike as they should normally, it can be determined whether or not the suspension has become mispositioned. Figure 9–34 shows a damaged truck front suspension.

Torsion Bar Suspension

There are no coil springs in a torsion bar independent front suspension system. In this system, a steel bar that twists lengthwise is used to provide the spring action (instead of a coil spring). As this bar twists, it resists up-and-down movement. One end of the bar attaches to the frame; the other end attaches to the lower control arm.

In a conventional torsion bar suspension, the torsion bar runs from front to rear (Figure 9–35). Either an A-arm or a single inner bushing control arm with a strut rod is used for the lower control arm. Torsion bars can be used to adjust vehicle riding height. They are not interchangeable from side to side, because the direction of the twisting (or torsion) is different between the left and right sides. An adjustable cam, also called an anchor cam, can be located at the forward or rear end of the bar. It is adjusted to place a twist on the torsion bar in order to control the riding height of the car.

In a transverse torsion bar system—the latest in torsion bar suspension—the bar is mounted from side to side and runs across the width of the chassis back to the control arm (Figure 9–36). This eliminates the need for a strut rod or A-arm.

Other than the difference between the torsion bar and the coil spring, this type of suspension is the same as the coil spring variety. If a torsion bar or adjustable cam is damaged, it must be replaced. All other damage is handled in the same manner as it is on the coil spring suspension.

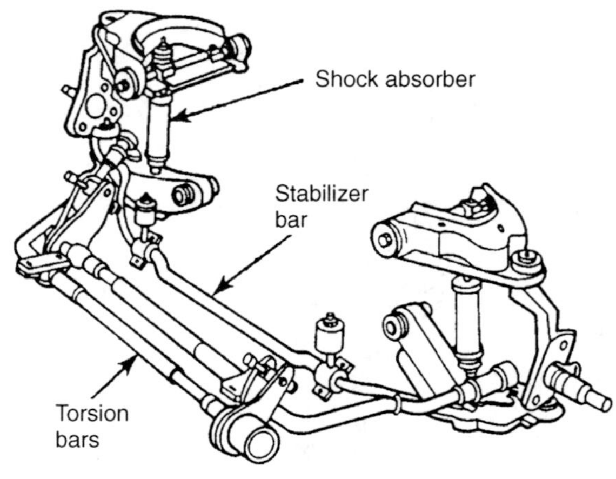

FIGURE 9–36 Transverse torsion bar.

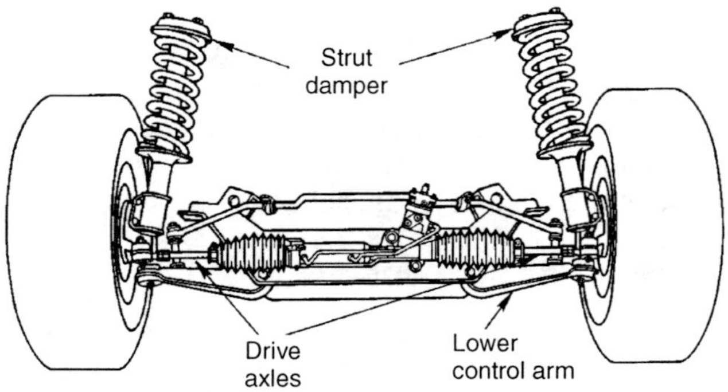

FIGURE 9–37 MacPherson strut suspension.

MACPHERSON STRUT SUSPENSION

This type of suspension system is used exclusively on vehicles with unitized construction. The MacPherson strut suspension is dramatically different in appearance from the traditional independent front suspension (Figure 9–37), but the similar components operate in the same way to meet suspension demands.

Its most distinctive feature is the combination of the main elements into a single assembly. The MacPherson strut assembly typically includes the spring, upper suspension locator, and shock absorber and is mounted vertically between the top arm of the steering knuckle and the inner fender panel. There is no upper control arm or upper ball joint.

When an automobile with MacPherson strut suspension is involved in a front-end collision, obtaining the correct front-end alignment is much more difficult, since this suspension type lacks all of the adjustments that are necessary to make proper front-end alignment. Thus the old techniques of obtaining proper body and frame alignment are outdated. A satisfactory job will no longer result merely by taking cross diagonal measurements, straightening the frame, hanging on the sheet metal, and sending the vehicle to an alignment shop. Because the suspension is attached to the body itself, the body as well as the frame and underbody must be aligned to factory specifications if correct wheel alignment is to be obtained. Figure 9–38 shows a damaged MacPherson strut.

FIGURE 9–38 Obvious MacPherson strut damage.

When engine or bodywork is performed on vehicles equipped with engine cradles, the cradle may become misaligned with the chassis. The alignment holes in the cradle must be directly in line with the corresponding holes in the body or steering pull because increased torque steer or an off-center steering wheel can result. Check cradle alignment during front-end alignment.

The need for precise front end and often rear-end alignment has led to a dramatic increase in the number of collision repair facilities doing their own alignment work. Relying on outside specialty shops for alignment often slows the repair process and increases costs if more than one trip to the alignment shop is needed. Shops that are equipped to perform their own alignment work increase their share of the repair bill and eliminate one of the most common customer complaints and comeback problems.

It should be noted that caster and camber adjustment kits are available for certain foreign vehicles that have nonadjustable caster and camber. These kits make proper wheel alignment an easier task.

TWIN I-BEAM SUSPENSION

Twin I-beam suspensions can be found on both two wheel drive and four wheel drive trucks and vans (Figure 9–39), although they are used almost exclusively on Ford vehicles. Twin I-beams are used on two

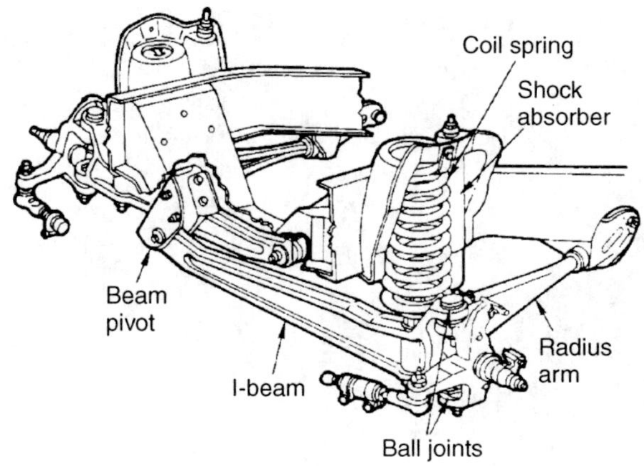

FIGURE 9–39 Twin I-beam suspension.

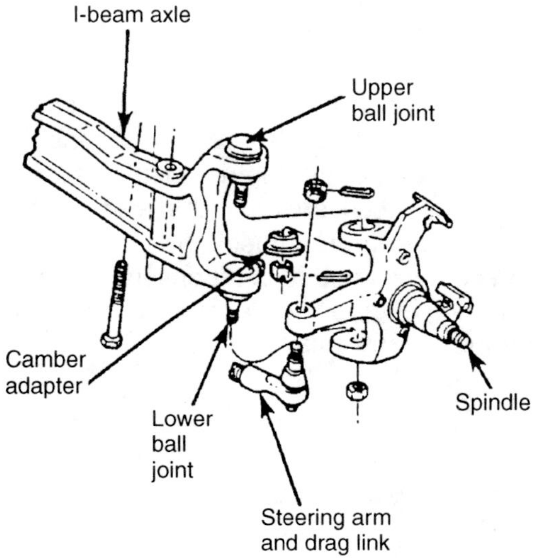

FIGURE 9–40 Twin I-beam axle with ball joints.

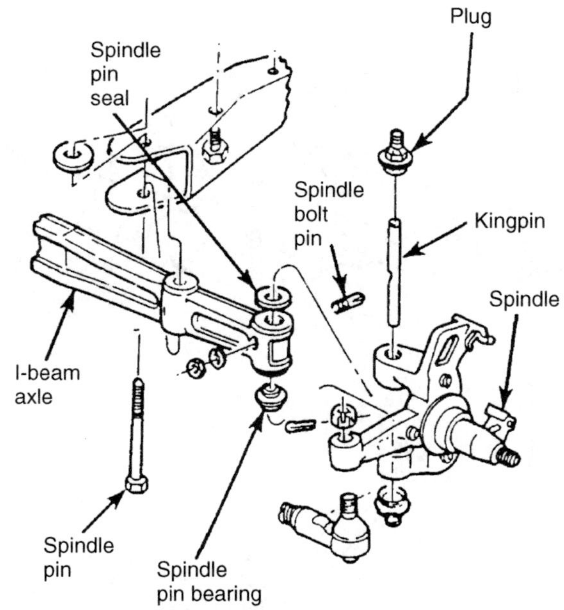

FIGURE 9–41 Twin I-beam axle with kingpin and spindle pin.

FIGURE 9–42 Bent I-beam on a Ranger pickup.

wheel drive vehicles; twin traction beams are used on four wheel drive vehicles. They are independent front suspensions.

Typical suspension components are the beams, springs, shock absorbers, radius arms, and ball joints or kingpins. Ball joints are used with traction beams, ball joints or kingpins with twin I-beams. Coil springs are used with twin I-beams, and either leaf springs or coil springs can be found on vehicles with traction beam suspensions.

Bushings, which are subject to wear, are used on both suspensions at the frame end pivot of the beams and at the frame end of radius arms on suspensions with coil springs. The method of attachment at the spindle varies with the vehicle application. First, there are the I-beam axles that have upper and lower ball joints to attach them as shown in Figure 9–40. Second, there are the I-beam axles that are secured to the spindle with a spindle pin (Figure 9–41). Figure 9–42 shows a damaged I-beam.

9.9 OTHER SUSPENSION SYSTEMS

A collision repair facility estimator or insurance company appraiser will encounter a variety of suspension systems. Some of the more popular rear suspensions are shown in Figure 9–43. The type of rear suspension used on a vehicle is determined by several factors: the weight of the vehicle; whether the vehicle is a sports, luxury, or off-road vehicle; and whether the vehicle is front or rear drive. Figure 9–44 shows a damaged rear suspension.

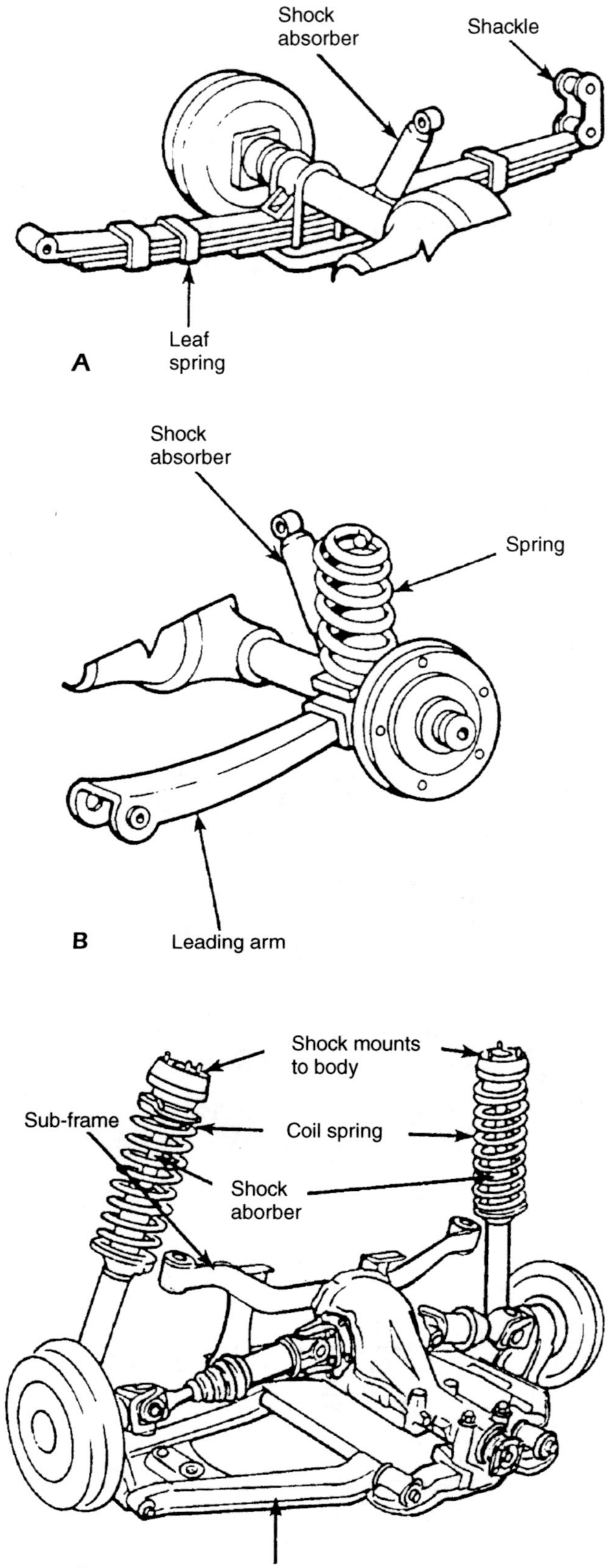

FIGURE 9–43 Examples of rear suspension systems: (A) leaf spring, (B) coil spring, and (C) independent MacPherson strut.

FIGURE 9–44 Damaged rear suspension.

The most recent development in suspension systems is electronically controlled air suspensions. The height of the vehicle and the dampening action of the springs are varied according to the speed of the vehicle and the type of terrain being traveled. Some vehicles are even designed to compensate for yaw during hard cornering. Figure 9–45 shows the components of an electronic air suspension system.

9.10 WHEELS

The tire is mounted on a wheel rim that is then mounted on the car with studs and nuts or lug bolts. The wheels are made of either a circular shape or a die-cast or forged aluminum or magnesium rim or disc. The latter are commonly referred to as "mag" wheels, although they are more commonly an aluminum alloy. Aluminum wheels are lighter in weight when compared with the stamped steel type.

All wheel rims on passenger cars have a dropped or lowered center (Figure 9–46). This type of rim is called a dropped center safety rim because it is designed to prevent the tire from accidentally coming off the rim. Tire-wheel separation due to extremely hard cornering or a blowout is prevented by these safety humps.

Typically, there are three types of wheel damage that can occur, as shown in Figure 9–47. Usually, an inward or outward bend of the wheel lip can be straightened to restore the wheel. However, if the bend extends past the first step, it will cause the wheel to wobble excessively as it rotates. A wheel damaged in this manner must be replaced.

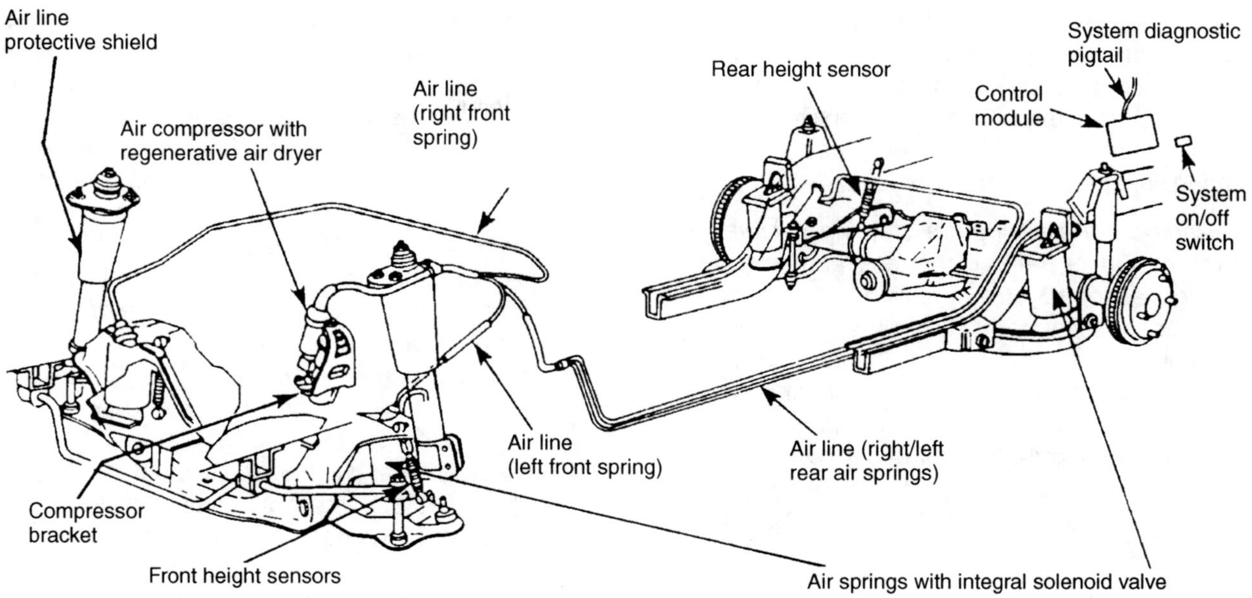

FIGURE 9–45 Electronic air suspension control.

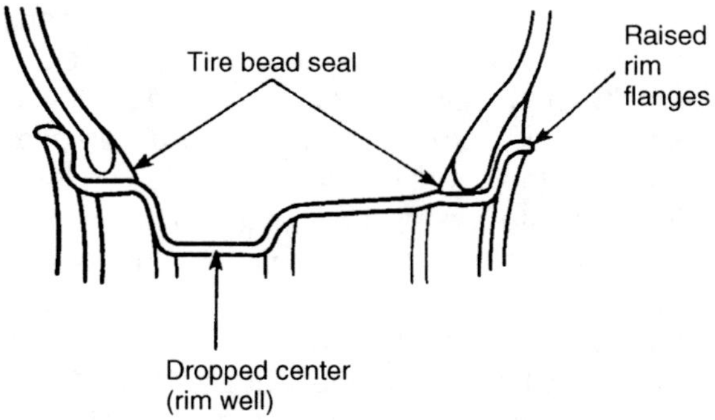

FIGURE 9–46 Dropped center safety rim.

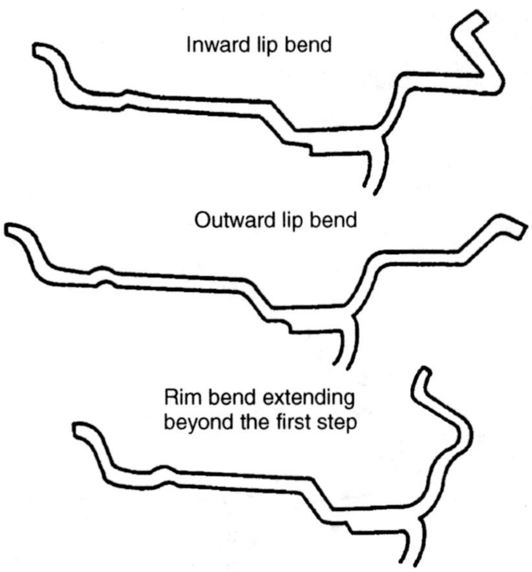

FIGURE 9–47 Three types of wheel damage.

Replacement wheels must be equal to the original equipment wheels in load capacity, diameter, width, offset, and mounting configuration. An improper wheel can affect wheel and bearing life, ground and tire clearance, or speedometer and odometer calibrations. Balancing, weights, and valve stem must be added separately to the damage report. They are not included with the wheels.

9.11 STEERING SYSTEMS

The typical steering system (Figure 9–48) consists of the following:

- *Steering Wheel.* Allows the driver to control the direction of the vehicle—the motion of the steering wheel goes through the steering system to the wheels.
- *Steering Shaft.* Connects the steering wheel to the steering linkage. It is attached to the steering wheel on one end and the gearbox on the other.
- *Steering Gear Assembly.* Allows the driver to turn the wheels by mechanical gear reduction.
- *Steering Linkage.* Connects the steering system to the wheels. Steering motion goes from the steering wheel and steering shaft, through the steering gear, and into the steering linkage, which sends movement to the wheels.

Steering systems can be divided into two types: manual steering and power assisted. They can also be divided according to the type of steering linkage used. The basic linkage components for both manual and power systems are the same. Three main types of steering linkage are in common use.

- *Standard or Parallelogram Steering System.* This system (Figure 9–49) is the most common type on conventional frame cars. A pitman arm attaches to the steering box either with a ball socket assembly or bushing. Steering action is relayed via the center link, again attached by either ball sockets or bushings. The idler arm supports the center link at the opposite end holding the system parallel, transmitting horizontal steering action (Figure 9–50). If up and down movement is excessive, toe change might exceed the manufacturer's limits, thereby creating premature and rapid tire wear. Tie rod ends are the final wearable pivots of the system. Looseness in the ball sockets indicates loss of tension in the internal spring. Even if slight tire motion is evident during the inspection (car off the ground, hand pressure only) when the vehicle is on the ground, moving and subjected to road forces, this motion might be dramatically more substantial. Figure 9–51 shows a damaged truck steering system.
- *Twin I-Beam Steering Linkage (Figure 9–52).* This system is often used with twin I-beam suspension and is frequently used in light trucks, vans, and four wheel drive vehicles. Its operation is similar to the parallelogram system.
- *Rack-and-Pinion Steering System (Figure 9–53).* This system is fast becoming a standard system for unibody vehicles. It gets its name from the pinion gear attached to the steering column and the rack gear in the steering gear housing. This rack gear is moved right to left within the housing by the rotation of the pinion gear. The ends of the steering rack are attached to the front wheel spindles by tie rods. The rack-and-pinion gear provides the same combined function as the intermediate rod, idler arm, pitman arm, and steering gear in a conventional steering system. In unibody construction, the rack-and-pinion steering gear assembly on some cars is mounted to the cowl panel. In other cases, the rack-and-pinion steering gear is mounted to the front suspension cross member or the engine cradle assembly. The rack-and-pinion steering gear must be mounted securely because any movement will cause the car to wander as it travels down the road. Figure 9–54 shows a damaged rack and pinion.

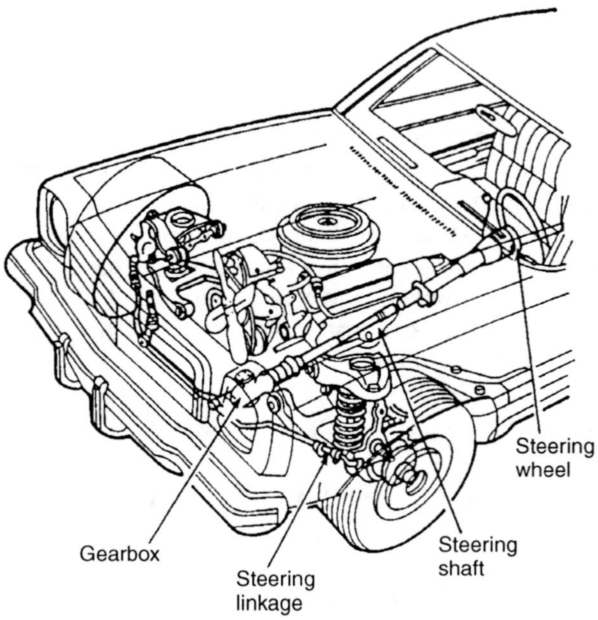

FIGURE 9–48 Rear wheel drive steering system.

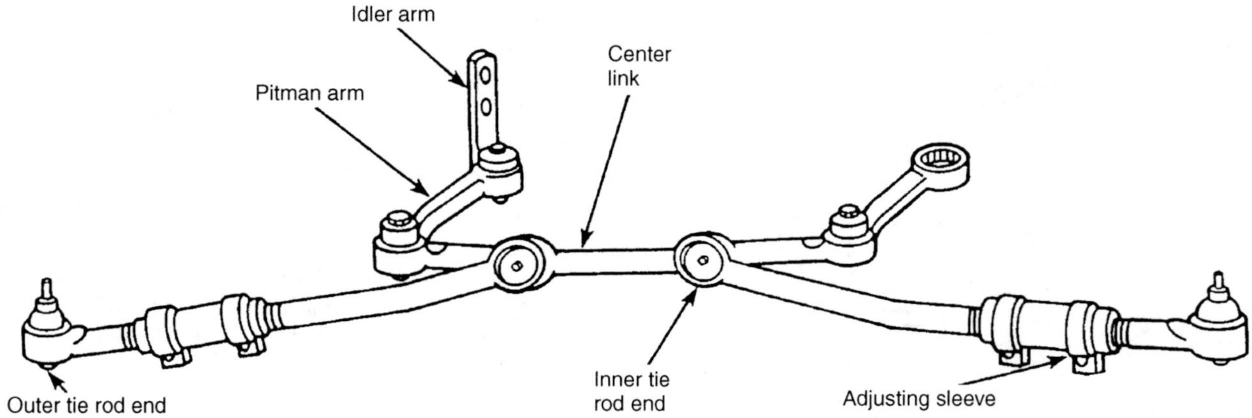

FIGURE 9–49 Parallelogram steering system.

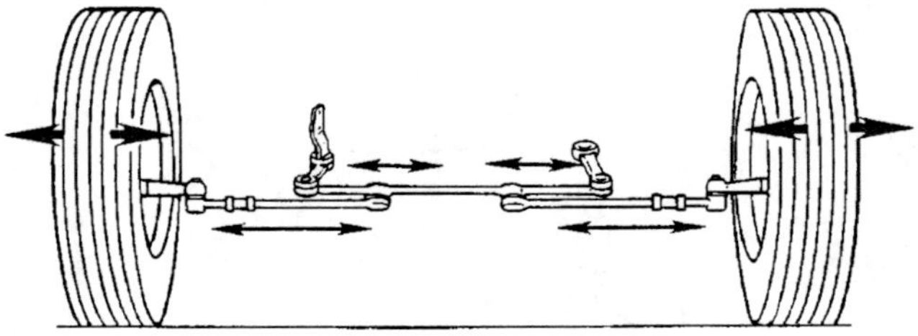

FIGURE 9–50 Linkage force diagram.

FIGURE 9–51 Pickup truck steering system damage. Notice that the tie rod is broken in two.

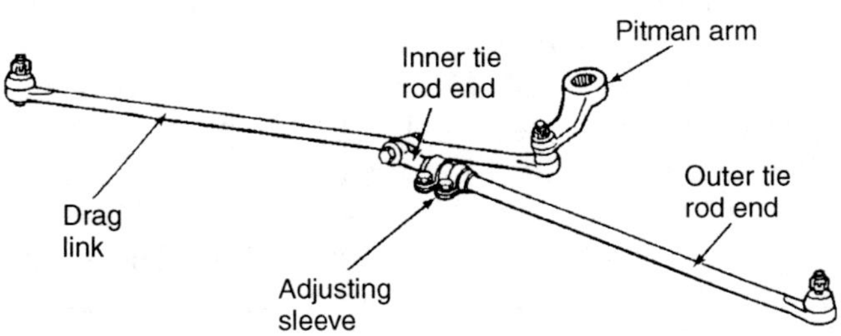

FIGURE 9–52 Typical twin I-beam steering linkage.

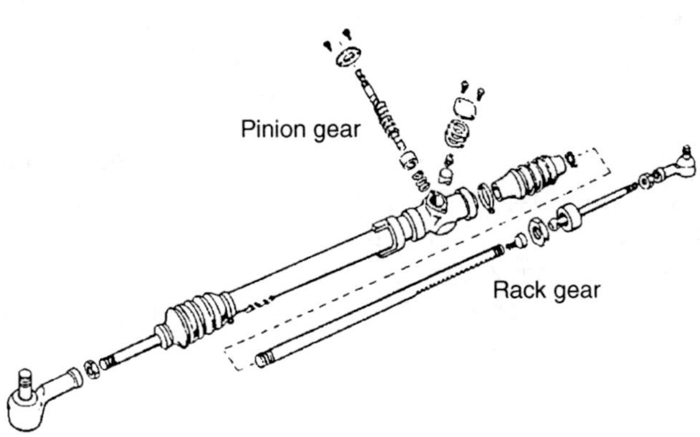

FIGURE 9–53 Rack-and-pinion steering system.

FIGURE 9–54 Damaged rack and pinion.

The position of the rack is critical and must be to factory specifications. If the rack and pinion are not mounted to factory specifications, changes in steering will take place as the suspension moves up and down. Road vibrations and impact forces are transmitted to the linkage from the tires, causing wear and looseness in the system. This looseness permits intermittent changes in the toe setting of the front wheels, allowing further tire wear. Since these points could be considered weak points in the steering systems, they should be carefully checked after an accident.

Whenever estimating front-end collision damage, be sure to inspect all parts and units that make up the manual or power steering gear assembly. In most cases, when damage to the steering gear has occurred, it can be corrected by removing and installing the pitman shaft and making the proper adjustment. If the damage extends beyond the pitman shaft, overhaul is usually necessary.

Perhaps the most frequently encountered complaints heard at a repair shop regarding a vehicle's performance are steering problems. These include wheel vibration, uneven tire wear, high efforts/poor returnability, the vehicle wandering from side to side, the vehicle pulling or drifting to one side, the steering being too loose or too tight, and specific noises, such as hissing, whistling, squawks, or rattles. It is important to become familiar with these common problems and their usual related repairs in order to avoid including prior damage and wear on the damage report.

Naturally, many vehicle owners are much more sensitive to steering performance after a collision and are likely to wrongly attribute all problems to the accident. Therefore, the adjuster must perform a careful and thorough inspection and exercise good judgment as to whether or not certain repairs are warranted.

STEERING ADJUSTMENTS

In some cases, a vehicle might require minor adjustments to the steering gear as well as a front end alignment. One telltale sign of such a need might be when the car's clear vision is off. Clear vision can be measured as follows:

1. Turn the steering wheel to the extreme right position.
2. Place a mark on the steering wheel and count the number of turns until it reaches the extreme left position.
3. Divide the total number of turns by 2 and turn the steering wheel that many times back toward the center.
4. The steering wheel is in the clear vision position and the road wheels should be in the straight-ahead position.

When the vehicle is in this position, the main shaft gear and pitman shaft gear should also be on the high spot, which is machined into the gears at the factory. The high spot helps hold the vehicle in the straight-ahead position and increases ease of handling.

If the vehicle's steering must be realigned to the clear vision position, there are two adjustments that can be made. The preload adjustment is made at the steering wheel nut, located under the horn pad. The center mesh-load adjustment is made by adjusting the nut on the steering gearbox. Both adjustments are measured with an inch-pound torque wrench.

These adjustments can also correct the amount of free play; that is, the distance the steering wheel must be turned before the road wheels turn.

All other front end factors must be checked and brought up to specifications before any adjustments are allowed on the damage report. Before any steering gear adjustments are made, the tires, wheel balance, front end alignment, steering linkage, and shock absorbers should all be within specifications.

9.12 BRAKES

The brakes are hydraulically operated. When the driver pushes in on the brake pedal, brake fluid from the master cylinder causes the front disc brakes and rear drum brakes to operate and slow the vehicle. The disc brakes consist of a rotor, caliper, and pads (Figure 9–55). Some vehicles have disc brakes on all four wheels. The drum brakes consist of drum, wheel cylinder, adjusters, springs, and shoes (Figure 9–56). An antilock brake system (ABS) is used to regulate brake pressure to maintain control of the vehicle when stopping on a slick surface. The ABS can either be rear ABS only or three or four channel, depending on how many controls are involved. Systems can consist of a tone ring, speed sensor, computer, and master cylinder (Figure 9–57). Maximum braking action is obtained just before the wheel stops completely. A complete stop of the wheel would cause a skid and loss of control. The speed sensor detects how fast the tone ring—connected to the wheel—is rotating. The speed sensor then sends the information to the computer. The computer regulates the brake pressure through the master cylinder to achieve the best braking action without causing a skid. In a rear ABS system, the rear wheels are controlled together. This system is found on pickup trucks. A three-channel system has the front wheels controlled individually and the rear wheels together. A four-channel system controls each wheel separately.

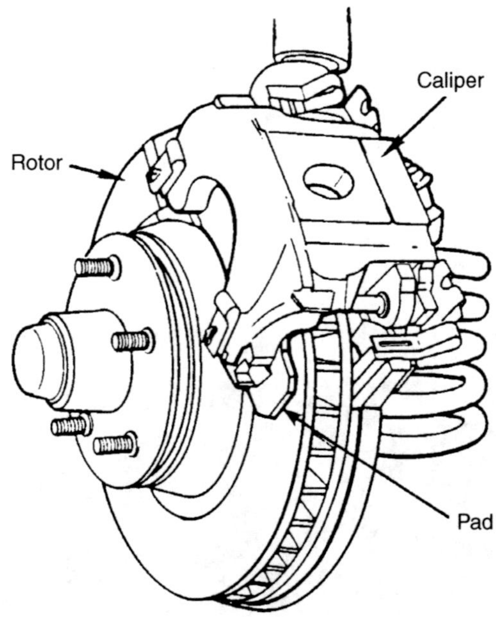

FIGURE 9–55 Disc brake parts.

An impact on a wheel can damage the brake system (Figure 9–58). Check for brake operation. A pedal that sinks to the floor under pressure indicates a broken brake line. A glowing ABS light in the dashboard indicates damage to the system. Always repair system damage by replacing parts.

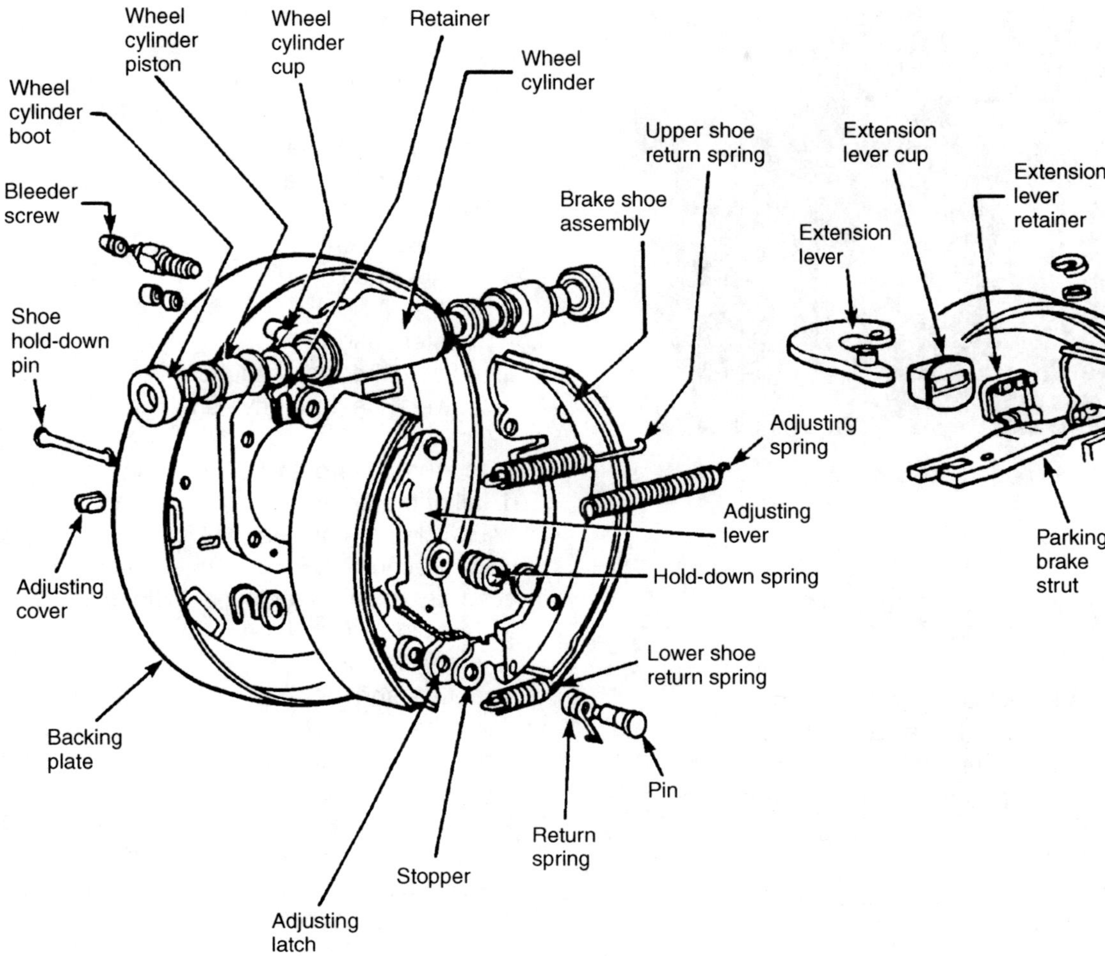

FIGURE 9–56 Drum brake parts.

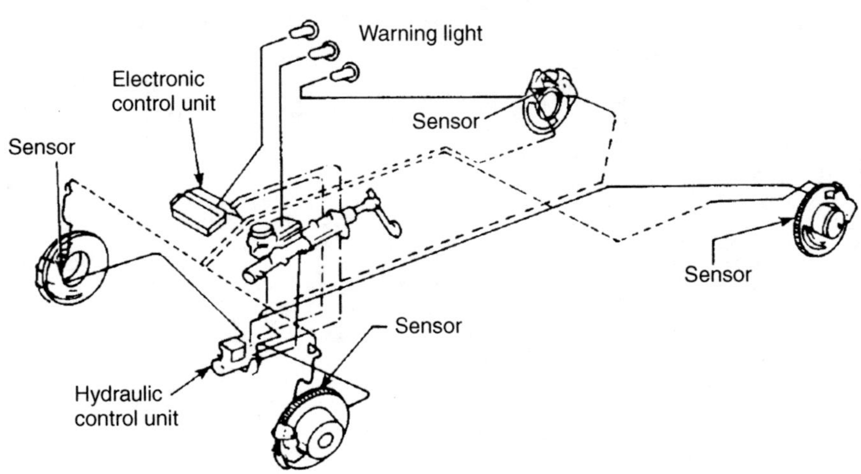

FIGURE 9–57 Four-channel ABS.

FIGURE 9–58 *Damaged ABS system.*

REVIEW QUESTIONS

1. Technician A says that the A/C system's receiver/dryer should be replaced if the A/C system has been open for one hour. Technician B says that the refrigerant boils in the evaporator. Who is right?
   a. technician A
   b. technician B
   c. both A and B
   d. neither A nor B
2. Radiator R&R time includes draining the coolant.
   True False
3. Technician A says that internal engine damage is rare, even on front wheel drive vehicles. Technician B says that the starter is often damaged on a front wheel drive vehicle. Who is right?
   a. technician A
   b. technician B
   c. both A and B
   d. neither A nor B
4. The most common type of transmission damage is a cracked ______.
5. Technician A replaces a CV joint boot if it is cracked. Technician B says that if the CV joint boot is cracked, the CV joint must be inspected and if needed, replaced. Who is right?
   a. technician A
   b. technician B
   c. both A and B
   d. neither A nor B
6. A transfer case
   a. powers the rear wheels.
   b. powers the front drive shaft on a four wheel drive.
   c. is found on front wheel drive vehicles.
   d. all of the above.
7. Technician A says that ball joints are subject to wear. Technician B says that a leaking shock absorber must be replaced. Who is right?
   a. technician A
   b. technician B
   c. both A and B
   d. neither A nor B
8. A bent MacPherson strut affects wheel camber.
   True False
9. Technician A says that an out-of-level rack and pinion causes bump steer. Technician B says that I-beam suspensions are found on front wheel drive vehicles. Who is right?
   a. technician A
   b. technician B
   c. both A and B
   d. neither A nor B
10. A three-channel ABS system controls
    a. all four wheels independently.
    b. rear wheels separately.
    c. front wheels independently, rear wheels together.
    d. front wheels together, rear wheels independently.

Chapter 10

Estimating Refinishing Costs

Objectives

After reading this chapter, you should be able to:

- **Describe the process involved in refinishing a vehicle.**
- **Explain how to obtain flat-rate refinishing times from the collision estimating guide.**
- **List the operations included in the refinishing times listed in the collision estimating guide.**
- **Figure reductions in refinishing times to compensate for overlap.**
- **Explain how to increase refinishing times when applying two- and three-stage paint systems.**
- **Determine the cost of refinishing materials per hour of refinishing time.**

Key Terms

Undercoat
Topcoat
Detailing
Single stage

Thus far in this book we have discussed the types of damage a vehicle can incur, how to inspect a vehicle for damage, how to determine the labor and parts costs involved in repairing a collision-damaged vehicle, and the processes involved in the repair. Once the vehicle has been restored to its pre-collision condition, the new or repaired panels must be refinished. This chapter covers the correct method of determining the cost of painting the repair area.

10.1 THE REFINISHING PROCESS

When making an estimate of the labor and materials involved in a refinishing operation, the estimator and adjuster should be familiar with the actual processes, tools, and supplies involved in painting a vehicle. This will enable them to write a fair and accurate estimate of the refinishing process

SURFACE PREPARATION

When the repaired vehicle leaves the metal shop and arrives in the paint shop, a variety of jobs must be performed to prepare it for refinishing.

Cleaning

The first step in refinishing is cleaning the repair area and adjacent panels (Figure 10–1). A pressure washer or compressed air can be used to dislodge accumulated dirt in cracks and under mouldings. The area to be painted should be washed with water and detergent, using a soft shop cloth or sponge. Then, the area must be thoroughly rinsed to remove any cleanser residue.

Dewaxing

Many waxes contain paraffins and silicones that can destroy a paint job if left on the panel surface. To remove these substances as well as road tars and grease, a wax and grease remover must be applied to

FIGURE 10–1 It is a good idea to thoroughly wash the vehicle prior to refinishing.

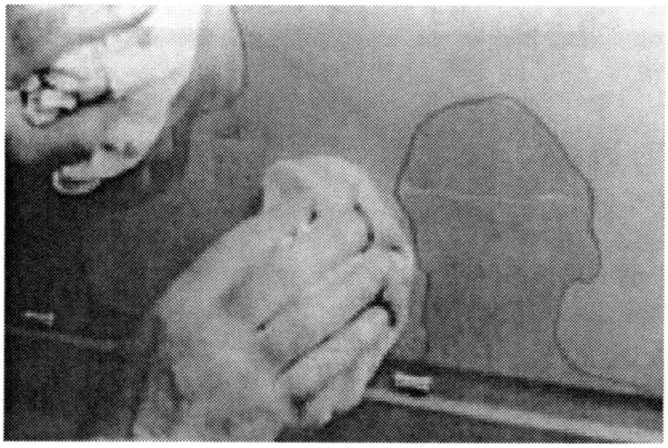

FIGURE 10–2 Road tar, waxes, and silicones must be removed from the repair area before sanding.

the repair area (Figure 10–2). The paint technician will apply the remover over a small area of the panel with a clean shop cloth; after waiting a moment to allow the remover to dissolve the wax, he or she will wipe the panel with another cloth to get rid of the wax and chemical remover. Very careful attention must be given to areas around door handles, along side mouldings and other body attachments. If wax is allowed to remain in these areas, the paint will eventually flake off.

Sanding

After chemical cleaning and dewaxing, the panel should be scuff sanded (Figure 10–3). Scuff sanding removes any ground-in contaminants and roughens the surface to promote adhesion. Any scratches or

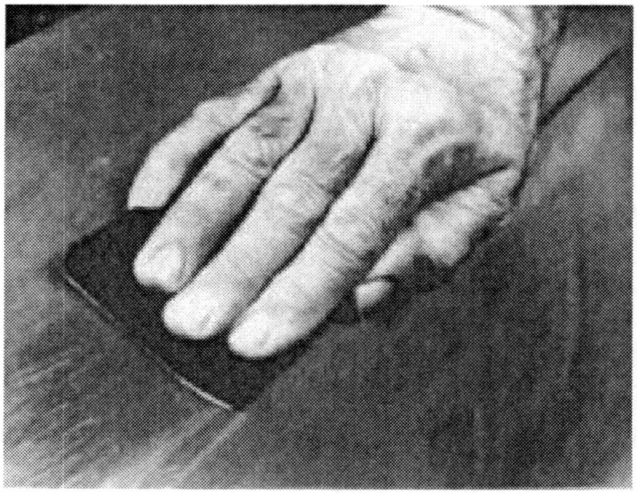

FIGURE 10–3 Scuff sanding removes ground-in dirt and wax and roughens the surface so that the surfacer or sealer will adhere well. (Courtesy of Maaco Enterprises, Inc.)

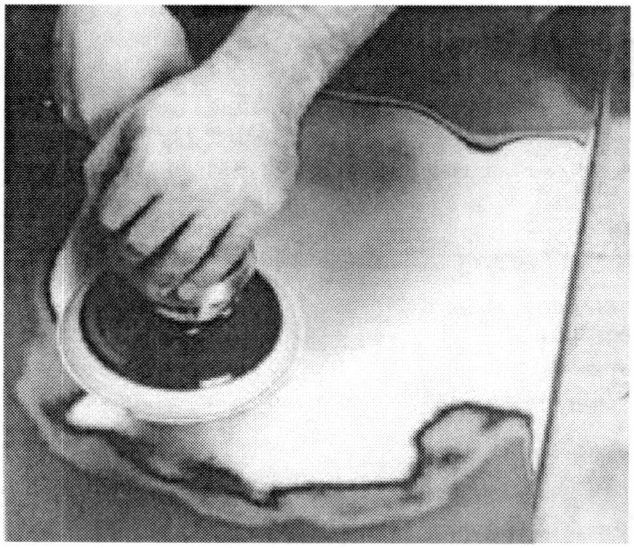

FIGURE 10–4 Proper featheredging, surfacer fill, and block sanding will make the repair edge invisible. (Courtesy of Maaco Enterprises, Inc.)

nicks in the existing paint surface must be featheredged. Featheredging tapers the edges of a break in the paint until the surface is smooth (Figure 10–4). Areas filled with plastic body filler must also be sanded level with the panel surface and carefully featheredged. After the panel or panels have been sanded smooth, compressed air is used to blow dust off the panel and out of crevices; then the panel is wiped clean with a tack rag.

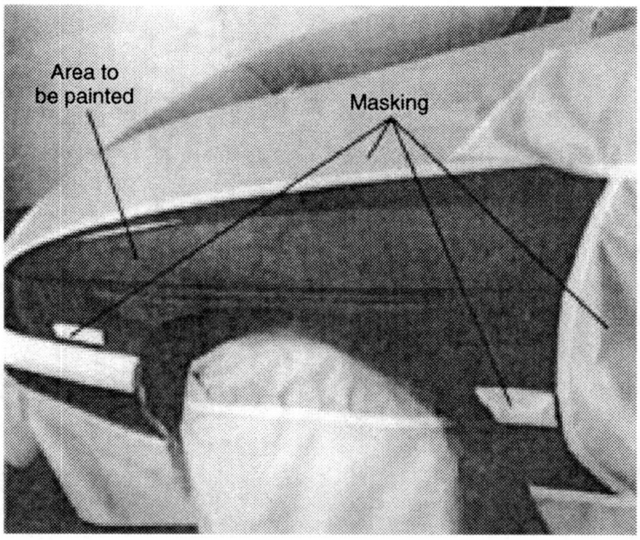

FIGURE 10–5 Masking tape and paper applied to the vehicle to prevent overspray damage.

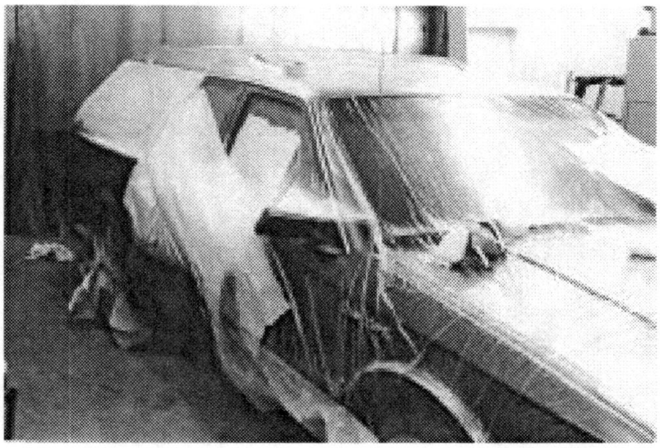

FIGURE 10–6 Plastic film can be used to cover the entire vehicle. Use eighteen-inch masking around the refinish area.

Masking

When the repair area is clean and sanded, the adjacent panels and trim must be protected from overspray. Some shops remove all obstructions on a panel. This means that they remove mouldings, locks, nameplates, and handles from all surfaces to be painted. This gives the best finish because there aren't any tape edges. Other shops use masking tape to keep paint off obstructions on the painted panel. To keep overspray off adjacent panels, shops use masking paper (Figure 10–5), plastic film (Figure 10–6), or liquid mask. The labor to mask adjacent surfaces to thirty-six inches is included in the refinish time. However, overspray will creep into any surface possible.

FIGURE 10–7 The application of metal conditioner neutralizes the surface rust. (Courtesy of Du Pont Co.)

The inside edges of the panels may also need to be masked. Proper masking will prevent many problems.

Metal Conditioning

If any of the repair panels have been sanded down to bare metal, the metal must be treated to prevent rust formation. Although the metal might appear rust-free, all bare metal has an invisible film of microscopic rust particles. If this film of rust is not neutralized, the rust will continue to react with the metal after the new paint has been applied and will eventually cause the finish to bubble and lift away from the metal in that area. To prevent this from occurring, the bare metal is treated with a metal conditioner and conversion coating (Figure 10–7). The conditioner neutralizes the rust and chemically etches the metal to increase adhesion between the primer and the metal. The conversion coating leaves negatively charged ions on the surface of the metal. These ions react with the positively charged paint, creating a bonding action that promotes adhesion.

APPLYING UNDERCOATS

Many shops treat bare metal with a primer. The two choices are self-etching primer and epoxy primer. Both can be sprayed on bare metal after metal conditioning or, if the bare metal has been exposed for a short time, alone. After primer, a urethane surfacer is applied to fill in featheredged areas. Surfacer is called high build, because it can fill scratches. A contrasting guide coat is dusted on the surface. After the surfacer cures, the guide coat is block sanded. The guide coat highlights scratches and low spots. When all of the guide coat is block sanded off, the surface is ready for topcoat. Some shops apply a sealer over all surfaces. The purpose of a sealer is to provide a uniform color surface and to promote adhesion of the topcoat. (Figure 10–8).

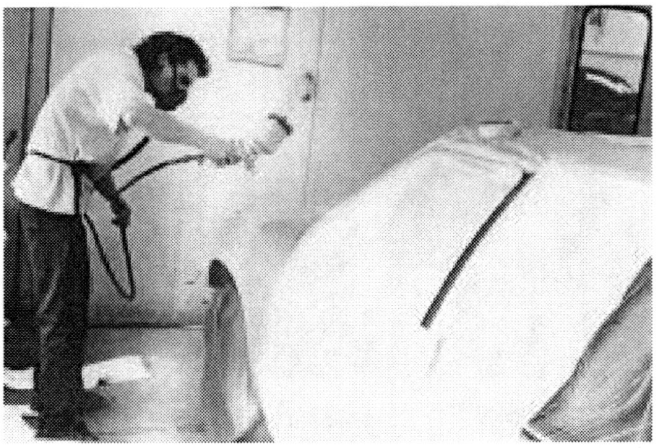

FIGURE 10–8 A sealer is sprayed on this OEM replacement fender to promote adhesion of the topcoat.

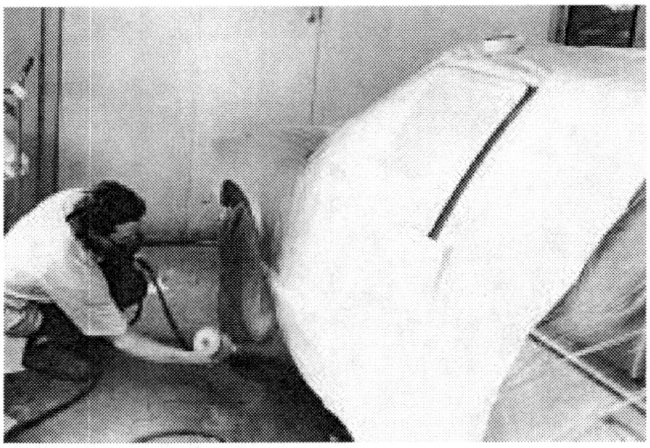

FIGURE 10–10 After the sealer flashes, the topcoat can be applied.

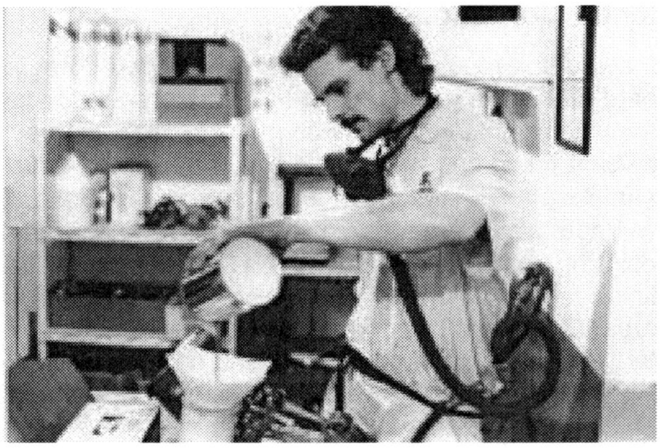

FIGURE 10–9 This shop uses a paint system consisting of several tints and a computer storing all OEM colors to mix the paint.

FIGURE 10–11 High production shops speed drying time by baking the paint. (Courtesy of Garmat USA, Inc.)

APPLYING TOPCOATS

Although no one step can be said to be most important to the success of the paint job, applying the color coat is probably the most complex. In some paint systems, the color coat is available premixed; other systems require mixing pigments with the paint base to achieve the desired color (Figure 10–9). The base paint must then be thinned, or reduced, and often additives are mixed with the paint to give it certain characteristics. For example, a hardener or catalyst is usually added to acrylic urethane paints to speed curing time and increase the hardness of the cured paint film. Other additives slow down the drying time, lower the gloss, minimize fish-eye cratering and wrinkling, or make the paint film flexible (the latter is used when refinishing flexible plastic panels).

When property tinted and mixed, the color coat is applied (Figure 10–10) in two or more coats, with a specific amount of flash time between coats. Single stage paint consists of color coat only. If the color coat is a two-stage system, clear urethane is then sprayed over the color basecoat. Three-stage paints require the application of a midcoat of clear urethane, containing flakes of mica, between the color basecoat and the clear topcoat. The color coats can be baked to speed the drying time (Figure 10–11).

DETAILING

The final step in any refinishing procedure is detailing the repaired panel and/or the entire vehicle. Detailing includes removing masking paper and overspray. If the paint surface has runs, dirt, or excessive orange

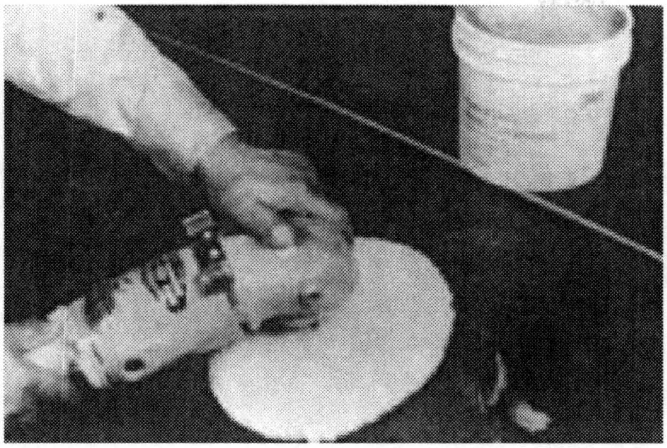

FIGURE 10–12 Buffing can eliminate minor surface defects.

peel, it should be wet-sanded with 1500 grit paper and buffed (Figure 10–12). Finally, the vehicle is cleaned inside and out, without any trace of compound, dust, or overspray remaining. The vehicle is now ready to be returned to the customer.

10.2 REFINISHING TIME

The estimator and adjuster must account for the preceding operations on the damage report and appraisal. The operation times given in the crash guides for R&R, R&I, and overhaul of panels do not include the time for painting those panels. The time required for painting is listed separately at the head of the section. The time shown in Figure 10–13 for painting the hood is 2.9 hours. The painting times listed are for one-color applications on the outer surface of new, undamaged panels only.

OPERATIONS INCLUDED

In a basic color coat application, the paint times include the following steps:

- Solvent wash
- Scuffing panel and cleaning
- Masking adjacent panels up to 36 inches
- Priming or sealing
- Final sanding and cleaning
- Mixing the paint
- Adjusting the spray equipment
- Applying the color coat
- Cleaning the paint equipment

Be sure to check whatever collision estimating guide is being used to see exactly what steps are included.

HOOD

Use Procedure Explanations 4 and 28 with the following text.

Refinish Hood Outside		**2.9**
Add for Underside		**1.5**
Refinish Hinge	R/L	**.3**
R&I Hood		**.5**
Add to F&R Hood w/Hood Insulator		**.3**

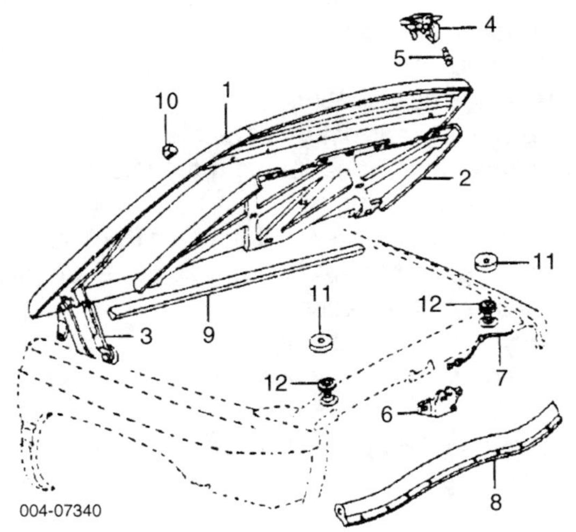

1 **Hood**		55274716	**1.3**	205.00
Emblem, Hood		55076512	**#.2**	14.25
#Included in R&R Hood				
2 **Insulator, Hood**		55216161	**.3**	41.00
3 **Hinge, Hood**	R/L	55255448-9	**#.7**	65.00
#Includes R&R Cowl Grille, w/Hood Removed				
Nut, Hinge (2/Side)	R/L	6100051		1.45
4 **Striker, Latch**		55255751	**.1**	23.00
5 **Screw, Striker**	R/L	6101455		.50
6 **Latch, Hood**		55255451	**.3**	23.50
7 **Cable w/Handle, Release**		55256496	**.5**	23.50
8 **Seal, Hood Front**		55036877	**.2**	25.00
9 **Seal, Hood Rear**		555255099	**.2**	17.25
10 **Bumper, Hood Side**	R/L	55256768		.85
11 **Bumper, Hood Front**	R/L	2276804		1.20
12 **Screw, Bumper**	R/L	6035724		1.45

FIGURE 10–13 Dodge Dakota pickup hood refinish times. (Courtesy of Mitchell International)

These painting times are not intended for calculating a complete vehicle refinish. Nor are these times accurate for refinishing repaired parts. Determining the costs of painting repaired or used parts and complete car paint jobs is covered later in this chapter.

OPERATIONS NOT INCLUDED

Other potential problems are not accounted for in the flat-rate refinishing times. For example, a damaged panel that had been recently painted with an enamel will have to be sealed before applying a new topcoat. Or if the finish is cracked, crazed, flaking, or otherwise unsound, the panel must be stripped, etched, and primed before applying the new finish. By not anticipating problems such as these and not adding time to cover the extras, an inaccurate damage report or appraisal will inevitably result. The following is a list of operations that are not figured into the basic color coat application times:

- Cost of paint or materials
- Matching and/or tinting color
- Grinding, filling, and smoothing welded seams
- Blending into adjacent panels
- Removal of protective coatings
- Spatter paint
- Custom painting
- Applying undercoating
- Applying anticorrosion materials
- Applying sound-deadening materials
- Painting panel edges
- Painting the underside of hood or trunk lids
- Covering entire vehicle, including interior, prior to refinishing, if necessary
- Applying protective coatings
- Additional time to produce custom, non-OEM finishes
- Spraying a test panel to check color match prior to painting

When painting repaired or used parts, the following operations are not included in the refinish time:

- R&I or masking of:
  —Glass
  —Handles
  —Antennas
  —Locks
  —Lights
- Panel repairs, including
  —Featheredging
  —Filling
  —Sanding
  —Block sanding

It should also be noted that painting times do not include the additional time required to correct finish imperfections caused by improper application, weather conditions, or environmental contamination such as dust, dirt, or grease. All parties should negotiate beforehand to determine a mutually acceptable solution in the case of such an event.

ADDITIONS TO PAINT TIMES

Paint times listed in the collision estimating guides allow for the application of one color coat to the outer surfaces of new replacement parts. However, the time required for painting the edges of the new parts is not included in that time. The edging time is listed separately at the top of each specific part section. No deductions from the edging times should be made for overlap. The times for painting the underside of the hood and trunk lid are also listed separately at the head of their part sections. Again, use the full painting time stated for these jobs without making any deduction for overlap.

FIGURE 10–14 Two-tone finishes require additional paint time.

Additional time must be allowed for multiple color applications (Figure 10–14), since the times listed are for one-color applications. Masking of interior edges is not part of refinish labor time. Figure 10–15 shows the amount of labor time that should be added for each panel, should it be necessary. Note the overlap deduction. The additions for clear coat and three-stage applications are covered later in this chapter. To add time for other surfaces, see Table 10–1.

REDUCTIONS IN PAINT TIMES

Reductions in the paint times should be made when more than one panel is being refinished. As is the case when figuring overlap for repair operations, the first major panel to be refinished should receive 100 percent of the time quoted in the collision estimating guides. (Major panels include header panel, hood, fender, door, rocker panel, roof, quarter panel, rear body panel, deck lid, tailgate, pickup cab corner, pickup cab back, pickup bed front, pickup bed side, van side, van rear corner, and rear gate.) Items such as fillers, inner panels, or valence panels are not considered major panels. However, for adjacent panels, deduct 0.4 hour and for extensions, deduct 0.2 hour.

When refinishing panels that are not adjacent, again list the full time as quoted for the first major panel. For each additional panel, deduct 0.2 hour. Although these panels are not connected, time has been included in the paint times of each for mixing paint, filling the spray gun, cleaning the gun, and so on. Some deductions for these overlapping operations are justified.

MASK INTERIOR ENTRYWAYS, ENGINE COMPARTMENT AND TRUNK OPENINGS

Interior masking prevents overspray damage from paint application to exterior surfaces. The performance of this operation is NOT INCLUDED in the Mitchell refinish labor time.

MASK INTERIOR OPENINGS, JAMBS

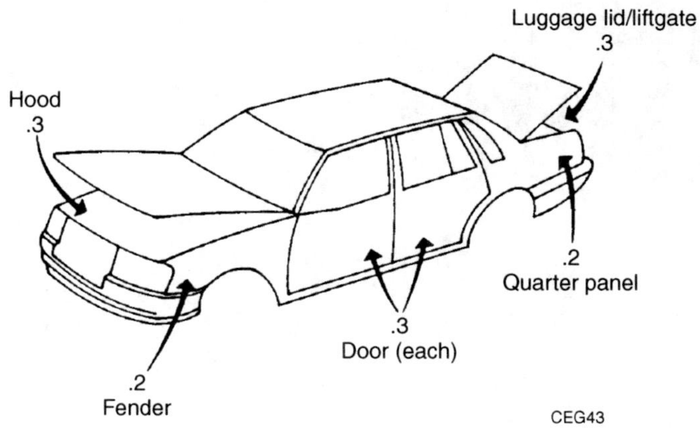

NOTE: The times shown in the illustration are for interior masking of that panel and/or opening. Labor time includes all pillars, jambs, weatherstrips, edges, entryways and openings as necessary. Deduct .1 hour overlap for each interior masked adjacent panel and/or opening.

FIGURE 10–15 Interior masking time from the P pages of a Mitchell Estimating Guide. (Courtesy of Mitchell International.)

TABLE 10–1: ADDITIONAL TIME FOR SURFACES*

Anticorrosion coating	Add 0.3 hour
Two-tone operations (unless otherwise specified in text)	
First panel	Add 0.5 hour
Each additional	Add 0.3 hour
Stone chip	
First major panel gravel guard	Add 0.5 hour
Each additional panel	Add 0.3 hour
Color sand and buff each panel per refinish hour	Add 0.3 hour

*Times might vary between collision estimating guides.

An elastomeric or flex additive may need to be added to the undercoat and topcoat when refinishing flexible parts, such as a fascia cover, fillers, and extensions. No labor overlap is taken when refinishing a flexible part, because it must be masked off and refinished separately from the other body panels not requiring the additive. The exception would be when refinishing more than one flexible item, in which case the adjacent and nonadjacent formula would apply.

Blending or spot painting is the partial painting of a panel. The minimum amount of time that should be charged for spot painting is one-half hour. No overlap is deducted on spot-repaired panels. The flat-rate guide should be used as a guide when allowing for spot refinishing.

TWO-STAGE FINISHES

The painting times listed in the collision estimating guides are based on the application of single stage paint. However, most factory finishes today are two-stage systems. The first stage of these systems is a base color coat. These base colors are usually very dense in metallic (or mica) and/or pigment, which gives them excellent "hiding" qualities. Thus a very thin film has the ability to uniformly cover the substrate, completely hiding the previous color. A flash time of ten to fifteen minutes must be allowed between each coat. As this paint flashes off, it loses all of its gloss. It dries to either a flat or satin finish. A drying time between thirty minutes and one hour must be allowed after the last coat of the basecoat is applied before beginning the second stage of the finish system. The second stage of the system is a clear coat that is applied over the basecoat. The clear coat provides gloss, distinctness of image, and protection against the elements.

FIGURE 10–16 *Areas to be blended and melted on a Chrysler LeBaron.*

To determine whether a panel has a clear coat, lightly sand the panel with #600 grit sandpaper. If a whitish dust collects on the paper, the topcoat is a clear finish. If the dust is the color of the car, the surface is not clear coated.

Problems can arise when refinishing a two-stage finish because the true shade of color is not known until the clear coat is applied. For this reason, manufacturers recommend spraying a test panel with the entire system before any topcoat is applied to the vehicle.

A blend is used to hide the difference in color between the vehicle paint and the refinish paint. A melt is used to make a clear coat edge invisible. For example, to hide a slight color mismatch when a quarter panel is replaced, the door and deck lid are blended. The color and clear coat are stepped out on the sail panel. The dry clear coat edge on the sail panel is melted (Figure 10–16).

Check the paint manufacturer's recommendations to see if an adhesion promoter is needed. If it is required, it is applied before the basecoat. To prepare for a blend, clean the panels with soap and water and wax and grease remover. Sand the clear coat with 800 grit dry paper on a dual action sander or 1000 grit wet on a sanding pad. The melt area is sanded with 2000 grit paper or rubbing compound. If adhesion promoter is required, it is sprayed at this time. On a melt, it is sprayed beyond where the melt will be made.

The first coat of basecoat is applied over the quarter panel. The second coat is applied over the quarter panel and stepped out about six inches onto the door and deck lid. At the sail panel, the basecoat extends about two inches beyond the primer. The third coat covers all surfaces that have received basecoat plus another six inches on the deck lid and door. The sail panel receives another two inches of color. After the base has dried to a dull finish, the clear coat can be sprayed. Start at the door edge and spray the entire door, quarter panel and deck lid. The dry edge of clear at the sail panel can be melted with a special clear. This melting clear is sprayed to flow out the dry clear edge to make the edge invisible. It can be sprayed after each coat of clear, two paint gun method, or after the last coat of clear, one gun method. Two or three coats of clear should be applied.

Because the melt is an extremely thin application, it lacks the durability of a normal clear coat finish. No paint manufacturer will warrant a melt area. To receive the paint manufacturer's warranty in the quarter panel replacement example, the sail panel, roof and opposite quarter panel should be prepared for clear coat. The basecoat is blended as described before and the clear is sprayed on the door, quarters, deck lid, and roof.

The labor time allowed to make a blend and melt is 0.5 hour per refinish hour. An example is the replacement of a quarter panel on a '95 Chrysler LeBaron coupe, the blending of the door and deck lid, and the melting of the sail panel. The blend time is calculated as follows:

	Refinish time	Blend time
Door	**2.8 hr × 0.5**	**1.4 hr**
Deck lid	**2.5 hr × 0.5**	**1.3 hr**
		2.7 hr.

Add 0.5 hour for the melt in the sail panel.

The collision estimating guides allow additional time for applying clear coats. However, the painting times listed for undersides and edges must be added before any adjustment for the clear coat time can be calculated. Once the time for the undersides and edges has been added, 50 percent of that total time is added to the total time to determine the painting time of the first major panel. For any additional panels, not only must the times for undersides and edges be added, but also overlap must be deducted before the clear coat time can be calculated. Once these two alterations are made, 25 percent (instead of 50 percent) of that total time is added to the total time to determine the painting time of each additional panel. Note that the percentages will vary from one guide to another.

Estimated time for the first major panel usually includes the following operations:

- Mixing the material
- Cleaning and tacking the surface
- Applying the material
- Cleaning the equipment

Estimated time for any additional panels includes time for cleaning and tacking the surface and applying the material.

FIGURE 10–17 This car has a three-stage or tricoat finish.

THREE-STAGE FINISHES

One of the newest OEM coating developments is the three-coat finish. These finishes are currently growing in domestic popularity (Figure 10–17), following the European trend.

These finishes consist of three distinct layers—a basecoat, a midcoat, and a clear coat—which produce a pearlescent appearance. Unlike the two-stage systems that change the value of a color (lightness/darkness) when viewed from different angles, the new three-stage finishes actually change the hue (color) as the angle of view changes. Thus a three-stage finish might appear to be red from one angle and blue from another.

When making a spot or panel repair on a vehicle with a three-coat finish, extra time and attention are needed to achieve a proper color match. The key to a precise color match lies in the application of the pearlescent midcoat. The color of the basecoat changes with each application of a midcoat.

To determine the correct number of midcoats, the technician must prepare a test panel before refinishing the repair area (Figure 10–18). The test panel is primed and sprayed with the basecoat. After the panel dries, it is masked into four equal sections. The top one-quarter of the panel is then sprayed with a coat of mica midcoat. After the first mica coat has flashed, masking is removed to expose the top half of the test panel. Another coat of midcoat is applied over the exposed panel. When this coat flashes, the masking is moved to expose three-quarters of the panel. After flashing, the masking is removed from the panel and the whole panel is given another coat of mica midcoat. Now the panel has four equal areas, each with a different number of midcoats (1, 2, 3, and 4). Now the test panel is masked in half lengthwise and sprayed with the manufacturer's recommended number of clear coats to complete the test panel. After it dries, the test panel can be compared to the actual finish on the vehicle to determine the number of mica midcoats needed to achieve a precise color match.

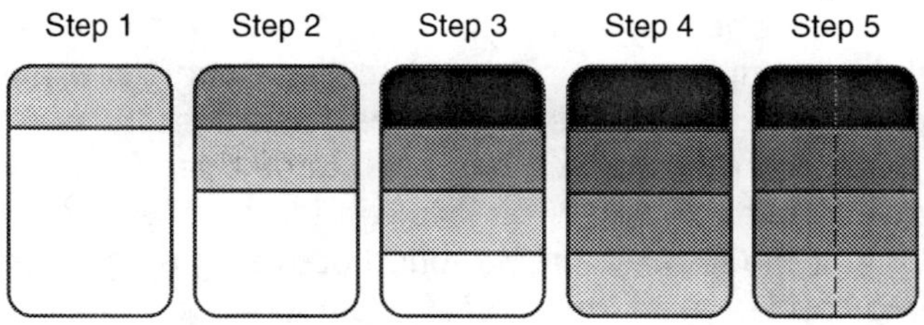

FIGURE 10–18 When spraying a three-stage paint, use a test panel to achieve a proper color match.

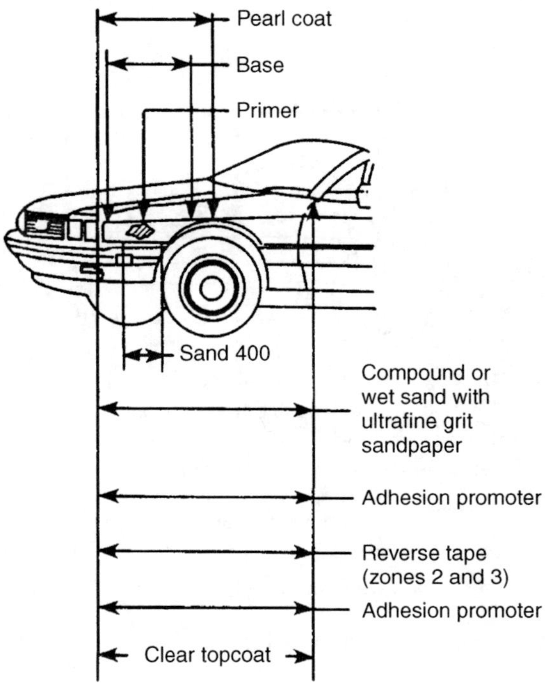

FIGURE 10–19 Spot repair.

Preparing a test panel takes time that is not necessary when applying a one-stage paint. This is one of the reasons why the estimating guide publishers allow extra time to be added to the given paint times when applying a three-coat system.

Once the surface has been prepared as it should be for any type of paint job, the adhesion promoter is applied. Figures 10–19, 10–20, and 10–21 illustrate the different application ranges for three typical repairs: spot, full panel, and deck lid. Two or three medium-wet coats of basecoat are applied to achieve hiding. After the basecoat is allowed to dry twenty minutes, the midcoat, thinned to the proper viscosity, is applied. As many as five coats might be needed to achieve a proper match and color uniformity. The midcoat is then allowed to air dry thirty to sixty minutes before adhesion promoter is applied over it. After thirty minutes of flash time, two clear coats are applied, with ten to fifteen minutes flash time between coats. After the surface has been force dried for forty-five minutes at 140 °F or allowed to dry overnight or longer, the repair area is rubbed out using a light cutting hand or machine polishing compound.

Typical time estimates for the application of three-stage finishes usually have built-in time allowances for the following first panel operations:

- Mixing the pearl/mica toner
- Applying the toner to the test panel
- Mixing the clear urethane
- Cleaning and tacking the surface
- Applying the clear urethane to the test panel
- Repeating the application to the surface being refinished
- Cleaning the equipment

For any additional panels being painted, the time allowance covers applying the pearl/mica toner, cleaning and tacking the surface, and applying the clear.

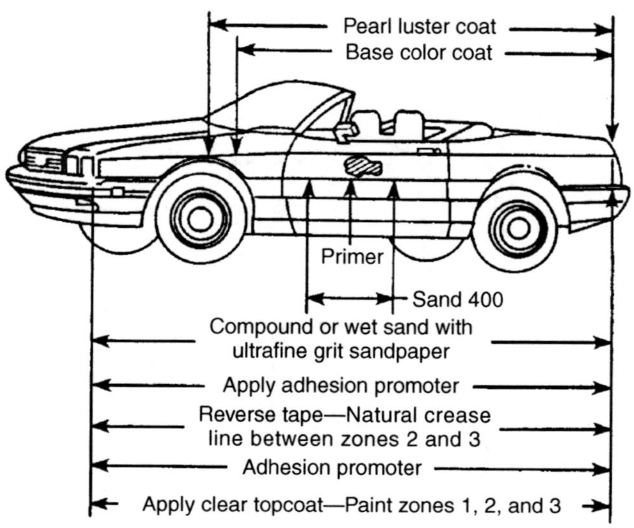

FIGURE 10–20 Full panel repair.

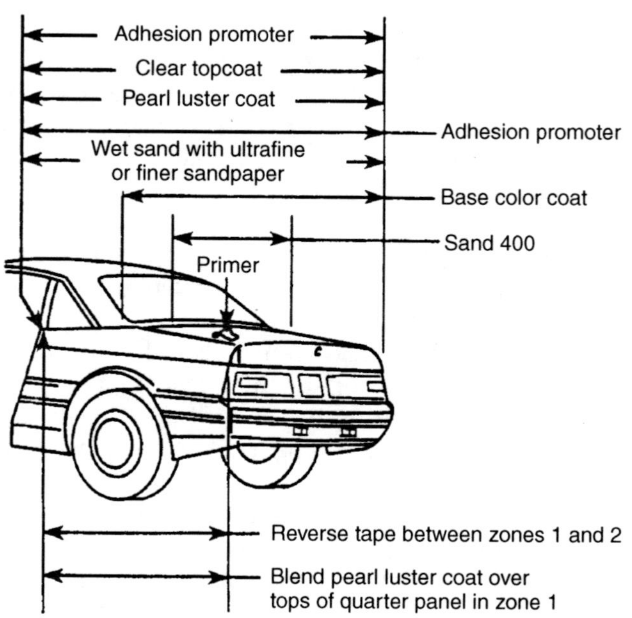

FIGURE 10–21 Deck lid repair.

10.3 COSTS OF REFINISHING MATERIALS

In many cases, the estimator multiplies the paint hours times a predetermined dollar amount to determine the cost of the paint material. For example, if the dollar amount is $20 per paint hour, the paint material cost for the door and quarter panel is 4.6 hours × $20

MFG. PAINT CODE	COLOR	PAGE				
		Clear	Acry Enam	Acry Ureth	Nat'l VOC	CA VOC
	1988-97 ALFA ROMEO					
022	Silver Pearl Met	*	3–2	3–8	3–14	3–20
024	White		3–1	3–8	3–14	3–20
130	Alfa Red Met	*	3–2	3–11	3–17	3–23
146	Bordeaux Met	*	3–4	3–10	3–16	3–22
230	Arctic White	*	3–1	3–7	3–13	3–19
258	Bright Yellow	*	3–2	3–8	3–14	3–20
270	Silver White Met	*	3–1	3–8	3–14	3–20
306	Foglia Green Met	*	3–1	3–8	3–14	3–20
364	Midnight Jade Green Met	*	3–1	3–7	3–13	3–19
370	Stellar Blue Met	*	3–1	3–7	3–13	3–19
371	Stellar Blue Met	*	3–1	3–7	3–13	3–19
375	Mediterranean Blue Met	*	3–1	3–8	3–14	3–20
382	Light Blue Met	*	3–1	3–9	3–15	3–21
477	Dark Blue Met	*	NA	3–9	3–15	3–21
530	Alfa Red		3–4	3–8	3–14	3–20
547	Cabernet Red Met	*	3–2	3–9	3–15	3–21
552	Bordeaux Red Met	*	3–3	3–8	3–14	3–20
555	Alfa Red	*	3–5	3–9	3–15	3–21
601	Onyx Black #2	*	3–1	3–7	3–13	3–19
602	Mystic Black Met	*	3–1	3–7	3–13	3–19
633	Graphite Met		3–1	3–7	3–13	3–19
643	Champagne Met	*	NA	3–8	3–14	3–20
676	Argento Silver Met	*	3–2	3–8	3–14	3–20
767	Sterling Silver Met	*	3–2	3–8	3–14	3–20
768	Light Gray Met	*	3–1	3–8	3–14	3–20
776	Anthracite Gray Met	*	3–1	3–8	3–14	3–20
856	Almond		3–1	3–7	3–13	3–19
913	Jet Black		3–1	3–7	3–13	3–19
914	Jet Black		3–1	3–7	3–13	3–19
945	Blue Gray Met	*	3–1	3–8	3–14	3–20

FIGURE 10–22 Find the paint code; read across to the page number. In this case, code 022 silver pearl metallic and the page number for California compliant VOC paint is 3–20.

per hour or $92 for paint materials. The dollar amount will have to be set by each shop using the local cost of materials plus a reasonable profit.

Another way to determine material cost is to use a refinish material guide (Figure 10–22). Each OEM color is assigned a code. Choices are given for acrylic enamel, acrylic urethane, national VOC compliant, and California VOC compliant. The code is used to determine paint materials cost (Figure 10–23). The guides are updated periodically to reflect price increases.

10.4 ENTIRE CAR PAINTING

There are variables involved in estimating the labor for an entire car paint job, primarily the condition of the existing paint coat and the type of system to be applied. A new car with an excellent finish might require ten hours of labor to be repainted another color.

NATIONAL VOC COST CHARTS

	SUGGESTED COST (Dollars)		
Estimated Refinish Hours	Color w/Materials	Color & Clear w/Materials	Three Stage w/Materials
0.0-1.0	$26.97	$28.23	$35.50
1.1-1.5	37.15	40.66	51.57
1.6-2.0	50.31	52.83	67.37
2.1-2.5	59.83	64.60	82.78
2.6-3.0	67.17	77.46	99.28
3.1-3.5	73.43	89.24	114.69
3.6-4.0	83.32	101.39	130.48
4.1-4.5	94.06	114.39	147.11
4.6-5.0	104.46	127.05	163.40
5.1-5.5	113.56	138.41	178.39
5.6-6.0	123.36	150.47	194.09
6.1-6.5	133.38	162.74	210.00
6.6-7.0	143.27	174.89	225.78
7.1-7.5	153.69	187.57	242.10
7.6-8.0	162.65	198.79	256.96
8.1-8.5	174.11	212.51	274.31
8.6-9.0	183.54	224.20	289.64
9.1-9.5	193.91	236.83	305.90
9.6-10.0	203.31	248.49	321.19
10.1-10.5	214.17	261.60	337.94
10.6-11.0	223.60	273.30	353.27
11.1-11.5	233.97	285.92	369.53
11.6-12.0	243.36	297.57	384.81
12.1-12.5	254.50	310.97	401.85
12.6-13.0	263.88	322.61	417.12
13.1-13.5	273.90	334.89	433.04
13.6-14.0	282.91	346.16	447.95
14.1-14.5	292.51	358.02	463.44
14.6-15.0	301.57	369.33	478.39
15.1-15.5	311.54	381.56	494.25
15.6-16.0	320.54	392.82	509.15
16.1-16.5	329.85	404.39	524.35
16.6-17.0	338.67	415.47	539.07
17.1-17.5	347.21	426.26	553.50
17.6-18.0	355.77	437.08	567.95
18.1-18.5	364.31	447.88	582.39
18.6-19.0	372.80	458.64	596.78
19.1-19.5	381.25	469.34	611.11
19.6-20.0	389.66	480.02	625.43

ADDITIONAL MATERIAL COSTS FOR BLENDING ADJACENT PANEL
Add according to procedure on page P-4

Hours	Cost	Hours	Cost
0.0-1.0	$10.62	3.1-3.5	$54.31
1.1-1.5	18.65	3.6-4.0	62.35
1.6-2.0	26.69	4.1-4.5	70.38
2.1-2.5	34.72	4.6-5.0	81.94
2.6-3.0	46.28	5.1-5.5	89.97

NOTE: Additional material may be required for special circumstances.

NATIONAL VOC COST CHARTS

	SUGGESTED COST (Dollars)		
Estimated Refinish Hours	Color w/Materials	Color & Clear w/Materials	Three Stage w/Materials
0.0-1.0	$28.16	$30.02	$39.08
1.1-1.5	38.81	43.35	56.94
1.6-2.0	52.70	56.41	74.54
2.1-2.5	62.69	69.08	91.74
2.6-3.0	70.27	82.84	110.03
3.1-3.5	76.76	95.51	127.23
3.6-4.0	87.14	108.56	144.81
4.1-4.5	98.35	122.45	163.23
4.6-5.0	109.23	136.01	181.32
5.1-5.5	118.80	148.26	198.10
5.6-6.0	129.08	161.22	215.59
6.1-6.5	139.57	174.39	233.29
6.6-7.0	149.94	187.43	250.87
7.1-7.5	160.83	201.01	268.97
7.6-8.0	170.28	213.13	285.63
8.1-8.5	182.21	227.74	304.77
8.6-9.0	192.12	240.33	321.89
9.1-9.5	202.96	253.85	339.94
9.6-10.0	212.84	266.40	357.03
10.1-10.5	224.17	280.41	375.57
10.6-11.0	234.09	293.00	392.69
11.1-11.5	244.93	306.52	410.74
11.6-12.0	254.79	319.07	427.81
12.1-12.5	266.41	333.36	446.64
12.6-13.0	276.27	345.90	463.71
13.1-13.5	286.77	359.07	481.41
13.6-14.0	296.26	371.24	498.11
14.1-14.5	306.33	384.00	515.40
14.6-15.0	315.86	396.21	532.14
15.1-15.5	326.31	409.33	549.79
15.6-16.0	335.79	421.49	566.48
16.1-16.5	345.57	433.95	583.48
16.6-17.0	354.87	445.92	599.98
17.1-17.5	363.38	457.62	616.21
17.6-18.0	372.92	469.33	632.45
18.1-18.5	381.94	481.03	648.68
18.6-19.0	390.91	492.68	664.86
19.1-19.5	399.83	504.28	680.99
19.6-20.0	408.73	515.85	697.09

ADDITIONAL MATERIAL COSTS FOR BLENDING ADJACENT PANEL
Add according to procedure on page P-4

Hours	Cost	Hours	Cost
0.0-1.0	$11.07	3.1-3.5	$59.24
1.1-1.5	20.00	3.6-4.0	68.17
1.6-2.0	28.93	4.1-4.5	77.10
2.1-2.5	37.86	4.6-5.0	89.55
2.6-3.0	50.31	5.1-5.5	98.49

NOTE: Additional material may be required for special circumstances.

FIGURE 10–23 In this example, there are four hours of paint time for this BC/CC color. The cost of materials would be $100.78.

CALFORNIA VOC COST CHARTS

Estimated Refinish Hours	SUGGESTED COST (Dollars) Color w/Materials	Color & Clear w/Materials	Three Stage w/Materials
0.0-1.0	$22.27	$26.76	$33.00
1.1-1.5	30.33	37.99	47.35
1.6-2.0	39.97	48.95	61.43
2.1-2.5	47.37	59.52	75.12
2.6-3.0	55.85	77.29	99.13
3.1-3.5	63.25	87.86	112.82
3.6-4.0	71.02	98.80	126.88
4.1-4.5	79.81	110.76	141.96
4.6-5.0	91.80	128.34	165.78
5.1-5.5	98.77	138.48	179.04
5.6-6.0	106.46	149.34	193.02
6.1-6.5	114.36	160.41	207.21
6.6-7.0	122.13	171.35	221.27
7.1-7.5	134.13	188.94	245.10
7.6-8.0	140.97	198.95	258.23
8.1-8.5	150.48	211.63	274.03
8.6-9.0	157.80	222.12	287.64
9.1-9.5	166.05	233.54	302.18
9.6-10.0	177.03	250.11	324.99
10.1-10.5	185.77	262.02	340.02
10.6-11.0	193.09	272.51	353.63
11.1-11.5	201.34	283.93	368.17
11.6-12.0	208.60	294.36	381.72
12.1-12.5	217.79	309.72	400.20
12.6-13.0	225.06	320.16	413.76
13.1-13.5	232.96	331.23	427.95
13.6-14.0	239.85	341.29	441.13
14.1-14.5	247.34	351.95	454.91
14.6-15.0	257.97	368.17	477.37
15.1-15.5	265.83	379.20	491.52
15.6-16.0	272.71	389.25	504.69
16.1-16.5	279.90	399.61	518.17
16.6-17.0	286.60	409.48	531.16
17.1-17.5	296.72	425.19	553.11
17.6-18.0	303.16	434.80	565.84
18.1-18.5	309.58	444.39	578.55
18.6-19.0	315.95	453.93	591.21
19.1-19.5	322.28	463.43	603.83
19.6-20.0	328.57	475.89	619.41

ADDITIONAL MATERIAL COSTS FOR BLENDING ADJACENT PANEL
Add according to procedure on page P-4

Hours	Cost	Hours	Cost
0.0-1.0	$12.14	3.1-3.5	$58.97
1.1-1.5	20.66	3.6-4.0	67.49
1.6-2.0	29.19	4.1-4.5	76.02
2.1-2.5	37.72	4.6-5.0	88.75
2.6-3.0	50.44	5.1-5.5	97.27

NOTE: Additional material may be required for special circumstances.

CALFORNIA VOC COST CHARTS

Estimated Refinish Hours	SUGGESTED COST (Dollars) Color w/Materials	Color & Clear w/Materials	Three Stage w/Materials
0.0-1.0	$23.32	$27.20	$33.88
1.1-1.5	31.80	38.65	48.67
1.6-2.0	42.07	49.83	63.19
2.1-2.5	49.89	60.62	77.32
2.6-3.0	58.79	78.83	102.21
3.1-3.5	66.61	89.62	116.34
3.6-4.0	74.80	100.78	130.84
4.1-4.5	84.01	112.96	146.36
4.6-5.0	96.84	130.98	171.06
5.1-5.5	104.23	141.34	184.76
5.6-6.0	112.34	152.42	199.18
6.1-6.5	120.66	163.71	213.81
6.6-7.0	128.85	174.87	228.13
7.1-7.5	141.69	192.90	253.02
7.6-8.0	148.95	203.13	266.59
8.1-8.5	158.88	216.03	282.83
8.6-9.0	166.62	226.74	296.88
9.1-9.5	175.29	238.38	311.86
9.6-10.0	187.11	255.39	335.55
10.1-10.5	196.27	267.52	351.02
10.6-11.0	204.01	278.23	365.07
11.1-11.5	212.68	289.87	380.05
11.6-12.0	220.36	300.52	394.04
12.1-12.5	229.97	316.10	412.96
12.6-13.0	237.66	326.76	426.96
13.1-13.5	245.98	338.05	441.59
13.6-14.0	253.29	348.33	455.21
14.1-14.5	261.20	359.21	469.43
14.6-15.0	272.67	375.87	492.77
15.1-15.5	280.95	387.12	507.36
15.6-16.0	288.25	397.39	520.97
16.1-16.5	295.86	407.97	534.89
16.6-17.0	302.98	418.06	548.32
17.1-17.5	313.94	434.21	571.15
17.6-18.0	320.80	444.04	584.32
18.1-18.5	327.64	453.85	597.47
18.6-19.0	334.43	463.61	610.57
19.1-19.5	341.18	473.33	623.63
19.6-20.0	347.89	486.01	639.65

ADDITIONAL MATERIAL COSTS FOR BLENDING ADJACENT PANEL
Add according to procedure on page P-4

Hours	Cost	Hours	Cost
0.0-1.0	$12.25	3.1-3.5	$60.24
1.1-1.5	21.01	3.6-4.0	69.00
1.6-2.0	29.77	4.1-4.5	77.75
2.1-2.5	38.52	4.6-5.0	90.71
2.6-3.0	51.48	5.1-5.5	99.47

NOTE: Additional material may be required for special circumstances.

FIGURE 10–23 (Continued)

An older car with a deteriorated finish might require removing the moulding and trim to access areas for repair and minor metalworking. This kind of paint job can require twenty-four hours to perform.

Because of the variables involved, no standard figures can be given for entire car paint jobs. Estimating the labor involved depends largely on the experience of the estimator, who makes allowances for washing and cleaning the car, repairing any defects, masking, priming, and painting. Since insurance companies are not involved in whole car paint jobs, these estimates are determined only by the collision repair facility.

REVIEW QUESTIONS

1. Technician A washes the surface with soap and water before any repairs. Technician B says that wax and grease remover dissolves substances that soap and water cannot. Who is right?
   a. technician A
   b. technician B
   c. both A and B
   d. neither A nor B

2. A ______________ is used to fill in scratches in the body filler.
3. Technician A says that masking of inner surfaces is included in the refinish time. Technician B says that inner surface masking labor must be added to the total labor time. Who is right?
   a. technician A
   b. technician B
   c. both A and B
   d. neither A nor B
4. A blend can hide a slight mismatch between the vehicle color and the refinish color.
   True False
5. Technician A says that clear coat labor time is included in the blend time. Technician B says that the time to clean up the equipment is not included in the refinish time. Who is right?
   a. technician A
   b. technician B
   c. both A and B
   d. neither A nor B
6. To prevent spraying clear coat on a roof and the opposite quarter panel, the clear coat edge is ______________ into the sail panel.
   a. blended
   b. sprayed
   c. melted
   d. none of the above
7. Technician A says that most new vehicles have a basecoat/clear coat finish. Technician B says that overlap must be deducted when spraying a fender and a urethane bumper. Who is right?
   a. technician A
   b. technician B
   c. both A and B
   d. neither A nor B
8. Color sanding is used to remove
   a. dirt.
   b. runs.
   c. sags.
   d. all of the above.
9. Technician A says that tricoat finishes are matched by variation in the number of midcoats. Technician B says that a spot repair cannot be made with tricoat. Who is right?
   a. technician A
   b. technician B
   c. both A and B
   d. neither A nor B
10. One way that refinish material cost can be calculated is to multiply the ______________ by a set dollar amount.

Chapter 11

Preparing Damage Reports/Appraisals

Objectives

After reading this chapter, you should be able to:

- Explain the importance of thorough, fair, and financially responsible damage reports.
- Define a vehicle's actual cash value, how it is determined, and how it is used to determine the "total" value of the vehicle.
- Define the terms **total loss ceiling, open bid,** and **teardown;** explain how they are used in damage report writing.
- Describe the key factors in the decision process used when writing a damage report.
- Identify the key information that must be listed on damage report forms, including VIN and paint code data.
- Describe a logical sequence for gathering key information and determining previous damage, depreciation, and betterment.
- Describe a logical sequence for inspecting a vehicle and assessing the actual damage to it.
- Describe the manner in which a handwritten damage report or appraisal is completed, line by line and operation by operation.
- Explain the importance of clear penmanship, detailed descriptions, and accurate calculations when writing a damage report.

Key Terms

ACV
VIN

A good damage report is more than a set of parts and labor prices accurately filled in and tabulated on a clean form or computer printout. A good damage report blends the knowledge, experience, and sound judgment of its writer with up-to-date pricing and labor data. Only then does it become an effective blueprint for the repair of the vehicle.

The damage report/appraisal must be fair to all parties involved.

1. It must guarantee the customer a safe, quality repair. All operations and parts needed to assure a complete repair must be listed, including any needed straightening, alignment, or adjustment times.
2. It must assure that the collision repair facility is fairly and accurately compensated for the labor and materials needed to repair the vehicle.
3. It must show financial responsibility to the insurance company and/or customer. The parts and repair methods selected should be the most economical way possible of returning the vehicle to its preaccident condition without compromising safety or quality (Figure 11–1).

FIGURE 11–1 Although the damage to this Sunfire appears minor, the repair cost was over $500.

11.1 FACTORS INFLUENCING THE REPAIR

In theory, any damaged vehicle can be restored to its preaccident or even "showroom-new" condition. But this is not always economically possible or desirable.

ACTUAL CASH VALUE (ACV) OF THE VEHICLE

Any vehicle is considered a "wear" item, with a definite lifespan. The average lifespan of American passenger cars, light trucks, and vans is twelve years. The average age of cars on the road today is about six years, and 80 percent of all vehicles are over three years of age.

In many cases, the value of a vehicle can significantly decrease with age and/or miles driven. This is even true when the vehicle has been properly maintained and serviced. For example, a vehicle that sold new for a list price of $15,000 several years ago can now have an average retail value of $8,000 if it is still in good condition without excessive mileage. This $8,000 figure represents the actual cash value, or ACV, of the vehicle, and becomes a very important figure for determining if and how the repair will be made, particularly in insurance claims.

Excluding certain types of vehicle restoration, investing substantial sums of money into a vehicle will not significantly increase its market value. Investing $3,000 into the collision repair of a $2,000 vehicle will not result in a $5,000 vehicle, or even a $3,000 vehicle. It will result in a vehicle having more or less its preaccident value, in this case $2,000.

24 FORD 1997–96

DOMESTIC CARS

Trade-in	BODY TYPE	Model No.	Loan	Retail
1997 THUNDERBIRD-AT-PS-AC-Cont.				
300	**Add** CD Player		300	300
200	**Add** CD Player		200	200
300	**Add** Custom Wheels		300	300
400	**Add** Leather Seats		400	400
225	**Add** Anti-Lock Brakes		225	225
300	**Add** Theft Deterrent		300	300
1996 ASPIRE-4 cyl.-5 Spd.-AC-FWD				
5525	H'back 2D	05	4975	7025
5825	H'back 4D	06	5250	7350
200	**Add** CD Player		200	200
175	**Add** Anti-Lock Brakes		175	175
250	**Add** Theft Deterrent		250	250
400	**Add** Automatic Trans		400	400
125	**Add** Power Steering		125	125
500	**Deduct** W/out Air Cond		500	500
1996 ESCORT-4 Cyl.-AT-PS-AC-FWD				
6175	H'back 2D	10	5575	7725
6925	H'back 2D LX	11	6250	8550
7125	H'back 4D LX	14	6425	8800
7225	Sedan 4D LX	13	6525	8900
7725	Sta Wgn 4D LX	15	6975	9450
8225	H'back 2D GT	12	7425	10025
75	**Add** Power Door Locks		75	75
125	**Add** Cruise Control		125	125
400	**Add** Power Sunroof		400	400
200	**Add** CD Player		200	200
125	**Add** Power Windows		125	125
75	**Add** Tilt Wheel		75	75
75	**Add** Luggage Rack (S/W)		75	75
200	**Add** Custom Wheels		200	200
175	**Add** Anti-Lock Brakes		175	175
250	**Add** Theft Deterrent		250	250
450	**Deduct** Manual Trans.		450	450
125	**Deduct** Convent Steer		125	125
550	**Deduct** W/out Air Cond		550	550
1996 CONTOUR-AT-PS-AC-FWD				
CONTOUR-4 Cyl.				
8950	Sedan 4D GL	65	8075	10825
9550	Sedan 4D GL	68	8600	11500
CONTOUR-V6				
9500	Sedan 4D GL	65	8550	11450
10100	Sedan 4D LX	66	9100	12100
10600	Sedan 4D SE	67	9550	12650
500	**Add** Power Sunroof		500	500
250	**Add** CD Player		250	250
100	**Add** Power Door Locks		100	100
175	**Add** Power Windows		175	175
175	**Add** Power Seat		175	175
175	**Add** Cruise Control		175	175
250	**Add** Custom Wheels		250	250
350	**Add** Leather Seats		350	350
175	**Add** Anti-Lock Brakes		175	175
250	**Add** Theft Deterrent		250	250

Trade-in	BODY TYPE	Model No.	Loan	Retail
550	**Deduct** Manual Trans		550	550
650	**Deduct** W/out Air Cond		650	650
1996 MUSTANG-AT-PS-AC				
MUSTANG-V6				
11525	Coupe 2D	40	10375	13675
14925	Convertible 2D	44	13450	17325
MUSTANG-V8				
13925	Coupe 2D GT	42	12550	16250
17325	Convertible 2D GT	45	15600	19925
19325	Coupe Cobra 5 Spd	47	17400	22025
22725	Conv Cobra 5 Spd	46	20475	25675
250	**Add** CD Player		250	250
200	**Add** MACH Stereo		200	200
100	**Add** Power Door Locks*		100	100
175	**Add** Power Windows*		175	175
175	**Add** Pwr Seat(Std.Cobra).		175	175
175	**Add** Cruise (Std.Cobra)		175	175
250	**Add** Custom Wheels		250	250
350	**Add** Leather Seats		350	350
175	**Add** ABS (Std. Cobra)		175	175
250	**Add** Theft Deterrent		250	250
450	**Deduct** Manual (V6)		450	450
650	**Deduct** W/out Air Cond		650	650
	*Std. Conv. Cobra			
1996 PROBE-4 Cyl.-AT-PS-AC-FWD				
9975	H/back 2D	20	9000	12025
11925	H'back 2D GT (V6)	22	10750	14100
250	**Add** CD Player		250	250
100	**Add** Power Door Locks		100	100
175	**Add** Power Windows		175	175
175	**Add** Power Seat		175	175
500	**Add** Power Sunroof		500	500
350	**Add** Leather Seats		350	350
175	**Add** Cruise Control		175	175
100	**Add** Tilt Wheel		100	100
250	**Add** Custom Wheels		250	250
350	**Add** SE Pkg		350	350
175	**Add** Anti-Lock Brakes		175	175
250	**Add** Theft Deterrent		250	250
450	**Deduct** Manual Trans.		450	450
650	**Deduct** W/out Air Cond.		650	650
1996 TAURUS-AT-PS-AC-FWD				
TAURUS-V6				
10050	Sedan 4D G	51	9050	12125
10425	Sedan 4D GL	52	9400	12525
11625	Sta Wgn 4D GL	57	10475	13775
12025	Sedan 4D LX	53	10825	14250
13225	Sta Wgn 4D LX	58	11925	15525
TAURUS SHO-V8				
16725	Sedan 4D	54	15075	19225
250	**Add** CD Player		250	250
200	**Add** JBL Stereo System		200	200
500	**Add** Power Sunroof		500	500
100	**Add** Power Door Locks*		100	100
175	**Add** Power Seat*		175	175

DEDUCT FOR RECONDITIONING

1998 FEBRUARY 1998

FIGURE 11–2 Page from NADA Official Used Car Guide that lists important market value and loan information concerning most vehicles. (Used with permission of National Automobile Dealers Used Car Guide Co.)

The ACV of a vehicle can be found using several sources. These include the NADA Official Used Car Guide, The Automobile Red Book published by Maclean Hunter Market Reports, Inc., and the Kelley Blue Book, among others. These guides contain similar information, but prices and exact data listed can vary slightly between guides.

Figure 11–2 illustrates a page from the NADA Official Used Car Guide. It lists the trade-in or wholesale value offered by car dealers for that particular make and model. It also gives the value of a loan that can be secured using the vehicle as collateral. The retail price given is the price that dealers would ask used car buyers to pay for that make and model. The retail value is the figure used in calculating the ACV.

The guide also lists the options available for the vehicle, including dollar amounts to add on or deduct for particular options. For example, power steering, air-conditioning, and other power and convenience features will increase the car's value. Features such as manual transmissions or diesel engines might reduce vehicle resale value. The guides also contain charts for determining value adjustments on vehicles with exceptionally high or low mileage.

Used car guides are continually updated and are normally obtained on a subscription basis. Because vehicle values may vary in different parts of the country, guide publishers often prepare regional editions reflecting these differences. In some cases, the average between two or more guides is used to determine the ACV.

Car dealership appraisals are also used as a source of determining ACV. Local dealers are often asked to determine a fair market value for vehicles. This is particularly true with "exceptional" vehicles. A good example would be an older vehicle with extremely low mileage that is in extraordinary condition. What would be the dealer's asking price if that car were on his or her lot? If the vehicle can easily be sold for greater than its "book" value, the insurance company is committed to settling at the higher ACV appraisal.

When determining the ACV of a vehicle, insurance companies do not normally consider recent mechanical repairs made at the expense of the vehicle owner. These are seen as normal operating and maintenance costs and do not increase ACV. For example, if a vehicle owner spends $1,000 on mechanical repairs to a vehicle with an ACV of $2,000, he or she might increase its lifespan, but the ACV remains at $2,000.

Total Loss Vehicles. A vehicle's ACV is often an important factor when the repair involves an insurance claim settlement. For obvious reasons, insurance companies do not pay for repairs that exceed the actual cash value of the vehicle. Instead, the vehicle is declared a total loss (Figure 11–3).

Individual insurance companies have their own guidelines and formulas for determining "total" situations. Three of the most common formulas are as follows:

1. When the total cost of repair equals or exceeds the ACV
2. When the total cost of repair equals or exceeds a fixed percentage of the ACV, such as 75 or 80 percent

FIGURE 11–3 The damage on this Ranger is so extensive that the cost of repair exceeds the actual cash value of the vehicle. This vehicle was declared a total loss.

3. When the total cost of repair, plus the salvage value of the vehicle, equals or exceeds the total ACV or a fixed percentage of the ACV

As you can see, both formulas 2 and 3 will actually "total" a vehicle at less than its full ACV. In all cases, an insurance company adjuster must still appraise the damage to confirm that the cost of all reasonable repair methods exceeds the company's total loss guidelines. Once this is proven, the vehicle owner receives a settlement check for the ACV amount. The title of the vehicle is transferred to the insurance company, who has effectively purchased the vehicle from the owner at the ACV price.

Any salvage value the vehicle might have now belongs to the insurance company. To minimize the time and effort needed to sell individual wrecks, some insurance companies sign yearly contracts with a single salvage yard in a given territory. This salvage yard agrees to purchase all wrecks from that company based on a certain percentage of their ACV (Figure 11–4). The range for late model wrecks is between 20 and 25 percent. With older models, the percentage is normally lower. Heavily damaged wrecks having less than three salvageable bolt-on parts are usually considered negotiable in terms of value. Such contracts are often awarded through competitive bidding.

TOTAL LOSS CEILING

Every vehicle has an ACV and a total loss ceiling based on the ACV. For minor to moderate repairs on late-model vehicles, the repair costs will be well below the total ceiling for the vehicle. There will also be a substantial money cushion available for engine or

FIGURE 11–4 *A salvage yard is the last resting place for total loss vehicles. It is an excellent source of used parts.*

transmission repairs that are often difficult to precisely price out until teardown and work begin. Having this cushion *does not* mean the damage report writer or adjuster can write an incomplete report. Always be as thorough and precise as possible, but recognize areas where hidden damage and supplements can occur.

Example 1:

A late-model passenger car equipped with an overhead cam engine is damaged in a front-end collision. There is moderate sheet metal damage, but the cover of the overhead camshaft has been cracked and loosened on its mounts. The likelihood of additional damage to the valve train and internal engine components is great, but until the system is disassembled, an exact appraisal cannot be made.

The total ceiling on this vehicle is $9,000. The insurance adjuster writes a complete appraisal on all sheet metal damage and other mechanical components, but excludes repairs to the camshaft and engine. The cost of repair to this point is $4,000. This leaves a repair cushion of $5,000 for the additional engine repairs needed to return the vehicle to its preaccident condition. The insurance company is certain the engine work can be completed for much less than $5,000, so they will authorize the repair.

In this case, the insurance company has left the bid or appraisal "open" for a future supplement. This does not mean that the body shop has $5,000 to make or sublet the engine repairs. As in any repair, the insurance company will require a detailed breakdown of the engine damage, including parts and labor. They might also send an adjuster to confirm the damage once engine teardown is complete.

Insurance companies are not comfortable with an open bid and will attempt to close every bid if at all possible. Rarely, if ever, will a bid be left open for straightening, sheet metal repair, panel replacement, or sectioning work. Both adjusters and collision repair facility personnel are confident in their ability to accurately determine costs in these areas. The same does not hold true for certain types of engine work, mechanical repairs, and electrical troubleshooting. Rapid advances in fuel injection, computerized engine control, antilock brakes, computer-assisted suspensions, and other new systems are revolutionizing how vehicles operate. They are also causing confusion in troubleshooting and repair. Pinpointing repair costs in these areas is difficult for skilled mechanics, let alone adjusters and shop damage writers. Until everyone in the auto repair industry becomes trained and experienced in these repairs, accurate costing of these repairs will be difficult.

In the preceding example, the decision to authorize the repair is rather clear-cut. The decision to repair becomes increasingly more difficult as the cost of repair nears the total figure.

Example 2:

A four-year-old vehicle has received collision damage to lower body panels. In addition to steering and suspension damage, the transmission casing is cracked.

The total ceiling on the vehicle is $8,000. The adjuster writes a complete appraisal on all body panel, steering, and suspension components. The cost of repair to this point is $7,500, which does not include the transmission repairs.

In the eyes of the insurance company, this vehicle has entered a gray area. The adjuster is not certain what the transmission repairs will cost but knows that all but the most minor problems will push the repair cost above the total. And once the insurance company authorizes repair, it is legally bound to pay for all costs, even if they exceed the total ceiling. So if the hidden costs of fixing the transmission exceed $500, the company stands to lose money.

In a case like this, the insurance company often authorizes a partial teardown of the transmission so a precise appraisal can be made. By paying this small labor charge, the insurance company can save hundreds of dollars in overruns. And if the company is paying a storage charge for each day the vehicle sits on the shop lot, resolving the claim quickly is also desired.

Based on the teardown findings, the insurance company will authorize the repair or declare a total. If the transmission repairs cost $200, the final appraisal is $7,700 or below total. Any cost over $500 results in a clear-cut "total" situation. Costs in the $400 to $500 range will again bring judgment and experience into play. In close calls, factors such as the overall condition of the vehicle, the possibility of other supplements, and the likelihood of comebacks and customer dissatisfaction might result in the vehicle being declared a total, even if it is technically several dollars short of the ceiling set by the company. This is an insurance company decision often based on the advice of the field adjuster familiar with the case.

BODY SHOPS, ACV, AND TOTALING

The vehicle's ACV and total ceiling is also important to the collision repair facility damage report writer. Remember, 75 percent of all collision repairs are settled through insurance claims. In three out of four cases, the shop and insurance company must reach an agreement on price. While collision repair facilities must always write their reports to include the work and parts they feel are needed to do the job, disregarding the total ceiling in the write-up process can lead to major disagreements.

Neither the collision repair facility nor the insurance company benefits from a total. A total brings no work to the shop. It also forces the insurance company to make its maximum possible payout. This is why insurance companies and collision repair facilities must work together to save as many vehicles as practically possible from the salvage yard.

THE DECISION PROCESS

The overall approach to damage report/appraisal writing should be "how can this vehicle be saved," not "how can this vehicle be totaled."

Saving a vehicle means making logical decisions on whether to repair or replace items. When replacement is the logical choice, it then means deciding if OEM, salvage, or aftermarket parts provide the like kind and quality needed.

Repair/Replace. Insurance companies consider repair/replace decisions from an almost entirely economic standpoint. If the labor and material cost of returning a part to its preaccident condition is considerably less than the cost of installing a new or salvaged part, the adjuster will specify the repair. But if repair cost is very close or equal to replacement cost, replacement is often selected for the following reasons:

1. There is less chance of problems and supplements.
2. The job is less labor intensive. It can be in and out of the shop quicker and does not tie up the shop's labor force.
3. The likelihood of higher repair quality and increased customer satisfaction and goodwill is increased.

Parts. Parts selection is probably the biggest factor influencing the overall cost of repair (Figure 11–5, OEM part prices are obviously higher than salvage parts. Certain aftermarket parts and components can also be less expensive than their OEM counterparts.

If every damage report and appraisal were written specifying new OEM parts, the number of totaled vehicles would rise dramatically. Use of salvage parts or aftermarket parts of like kind and quality can significantly reduce parts costs. OEM, salvage, and aftermarket parts all have their place in the collision repair industry.

Aftermarket Parts Disclosure. The availability and coverage of aftermarket collision parts such as bumpers, fenders, and hoods has increased greatly. A heated and ongoing debate among OEMs, aftermarket suppliers, insurance companies, and collision repair facilities has erupted concerning the quality, fit, and appropriate use of aftermarket parts. As in any competitive market, quality can vary between manufacturers and collision repair facilities, and insurance companies must research and choose any aftermarket supplier carefully. The Certified Automotive Parts Association (CAPA) has developed a program of testing and certification for aftermarket sheet metal parts. Aftermarket manufacturers voluntarily use the testing certification program, and parts that pass the test are marked with the CAPA stamp of approval, as well as the name of the supplier. Many aftermarket suppliers also offer lifetime warranties on their products.

A significant number of states have passed legislation requiring body shops and/or insurance companies to inform their customers in writing whenever non-OEM parts are used in the repair of their vehicles. A written estimate is needed on non-OEM parts clearly identifying each major part. Many states also require a written estimate disclosure statement. The following is a typical example of such a disclosure statement:

> "This estimate has been prepared based on the use of automobile parts not made by the original manufacturer. Parts used in the repair of your vehicle by other than the original manufacturer are required to be at least equal in like kind and quality in terms of fit, quality, and performance to the original manufacturer parts they are replacing. All aftermarket parts installed on the vehicle shall be clearly identified on the estimate of such repair."

REPLACE	REPAIR	DESCRIPTION	PARTS	LABOR	REFINISH	SUBLET
✓		front bumper	248.81	1.6		
✓		license bracket	8.88	0.2		
✓		bumper pad	35.67	0.3		
✓		right bumper reinforcement	27.20	IOH		
✓		right mounting arm	24.21	IOH		
✓		right insulator reinforcement	18.67	IOH		
	✓	stone deflector		1.5	1.4	
✓		right bumper filler	11.66	0.1		
	✓	clear coat			0.6	
		TOTALS	375.10	3.7	2.0	

The above is an estimate based on our inspection and does not cover any additional parts or labor which may be required after the work has been started. Occasionally, worn or damaged parts are discovered which may not be evident on the first inspection. Because of this, the above prices are not guaranteed. Quotations on parts and labor are current and subject to change.

AUTHORIZATION FOR REPAIR. You are hereby authorized to make the above repairs:

SIGNED: ________

DATE: ________

TOTAL PARTS	$ 375.10
TOTAL LABOR	$ 125.80
TOTAL REFINISH	$ 68.00
TOTAL ~~SUBLET~~ paint mat	$ 40.00
TAX	$ 25.94
FRAME REPAIR	$
TOTAL	$ 634.84

FIGURE 11–5 *As shown on this damage report, the cost of parts often exceeds the labor cost of a repair.*

Some disclosure laws require that the consumer be informed as to cost comparisons between OEM and aftermarket parts. Others require that any aftermarket part warranty be provided upon customer request. The use of an aftermarket part might invalidate any remaining warranties of the original manufacturer on that part.

States requiring non-OEM disclosure

Alabama
Alaska
Arizona
Arkansas
California
Colorado
Connecticut
Florida
Georgia
Hawaii
Idaho
Illinois
Indiana
Iowa
Kansas
Louisiana
Maine
Maryland
Massachusetts
Minnesota
Mississippi
Missouri
Nevada
New Hampshire
New Jersey
New York
North Carolina
Ohio
Oklahoma
Oregon
Pennsylvania
Rhode Island
South Dakota
Texas
Utah
Vermont
Virginia
Washington
West Virginia
Wisconsin

It is vital that every body shop worker and insurance company adjuster know and follow the guidelines in his or her state.

PARTS SELECTION FACTORS

Each repair situation requires close attention and sound judgment in determining which type(s) of part(s) to use. Quality, fairness, and financial responsibility are all concerns that must be addressed. Factors entering into the decision will be:

1. Actual cash value and total ceiling of the vehicle
2. Age, condition, and mileage of the vehicle
3. Availability of OEM, salvage, and aftermarket parts
4. Condition of the salvage parts
5. Wishes of the customer

Example 1:

A seven-year-old vehicle with an ACV of $1,900 has received moderate to light front-end damage. Vehicle mileage is low (58,000) and the vehicle is in good, but not extraordinary shape. The owner is anxious to keep the vehicle. The repair is to be settled through the customer's insurance company.

The customer's insurance company totals vehicles at 80 percent of their ACV, so the total ceiling on this job is $1,520. This vehicle is a prime candidate for salvaged parts. New parts will likely push the cost of repair over $1,520, a fact that can easily be confirmed once the damage report and/or appraisal has been written.

In addition to the cost savings, structurally sound salvage parts are probably the closest in like kind and quality to those damaged. The customer is entitled to have the vehicle returned to its preaccident condition, not better than preaccident condition. With salvage parts, the customer does not face betterment charges if the damaged parts are slightly corroded or show dings and dents not caused by this accident. If a new part was used, the customer would pay a percentage of the cost based on the extent of the previous damage or corrosion. Such is not the case with salvage parts. And because it is unethical to install a part of lesser quality than the panel being replaced, the customer is actually receiving betterment at no charge. And at this point in the car's lifespan, OEM parts can actually be more difficult to locate than salvaged parts.

Example 2:

A late-model-year vehicle with less than 18,000 miles receives moderate damage to the rear end and trunk. ACV is $11,500.

The total ceiling will obviously not come into play in this case. This vehicle is a prime candidate for new replacement parts for many of the reasons salvage parts were ideal in Example 1. The likelihood of the customer having to pay betterment is low. New late-model parts are readily available, while salvage pool stockpiles of late-model parts are low. And most importantly, a new part matches like kind and quality guidelines.

CUSTOMER PAID REPAIRS

Not all repairs are insurance paid. A vehicle owner without collision or comprehensive coverage must personally pay for damage repairs to his or her vehicle (when he or she is at fault). Even motorists with insurance might want to settle small accidents out of their own pockets to prevent an increase in their insurance rates.

In an insurance claim, both the collision repair facility and the adjuster understand the options available for vehicle repair. They might not always be in agreement, but they do act as a series of checks and balances. The novice customer must rely on the collision repair facility to accurately explain the options available. And the shop should approach the repair with the same respect for quality and financial responsibility.

While it is unethical to approach the uneducated customer as an easy opportunity to select the repair route that best profits the shop, it is also poor business practice to bow to customer requests for quick, cheap repairs.

Every car the shop repairs is a traveling billboard of the shop's quality and workmanship. A customer might say he is looking for a shop to bang out the buckle and make it "presentable." He might not even complain if this is exactly what the shop does. He might pay and go happily on his way. What he probably will not do, however, is explain the entire story to the first person who asks him where he had his car repaired. The person probably thinks that the slight wrinkle or less-than-perfect paint match is the shop's fault, not the customer's request.

11.2 PREPARING A DAMAGE REPORT/ APPRAISAL

In estimates of any type of damage to a vehicle, whether minor or major, a logical sequence must be followed. The analyzing of damage procedures given in the earlier chapters of this book, particularly Chapter 5, must be followed by both collision repair facility and insurance company personnel.

But before beginning the actual listing of damaged parts and labor needs, the damage report writer must gather vital information about the vehicle and make several important inspections.

The forms used by collision repair facilities, insurance companies, and independent adjusters will vary slightly. Figure 11–6 illustrates a typical form used by a collision repair facility when writing up a damage report. A typical insurance company appraisal form is shown in Figure 11–7. As you can see, both forms record much of the same vital data concerning the owner, vehicle, and the insurance claim.

NAME, ADDRESS, AND PHONE NUMBER

This vital information is so basic many collision repair facility and insurance company personnel fail to see its real value. Of course it informs the shop or insurance company who the customer is, where he or she lives, and how he or she can be contacted at any time during the repair. But far more importantly, it allows you to address the customer by name. People appreciate being addressed by name. It encourages them to be more open and friendly and places the damage report writer in a more personal position.

Of course, not all reports will be written with the customer present, but most will. When taking down this basic information, use this minute or so to establish a friendly, but professional tone with the customer. The customer has suffered a shock and probably wants someone to hear his story and listen to his concerns. Be a good listener, be sympathetic, ask pertinent questions when needed, and provide answers to the customer's questions and concerns.

INSURANCE INFORMATION

The insurance company name, policy number, claim number, date of accident or loss, and customer-paid deductible amount are all vital to the fast, accurate processing of the claim. Once again, the collision repair facility can use this information to its advantage. If the collision repair facility has a good working relationship with the insurance company involved in the claim, use this fact as a selling tool. Tell the customer he or she has a good insurance company that works to see that their claims are settled quickly and fairly.

The collision repair facility should never put down an insurance company, even if their working relationship has a history of problems. As you will see, the best time to build customer support concerning a potential shop/adjuster disagreement is after the damage report has been completed. Review the report line by line, explaining how the facility plans to repair the vehicle and the parts, labor, and material needed. Then, if a disagreement arises, the fact that the customer clearly understood the facility's position from the beginning might work in the facility's favor.

Obviously, experienced adjusters will use the same tactic, particularly when the vehicle is appraised at a drive-in claim center before being taken to a facility. The point is that *both* collision repair facility estimators and insurance adjusters must write reports that fully explain the job to the customer.

As the insurance information is being recorded, be sure the customer understands how the insurance process will work. Explain any important laws concerning collision repair, such as aftermarket parts disclosure. Be sure the customer knows his or her rights and obligations in the repair process (see Chapter 1). Record any deductible amount involved in the settlement and clearly explain the shop's policy in arranging for payment.

Once the needed customer and insurance information is recorded, move on to the vehicle inspection and information.

VIN NUMBERS

Beginning with the 1981 model year, the National Highway Traffic Safety Administration, Department of Transportation, required all manufacturers selling over-the-road vehicles in the United States to produce the vehicles with a seventeen-character vehicle identification number (VIN), which can be used to determine valuable information about the vehicle and its repair.

The VIN number is stamped into a small metal plate that is riveted to the upper left corner of the instrument panel, visible through the windshield (Figure 11–8).

This standard establishes a fixed VIN format, including a check digit, and applies to all passenger cars, multipurpose passenger vehicles, trucks, buses, trailers, incomplete vehicles, and motorcycles with a gross vehicle weight of 10,000 pounds or less (Figure 11–9).

The first three characters of the VIN are designated the WMI, or World Manufacturers Identification, which uniquely identifies the nation of origin, manufacturer, make, and type of vehicle.

The second section has five characters and has been designated the VDS, or Vehicle Description Section. The VDS uniquely identifies the attributes of the vehicle such as model, model style, and engine.

The ninth character in the VIN is the check digit, which can be used to confirm the validity of the VIN if there is reason to suspect that the VIN plate has been tampered with, switched, or is otherwise out of order.

The third section of the VIN is located after the check digit. It is eight characters in length and is called the VIS, or Vehicle Identification Section. The first character represents the vehicle model year as indicated in Table 11–1. The second character repre-

DAMAGE REPORT

Name __________ Date __________ Work Phone __________ Home Phone __________

Address __________ City __________ State __________ Zip __________

Year __________ Make __________ Model __________ I.D. No. __________

Paint Code _____ Prod. Date _____ Trim _____ Mileage _____ License No. _____ Date of Loss _____

Written By _____ Ins. Co. _____ File No. _____ Claim No. _____ P.O. No. _____

Adjuster __________ Lic. No __________ Phone __________ Deductible/Betterment __________

Line No.	Re-pair	Re-place	Details of Repair N = New U = Used R = Repair S = Straighten RC = Recycle/Rechrome/Recore	Parts	Labor	Paint	Sublet/ Misc
1							
2							
3							
4							
5							
25							
26							
27							
28							

I hearby authorize the above work and acknowledge receipt of copy. **Totals**

Signed X __________ Date __________

This estimate is based on our inspection and does not cover additional parts or labor which may be required after the work has been started. After the work has started, worn or damaged parts which are not evident on first inspection may be discovered. Naturally this estimate cannot cover such contingencies. Parts prices subject to change without notice. This estimate is for immediate acceptance.

ESTIMATE GOOD FOR 30 DAYS.

Parts *Prices subject to invoice*	$ ______
Labor _____ hrs. @ _____	$ ______
Shop Supplies	$ ______
Paint _____ hrs. @ _____	$ ______
Paint Supplies	$ ______
Towing/Storage	$ ______
Sublet/Miscellaneous	$ ______
EPA/Waste Disposal Charge	$ ______
______	$ ______
Subject	$ ______
______	$ ______
Tax	$ ______
Total	$ ______

FIGURE 11–6 Typical collision repair facility damage report form.

DAMAGE APPRAISAL

Name Insured		Claim No.	Date of Loss
Registered Owner	Date Assigned	Date Inspected	Inspected At
Address		Phone Number	

Year	Make	Model	Body Style	Vehicle ID No.	License No.	Mileage	Color

Engine: ____ # Cyl. ☐ Auto Trans. ☐ P/S ☐ P/B ☐ A/C ☐ T/Glass ☐ Vinyl Roof Condition
☐ Radio ☐ AM/FM ☐ Stereo ☐ Tape Deck ☐ Other: ☐ G ☐ F ☐ P

Repair	Replace	Details of Repair and/or Replacement	Labor Hours	Parts & Material	Sublet

Deductions:

Estimate Total* . . . . $ ______

Minus Applicable Deductible. . . . . $– ______

Minus Old Damage (See Itemization Att.) . . . $– ______

Minus Betterment (See Itemization Att.) . . . . $– ______

Net Total $ ______

This Is Not an Authorized Repair Order
Deductibles, betterment and previous damage, if included in the estimate, will be deducted from any settlement agreement with the owner. Nothing is to be charged to the Insurance Company.

The Undersigned agrees to complete repairs as:
☐ Itemized ☐ Contracted for the total price of $

Repair Shop	Phone No.
Address	
Shop Representative	Registration No.

Totals

Labor Hours	@		$
Parts	Less	%Disc	$
Tax	%		$
Sublet			$

***Note:** Estimate Total does not include Betterment or Old Damage

*Estimate Total $

NOTICE: Choose your repair shop carefully. New auto designs, such as unibody construction, and materials may require special welding and measuring equipment to safely repair.

Settlement communicated to vehicle owner:
☐ Yes ☐ No

Was Draft Issued? ☐ Yes ☐ No Draft No.

Claim Representative Signature

Claim Representative License No.

FIGURE 11–7 Typical insurance company appraisal form.

FIGURE 11–8 *VIN location on upper left corner of the instrument panel, visible through the windshield.*

sents the plant of manufacture, while the last six characters represent the sequential production number.

The collision estimating guides and manuals list all applicable VIN information for the vehicles covered in those guides. They will also list the location and codes for vehicle identification numbers prior to the standardization in 1981.

Ford Motor Co. Trucks & Vans

1 F D E E 1 4 H 4 M H A 1 6 6 6 8

(1)(2)(3)(4)(5)(6)(7)(8)(9)(10)(11)(12)(13)(14)(15)(16)(17)

1 Manufacturer
1 • U.S.A.
2 • Canada
4 • U.S.A.
J • Japan

2 Make
C • Courier
F • Ford

3 Type
B • Bus
C • Basic Chassis
D • Incomplete Vehicle
M • Multi-Purpose Vehicle
T • Truck

4 Brake System/Gross Vehicle Weight
A • Hydraulic Brakes/Under 3000 lbs.
B • Hydraulic Brakes/3001-4000 lbs.
C • Hydraulic Brakes/4001-5000 lbs.
D • Hydraulic Brakes/5001-6000 lbs.
E • Hydraulic Brakes/6001-7000 lbs.
F • Hydraulic Brakes/7001-8000 lbs.
G • Hydraulic Brakes/8001-8500 lbs.
H • Hydraulic Brakes/8501-9000 lbs.
J • Hydraulic Brakes/9001-10,000 lbs.
K • Hydraulic Brakes/10,000-14,000 lbs.
L • Hydraulic Brakes/14,001-16,000 lbs.
M • Hydraulic Brakes/16,001-19,500 lbs.

5 Model or Line
A • Aerostar
A • Windstar
E • Econoline Van
F • Truck Standard Cab
R • Ranger
S • Super Van
U • Bronco II
U • Bronco
U • Explorer
V • Villager
W • Truck Crew Cab
X • Truck Super Cab

6–7 Chassis & Body Type
Aerostar
11 • Wagon (Standard)
14 • Cargo Van (Standard)
15 • Window Van (Standard)
21 • Wagon 4WD (Standard)
24 • Cargo Van 4WD (Standard)
25 • Window Van 4WD (Standard)
31 • Wagon (Extended)
34 • Cargo Van (Extended)
35 • Window Van (Extended)
41 • Wagon 4WD (Extended)
44 • Cargo Van 4WD (Extended)
45 • Window Van 4WD (Extended)
Windstar
51 • Wagon
54 • Van
Econoline Van
01 • 1/2 Ton Club Wagon
04 • 1/2 Ton Cargo Van
05 • 1/2 Ton Window Van
06 • 1/2 Ton Display Van
11 • 1/2 Ton Club Wagon
14 • 1/2 Ton Cargo Van
15 • 1/2 Ton Window Van
16 • 1/2 Ton Display Van
21 • 3/4 Ton Club Wagon
24 • 3/4 Ton Cargo Van
25 • 3/4 Ton Window Van
26 • 3/4 Ton Display Van

6–7 Chassis & Body Type (Cont)
Econoline Van
31 • 1 Ton Club Wagon
34 • 1 Ton Cargo Van
35 • 1 Ton Window Van
36 • 1 Ton Display Van
Truck Standard Cab
07 • 1/2 Ton Pickup 2WD Flareside (97)
08 • 1/2 Ton Pickup 4WD Flareside (97)
17 • 1/2 Ton Pickup 2WD Styleside (97)
18 • 1/2 Ton Pickup 4WD Styleside (97)
10 • 1/2 Ton Pickup
14 • 1/2 Ton Pickup 4WD
15 • 1/2 Ton Pickup
17 • 1/2 Ton Cab & Chassis
18 • 1/2 Ton Cab & Chassis 4WD
25 • 3/4 Ton Pickup
26 • 3/4 Ton Pickup 4WD
27 • 3/4 Ton Cab & Chassis
28 • 3/4 Ton Cab & Chassis 4WD
35 • 1 Ton Pickup
36 • 1 Ton Pickup 4WD
37 • 1 Ton Cab & Chassis
38 • 1 Ton Cab & Chassis 4WD
47 •1 Ton Cab & Chassis Super Duty
Ranger
10 • Standard Cab
11 • Standard Cab 4WD
14 • Super Cab
15 • Super Cab 4WD
Super Van
11 • 1/2 Ton Club Wagon
14 • 1/2 Ton Cargo Van
15 • 1/2 Ton Window Van
16 • 1/2 Ton Display Van
21 • 3/4 Ton Club Wagon
24 • 3/4 Ton Cargo Van
25 • 3/4 Ton Window Van
26 • 3/4 Ton Display Van
31 • 1 Ton Club Wagon
34 • 1 Ton Cargo Van
35 • 1 Ton Window Van
36 • 1 Ton Display Van
Bronco II
12 • Bronco II
14 • Bronco II 4WD
Bronco
15 • Bronco 4WD
Explorer
22 • 2-Door
24 • 2-Door 4WD
32 • 4-Door
34 • 4-Door 4WD
35 • 4-Door Full Time 4WD
Mountaineer
52 • 4-Door
54 • 4-Door 4WD
55 • 4-Door Full Time 4WD
Truck Crew Cab
25 • 3/4 Ton Pickup
35 • 1 Ton Pickup
36 • 1 Ton Pickup 4WD
Truck Super Cab
07 • 1/2 Ton Pickup 2WD Flareside (97)
08 • 1/2 Ton Pickup 4WD Flareside (97)
17 • 1/2 Ton Pickup 2WD Styleside (97)
18 • 1/2 Ton Pickup 4WD Styleside (97)
14 • 1/2 Ton Pickup 4WD
15 • 1/2 Ton Pickup
25 • 3/4 Ton Pickup
26 • 3/4 Ton Pickup 4WD
35 • 1 Ton Pickup
37 • 1 Ton Pickup

6–7 Chassis & Body Type (Cont)
Villager
11 • Wagon
14 • Cargo Van

8 Engine Type
A • 2.3L 4-Cylinder Gas
C • 2.0L 4-Cylinder Gas
C • 7.3L V8 Diesel Turbo
D • 4.2L V8 Gas
E • 2.3L 4-Cylinder Diesel Turbo
E • 4.9L 6-Cylinder Gas
F • 5.0L V8 Gas
F • 7.3L V8 Diesel
G • 5.8L V8 Gas 80-85
G • 7.5L V8 Gas 87-95
H • 5.8L V8 Gas
K • 7.3L V8 Diesel
L • 7.5L V8 Gas
M • 7.3L V8 Diesel
N • 5.0L V8 Gas
P • 2.2L 4-Cylinder Diesel
R • 5.8L V8 Gas
R • 5.8L V8 Gas
S • 2.8L V6 Gas
T • 2.9L V6 Gas
U • 3.0L V6 Gas
W • 3.0L V6 Gas
W • 5.8L V8 Gas
X • 4.0L V6 Gas
Y • 4.9L 6-Cylinder Gas
Z • 2.3L 4-Cylinder Gas
Z • 4.9L V6 Gas
Z • 6.6L V8 Gas
1 • 6.9L V8 Diesel
3 • 3.8L V6 Gas
4 • 3.8L V6 Gas
7 • 5.8L V8 Gas
9 • 4.9L 6-Cylinder Gas

9 VIN Check Digit
Manufacturer's Internal Code

10 Vehicle Model Year
A • 1980
B • 1981
C • 1982
D • 1983
E • 1984
F • 1985
G • 1986
H • 1987
J • 1988
K • 1989
L • 1990
M • 1991
N • 1992
P • 1993
R • 1994
S • 1995
T • 1996
V • 1997

11 Assembly Plant Code
B • Oakville, Ontario
C • Ontario Truck
D • Ohio Truck
E • Kentucky
H • Lorain
I • Highland Park
J • Monterey, Mexico
K • Kansas City
L • Michigan Truck
M • Cuautitlan (Mexico)
N • Norfolk
P • Twin Cities
S • Pilot Plant
T • Edison
U • Louisville
Z • St. Louis

FIGURE 11–9 *Typical VIN decoding information as found in collision estimating guides. (Courtesy of Mitchell International.)*

TABLE 11–1: VIN MODEL YEAR CODES

L	1990	X	1999
M	1991	Y	2000
N	1992	I	2001
P	1993	2	2002
R	1994	3	2003
S	1995	4	2004
T	1996	5	2005
V	1997	6	2006
W	1998	7	2007

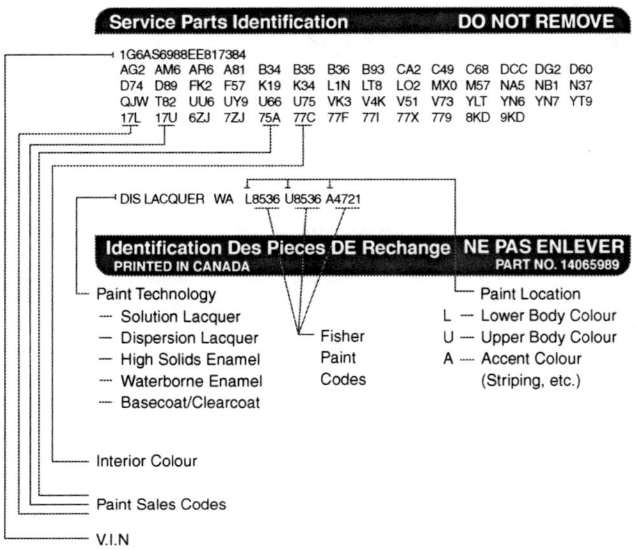

FIGURE 11–10 Typical General Motors service parts identification label. (Courtesy of Du Pont Co.)

IDENTIFYING PAINT SYSTEMS AND CODES

It is important to determine the type of paint system used on the vehicle and the exact vehicle paint code. Note the use of metallic, two-stage (basecoat/clear coat) or three-stage (basecoat/mica midcoat/clear coat) finishes. A quick test to determine if the finish uses a clear coat is to lightly sand the damaged painted surface. If the dust is whitish, the finish is clear coat. If the sandpaper shows color, the finish is single stage. In spot or panel repairs, the paint code is used to ensure that the color used in the repair will accurately match the original color of the vehicle.

The paint code(s) are listed on the vehicle's service identification label (Figure 11–10). The location of the service identification label is not standardized throughout the industry. Figure 11–11 illustrates common locations used by vehicle manufacturers as keyed to Table 11–2. The collision estimating guides list paint code location and which colors are clear coat (Figure 11–12).

LICENSE AND MILEAGE

Record the vehicle license number. Also record the odometer or mileage reading. The used car guide book values are based on vehicles with average mileage. As mentioned earlier, extremely high or low mileage will decrease or increase a vehicle's ACV. Figure 11–13 illustrates an example of one type of high mileage table. Similar tables are included for low mileage appraisals.

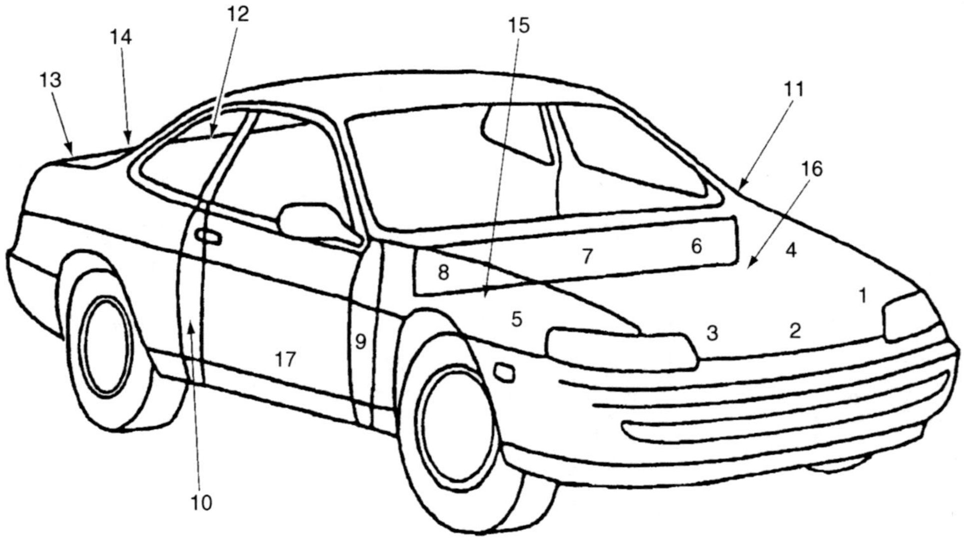

FIGURE 11–11 Common locations of paint codes on various vehicles.

TABLE 11–2: PAINT CODE LOCATION

Model	Position	Model	Position
Acura	9	Honda	8, 10
Alfa Romeo	4, 13	Hyundai	6, 7
AMC	9, 10	Isuzu	2, 10
Audi	12, 13	Lexus	7, 8
Austin Rover	17	Mazda	1, 2, 3, 4, 6, 8
BMW	4, 5	Mercedes	2, 7, 9
Chrysler	3, 5, 16	Mitsubishi	7
Chrysler Corp.	3, 5	Montero/Pickup	3
Caravan/Voyager/ Ram Van	6	Cordia/Tredia	4
Chrysler Imports	1, 2, 4	Others	1, 2, 3
Colt Vista	16	Nissan	1, 3, 4, 6, 8, 15,*
Conquest	7	Peugeot	2, 3, 4, 5, 8
Diahatsu	1, 6, 7	Porsche	9
Datsun	2	Renault	1, 3, 4, 5, 8
Dodge D50	3	Rover	1, 3, 4, 5
Ford	10	Saab	5, 6, 8
Ford Motor Co.	10	Subaru	2
General Motors		Suzuki	7, 11
A, J and L Bodies	14	Toyota Passenger	7, 8, 14
E and K Bodies	12	Truck	4
B,C,H and N Bodies	13	Volkswagen	2, 11
GM Imports	2, 12, 13, 14	Volvo	6, 7, 8
		Yugo	12

PAINT CODE LOCATION

Paint code located on rear face of left front door or on left pillar.

CLEAR COAT IDENTIFICATION

1993-94
All colors except YC Black, YO White, E7 Dark Ruby, DH Mocha, and EY Performance Red are clear coat.

1995-97
All colors are clear coat.

FIGURE 11–12 Paint code location and clear coat information from an estimating guide.

HIGH MILEAGE TABLE

VALUES SHOWN BELOW TO BE DEDUCTED FROM BASE GUIDEBOOK VALUES

See beginning of each manufacturer for mileage classifications.

Estimated values due to High and Low Mileage not applicable to Kentucky Usage Tax Valuations

MILEAGE	CLASS	1997	1996	1995	1994	1993	1992	1991
55001 to 60000	I	1150	725	375				
	II	1525	975	500				
	III	2275	1450	750				
	IV	2650	1700	875				
60001 to 65000	I	1275	875	525	150			
	II	1700	1150	700	200			
	III	2550	1725	1025	300			
	IV	2950	2025	1200	350			
65001 to 70000	I	1400	1000	650	300			
	II	1875	1350	875	400			
	III	2800	2000	1300	600			
	IV	3250	2350	1525	700			
70001 to 75000	I	1525	1150	800	450	150		
	II	2050	1525	1050	600	200		
	III	3050	2275	1575	900	300		
	IV	3550	2650	1850	1050	350		
75001 to 80000	I	1650	1275	925	600	300	125	
	II	2200	1700	1250	800	400	175	
	III	3300	2550	1850	1175	600	275	
	IV	3850	2950	2175	1375	700	325	
80001 to 90000	I	1850	1475	1125	800	525	350	100
	II	2450	1950	1500	1075	700	475	125
	III	3675	2925	2275	1600	1025	725	250
	IV	4275	3400	2650	1850	1200	850	325
90001 to 100000	I	2075	1725	1400	1075	800	650	400
	II	2775	2300	1850	1425	1075	850	550
	III	4150	3425	2800	2150	1600	1300	825
	IV	4825	4000	3250	2500	1850	1500	1000
100001 to 115000	I	2375	2025	1700	1400	1150	975	775
	II	3150	2700	2275	1875	1525	1325	1000
	III	4725	4025	3425	2800	2275	1975	1525
	IV	5500	4700	4000	3250	2650	2325	1825
115001 to 130000	I	2700	2375	2075	1775	1525	1375	1175
	II	3575	3150	2775	2375	2050	1850	1550
	III	5375	4725	4150	3550	3050	2775	2350
	IV	6250	5500	4825	4150	3550	3250	2750
130001 to 150000+	I	3050	2750	2450	2200	1950	1825	1625
	II	4050	3650	3300	2925	2600	2425	2175
	III	6075	5475	4925	4375	3900	3650	3225
	IV	7075	6375	5750	5100	4550	4250	3800

NOTE: DEDUCTION FOR HIGH MILEAGE SHOULD NOT EXCEED 40% OF TRADE-IN VALUE

FEBRUARY 1998 **V**

FIGURE 11–13 Typical high mileage table. Values listed are deducted from the ACV of the vehicle. (Used with permission of National Automobile Dealers Used Car Guide Co.)

IMPORTANT VEHICLE OPTIONS

Certain vehicle options will increase the ACV of the vehicle and should be noted on the damage report or appraisal form. These options might include:

- Specific engine sizes or options
- Sunroofs
- AM/FM stereo, tape player, compact disc player (OEM only)
- Power door locks, windows, and seats
- Cruise control, tilt steering wheel
- Leather seats, special wheel covers, luggage racks (OEM), and special accessory packages

In addition to listing the dollar value of ACV-raising options, the used car guide books also list dollar values for ACV deductions. Features such as diesel engines, manual transmissions, or the absence of air conditioning, option packages, or power steering may require a lowering of the ACV.

PRIOR DAMAGE

The vehicle must be thoroughly inspected inside and outside for all prior damage not related to this claim. Examples of such damage are:

- Old scratches and dents (Figure 11–14).
- Rust, corrosion, or chipped or cracked paint finishes
- Cracked or dented plastic or rubber parts on bumpers, valances, shields, etc.
- Ripped or split seams on seats or upholstery
- Stains or damage to seats, carpets, and interior surfaces
- Broken or cracked glass or mirrors
- Damaged or missing wheel covers or moulding strips
- Cracked light lenses or broken or burned-out bulbs
- Broken optional equipment, such as air-conditioning and heating units, rear defrosters, etc.

Determining previous damage, recording it on the damage report or appraisal, and informing the customer of this damage is extremely important. It eliminates the possibility that the customer might

1. Accuse the collision repair facility of making an incomplete repair
2. Accuse the collision repair facility of damaging their vehicle during the time it is in the shop.

Adjusters must be very thorough in appraising previous damage not caused by this claim. Insurance companies obviously do not want to pay for prior damage repairs. Their responsibility is to pay for damages caused by this accident or loss.

FIGURE 11–14 On an insurance claim settlement, old scratches and dents that were not caused by this accident or loss must be identified and carefully noted.

Some insurance companies alter their payment method if their field adjusters spot significant damage from a previous accident that has not been repaired. The insurance company sees this as a signal that the vehicle owner might settle the claim, but not have the repair made if the vehicle is still driveable. In such cases, the insurance company might make only a partial payment up-front and hold full payment until after the customer proves the repair has been made. Usually a visual inspection by the adjuster is needed.

Determining whether vehicle damage was caused by past or recent accidents is not always easy, particularly if the damage is in the same general area of the vehicle. Rust and dirt on surfaces that have been scratched or gouged to the bare metal are signals of old damage. If plastic or rubber pieces have a weathered or dull appearance on surfaces that have only "recently" been exposed to the weather, the damage is probably old.

Be suspicious of two different areas of damage separated by a long stretch of undamaged area: for example, a ding or scratch on the fender, no damage on the door, and a scratch or damage on the rear quarter panel.

Slight depressions or craters in the hood or trunk panels can be caused by overzealous closing. Do the depressions match the typical hand positions for closing the hood or trunk?

An inoperative air conditioner or heater should show signs of recent physical damage. On older vehicles these items might have broken or worn out long before the accident.

Investigate the details of the accident. How did it happen? At what speed was the vehicle traveling? What was the angle of impact? Did the vehicle strike one object, then slide or bounce into another?

Once the estimator knows these facts, he or she can begin to make logical conclusions about the pattern of damage. It also becomes easier to distinguish between old damage and damage caused by this accident or loss. Do the facts support the likelihood that specific damage was caused by this accident? Was the impact strong enough to crack the taillight housing, side view mirror, or window glass on the opposite side of the vehicle? Or is it previous damage that the vehicle owner chose to live with rather than have repaired? Does the owner's statement that the missing wheel cover was knocked free and lost make sense? Is there damage to the wheel? Are all other wheel covers present? Do you know if this model vehicle has a reputation for losing wheel covers during normal operation?

For this reason, insurance companies have found that former police officers and officials make excellent adjusters due to their investigative experience and ability to draw sound conclusions from an often odd assortment of facts.

Experience and a good memory are the two best tools any damage report writer or adjuster can have. A full-time adjuster or damage report writer might write up 1,000 to 2,000 vehicles per year. This gives the writer a large bank of data to draw on. Does this vehicle normally sustain frame damage in front-end hits of this magnitude? If the writer knows that in four out of five similar cases she has seen in the past year, frame damage was present on this model vehicle, she will undoubtedly check the side rails and cross members very carefully.

DEPRECIATION AND BETTERMENT

For insurance companies, identifying previous damage does more than define what is to be repaired under this claim. It also serves as a guideline for determining depreciation values on damaged parts.

If body panels involved in this settlement have previous damage, such as rust, dents, or buckles in other areas, or missing mouldings or decorative trim, the cost of repairing this previous damage is deducted from the cost of the new replacement panel (Figure 11–15).

The damage report writer or adjuster must accurately determine the cost of repairing this previous damage using the methods outlined in Chapters 5 and 6. For example, a new trunk lid lists for $300. The damaged lid being replaced has previous rust damage. The adjuster determines two hours of shop time at $25 per hour would be needed to repair the rust damage. The insurance company would then pay $250 toward the purchase of a new trunk lid. The customer would pay $50 for the betterment of the vehicle.

As described in Chapter 6, mechanical components, tires, batteries, interior upholstery, and many other items are depreciated according to their age, mileage, or a combination of both. During the inspection, the adjuster or report writer must note all components that must be depreciated.

11.3 WRITING UP THE REPORT

Once the preliminary information and inspection have been completed, the collision repair facility writer or insurance adjuster can begin listing individual part and labor entries on the report from a set of abbreviated notes taken during the inspection. Most experienced writers try to write as complete a report as possible the first time pencil is put to paper.

On small to moderate write-ups, working from notes can be a waste of time. It is just as easy to be complete and precise the first time through. On larger estimates, working from notes might be an advantage in organizing and double-checking all operations. Using a microcassette recorder to dictate detailed damage descriptions and parts lists might also be helpful for larger write-ups (Figure 11–16). The information can then be transcribed and organized on the report form or entered into a computerized estimating system.

When inspecting slight damage on vehicles they are familiar with, experienced writers might not require the collision guides at their sides. They can accurately list the part names in a logical order from memory.

FIGURE 11–15 If a damaged panel has unrelated rust damage, the insurance appraiser must calculate betterment if replacing it with a new panel.

FIGURE 11–16 Dictating damage information into a microcassette recorder is an excellent method of documenting vehicle damage.

Once the inspection is complete, they then return to the office, sit down with the guides, and fill in the needed part numbers and labor and paint times.

When manually writing moderate to involved damage reports, it is always best to have the appropriate, up-to-date collision guide at your side. Open the guide to the appropriate listing and refer to it if there is any question as to part names or configurations. Having the guide at the inspection site also allows labor times and paint times to be listed as each entry is made.

Be Fully Prepared

A complete list of tools needed to perform a damage report inspection appears in Chapter 1. Gather all tools, forms, collision guides, and measurement specification guides before beginning the write-up. Having all needed tools and materials on hand will reduce the likelihood of interruptions. Stopping work in the middle of a write-up can lead to mix-ups, omissions, and repeated entries.

Follow a Logical Sequence

The importance of following a logical sequence when assessing damage or labor needs has been constantly stressed throughout this book. The damage report form is where this attention to detail and logic are seen in written form.

Always remember the zone method of inspection covered in detail in Chapter 5. A quick summary and review of the five basic zones are given below.

Zone 1: The immediate impact area containing the primary damage
Zone 2: Remaining body damage defined as secondary damage
Zone 3: Mechanical damage, including any damage to drivetrain and accessories
Zone 4: The passenger compartment, including all damage to the compartment, trim, lights, accessories, controls, and finish
Zone 5: The exterior components and trim

Begin the inspection at the immediate point of impact and work inward through the damaged area, listing all affected parts. Follow the same path the impact force followed as it moved through the vehicle. In front-end impacts, the most common type of collision, the inspection procedure involves starting at the front end of the vehicle and working back to the rear. Mentally divide the vehicle into major subassemblies or groups of related parts, and list the damaged parts within each group, working from the outside in.

For example, consider a vehicle that has suffered damage to the front left corner. Typical subassemblies damaged in such a collision are (in order from front to back): the bumper, grille, left headlight, radiator, hood, and fender. Starting at the front of the vehicle, inspect each group for damaged components, listing the parts in order from outside to inside. For example, the damaged fender might need the following parts replaced: the fender, strut mount, strut tower, rear extension, splash shield, and skirt. Attachments to the fender, such as the body mouldings and lights, might also need to be replaced.

To make this front-to-back and outside-to-inside process easier, the collision estimating guides follow this sequence when listing parts and operation times. Both major manuals list the front bumper as the first major subassembly of parts for each vehicle listed in the manual. Both manuals then work toward the rear of the vehicle, defining major subassemblies and part groups until they reach the rear bumper assembly. There are slight differences in parts breakdown and exact group order, but both follow a logical front-to-back, outside-to-inside breakdown of the vehicle.

Figure 11–17 is an exploded view of a front bumper subassembly as illustrated in a typical estimating guide. It clearly demonstrates how guides list parts from the outside of the assembly to the inside. By following this sequence, collision repair facility personnel and adjusters will write more accurate reports and find the guides much easier to use when looking up parts prices, numbers, and operation times.

Making Entries

Each entry line on the damage report form should list an individual action or operation. Do not lump several operations under a general entry such as "Fix front end damage," or simply list damaged parts without explanation.

Use Action Words. Begin each entry with an action word such as *Repair, Replace, Straighten, Refinish, Overhaul,* etc. These words explain the action to be taken to anyone—customer, adjuster, shop worker—who reads the report.

Avoid Abbreviations. Avoid excessive use of abbreviations unless the abbreviations are defined by a key on the estimate. When a quarter panel must be straightened, do not write "str. qtr. pnl." Write "straighten quarter panel." And be as specific as possible. Distinguish between left and right side parts. Write "straighten *left* quarter panel" if such is the case. Using repeat signs might save a minute or two when writing up the report, but overuse can lead to a sheet that is confusing and difficult to read. Spell out each entry.

Write Neatly. Customers and adjusters will be impressed by clean, neat sheets. They are much easier

FRONT BUMPER

DR2MX87

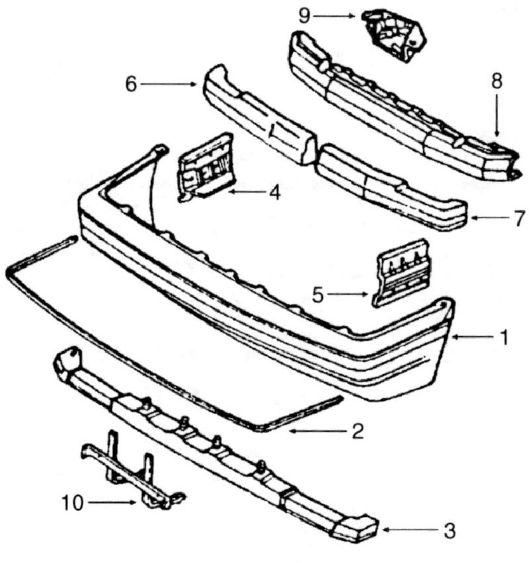

FRONT BUMPER
REFINISHING NOTES

Deduct for overlap when necessary. See GUIDE to ESTIMATING pages.

Add for edging				
cover			**0.5**	
Add for clear coat if required				
cover			**1.3**	

PARTS & LABOR

Item	Side	Price		Time	Part No.
Assembly R&I				**0.8**	
Overhaul			#	**2.0**	
1 **Cover** (p2.5)		335.57	#	**2.0**	E7GZ 17D957 C
2 moulding		44.57		**0.3**	E7GZ 17C829 A
Reinforcement					
3 center		129.05			E7GZ 17A792 A
4 side	R	2.85			E7GZ 17C861 A
5	L	13.67			E7GZ 17C861 B
6 **Absorber, side**	R	81.35			E7GZ 17C947 A
7	L	81.35			E7GZ 17C947 B
8 **Bar**		308.95	§	**2.0**	E7GZ 17757 A
9 bracket					
hatchback	R	27.55			E7GZ 17C886 A
	L	26.77			E7GZ 17C887 A
wagon	R	22.33			E8GY 17C897 A
	L	22.33			E8GY 17C897 B
10 **License bracket**		6.83		**0.3**	E7GZ 17A385 A

#Time is for overhaul, after grille and headlamps are removed.
§Time is for overhaul.

FIGURE 11–17 Exploded view of the front bumper subassembly. (Courtesy of Motor Information Systems.)

to understand and work from. A clean, accurate report with no scratched-out entries, smudges, or unreadable words or figures signals to everyone that this shop or adjuster is a professional.

But sloppy or illegible handwriting does more than reflect poorly on the professionalism of the writer (Figure 11–18). Customers will likely take their business to the shop whose damage reports they can easily read and understand. Difficult-to-read numbers can lead to pricing mistakes.

Major problems occur when the part price of $92.50 is read and calculated at $42.50. The shop is out $50.00 on the cost of the part. Such major mistakes are not always obvious. It is surprising how many people cannot read their own handwriting or forget what they have written down several minutes later.

Also, do not count on the customer, adjuster, or others involved in the repair process to catch and point out mistakes and inaccuracies. That is not their job. It is the responsibility of the person filling out the sheet.

Record All Needed Data. After each concise entry description, record the price of any needed part and the labor involved in the operation, if any. Figure 11–19 illustrates several typical entries on a moderate front-end collision. The part names listed are those given in the estimating manual. Because part numbers are not normally listed on a handwritten damage report, using the correct name for the part will make it easier to look up these numbers when it is time to order parts. It also prevents confusion when another party reviews the damage report or uses it to generate work orders, part orders, or part invoices.

Specify the use of parts other than OEM. If the part is a salvage part, note this on the entry line. It is also wise to list the salvage yard name and telephone number. Include any repairs to the salvage part.

Note any aftermarket panels or parts that are to be used. Remember, it is now law in some states that disclosure of aftermarket sheet metal parts be made on the damage report.

Operation and Paint Times. Accurately record all labor time needed for each entry. Remember that only specific operations are included in the operation times listed in the estimating manuals.

Steps and activities needed for the complete repair of the vehicle that are not standardized and listed in the manuals must be recorded as separate entries. Such jobs include removing broken glass, installing weld-on hinges, aiming headlights, and many others. Refer to the collision estimating guide being used for a complete listing and work accordingly.

Also check the footnotes that precede the operation times in the estimating guides. The footnotes might also define additional operations that are not included in the operation times. For example, the # beside the operation time for head lamp mounting panel in Figure 11–20 will direct you to a footnote that says "with necessary bolted parts and assemblies removed." The footnote in Figure 11–21 explains that the R&R times do not include alignment or brake system bleeding.

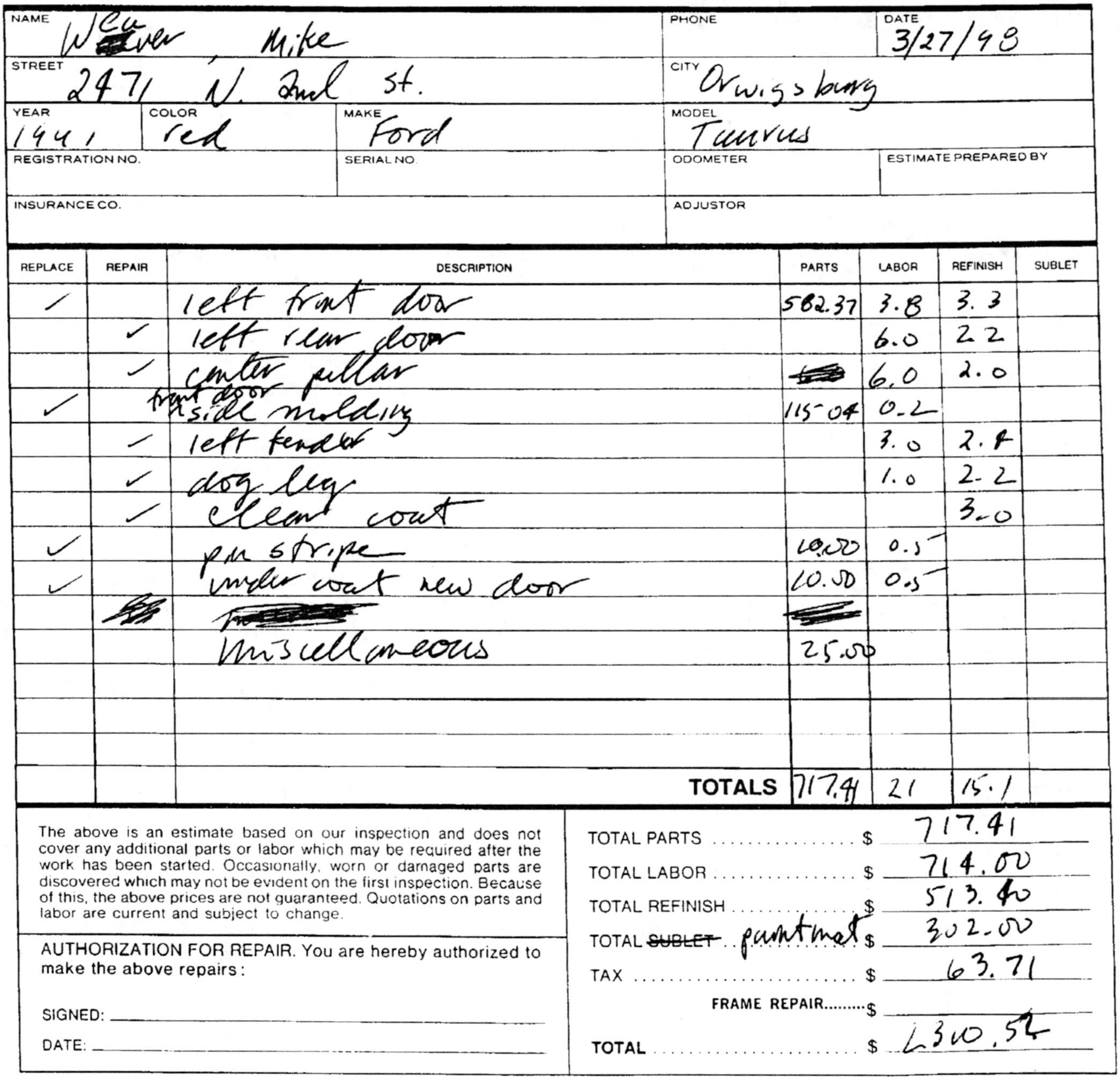

NAME Weaver, Mike	PHONE	DATE 3/27/98
STREET 2471 N. 2nd St.	CITY Orwigsburg	
YEAR 1991 / COLOR red / MAKE Ford	MODEL Taurus	
REGISTRATION NO. / SERIAL NO.	ODOMETER	ESTIMATE PREPARED BY
INSURANCE CO.	ADJUSTOR	

REPLACE	REPAIR	DESCRIPTION	PARTS	LABOR	REFINISH	SUBLET
✓		left front door	582.37	3.8	3.3	
	✓	left rear door		6.0	2.2	
	✓	center pillar	[illegible]	6.0	2.0	
✓		front door side molding	115.04	0.2		
	✓	left fender		3.0	2.4	
	✓	dog leg		1.0	2.2	
	✓	clear coat			3.0	
✓		pin stripe	10.00	0.5		
✓		under coat new door	10.00	0.5		
[illegible]		[illegible]	[illegible]			
		miscellaneous	25.00			
		TOTALS	717.41	21	15.1	

The above is an estimate based on our inspection and does not cover any additional parts or labor which may be required after the work has been started. Occasionally, worn or damaged parts are discovered which may not be evident on the first inspection. Because of this, the above prices are not guaranteed. Quotations on parts and labor are current and subject to change.

AUTHORIZATION FOR REPAIR. You are hereby authorized to make the above repairs:

SIGNED: ____________

DATE: ____________

TOTAL PARTS	$ 717.41
TOTAL LABOR	$ 714.00
TOTAL REFINISH	$ 513.40
TOTAL ~~SUBLET~~ paint mat	$ 302.00
TAX	$ 63.71
FRAME REPAIR	$
TOTAL	$ 2310.52

FIGURE 11–18 *Sloppy, unreadable reports lose jobs and cause errors.*

Page of Pages

estimate

SMASH PALACE
13105 N. Jackson Rd.
BRIMFIELD, IL 61517
(309) 446-3162
Rebuilder License #UDL 10439
Repairer License #UDL 10455

NAME			PHONE	DATE
Kyle Bauman			348-1085	3/29/98
STREET			CITY	
506 E. Union St.			Delano	
YEAR	COLOR	MAKE	MODEL	
1997	code FU	Ford	Mustang	
REGISTRATION NO.		SERIAL NO.	ODOMETER	ESTIMATE PREPARED BY
228-346		1FASP12T3VF675455	15828	MSR
INSURANCE CO.			ADJUSTOR	
Ideal			B. Defftey	

REPLACE	REPAIR	DESCRIPTION	PARTS	LABOR	REFINISH	SUBLET
✓		left headlamp assbly F4ZZ13008F	161.60	0.3		
		aim headlamps		0.4		
✓		left park lamp F4ZZ13201A	50.01	inc		
✓		left fender F6ZZ16006AA	234.38	1.8	2.5	
✓		left splash shield F4ZZ16103B	28.43	0.4		
	✓	left door		4.0	2.2	
✓		wheel, cast aluminum used	150.00	0.3		
		Robertson 697-5112				
		front end alignment		1.8m		
		blend paint on bumper, hood and left			3.7	
		quarter panel				
		R+I moldings on fender and door		0.7		
	✓	Clear coat			1.4	
		"This business is required to be licensed by the Secretary of State, pursuant to Illinois Revised Statutes, Chapter 95½ Sec. 5-301 ANY complaints as to the quality of service obtained here may be brought to the attention of the Attorney General"				
		TOTALS	624.42	7.9	9.8	

The above is an estimate based on our inspection and does not cover any additional parts or labor which may be required after the work has been started. Occasionally, worn or damaged parts are discovered which may not be evident on the first inspection. Because of this, the above prices are not guaranteed. Quotations on parts and labor are current and subject to change.

AUTHORIZATION FOR REPAIR. You are hereby authorized to make the above repairs:

SIGNED: ______

DATE: ______

TOTAL PARTS	$ 624.42
TOTAL LABOR	$ 268.60
TOTAL REFINISH	$ 333.20
TOTAL ~~SUBLET~~ paint mat	$ 196.00
TAX	$ 51.27
~~FRAME~~ m REPAIR	$ 81.00
TOTAL	$ 1554.49

FIGURE 11–19 Example of complete, detailed entries on a handwritten damage report. Salvage and aftermarket parts must be noted on the report. Include the salvage yard sources.

FRONT INNER STRUCTURE

Use Procedure Explanations 8 and 28 with the following text.

FRONT

Operation		Time
Refinish Radiator Support Complete		1.5
Refinish Upper Tie Bar		.5
Refinish Lower Tie Bar		.5
Refinish Side Support Panel	R/L	.5
R&R Radiator Support Complete		#7.7

#Includes R&R Radiator Fan Assy, Horns, Hood Latch, Prop Rod, Coolant Reservoir, W/S Washer Assy, R&I/R&R H/Lamp Assemblies, Air Cleaner Assy & Disconnect & Connect Hood Release Cable, Front Eng Mount, Wiring and/or Harness

Add to R&R Radiator Support Complete

Operation		Time
w/ A/C	m	#.4

#Includes R&I/R&R Condenser & Receiver/Drier

Operation		Time
w/Trans Cooler		#.1

#Includes R&R Trans Oil Cooler

Operation		Time
To Aim Lamps		.4
Evacuate & Recharge System	m	1.4
Refrigerant Recovery	m	.3

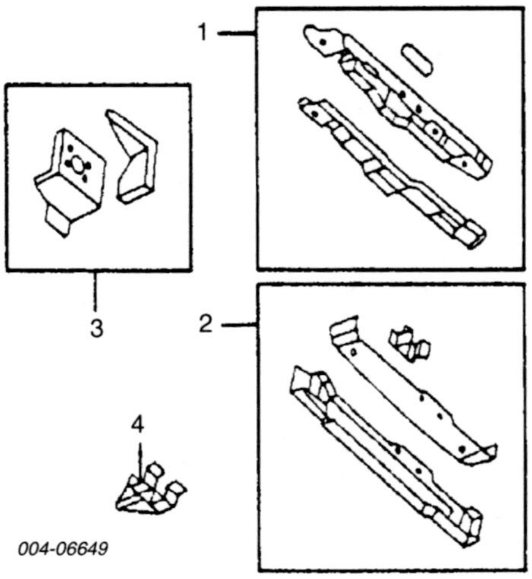

Part		Part No.		Time	Price
1 **Tie Bar, Upper**		4860194	s	#2.0	105.00
2 **Tie Bar, Lower**		4716500	s	#3.0	94.50
Seal, Tie Bar/Radiator		4677365			14.50
3 **Panel, H/Lamp Mtg**	R	4798352	s	#2.5	130.00
	L	4798353	s	#2.5	133.00

#w/Necessary Bolted Parts & Assemblies Removed

Part		Part No.		Time	Price
4 **Hook, Tiedown**		4716833			12.25

SIDE

Operation		Time
Refinish Apron/Side Rail	R/L	2.0
Refinish Front Apron	R/L	1.0
Refinish Apron Reinf	R/L	.5
Refinish Strut Tower	R/L	.5
Refinish Side Rail	R/L	1.0

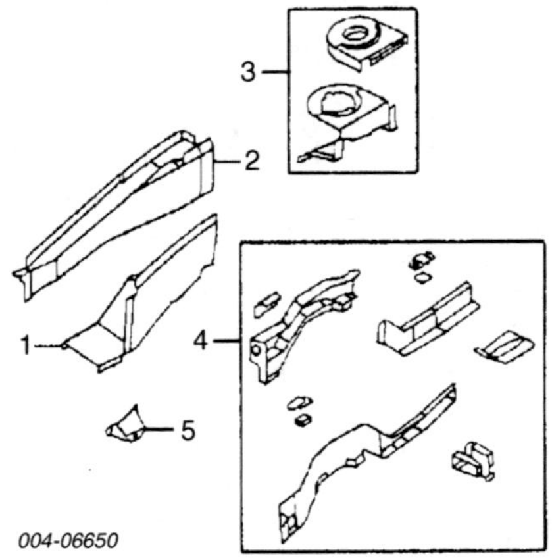

Part		Part No.		Time	Price
1 **Apron, Front**	R	4797814	s	#6.0	43.75
	L	4797815	s	#6.0	46.25

#Includes R&R Apron Reinforcement, w/Side Supt Panel Removed Deduct 1.0

Part		Part No.		Time	Price
2 **Reinf, Apron**	R/L	4797904-5	s	2.0	42.25
3 **Tower, Strut**	R	4886046AA	s	#8.0	77.50
	L	4797807	s	#8.0	77.50

#Includes R&R Front Apron, w/Side Rail Removed Deduct 1.0

Part		Part No.		Time	Price
4 **Rail, Side**	R/L	4716498-9	s	#12.0	220.00

#w/Side Supt Panel Removed Deduct 1.0, w/Front Apron Removed Deduct 2.5

Part		Part No.		Time	Price
5 **Reinf, Fender/Rad Panel**	R/L	4797858-9		2.0	40.00

SUB-FRAME

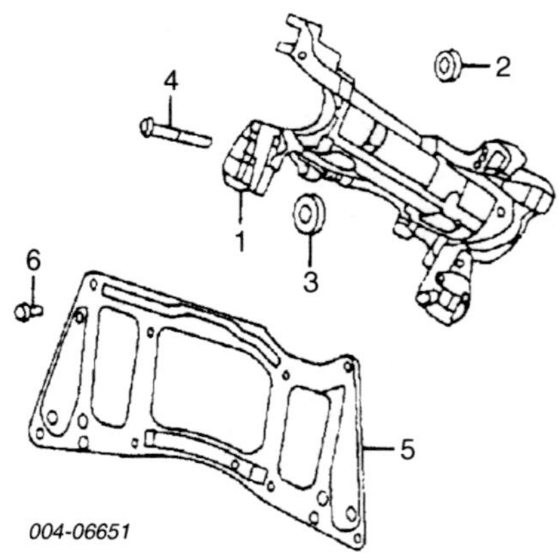

Part		Part No.		Time	Price
1 **Sub-Frame, Susp**		4694725	m	#4.5	535.00

#Includes R&R Stabilizer Bar, Links & Lower Control Arms, Does Not Include R&I or Suspended Eng Assy

Part		Part No.		Time	Price
2 **Isolator, Sub-Frame Front**					
Upper	R/L	4684286	m		9.35
Lower	R/L	4684287	m		12.50
3 **Isolator, Sub-Frame Rear**					
Upper	R/L	4684284	m		9.35
Lower	R/L	4684285	m		11.50
4 **Bolt, Mtg (2/Side)**	R/L	6504155	m		3.00
5 **Reinf, Sub-Frame**		4684659	m		79.50
6 **Bolt, Reinf Mtg (10)**		6504013	m		2.00

FIGURE 11–20 Example of head notes that contain vehicle information on the repair process.

REAR SUSPENSION

Use Procedure Explanation 23 with the following text.

FRONT WHEEL DRIVE

R&I Suspension/Crossmember Assy.......................................m 1.7
O/H Suspension
One Side...R/Lm 3.1
Both Sides...m 6.0
Adjust Four Wheel Alignment.......................................m 2.1
Adjust Rear Alignment.......................................m 1.6
Bleed Brake System
w/ABS...m .6
w/o ABS...m .3

NOTE 1: R&I/R&R or O/H Does Not include alignment or bleed brake system.
NOTE 2: All Parts in this section are included in overhaul unless noted otherwise.

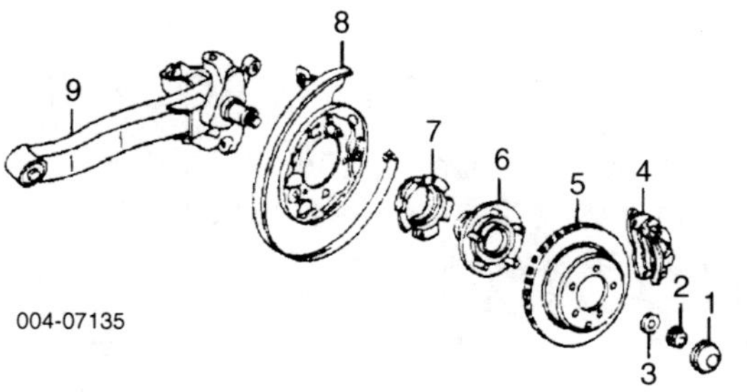

Part	Year	Side	Part No.		Labor	Price
1 **Cap, Dust**		R/L	MB515405	m		3.25
2 **Nut, Axle**		R/L	MB631529	m		3.85
3 **Washer, Spindle**		R/L	MB631528	m		6.70
4 **Caliper, Brake**		R/L	MB857332-1	m	#.5	375.00
#w/Brake Rotor Removed .2						
Hose, Brake		R/L	MR129750	m	#.3	13.00
#w/Brake Caliper Removed Deduct .2						
5 **Rotor, Brake**		R/L	MB587579	m	#.8	149.00
#R&R Both 1.4						
6 **Hub**		R/L	MB633512	m	#1.0	135.00
7 **Rotor, ABS Sensor**		R/L	MB668632	m	#1.0	96.00
#R&R One Side Complete, R&R Both Sides 1.8, Includes R&R Brake Rotor						
Sensor, Rear ABS-See Antilock Brake System Section						
8 **Shield, Splash¶**						
w/ABS	91-95	R	MB857294	m	#1.3	83.50
		L	MB857266	m	#1.3	91.00
	96	R	MR205822	m	#1.3	103.00
		L	MR205821	m	#1.3	103.00
w/o ABS	91-95	R	MB857265	m	#1.3	83.50
		L	MB857264	m	#1.3	91.00
	96	R/L	MR205820-19	m	#1.3	103.00
¶Order by Application						
#Includes R&R Hub & Brake Rotor						
9 **Arm, Trailing**						
w/ABS		R/L	MB663211-0	m	#1.6	520.00
w/o ABS						
Base Model		R/L	MB844156-5	m	#1.6	520.00
ES, R/T Model		R/L	MB631358-7	m	#1.6	520.00
#Includes R&R Brake Rotor, Hub & Splash Shield, R&R Both Sides 3.0						

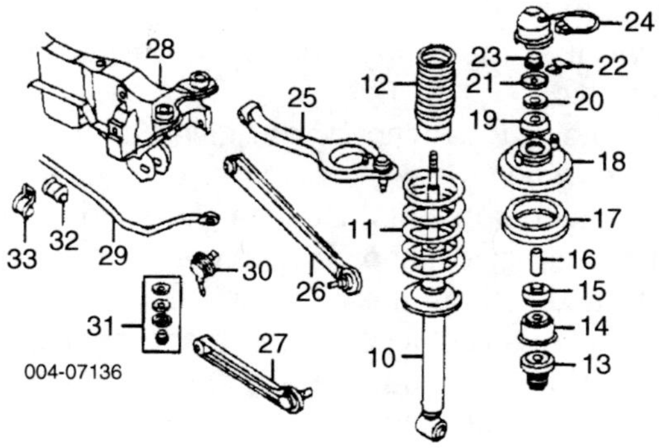

Part	Year	Side	Part No.		Labor	Price
10 **Absorber, Shock**						
w/Elec Susp	91	R/L	MR179070	m	#1.3	255.00
	92-96	R/L	MB871464	m	#1.3	172.00
w/o Elec Susp		R/L	MB631741	m	#1.3	95.00
11 **Spring, Coil**						
w/Elec Susp	91-93	R/L	MB631663	m	#1.3	70.50
	94-96	R/L	MB949448	m	#1.3	50.00
w/o Elec Susp	91-93	R/L	MB631740	m	#1.3	76.50
	94-96	R/L	MB949447	m	#1.3	48.00
12 **Boot, Dust**						
w/Elec Susp		R/L	MB631670	m	#1.3	8.00
w/o Elec Susp		R/L	MB518025	m	#1.3	7.50
13 **Damper**						
w/Elec Susp		R/L	MB663671	m	#1.3	4.95
w/o Elec Susp		R/L	MB663524	m	#1.3	7.20
14 **Cup, Damper**						
w/Elec Susp		R/L	MB631672	m	#1.3	7.75
w/o Elec Susp		R/L	MB338614	m	#1.3	5.55
15 **Bushing**						
w/Elec Susp		R/L	MB663671	m	#1.3	4.95
w/o Elec Susp		R/L	MB338618	m	#1.3	4.25
16 **Collar**						
w/Elec Susp	91	R/L	MB663672	m	#1.3	3.85
	92-96	R/L	MB871465	m	#1.3	4.50
w/o Elec Susp		R/L	MB338620	m	#1.3	2.35
17 **Pad, Spring**						
w/Elec Susp		R/L	MB631668	m	#1.3	9.75
w/o Elec Susp		R/L	MB338613	m	#1.3	13.00
18 **Insulator, Strut**						
w/Elec Susp		R/L	MB631669	m	#1.3	65.50
w/o Elec Susp		R/L	MB338610	m	#1.3	24.75
19 **Bushing, Upper**						
w/Elec Susp		R/L	MB663670	m	#1.3	5.00
w/o Elec Susp		R/L	MB242358	m	#1.3	3.00
20 **Washer**						
w/Elec Susp		R/L	MB663043	m	#1.3	6.85
w/o Elec Susp		R/L	MB242360	m	#1.3	1.95
21 **Spacer**		R/L	MB844349	m	#1.3	3.70
22 **Clip, Elec Susp**		R/L	MB663061	m	#1.3	3.10
23 **Nut, Lock**		R/L	MB573244	m	#1.3	1.50
24 **Cap, Strut**						
w/Elec Susp		R/L	MB633391	m	#1.3	12.50
w/o Elec Susp		R/L	MB515840	m	#1.3	6.75
#R&R One Side Complete, w/Trailing Arm Removed .4, R&R Both Sides 2.4						
25 **Arm, Upper Control**		R/L	MB698979	m	#1.5	172.00
26 **Arm, Rear Lower**		R/L	MB698980	m	#1.5	121.00
27 **Link, Lower**		R/L	MB584857	m	#1.5	99.50
#w/Trailing Arm Removed 1.3						
28 **Crossmember, Susp**			MB631660	m	#3.5	310.00
#Includes R&R Stabilizer Bar, Not Included in O/H						
29 **Bar, Stabilizer**			MB809547	m	#.8	62.00
30 **Lick, Bar**	91-93	R/L	MB584467	m	#.8	37.25
	94-96	R/L	MR131733	m	#.8	26.25
31 **Link, Kit**		R/L	MB185476			10.25
32 **Bushing, Bar**		R/L	MB809548	m	#.8	3.10
33 **Bracket, Bar**		R/L	MB573053	m	#.8	5.10
#R&R Complete, Not Included in O/H						

ALL WHEEL DRIVE

R&I Suspension/Crossmember.......................................m 4.2
O/H Suspension
One Side...R/Lm 3.5
Both Sides...m 6.7
Adjust Four Wheel Alignment.......................................m 2.1
Adjust Rear Alignment.......................................m 1.6
Bleed Brake System
w/ABS...m .6
w/o ABS...m .3

NOTE 1: R&I/R&R or O/H Does Not include alignment or bleed brake system.

FIGURE 11–21 Example of labor identified with an **m** for mechanical. Mechanical repairs can be charged at a higher rate than body repairs.

If footnotes are not read when an estimate is written, time for operations that are not included in the recorded flat rate will be omitted. In effect, the shop will then be doing part of the repair without compensation.

Calculate any overlap for the given operation and paint times when the work involves adjacent panels or interrelated components. Reduce these times based on the overlap guidelines for clear coat and three-coat systems and include the allowed amount of time for color blending on adjacent panels if applicable.

Materials. Make separate entries for both paint material and miscellaneous material costs. Itemize miscellaneous materials.

Straightening Times. Specify straightening times by component part and problem. Do not enter a catch-all time estimate for all pulls and setups. Include separate entries for setup times.

Sublet Operations. Specify any portion of the repair that is being sublet to other facilities or independent mechanics. While more and more larger collision repair facilities are doing repair jobs such as wheel alignment, rustproofing, and tire replacement, smaller operations still sublet these jobs to specialty organizations. When this is done, the specialty shop bills the collision repair facility for the work. Generally, this is done at a rate less than the normal retail labor cost for work. In this way, the collision repair facility can charge the normal retail cost and still make a small profit. Shops that farm out work usually have a column—Sublet—where the retail labor cost is marked. If final cost of sublet operations is not known, leave the price open.

Special Instructions. Any special instructions should be clearly stated on a separate entry. They may warn of suspected hidden damage, give tear-down instructions, and specify the details of a sectioning repair. It is very important that everyone involved in a sectioning repair—the insurance company, collision repair facility, and customer understand what is involved in the repair, how the repair is to be performed, and the importance of quality work.

Customer Requests. If the vehicle owner wishes to have additional work performed that is not covered by this claim (such as repair to damage that occurred prior to the collision), this should be noted as "customer-requested (C/R)" repairs. As a general rule, the insurance company will not pay for customer-requested repairs. Often a separate estimate must be made for the customer in such cases.

TOTALING LABOR AND PARTS COSTS

Labor rates for metal, straightening, mechanical, and refinishing operations often differ, so be certain to use the applicable rate when totaling labor costs. Total all hours in each labor category and multiply these figures by the applicable labor rate.

For example, if the repair requires hours of metal labor based on a rate of $28 per hour, the cost of metal repair labor is $406. Five hours of mechanical labor at $32 would be $160.

Add together the cost of all individual component and assembly parts to determine the total part cost of the job. Additional costs might include towing costs, vehicle storage costs, and state and local sales taxes.

REVIEWING THE DAMAGE REPORT

Figure 11–22 illustrates a completed damage report having many of the features described previously. Once the report is complete and all figures totaled and checked, review the report with the customer. Go over the report entry by entry, explaining what is to be done, how it will be done, and why certain types of parts and repair methods are best for the vehicle.

PHOTOGRAPHIC RECORDS

Virtually all insurance companies require a photographic record of the damage for their files (Figure 11–23). Many collision repair facilities also take photographs of vehicles to document damage, or to create a "before-and-after" file of repairs. Before-and-after files of select repairs can be a great help in building prospective customer confidence and showcasing the facility's commitment to quality.

Most insurance companies issue instant cameras to their adjusters so separate developing and printing is not required. This saves time while giving a color or black-and-white photo of adequate quality. To reduce the possibility that the photos will be forgotten, adjusters should develop the habit of taking these cover shots at the same time in every inspection, such as immediately after assessing previous damage, but before beginning the actual write-up. Facilities interested in before-and-after files should invest in better quality 35 mm camera and flash equipment.

While photos are excellent for documenting the general extent of the damage, they might or might not show hidden damage, frame damage, slight misalignments, and small scratches and dings. For these reasons many insurance companies will not always accept a photograph as proof of supplemental damage. However, since stopping work on a vehicle until the adjuster can survey the hidden damage can cost the facility valuable time, many facilities photograph the additional damage, resume the work, and count on the majority of settlements being made in their favor. Low-light video equipment is another method some facilities

Page 1 of 2 Pages

estimate

SMASH PALACE
13105 N. Jackson Rd.
BRIMFIELD, IL 61517
(309) 446-3162
Rebuilder License #UDL 10439
Repairer License #UDL 10455

NAME	PHONE	DATE
Anne Henkel	342 - 4168	6/24/98

STREET	CITY
RR#2	Tamaqua

YEAR	COLOR	MAKE	MODEL
1997	KS	Ford	Escort

REGISTRATION NO.	SERIAL NO.	ODOMETER	ESTIMATE PREPARED BY
353 666	1FASP09K6VG457535	8347	MSC

INSURANCE CO.	ADJUSTOR
Interstate	S. Donovan

REPLACE	REPAIR	DESCRIPTION	PARTS	LABOR	REFINISH	SUBLET
✓		front bumper	186.47	inc	2.7	
✓		right head lamp assbly	209.69	inc	—	
✓		hood	143.85	1.0	4.4	
✓		hood labels	10.00	0.1	—	
✓		radiator aftermarket	180.00	inc	—	
✓		antifreeze	10.00			
✓		condensor aftermarket	120.00	0.5m		
		AC recharge R134a	50.00	1.4m		
✓		radiator support	291.84	7.6	1.5	
	✓	unibody 1.5hr set up 3.0hr align		4.5F		
	✓	right lower front frame rail		2.0	1.0	
✓		right fender	64.25	1.4	2.5	
✓		right fender liner	26.16	inc		
✓		right lower control arm assbly	170.62	1.6m		
✓		right strut	74.95	1.4m		
		4 wheel alignment		1.9m		
		"This business is required to be licensed by the Secretary of State, pursuant to Illinois Revised Statutes, Chapter 95½ Sec. 5-301 ANY complaints as to the quality of service obtained here may be brought to the attention of the Attorney General"				
		TOTALS				

The above is an estimate based on our inspection and does not cover any additional parts or labor which may be required after the work has been started. Occasionally, worn or damaged parts are discovered which may not be evident on the first inspection. Because of this, the above prices are not guaranteed. Quotations on parts and labor are current and subject to change.

AUTHORIZATION FOR REPAIR. You are hereby authorized to make the above repairs:

SIGNED: ______________________

DATE: ______________________

TOTAL PARTS .... $ ________
TOTAL LABOR .... $ ________
TOTAL REFINISH .... $ ________
TOTAL SUBLET .... $ ________
TAX .... $ ________
FRAME REPAIR........ $ ________
TOTAL .... $ ________

FIGURE 11–22 A professional damage report.

Page 2 of 2 Pages

estimate

SMASH PALACE
13105 N. Jackson Rd.
BRIMFIELD, IL 61517
(309) 446-3162
Rebuilder License #UDL 10439
Repairer License #UDL 10455

NAME		PHONE	DATE
Anne Henkel			
STREET		CITY	

YEAR	COLOR	MAKE	MODEL	
REGISTRATION NO.		SERIAL NO.	ODOMETER	ESTIMATE PREPARED BY
INSURANCE CO.			ADJUSTOR	

REPLACE	REPAIR	DESCRIPTION	PARTS	LABOR	REFINISH	SUBLET
	✓	blend paint on right front door				
		and left fender			2.2	
	✓	clear coat			4.0	
	✓	R+I door molding		0.2		
		paint overlap			-0.4	
		"This business is required to be licensed by the Secretary of State, pursuant to Illinois Revised Statutes, Chapter 95½ Sec. 5-301 ANY complaints as to the quality of service obtained here may be brought to the attention of the Attorney General"	F	4.5		
			m	6.8		
		TOTALS	1542.83	12.3	17.9	

The above is an estimate based on our inspection and does not cover any additional parts or labor which may be required after the work has been started. Occasionally, worn or damaged parts are discovered which may not be evident on the first inspection. Because of this, the above prices are not guaranteed. Quotations on parts and labor are current and subject to change.

AUTHORIZATION FOR REPAIR. You are hereby authorized to make the above repairs:

SIGNED: ______________________

DATE: ______________________

TOTAL PARTS	$ 1542.83
TOTAL LABOR	$ 418.20
TOTAL REFINISH	$ 608.60
TOTAL ~~SUBLET~~ paint material	$ 358.00
TAX	$ 118.80
FRAME REPAIR	$ 202.50
m	$ 306.00
TOTAL	$ 3554.93

FIGURE 11–22 *(Continued)*

FIGURE 11–23 Damage is often documented visually using a camera.

are using to accurately record vehicle damage. Follow these tips when using camera or video equipment:

1. Know your equipment. Read all operating instructions.
2. Analyze what needs to be shown. Frame the action or shot squarely in the lens or viewfinder. For high-quality photos or videos a tripod should be used.
3. Never shoot into direct sunlight. Photos will be washed out. Pointing tube-type video cameras into the sun, bright lights, open flames, or welding equipment arcs will *severely damage* the camera tube. Chip video cameras have no tubes and eliminate this concern.
4. Eliminate shadows by using even, overhead lighting, flash equipment, or handheld or camera-mounted video fill lights. Slightly overcast days are ideal for outdoor shooting. When indoors, remember that certain types of fluorescent, mercury, metal halide, and sodium vapor lights can affect exact color reproduction. This can be a problem when exact color reproduction is desired for high-quality shots. For example, the blue-green light of mercury and metal halides can turn a red car a dull brown. Fluorescents can add blue or green tints. Sodium vapor lamps add yellow and orange. Special filters can help.
5. When using an instant camera, avoid extreme closeups. Stay at least three feet back from the subject.
6. Load and unload 35 mm film in low-light areas.
7. If needed, take shots from several angles to completely document the damage.
8. Protect your equipment. Return it safely to its case when not in use.

SUMMARY AND REVIEW

The following list summarizes many of the points that have been made in this chapter and throughout the text.

1. Be pleasant and professional. Respect the rights and concerns of the customer, collision repair facility, and insurance company.
2. Write clearly and legibly. Good-looking reports land jobs and avoid confusion.
3. Develop a pattern of actions. Methodically list each and every item or service provided. Leave a blank entry every five lines or so in case additional entries must be made.
4. Examine the damage from all angles. Inspect the inside and underside of the area. Pull up trunk and passenger compartment carpets as needed. Use a creeper, jack stands, or a lift as needed to thoroughly inspect the vehicle.
5. Refer to the estimating guides to accurately determine the subassemblies involved and exact part names and prices.
6. Inspect carefully for secondary damage remote from the immediate impact area. Measure distances between the manufacturer's check points to confirm frame or alignment damage.
7. Check alignment and fit of major body panels.
8. Check the electrical system. Do the horn, lights, turn signals, wipers, and other accessories work? Is the battery leaking acid or otherwise damaged?
9. Inspect for bent suspension components, steering parts, cracked or bent wheels, or cut tires. If the vehicle is driveable, have an assistant drive it slowly in a straight line. Do the rear wheels track properly? Are there signs of wobble and shimmy?
10. Check the vehicle's mechanical functions when possible. Does it start and run properly? Is there clearance for pulleys and belts and do the pulleys line up with the V-belts? Is the fan blade centered in the fan shroud? Are the motor mounts secure? Does the transmission shift properly? Is there evidence of cracked casings or leaking fluids?
11. Inspect the glass alignment. Do the windows open and close smoothly? Do all door, hood, and trunk latches work properly?
12. Inspect the interior upholstery and trim. Do the seats run smoothly on their tracks? Have the seat and shoulder belts been damaged or fatigued by the accident?
13. Check all fluid levels, coolants, and oils. Are there signs of leakage?

14. Does the steering wheel turn properly? Is the steering column centered in the dash opening?
15. Make logical economically sound decisions concerning the use of OEM, aftermarket, salvage, or rebuilt parts.
16. Fill in all information required on the damage report.
17. Attempt to "close" all replacement or repair costs. Open bids invite abuse.
18. Be precise with ACVs, calculations, and figures. Recheck all totals, agreed upon discounts, taxes, deductibles, etc.
19. Review all aspects of the damage report with the customer. Be prepared to negotiate in good faith with the adjuster or collision repair facility representative.
20. Document all claims with clear photos and descriptions.

REVIEW QUESTIONS

1. Technician A says that some insurance companies include the salvage value of a vehicle when deciding if a vehicle should be totalled. Technician B says that the VIN will indicate the model year of the vehicle. Who is right?
   a. technician A
   b. technician B
   c. both A and B
   d. neither A nor B
2. OEM parts fit better than aftermarket parts.
   True False
3. Technician A uses aftermarket parts when OEM have been specified. Technician B says that CAPA parts have been tested. Who is right?
   a. technician A
   b. technician B
   c. both A and B
   d. neither A nor B
4. The paint code for a Ford is found
   a. inside the driver's door.
   b. inside the trunk.
   c. inside the glove compartment.
   d. under the hood.
5. Technician A checks the entire vehicle for scratches when writing an estimate. Technician B looks at the damaged area only. Who is right?
   a. technician A
   b. technician B
   c. both A and B
   d. neither A nor B
6. When a damaged vehicle is declared a total loss, the insurance company pays the owner the ______ price of the vehicle.
   a. retail
   b. wholesale
   c. loan
   d. new
7. Technician A says that betterment must be deducted, if a tire is damaged and replaced. Technician B says that betterment must be deducted if a battery is replaced. Who is right?
   a. technician A
   b. technician B
   c. both A and B
   d. neither A nor B
8. Which of the following features usually does not increase the value of the vehicle?
   a. air-conditioning
   b. power steering
   c. manual transmission
   d. ABS
9. Technician A checks for engine pulley wobble with the engine running. Technician B says that in some collisions, this is impossible to do. Who is right?
   a. technician A
   b. technician B
   c. both A and B
   d. neither A nor B
10. The VIN plate is located on the upper left corner of the instrument panel, visible through the windshield on all cars manufactured after 1971.
    True False

Chapter 12

Computerized Damage Reports and Shop Management

Objectives

After reading this chapter, you should be able to:

- List at least five different ways computer-assisted damage report writing helps increase speed and productivity.
- Define the term **job costing** and explain its importance in collision repair facility operations.
- Explain how a computer system can aid the collision repair facility in such areas as damage report supplements, job control, inventory control, parts ordering, and paint formulation.
- Define the term **business analysis,** explain its importance to collision repair facilities, and list three examples of computer-generated analysis reports.
- List and explain the functions of the major components of a typical computer system.
- List and explain the information an automated estimating system can provide.
- List the advantages of an automated estimating system over a manual estimating system.

Key Terms

modem
mouse
printer
CD

Computers provide immeasurable benefits in almost every facet of our daily lives. More and more, the computer is moving into the work environment, including the collision repair industry. For many years computerized systems have been a mainstay in the insurance industry. Insurance companies rely on the accurate processing of millions of bits of data to compute probabilities and projected returns on investments. Because almost every other phase of the insurance business is data intensive, it is easy to understand why insurance companies, not collision repair facilities, were the first to move heavily toward computerized estimating programs and equipment.

But computerized estimating and shop management systems are being used by an ever-increasing number of facilities, both large and small (Figure 12–1). Regardless of size, collision repair facilities face concerns that demand increased efficiency and speed. A dramatic increase in the number of individual vehicle makes and models has led to a proliferation of sheet metal and mechanical components. Prices are constantly updated and changed. Calculating the needed overlaps and deductions takes precision and accuracy. The time needed to do these jobs is time not spent selling customers or repairing vehicles.

In addition, studies conducted by industry financial analysts have concluded that the average gross profit on a collision repair job has decreased significantly in

FIGURE 12–1 An estimator for a collision repair facility at work on a computer.

the past ten years. The lighter gauge, high-strength steels used in unibody design result in most unibody repairs being parts intensive, not labor intensive. This means more components are simply replaced, not repaired. Because body shops earn an average of 47 percent gross profit on labor versus 21 percent gross profit on parts, overall profit margins have shrunk quite noticeably.

As a result of these tighter profit margins, collision repair shops must run their businesses and manage their financial resources more efficiently. The average collision repair shop may realize a 5 percent net profit on sales. The profit potential, if managed properly, can be as high as 10 percent.

As explained throughout this text, collision repair shops have a ceiling on the amount they can charge for repair work due to the following factors:

1. Shops must negotiate an agreed price of repair with insurance company representatives.
2. Parts prices and labor hours are obtained from collision estimating guides.
3. Insurance companies use these guides to determine repair costs they will pay for.
4. Competition among local body shops tends to keep labor rates stable within a given geographic area.

With a shrinking profit margin, controlled pricing on parts and labor, and increased competition, many collision repair facilities are facing a profit squeeze. A computerized system of damage report writing and shop management can simplify and quicken almost all areas of shop operation. It increases productivity in the following ways.

12.1 DAMAGE REPORT WRITING

A number of advantages exist for collision repair facilities using an automated estimating system. Customers find computer-generated reports easier to understand and read (Figure 12–2). Customers have more confidence in a shop issuing computer damage reports. Shops can be sure that all calculations are correct and (with some systems) all overlaps and deductions have been included in the calculations.

Computers also streamline the estimating process. With computer damage report writing, many items can be reduced to select and click, and some computer programs reduce typing to a minimum through a code system that is easily understood by the user (often called user-friendly). An additional benefit is that the writer or adjuster does not need to take his or her guides, notes, and calculator into a quiet back room to grind out the numbers and figures. The process is much faster.

Instead of abandoning the customer, a computer system can keep the customer involved in the creation of the report. Some shops have found that allowing the customer to watch as the estimate is produced is a way of increasing shop/customer interaction. The customer sees the parts and procedures being used to repair the vehicle and is presented with a clean, crisp printout. This is often the difference between selling the job and seeing it taken elsewhere.

The second major advantage of computerized reports is found in dealing with insurance companies. As mentioned above, insurance companies depend upon computers very heavily. They tend to accept computer-generated information over handwritten information as being more accurate, complete, and "honest." And once the shop and insurance company begin to speak the same language and meet on common ground, negotiations become easier. The more difficult to settle charges, such as color blending, vehicle prep, and detailing, are far more likely to be collected when presented on a computer-generated damage report.

Computer-generated sheets can be easily modified, another plus in the shop/adjuster negotiating process. Data stored in the computer can be revised in minutes to reflect agreed-upon changes. With several keystrokes the old hood or trunk will be repaired rather than replaced, the labor cost can be adjusted to reflect $28 per hour instead of $30. Material and additional costs can be altered just as easily.

Description: 1994 Ford Escort LX
Body Style: 4D Sed Drive Train: 1.9L Inj 4 Cyl 5M
VIN: 3FARP13J8RR155203

Line Item	Entry Number	Labor Type	Operation	Line Item Description	Part Type/ Part Number	Dollar Amount	Labor Units
1	700590	BDY	REMOVE/INSTALL	FRT BUMPER ASSY			0.8
2	705580	BDY	REMOVE/REPLACE	R FENDER PANEL	Qual Recycled Part	120.00*	1.4 #
3	AUTO	REF	REFINISH	R FENDER OUTSIDE			C 2.4
4	AUTO	REF	REFINISH	R FENDER EDGE			C 0.5
5	705632	BDY	REMOVE/REPLACE	R FENDER SPLASH SHIELD	Qual Recycled Part	10.00*	0.1 #
6	706170	BDY	REMOVE/REPLACE	R FENDER ADHESIVE MOULDING	F2CZ 16003 AA	35.49	0.1
7	700240	BDY	REMOVE/REPLACE	R FRT DOOR SHELL	Qual Recycled Part	400.00*	3.5
8	AUTO	REF	REFINISH	R FRT DOOR OUTSIDE			C 1.8
9	AUTO	REF	REFINISH	R FRT ADD FOR JAMBS & INSIDE			C 1.0
10	719240	BDY	REMOVE/INSTALL	R FRT BELT MLDG			INC
11	719500	BDY	REMOVE/REPLACE	R FRT DOOR ADHESIVE MOULDING	ORDER FROM DEALER	42.39	0.1
12	719720	BDY	REMOVE/REPLACE	R FRT DOOR POWER MIRROR	Qual Recycled Part	40.00*	0.3
13	720260	BDY	REMOVE/INSTALL	R FRT DOOR HANDLE & LOCK CYL			INC #
14	721980	BDY	REPAIR	R REAR DOOR REPAIR PANEL	Existing		4.0*
15	AUTO	REF	REFINISH	R REAR DOOR OUTSIDE			C 1.8
16	722180	BDY	REMOVE/REPLACE	R REAR DOOR ADHESIVE MOULDING	F2CZ 5425532 AA	30.03	0.2
17	900500	BDY*	REMOVE/INSTALL	R REAR DOOR TRIM & NECC. PARTS	Existing		1.3*
18	728250	BDY	REPAIR	R QUARTER OUTER PANEL	Existing		2.5*#
19	AUTO	REF	REFINISH	R QUARTER PANEL OUTSIDE			C 2.1
20	700437	BDY	REMOVE/INSTALL	R REAR COMBINATION LAMP			0.2
21	900500	BDY*	REMOVE/REPLACE	RESTRIPE REPAIRS	**Qual Repl Part	14.00*	0.8*
22	900500	BDY*	ADD'L LABOR OP	LKQ CLEAN UP TIME	Existing		2.0*
23	AUTO	REF	ADD'L OPR	CLEAR COAT			2.5
24	AUTO		ADD'L COST	PAINT/MATERIALS		254.10*	
25	AUTO		ADD'L COST	HAZARDOUS WASTE DISPOSAL		5.00*	

* Judgment item
\# Labor note applies
C Included in clear coat calc

I. Labor Subtotals	Units	Rate	Add'l Labor Amount	Sublet Amount	Totals
Body	17.3	36.00	0.00	0.00	622.80
Refinish	12.1	36.00	0.00	0.00	435.60
	Nontaxable Labor				1,058.40
Labor Summary	29.4				1,058.40

II. Part Replacement Summary		Amount
Taxable Parts		691.91
Sales Tax	@ 6.500%	44.97
Total Replacement Parts Amount		736.88

III. Additional Costs		Amount
Taxable Costs		259.10
Sales Tax	@ 6.500%	16.84
Total Additional Costs		275.94

IV. Adjustments	Amount
Customer Responsibility	0.00

I. Total Labor:	1,058.40
II. Total Replacement Parts:	736.88
III. Total Additional Costs:	275.94
Gross Total:	2,071.22
IV. Total Adjustments:	0.00
Net Total:	2,071.22

This is a preliminary estimate.
Additional changes to the estimate may be required for the actual repair.

ESTIMATE RECALL NUMBER: 6/15/98 07:43:27 288
UltraMate is a Trademark of Mitchell International
Mitchell Data Version: JUN_98_A Copyright © 1994–1997 Mitchell International
All Rights Reserved

FIGURE 12–2 Computer-generated estimate.

Once all entries are satisfactory to all parties, the data can instantly be shuffled into any number of useful formats and printed out. A final damage report might be desired. Or the data can be turned into a repair order, printed in the industry standard six-column format listing a complete breakdown of charges.

In addition, many systems can also break down the total damage report in detailed subreports listing only insurance payable items, customer payable items, labor, or parts costs.

Parts orders to dealerships, jobbers, or other parts sources can also be generated from the data base. The parts or shop manager does not have to spend time writing out handwritten orders. Clean, clearly typed reports can easily be transmitted by fax machine, which decreases telephone time and the possibility of verbal mix-ups.

12.2 JOB COSTING

Job costing involves making a detailed analysis of all parts, labor, sublet, material, and miscellaneous expenses involved in performing a repair. It gives the shop owner or manager a clear indication of where profits are being made or lost. It is estimated that no more than one-half of all shops do job costing on a regular basis.

There is no way a collision repair facility can lose money on every job and make up for it in volume. Every job taken on by the shop must be a profit maker. While job costing, like damage report writing, can be done manually, a computerized system can help maintain a shop's desired profit margin by providing shop owners with a way to spot missed parts that might have been purchased and were not included in the insurance company damage appraisal. In addition, job costing allows shop owners to compare actual labor required to complete a job to the labor hours allowed by the insurance company appraisal and agreed price of repair. This analysis helps shop owners to fine-tune their estimating skills as well as monitor the performance of their repair technicians. Job costing also enables report writers and managers to spot those miscellaneous items and expenses easily overlooked until the vehicle is paid for and out of the shop.

Some computer management systems can also project profits (or lack of profits) on incoming jobs by comparing the damage report with the previous month's profit-by-department statistics, or the shop's profit history on similar models and situations. This can be a real benefit when the shop is negotiating an agreed price. Areas where the shop can bend slightly, stand firm, or demand more become clearly defined. The decision to accept a job or reject it based on profit becomes much easier with accurate figures and predictable trends.

12.3 SUPPLEMENTS AND JOB CONTROL FUNCTIONS

Often the time and effort needed to generate a supplement are worth more than the supplement itself. On computerized systems, an automatic supplementing function makes this process fast and simple. With a computerized system, it is now worthwhile to track down money for parts price increases and unforeseen work.

Many other reports and forms can be generated and printed to allow shop owners to maximize efficiency and profit through better communication and internal organization. Depending on the system, these job control features might include

- User-defined form letters, such as follow-ups to damage reports or completed jobs, checklists, birthday greetings, or notification of sales to select customer groups
- Power of attorney forms so the customer can conveniently authorize the shop to legally act on his/her behalf
- Work authorization forms outlining shop/customer responsibilities and the terms of the repair
- Work schedule, open estimates, work in progress, work delivery dates, invoiced jobs, and many other specialized reports

Time Control. Some systems offer computerized time sheets in which a technician logs on and off particular jobs using a computer terminal placed in the shop. The information is compiled and charged to the job. The shop manager or owner can then retrieve the data for progress reports or to determine if the job is exceeding its allotted time and requires a supplement or investigation as to the cause.

Inventory Control. Inventory control is another area in which the computer can save time and money. The parts manager can log in all incoming parts so that the job can be scheduled into the metal shop as soon as practical. The system can also monitor supplies of "staple" parts and materials common to all jobs such as fasteners, clips, and refinishing supplies. Facilities can lose time and money if they hold a job because no one knows all parts are in. More than one facility has put a job on hold while waiting for a part that has already arrived and is sitting off in a corner. Job delays caused by running out of everyday supplies can be even more frustrating. An inventory program can help eliminate these money-wasting situations.

Paint Shop Operations. Paint companies have developed a number of systems to help streamline paint shop operations. As a job moves into the refinishing sequence, the technician can call up on the terminal the vehicle color code and the overall amount of paint needed. The computer program will then determine how much of each of the tints to mix together to achieve the right match and the right amount. Waste is kept to a minimum, a major concern, since unused paint is considered a hazardous waste.

If a formula is slightly off, the technician can make the correct tint adjustment and enter this information into the data base. The next time that a color code is called up, the mistake will be noted and the correction made before the paint is mixed and sprayed. Again, waste is avoided.

12.4 ACCOUNTING AND BUSINESS ANALYSIS

A computer can drastically cut the number of hours required to perform basic (and tedious) bookkeeping and accounting tasks. The payroll and accounts receivable and payable can be handled much more quickly by computer. A payroll that was an all day task now becomes a thirty-minute job. Cash flow increases with automatic generation of customer statements, aged balance reports, and automatic addition of finance charges on delinquent accounts.

A computer can also help a shop owner organize payables in a planned, orderly manner. As a result, a good credit rating is easier to maintain, and the shop is also able to take advantage of any discount incentives and special payment opportunities offered by suppliers.

A computer also allows the collision repair facility to perform invaluable sales and business analysis. Its ability to quickly process data allows the user to spot business trends and probabilities that can be used to the facility's advantage.

For example, the system can break down all insurance claim repairs by individual insurance companies. A collision repair facility owner might be surprised to find that the insurance company he or she thought represented 40 percent of all business actually accounts for only 15 percent. Some insurance companies might be highly profitable to work with, while others are not. A program that allows the owner to call up such information can help determine which accounts and companies are most valuable to the shop.

FIGURE 12–3 Digital camera shown with the computer it is used with.

Analysis can show what type of vehicles turn the highest profit. The facility might turn a larger profit on imports, domestics, new models, or older vehicles. Profitability by make and model might help the collision repair facility decide on a field of specialization, or mold the shop's advertising campaign. Entering some simple facts about customers into the data base can also be useful. Where do your customers live? Why did they choose the shop? Where did they learn about the shop? Is that expensive billboard pulling in customers, or is the simple ad in the yellow pages doing the job? Knowing where the work comes from is also important when determining how promotional budgets are spent.

A shop may elect to pay employees through any number of methods (e.g., hourly, flat ratio, commission, or a combination of all of these methods). For this reason, determining the shop's most productive workers and what will motivate them is the best key to the shop's success. A computer can compile all of the pertinent information to determine which technicians are generating the most profits and which technicians are costing the shop money.

Digital Imaging

A computer can be used with a digital camera to record vehicle damage. Figure 12–3 shows a digital camera. The digital image is recorded on a disk that can be loaded into a computer. Once an image is

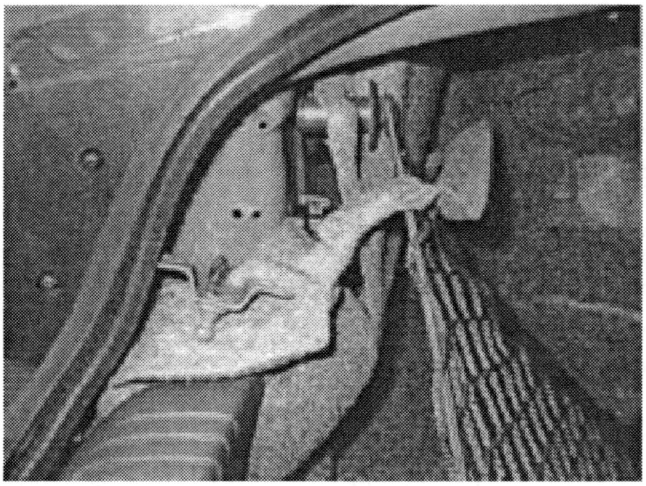

FIGURE 12–4 Digital image of the inside of a Cavalier trunk.

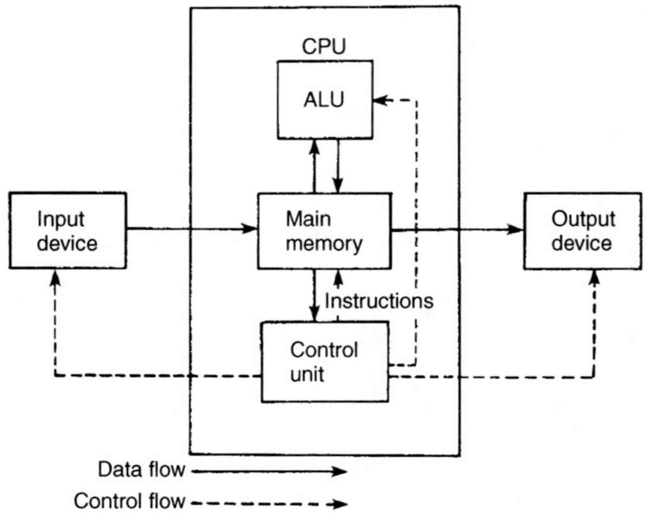

FIGURE 12–6 Basic components of a personal or business computer system.

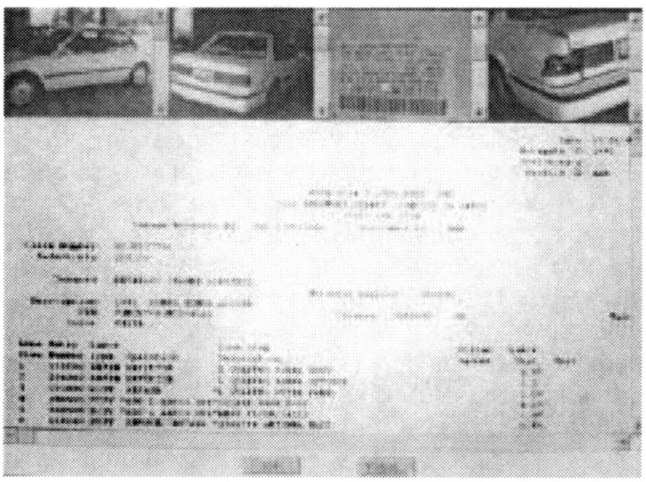

FIGURE 12–5 Digital images of a damaged vehicle with an estimate. (Courtesy of Mitchell International.)

stored in the computer, it can be edited, stored, printed, and even sent by phone to another computer. Figure 12–4 shows a digital image of trunk damage done to a Cavalier. Figure 12–5 shows how digital images can be stored with a damage report.

12.5 COMPUTER COMPONENTS

The computer is simply a machine. Just as a driver need not understand the various components of an automobile to drive it, the computer operator need not thoroughly understand the workings of a computer to use it. However, the driver who does have some knowledge of the mechanics of a car might notice early warning signals of a mechanical problem. And the person who understands the computer might detect problems sooner as well.

Like the auto, the computer is made up of various systems and components (Figure 12–6). The input/output devices are items such as cathode ray tubes (CRTs) and printers. The central processing unit (CPU) is where the work—processing data—actually takes place.

Most people have probably seen a CRT in the form of a screen. This device serves as both an input device where the data is keyed in and as an output device where the information is displayed on the screen. Information that is keyed in is sent to the CPU to be processed. Then the information is returned to the screen for viewing.

COMPUTER JARGON

An awareness of computer jargon is essential to understanding the computer. The following is a list of definitions to acquaint the damage report writer with computer terminology, increase an understanding of the computer, and serve as a reference guide for future questions.

- **ALU (Arithmetic and Logic Unit).** The portion of the CPU (Central Processing Unit) that performs arithmetic and makes logical comparisons. The CPU has two parts: a main memory that holds the computer program and the processing data, and an ALU that actually executes the program's instructions.

FIGURE 12–7 Diskette.

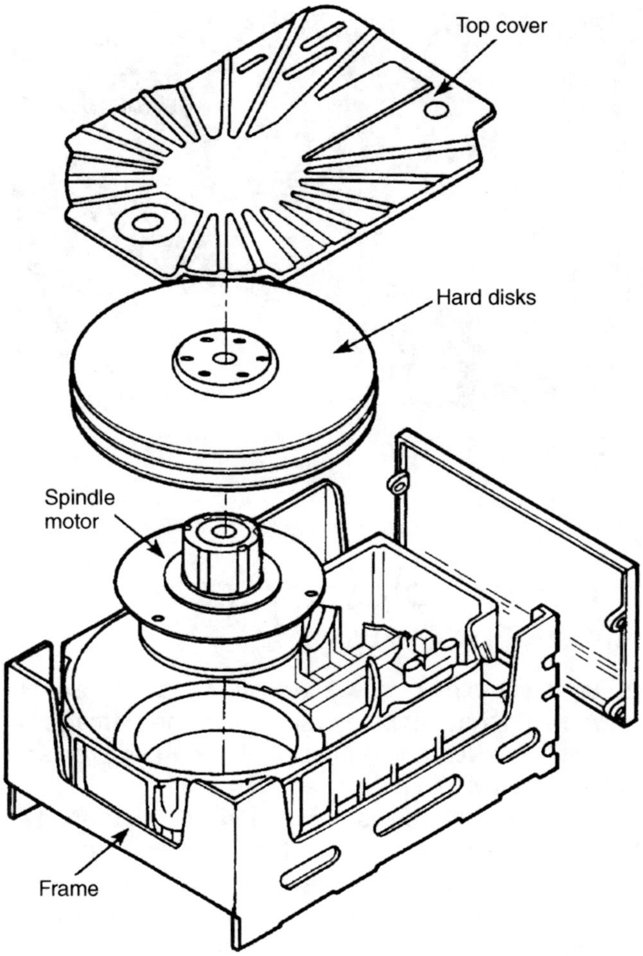

FIGURE 12–8 Hard disk.

- **CD.** Compact disc that can hold 650 megabytes of data.
- **Command.** An instruction to the computer to perform a predefined operation.
- **Control Unit (of CPU).** The section of the CPU that directs the sequence of operations by electrical signals and governs the actions of the units that make up the computer.
- **Cursor.** A movable, blinking marker on the terminal video screen that defines the next point of character entry or change.
- **Downtime.** The period of time when a device is not working.
- **Drive.** A device that holds a disk or diskette so that the computer can read data from and write data onto it.
- **Diskette.** A thin magnetic disk (in protective paper jacket) that stores data or programs. It is erasable and reusable and comes in several diameters (Figure 12–7).
- **Hard Disk.** A device to store data on magnetic disks that, unlike diskettes, is rigid and not readily interchangeable but can store and retrieve data faster (Figure 12–8).
- **Hardware.** The physical parts of a computer, such as keyboard, screen, disk drives, printer, and so on.
- **Mainframe Computer.** A very large, powerful computer capable of storing and processing extremely large amounts of data. Sometimes called the "host" computer in some systems.
- **Main Storage.** Also known as internal storage, memory, or primary storage unit; the section of the CPU that holds instructions, data, and intermediate and final results during processing.
- **Menu.** A displayed list of options from which the user selects an action to be performed.
- **Modem.** Telephone line hookup to the computer.
- **Mouse.** Device that allows the cursor to be moved.
- **Peripherals.** Hardware that can be added to a basic computer, such as a printer, disk drive, or modem, for transferring information by telephone (Figure 12–9).
- **Printer.** Laser, ink jet, or dot matrix device that types out printed reports.
- **Secondary Storage.** Also known as permanent storage; the components of the system, such as magnetic tape or disks that hold large amounts of raw data that can be called up to the CPU for processing.
- **Software (also Programs).** Instructions written in a computer language that tell the computer what the users wants to do. Most programs come on diskettes for small computers and on magnetic tape for larger ones.
- **Terminal.** An input/output device used to enter data into a computer and record the output. Terminals are divided into two categories: hard copy (e.g., printers) and soft copy (e.g., video terminals).

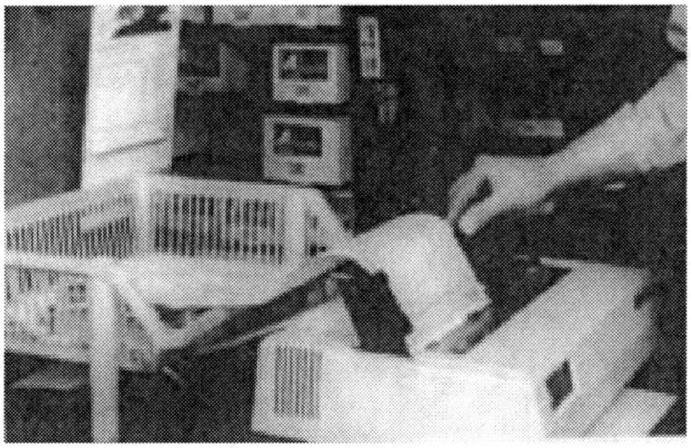

FIGURE 12–9 Printer.

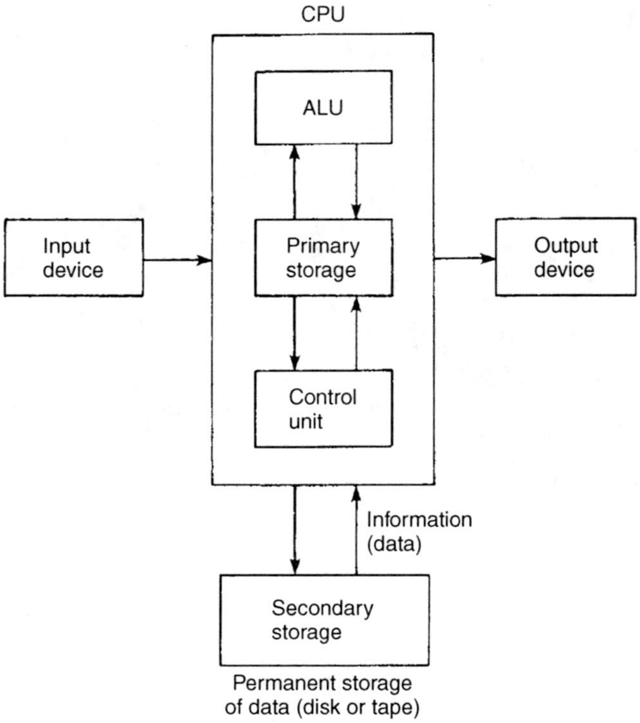

FIGURE 12–10 Basic components of a personal or business system, including a secondary storage system that can be on-site or remote.

INFORMATION STORAGE AND RETRIEVAL

An understanding of how information is stored and retrieved within a computer will enable the damage report writer to determine the cause of certain problems encountered within the system.

Figure 12–10 shows the general configuration of the workings of most computer systems. The major difference between this diagram and Figure 12–6 is the addition of the secondary storage block. When not in use, information is stored in secondary storage. It is permanent storage as compared to primary storage, which is temporary. Examples of devices used for secondary storage are magnetic tape and magnetic disk.

Information is retrieved from secondary storage and brought into the CPU for processing when requested by the user. For example, the damage report writer can access all pertinent operation and parts price information for a certain vehicle make and model by keying in the request on the CRT terminal. The secondary storage data base, either magnetic tape or magnetic disk, can be physically located at the CPU or accessed from a central remote location using telephone modem hookups.

The CRT terminal is the user's link with the CPU. The CRTs do no processing themselves but serve as display units for the dialog that the user holds with the CPU. The request for information is sent to the CPU, where the control unit takes over. Someone has already written a program that contains a series of instructions directing the computer what to do next.

THE PROCESS

The control unit issues an electronic command that causes the requested data to be found in secondary storage and copied into primary storage. After this transfer is complete, the information is available for use. Note that information is available for use only after it has been transferred into primary storage, never while it is in secondary storage. The program (instructions) that specifies what must be done to this data is also stored in primary storage.

User-Keyed Data. Rather than rely on magnetic tape or disk data banks compiled from information in the popular collision estimating guides, some computer software programs allow the user to key in data from the collision estimating guides into a spreadsheet-type program. Although not as fast as fully computerized systems, user-keyed data is more flexible. For example, it can be slightly altered to reflect different part prices and labor time projections. Again, once the information is keyed into the computer, it can be used throughout the rest of the job process to print various form letters and work reports.

The following selection discusses the features of the major computer-assisted damage report systems used in the collision repair industry.

COMPUTERIZED ESTIMATING SYSTEMS

History

The first automated estimating system in the United States and Canada was introduced by Audatex™ in the late 1970s. Automatic Data Processing (ADP), an independent computer services company, began actively marketing the Audatex™ system in 1980. Today, there are three major suppliers of automated estimating systems: ADP, CCC, and Mitchell International.

Function

Automated estimating systems access the tremendous amount of information required in producing damage reports. The massive data bases are stored on CD-ROMs or servers. The computer can be either a desktop or laptop and can generate the damage report in an office or at the repair site. The damage report can then be combined with imaging and sent via modem anywhere in the world.

An automated estimating system will provide the following important information:

- **Administrative Data.** Includes the name and address of the owner of the vehicle and other pertinent information such as claim and policy numbers, deductible, point of impact, place of inspection, etc.
- **Vehicle Data.** The user enters the VIN. From this number, the vehicle make, model, and other information is decoded by the computer. Vehicle options should also be noted, as they can have a direct impact on repair times.
- **Part Information.** Upon entering a part in an automated system, a number of things occur.
    - With a click of the mouse or a simple keystroke, current OEM part prices, part numbers, and labor times are listed on the estimate, simplifying the parts ordering process.
    - *Overlap.* The automatic application of overlap adjusts labor times on procedures that share common operations, per vehicle-specific labor notes or procedure pages.
    - *Included/Not Included.* Labor time or parts included in the replacement time of other parts are automatically shown as included on the estimate.
    - *Automatic Functions.* In the instance where a part requires refinishing (e.g., edges and jambs), the system will automatically add a refinish line. When the headlamp adjustment is affected by parts replacement, the system will instinctively add in the adjustment time. If the vehicle has options that increase the base labor time of the operation, the system will increase the time automatically.

In the same way a manual estimate is written, automated estimating systems have processes in place to specify labor for panel straightening, frame repair, and other judgment times. The judgment time is entered by the user. Automated estimating systems offer a wide variety of part types to choose from (new, used, aftermarket, etc.), labor operations (R&R, repair, sublet, etc.), and labor rates (body, frame, mechanical, etc.). This myriad of choices has been designed to accommodate any conceivable situation that might arise in the estimating process. Aftermarket and LKQ/used parts can be specified in place of OEM parts. Automated estimating systems offer two ways to enter aftermarket and LKQ/used parts:

- The user must enter the aftermarket or LKQ/used part price manually.
- The computer automatically searches for aftermarket or LKQ/used parts using criteria specified by the user. When a part is selected, the system provides the correct part number and price but also references the supplier who has the part in stock.

Finally, an automated estimating system will also permit the user to enter a line of parts information manually. This function is primarily used to enter accessories and other items not typically found in the systems data base. Automated estimating systems provide a tremendous amount of flexibility in processing an estimate that would otherwise have to be done manually. The computerized estimate also provides a more professional presentation of the information.

System Use

The creation of manually produced and computer-generated damage reports share a common factor in that the estimator is required to enter the information. An automated estimating system is basically a tool. As with any tool, the estimator must learn how to use it. Once the user acquires a minimum skill level, the system can significantly accelerate and simplify the process. Usually, the first step in producing an automated estimate is entering the administrative data. As in manually produced estimates, automated systems

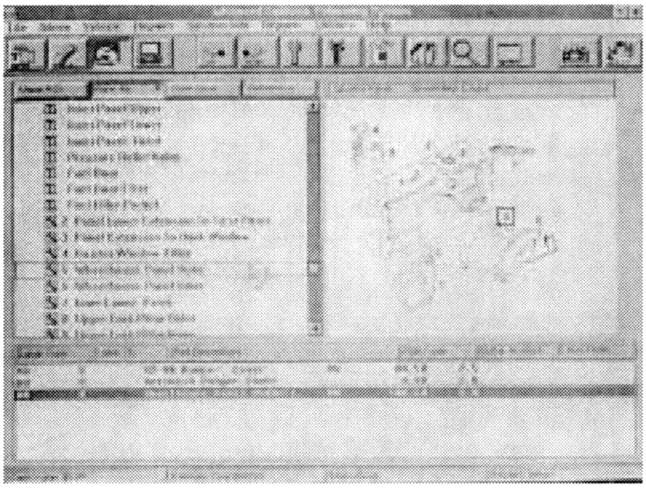

FIGURE 12–11 Screen from a computerized estimating system. (Courtesy of Mitchell International.)

FIGURE 12–12 Front hit of a Chrysler LeBaron.

allow for the administrative data to be entered at different times in the process. The name, address, and phone number of the customer can also be entered prior to the customer's arriving at the shop for their appointment. For drop-ins, the vehicle can first be inspected for damage and then the administrative data can be entered. There are two things the estimator should do before writing an automated damage report:

- *Provide a system identifier.* This could be a claim or plate number. If needed, all systems have the ability to generate their own identifying numbers. This number becomes the permanent identifier for the damage report and cannot be changed.
- *Provide the vehicle identification number (VIN).* When the VIN is entered, the system uses it to identify the vehicle and facilitate the selection of parts that relate to that specific year, make, and model. A thorough examination of the damaged vehicle is essential. A computer can be a useful tool for writing a report, but it cannot take the place of a complete and thorough evaluation of the damaged vehicle.

In most automated estimating systems, the damaged parts are entered on a series of screens (Figure 12–11). When the repair information has all been entered, the computer will total up the labor, parts, and materials. The report can be printed out at this time. Figure 12–12 displays a picture of a damaged 1991 Chrysler LeBaron. The corresponding automated estimate for the vehicle is shown in Figure 12–13.

Advantages

An automated estimating system has several advantages over a manually written estimate:

- **Speed.** Long damage reports can be produced faster by a computer than by manually writing them.
- **Accuracy.** As the data is entered, the system automatically supplies current, up-to-date information or prompts the user to add the needed information.
- **Automated Functions.** The computer performs time-consuming functions for the user.
- **Overlap.** The systems apply overlap per the procedure pages and/or labor notes.
- **Organization.** A computer-generated damage report is organized and easy to read.
- **Quality Review.** The on-screen data can be reviewed before printing.
- **Modifications.** If errors are found on the screen, they can be corrected easily before the report is printed.
- **Presentation.** The printed report provides a more professional presentation than a handwritten damage report.
- **Multiple Copies.** The estimating system can produce additional copies of a damage report if needed.
- **Management Information.** Data is easily transferred into the existing shop management system.

1991 CHRYSLER LE BARON PREMIUM LX 2 DR COUPE
CLAIM # XXX-XXXXXXX LOG 1609 -0 3-10-98 8:31 AM

VEHICLE
1991 CHRYSLER LE BARON PREMIUM LX 2 DR COUPE
6CYL GASOLINE 3.0

OPTIONS

TWO-STAGE - EXTERIOR SURFACES	HEATED REMOTE CONTROL MIRRORS
POWER DOOR LOCKS	HEATED BACK GLASS
TILT STEERING WHEEL	AIR CONDITIONING
AUTOMATIC TRANS	CRUISE CONTROL

BODY COLOR	MAROON	MILEAGE	111,188
CONDITION		VIN	1C3XJ5131MGXXXXXX
LICENSE #	XXXXXX	CODE	M251

REMARKS:

PART		GDE	OPERATION	PRICE		LABOR	RT
FRONT BUMPER							
Strip, Front Impact		0021	Quality Repl. Part	25.00		INC	SM
Cover, Front Bumper		0006	Quality Repl. Part	223.00		2.6	SM
Cover, Front Bumper		0006	Refinish			3.6	RF
			Surface		2.5		
			Two-stage		0.5		
			Two-stage setup		0.6		
FRONT END PANEL AND LAMPS							
Grille, Upper		0028	Quality Repl. Part	38.00		INC	SM
Panel, Front End		0030	Quality Repl. Part	138.00		3.4	SM
Panel, Front End		0030	Refinish			2.3	RF
			Surface		1.5		
			Edge		0.5		
			Two-stage		0.3		
Moulding,Grille 4334363		0031	Replace OEM	180.00		INC	SM
Cover, Headlamp	LT	0059	Refinish			0.7	RF
			Surface		0.6		
			Two-stage		0.1		
Cover, Headlamp	RT	0060	Refinish			0.7	RF
			Surface		0.6		
			Two-stage		0.1		
Headlamps Aim		0973	Additional Labor			0.6	SM
RADIATOR SUPPORT							
Panel Assembly,Rad Sup 07		0073	Repair			2.0*	SM
FRONT BODY AND WINDSHIELD							
Supt,Hood Lock Vert 4655703		0102	Replace OEM	9.35		0.7	SM
Fender, Front	LT	0103	Refinish			1.0*	RF
Fender, Front	RT	0104	Refinish			1.0*	RF
Panel, Hood 5253839		0083	Replace OEM	445.00		1.5	SM
Panel, Hood		0083	Refinish			4.6	RF
			Surface		2.8		
			Edge		1.2		
			Two-stage		0.6		

16 ITEMS

MC MESSAGE
07 STRUCTURAL PART AS IDENTIFIED BY I-CAR

FIGURE 12–13 Computer-generated estimate for the LeBaron. (Courtesy of ADP.)

FINAL CALCULATIONS & ENTRIES					
PARTS					
GROSS PARTS				$ 634.35	
OTHER PARTS				$ 424.00	
PAINT MATERIAL				$ 278.00	
ADJUSTMENTS		DISCOUNT	MARKUP		
PARTS TOTAL					$ 1,336.35
TAXON PARTS & MATERIAL @ 7.250%					$ 96.89
LABOR	RATE	REPLACE HRS	REPAIR HRS		
1-SHEET METAL	$ 36.00	8.2	2.6	$ 388.80	
2-MECH/ELEC	$ 45.00				
3-FRAME	$ 45.00				
4-REFINISH	$ 36.00	13.9		$ 500.40	
5-PAINT	$ 20.00				
LABOR TOTAL					$ 889.20
TAX ON LABOR @					
SUBLET REPAIRS					
TOWING					
STORAGE					
GROSS TOTAL					$ 2,322.44
LESS: NET DEDUCTIBLE					NONE-
NET TOTAL					$ 2,322.44

Owner To Pay All Damages

PXN Y/04/04/00.00.00 CUM 04/04/00.00.00 Geocode: 61554 PEORIA
ADP PENPRO W0330 ES LOG 1609 -0 03-10-98 08:43:20 REL 3.30 CD 02/98
COPYRIGHT, AUTOMATIC DATA PROCESSING, INC. 1997

2.2 HOURS WERE ADDED TO THIS ESTIMATE BASED ON ADP'S TWO-STAGE REFINISH FORMULA: 20% OF REFINISH HOURS, AFTER OVERLAP, PLUS 0.6 HOURS FOR THE FIRST MAJOR PANEL, WHERE NOTED.

THIS ESTIMATE HAS BEEN PREPARED BASED ON THE USE OF CRASH PARTS SUPPLIED BY A SOURCE OTHER THAN THE MANUFACTURER OF YOUR MOTOR VEHICLE. WARRANTIES APPLICABLE TO THESE REPLACEMENT PARTS ARE PROVIDED BY THE MANUFACTURER OR DISTRIBUTOR OF THESE PARTS RATHER THAN THE MANUFACTURER OF YOUR VEHICLE.

THIS IS NOT A REPAIR AUTHORIZATION, OWNER MUST AUTHORIZE REPAIRS TO THE SHOP. ADD ORDERS WILL NOT BE ALLOWED UNLESS AUTHORIZED AND INSPECTED BY THE ADJUSTER.
ILLINOIS LAW REQUIRES THAT VEHICLE REPAIRERS MUST BE LICENSED IN ACCORDANCE WITH SECTION 5-301 OF THE ILLINOIS VEHICLE CODE.

FIGURE 12–13 (Continued)

REVIEW QUESTIONS

1. Technician A says that job costing pinpoints profit or loss. Technician B says that most collision repair facilities do job costing on a regular basis. Who is right?
   a. technician A
   b. technician B
   c. technician A and Technician B
   d. neither Technician A nor Technician B
2. A high volume of work will make up for individual unprofitable jobs.
   True False
3. Technician A says that a computer system can be used for inventory control. Technician B says that a computer system can be used for paint shop operations. Who is right?
   a. technician A
   b. technician B
   c. technician A and Technician B
   d. neither Technician A nor Technician B

4. The physical parts of a computer are known as the _______.
5. Technician A says that labor time overlap is automatically deducted from an automated estimate. Technician B says that if the VIN is entered, the vehicle make and model will be provided by the computer. Who is right?
   a. technician A
   b. technician B
   c. technician A and Technician B
   d. neither Technician A nor Technician B
6. Judgment time is automatically supplied by the computer.
   True False
7. Technician A says that in an automated estimating system, aftermarket parts may need to be entered manually. Technician B says that in some automated estimating systems, used parts are automatically searched. Who is right?
   a. technician A
   b. technician B
   c. technician A and Technician B
   d. neither Technician A nor Technician B
8. The advantage of an automated estimating system over a manually written estimate is
   a. accuracy.
   b. multiple copies.
   c. speed.
   d. all of the above.
9. Technician A says that on-screen data can be reviewed and corrected before printing an estimate. Technician B says that in an automated estimating system, labor overlap must be entered manually. Who is right?
   a. technician A
   b. technician B
   c. technician A and Technician B
   d. neither Technician A nor Technician B
10. The device that allows the computer to be connected to a phone line is known as a ____________________.

GLOSSARY

Access Time Time required to remove extensively damaged collision parts by cutting, pushing, pulling, etc.
Actual Cash Value (ACV) Current market value of a standard production vehicle and its accessory options as determined by used car guidebook listings or car dealer assessments.
Adjuster The insurance company representative who is responsible for approving the collision repair bid and satisfying a customer's vehicle damage claim.
Aftermarket Parts produced by a company other than OEM.
Air Deflector A structure that increases air flow efficiency by directing air, usually located at the front of a vehicle.
Align To adjust a line or predetermined relative position.
Anchor To hold in place.
Appraisal Estimate written by an insurance adjuster.
Appraiser An individual who writes estimates for an insurance company.
Apron The panel that is welded to the strut tower and radiator support on unibody vehicles.
Arc Braze A flexible factory weld used to fill panel seams.
Assembly A number of auto body parts that are either bolted or welded together, forming a single unit.

Back Glass Vehicle window located behind the passengers.
Backup Light The white or clear portion of the taillight that comes on when the vehicle is in reverse.
Baffle Sheet metal panel used to direct air to the radiator.
Basecoat/Clear Coat A paint system in which the color effect is given by a highly pigmented basecoat. Gloss and durability are given by a subsequent clear coat. The basecoat can be either straight color or metallic.
Belt Line The horizontal moulding or crown along the side of the vehicle.
Belt Moulding A trim strip that goes around the vehicle slightly below the lower portion of the vehicle's windows.
Bench System A vehicle underbody measurement device; the vehicle is set on preadjusted pins to check its frame and suspension dimensions for damage.
Betterment Depreciation of vehicle parts due to age or wear, determined by the insurance adjuster.
Bezel A chrome or plastic rim surrounding a lamp or instrument.
Blend A gradual tapering of color used to hide a slight color mismatch.
Body Filler Polyester compound mixed with a catalyst to bring low areas up to the proper level.
Body Over Frame Vehicle construction that consists of a separate body bolted to a frame.
Body Panels Sheets of steel fastened together to form the car body.
Body Section A repair method that involves cutting the damaged vehicle in half and welding on a replacement half.
Body Trim Upholstery or body ornamentation.
Bolt-On Panels Auto body panels fastened with screws or bolts.
Bracket An attachment used to secure parts to the body or frame.
Brake Light That portion of the taillight assembly that lights up when the brake pedal is depressed.
Buckles High places associated with dents in a metal panel.
Bulge High crown or area of stretched metal.
Bumper Brackets Supports joined to the body frame to which the bumper is attached.
Bumper Isolator A mechanism designed to help contain inertial energy created in a collision involving the vehicle's bumper.
Bumper Reinforcement Box-shaped energy-absorbing structure located behind the bumper or bumper cover. It can be made of steel, aluminum, or aluminum/plastic.
Bumper Surround Panel The panel surrounding the face bar of the bumper; usually a combination of the filler, valance, and end caps.

Camber Inward or outward tilt of the wheel at the top. It is the tire wearing angle measured in degrees and the amount the centerline of the wheel is tilted from true vertical.
Capsule The headlight mounting bucket, retaining ring, adjusting screws, tension spring, screws, and sometimes the sealed beam purchased as a unit.
Caster The backward or forward tilt of the kingpin or spindle support arm. It is the directional control angle measured in degrees and is the amount the centerline of the spindle support arm is tilted from the true vertical.
Centerline Imaginary line that divides the car into equal left and right halves.
Center Pillar Four-door sedan and four-door hardtop; serves as a hinge pillar or post for the rear doors and as a latch pillar for the front doors. Same as center post.

Chassis The assembly of mechanisms that make up the major operating systems of the vehicle; usually assumed to include everything except the car body.
Clearance The amount of space between adjacent panels.
Clear Coat Top coating on a painted surface that is transparent so that the color coat beneath it is enhanced and protected.
Clip Section of a salvaged vehicle that contains more than one complete panel; usually obtained from a salvage yard.
Clip Moulding Moulding attached to a vehicle by spring clips, screws, nuts, or bolts.
Closed Item A damage report or appraisal in which every item has an amount listed for all repair parts.
Closed Structural Members Boxed-in sections, typically accessible only from the outside, such as rails, pillars, and so forth.
Collision Estimating Guide Manuals containing part prices and flat-rate times for panel replacement.
Collision Insurance Insurance that covers the repair of the insured's car if the insured causes the damage. Also, the insured's vehicle is covered for repairs by this insurance if it is hit by an uninsured motorist.
Compounding The action of using an abrasive material—either by hand or machine—to smooth and bring out the gloss of the applied lacquer topcoat.
Comprehensive Insurance Covers damages not covered by collision or liability insurance, such as hail, wind, fire, and theft.
Computer Bid An estimate that obtains parts costs and labor rates from a computer programmed for this purpose.
Cone Concept A way of looking at and thinking about the progress of damage in a collision.
Control Points Points on a vehicle, including holes, flats, or other identifying areas, used to position panels and rails during the manufacturing and redimensioning of vehicles.
Conventional Frame An auto body construction type in which the engine and body are bolted to a separate frame.
Cornering Lamp A lamp located on the lower front fender. It illuminates the area on the side of a vehicle when the direction lights are turned on.
Courtesy Bid Estimates written higher than the original estimate to enable the original bidder to be the lowest bidder. A form of price fixing.
Cowl and Dash Assembly Sheet metal panels welded together that form the front of the body shell or passenger compartment of the vehicle.
Cowl Panels Panels in front of the passenger compartment to which the fenders, hood, and dashboard are bolted.
CPU Central processing unit—the computer.
Crazing Small cracks in the paint surface.
Cross Member Major portion of the frame that holds the frame rails together and adds rigidity.
Crown A bend in the metal. The higher the crown, the higher the strength.

Damage Report An estimate.
Dash Partition between the engine and the operator; also known as the fire wall.
Datum Line An imaginary line that appears on frame blueprints or charts to help determine correct frame height.
Dedicated Fixture Measuring System A bench with fixtures that are set in specific points for body measurement.
Depreciation Loss of value due to age or wear.
Detailing Final cleanup of a vehicle before it is released to the customer.
Diamond Body damage in which one side of the vehicle has been moved to the rear or front, causing the frame/body to be out of square.
Direct Damage Damage to the area in direct contact with the damaging force or impact.
Dog Tracking Off-center tracking of rear wheels as related to the front wheels.
Door Skin Outer sheet metal panel of a door.
Drip Moulding Metal moulding that serves as a rain gutter over side glass.
Drivetrain Engine, transmission, differential, and drive axels.

Elastic Deformation Occurs when the force is less than the yield strength of the metal. The metal will return to its normal shape when the force is removed.
Energy Absorber Any of several impact-absorbing bumper attachments that lessen vehicle damages in front- or rear-end collisions.
Estimate A written report giving a detailed description of all parts and labor costs needed to restore a vehicle to its preaccident condition. Estimates made by collision repair facilities are often called damage reports. Insurance adjusters call their reports appraisals.
Estimator Person who prepared estimates or damage reports for the collision repair facility.

Face Bar A metallic bumper located at the front or rear of the vehicle. May be chrome plated, painted, or covered with a urethane valance.
Featheredging Tapering the edges of the damaged area with sandpaper or special solvents.
Fender Panel structure that encloses the outside wheel.
FF A front engine, front wheel drive vehicle.
Flat Rate The time listed in an estimating guide to perform a specific labor operation.

FR A front engine, rear wheel drive vehicle.
Frame The heavy metal structure that supports the auto body and other external components.
Frame Alignment The procedure by which the frame of a car, truck, or bus damaged in an accident is restored to the manufacturer's specifications.
Frame Horn Portion of an automobile frame used to support the front bumper assembly.
Front Clip A group of used parts including the hood, fenders, grille, bumper, radiator and support, cowl, and front half of the frame that can be purchased as an assembly to be installed on another car.
Front Wheel Drive Drive train that powers the front wheels.
Fuel Door The hinged fuel intake cover located on the vehicle's side or rear.
Full Body Section Simultaneous section repairs to both rocker panels, windshield pillars, and floor pan; required to join the undamaged front half of one vehicle to the replacement undamaged half of another vehicle.

Garnish Moulding Trim surrounding glass openings on the interior of the vehicle.
Gauge A measure of thickness of sheet metal.
Grille Body panel used to protect the radiator of the car. It allows air to pass through the front of the vehicle to the radiator.

Hammer Off Dolly Metal finishing technique used to level metal.
Hammer On Dolly Metal finishing technique used to stretch metal.
Hardtop An auto body style that has a rigid roof and convertible-type door.
Hatchback A vehicle body that has a large rear end door, which is accessible to the passenger compartment.
Hazardous Waste Leftover or consumed products produced by a collision repair facility that cannot be disposed of in a landfill, such as paint, oil, antifreeze, and spray booth filters.
Header Bar Framework or inner construction that joins the upper sections of the windshield and pillars, forms the upper portion of the windshield opening, and reinforces the panel.
Header Panel Headlight mounting panel, found on some vehicles.
Headlining The cloth or plastic material covering the ceiling area in the passenger compartment.
High-Strength Steel A low-alloy steel that is much stronger than hot-rolled or cold-rolled sheet steels normally used in the manufacture of car body frames.
Hinge Pillar Framework or inner construction to which the door hinges fasten.
Hookup Attachment to a vehicle to pull or hold.
HSLA High-strength low-alloy steel used in unibody construction.
HSS High-strength steel used in unibody construction.

Impact Absorber Bumper energy absorber mounted between the frame and reinforcement.
Indirect Damage Damage occurring somewhere other than at the point of impact.
Inner Panel Reinforcing metal panel connected to outer panels.
Insured The individual who has paid for insurance and is thereby represented by his or her insurance company in the settlement of damage claims.
Interior Trim All the upholstery and moulding on the inside of the vehicle.

Job Costing Process that evaluates the profit or loss on a repair job.

Ladder Frame BOF pickup truck frame, straight side rails.
Length The linear dimension of a vehicle component.
Liability Insurance Usually covers the insured against claims resulting from an accident caused by the operation of the insured's vehicle. It protects the insured's personal possessions from being lost through a lawsuit initiated by the owners of the vehicle or property that the insured damaged. This insurance also pays for the repair of the vehicle belonging to the other motorist who is not at fault in the collision.
Like Kind and Quality Salvage parts that are equal in value to the precollision condition of the damaged parts.
Lock Pillar The door post containing the lock striker.

MacPherson Strut A type of independent suspension without an upper control arm; supported on top by a reinforced inner panel.
Martensitic Steel High-strength steel used in door impact beams.
Mash A type of vehicle damage in which the length of any section or frame member is shorter than factory specifications.
Masking Using tape and paper to protect an area that will not be painted.
Melt An application of reducer used to dissolve the dry overspray at the edge of the clear coat.
Metal Bumping Returning damaged metal to its original shape.
Metal Conditioner A chemical cleaner that removes rust and corrosion from bare metal and helps prevent further rusting.
Metallic Paint Finish-paint colors that contain metallic flakes in addition to pigment.
MIG Gas shielded metal arc welding.

No-Fault Insurance A statute in some states that allows motorists to carry 100 percent no-fault insurance. This means the insured's vehicle and/or personal injury are paid by the insurance company regardless of who causes the accident. No effort is made to determine which motorist is at fault when settling the claim if both parties carry no-fault.

OEM Original equipment manufacturer.
OH Overhaul. Take the assembly off; remove all of the components. Reassemble and mount.
Open Item An item on a vehicle suspected of needing repair for which a bid price cannot be made until the damage is opened up.
Outer Panels Sheet metal sections that, when attached to the inner panels, form the auto's exterior.
Overlap When repairing adjacent panels, common labor operations can be deducted from one panel.

Perimeter Frame BOF, side rails not straight, has torque boxes at the corners of the center section.
Pillar A reinforced, boxed upright body structure portion to which doors and windows are attached or held.
Plastic Deformation Occurs when the force is greater than the yield strength of the metal. The metal will not return to its normal shape when the force is released.
Pressure Area A high spot in the metal.
Primary Damage Local damage that occurs at the point of impact on the vehicle.
Prime Coat The first coat in a paint system—its main purpose is to impart adhesion.

Quarter Glass Glass located in the upper quarter panel.
Quarter Panel The side panel extending from the lock pillar to the rear of the vehicle. This includes the wheel opening, rear door rear pillar, sail panel, and, on station wagons, the window openings.
Quarter Panel Extension A short bolt-on extension panel that closes the rear end of a quarter panel.

Radiator Support The front structural part of a unibody vehicle.
R&I Remove and install. When a part is in the way, take it off and then put it back.
R&R Remove and replace a damaged part.
Rear Clip A group of used parts—including two quarter panels, upper and lower body panels, deck lid, bumper, floor pan, and sometimes the rear half of the frame—that can be purchased as an assembly to be installed on another car.
Rear Wheel Drive Drivetrain that powers the rear wheels.
Rear Wheelhouse Panel An inner panel attached over the rear wheel that keeps out mud and water; part of the quarter panel.
Re-core A repair to the radiator in which the water passageways and air directional fins are replaced with a new unit using the original tanks, fillings, etc.
Resistance Spot Weld Structural electrical weld.
Retainer Ring A small steel ring that attaches to the headlight body and holds a sealed beam in position.
Reveal Moulding An exterior window trim strip that attaches to the edge of the glass.
Rocker Panel The narrow, outer panel attached below the car door. Same as a sill panel.
Roof Panel Forms the top of the passenger compartment.
Roof Rails Framework or inner construction that reinforces and supports the sides of the roof panel.

Sag A type of frame damage in which one or both side rails bend and sag at the cowl, causing buckles.
Salvage Used parts.
Sealer The undercoat used as an adhesion promoter and uniformer.
Secondary Damage Damage that occurs due to misplaced energy; causes stresses in suspension and/or body dimensions at areas other than the primary impact zone.
Section Repair A repair made by cutting a component, removing the damaged portion, and replacing it with an undamaged component.
Shrink A metal finish technique used to reduce the metal surface area.
Side Member A piece that runs parallel with the frame to add support, especially in the front and rear of the vehicle; side rail.
Sidesway Damage that results when either the front, center, or rear of the vehicle has been pushed sideways out of alignment, causing the structural linear alignment of the vehicle to be altered.
Single Stage Paint system that is top coat color only, no clear coat.
Space Frame Unibody design consisting of cosmetic outer panels attached to a rigid one-piece framework.
Spot Finish Repair Refinish repair job in which a section of a car smaller than a panel is refinished.
Stress A disruption in the grain structure of the metal.
Stress relief Returning the grain structure in the metal back to normal by tension, spring hammering, or heat.
Stretch A damage condition in which the metal surface area is greater and the panel thickness is less.
Stub Frame A unitized body frame with only front and rear stub sections of the side rail and no center side rail portions.

Sublet Someone else to whom the damaged vehicle is given to perform a service operation.
Supplement An addition to an estimate.
Supplemental Restraint System Air bag system.
Surfacer The undercoat used to fill scratches.

Technician A trained autobody repair professional, body person, frame person, mechanic, or painter.
Tension Area A low spot in the metal.
Tie-down Attachment to the vehicle to immobilize it.
Tinted Windshield A windshield with a blue, green, or reddish tint that reduces sunlight glare.
Toe Condition The rotation of the wheel in or out as seen at the front of the wheel looking from the side.
Tolerances The acceptable alteration limit of the vehicle dimensions as provided by the manufacturer; for example, +3 mm is typical.
Topcoat The final color coat.
Torque Box Frame side rail configurations designed to trap and hold damaging inertial forces; located in the lower cowl and front or rear suspension areas.
Tracking The ability of the rear wheels to follow the front wheels.
Tram Gauges Gauges used to accurately measure and diagnose body and frame collision damages for all conventional and unibody vehicles.
Transfer Time Time necessary to weld or rivet brackets, braces, or reinforcements from an old part to a new part.
Tricoat Paint system of a base color, pearl midcoat, and clear coat.
Twist A type of frame damage in which both side rails are bent out of alignment so that they do not run parallel to one another.

Undercoat A first coat: primer, sealer, or surfacer.
Unibody Unitized construction.
Unitized Construction Auto body construction in which the body and frame are welded together as one unit.
Universal Measuring System A measuring system that has devices mounted on a frame that can be adjusted for various vehicle bodies. This system can utilize laser systems, universal benches, or mechanical gauges.

Valence Panel An outer facial panel placed over or below the front or rear bumpers to cover the underbody.
Vehicle Identification Number (VIN) The number assigned to each vehicle by its manufacturer, primarily for registration and identification purposes.

Weathering The deterioration of a car's painted surface due to exposure to environmental conditions.
Wheel Alignment Positioning the suspension and steering components to assure a vehicle's proper handling and optimum tire wear.
Windshield Header Box-type reinforcement for the top of the windshield opening and front of the roof panel.
Work Hardening Stiffening of the metal from plastic deformation.

X Check Two diagonal measurements used to check for squareness.

Zone 1 Primary damage area.
Zone 2 Secondary damage area.
Zone 3 Mechanical components.
Zone 4 Interior components.
Zone 5 Exterior trim and finish.
Zone Inspection A method of systematically observing a damaged vehicle, including checking primary and secondary damage, the mechanical components, interior components, and, lastly, the exterior paint and trim of the vehicle.

Index

Page numbers followed by a t or a f indicate that the entry is included in a table or figure.

A

Access time, of repair operations, 138, 139f
Accounting, computerized, 283–284
Acrylic enamels, 100
Acrylic lacquers, 99
Acrylic urethane enamels, 100, 100f
Acts-of-nature damage, covered by comprehensive insurance, 6
Actual cash value (ACV), of vehicle, 254–255, 254f–256f, 257
  options affecting, 265
Acura NSX, all-aluminum unibody of, 34–35
Adhesion promoter, 98–99
Adhesives, 2, 39–41, 40f–41f
  for glass installation, 42–43, 43f, 176
Aftermarket parts, 6–7
  disclosure statement for, 257–258
  estimating prices for, 145, 145f
Air bags. *See also* Restraint systems, assessment of
  assessment of, 125, 201, 203f–205f
Air-conditioning, damage assessment of, 212–214, 213f, 215f
Alignment. *See also* Realignment
  of grain structure, of sheet metal, 58, 58f, 69, 70f, 131f
  on unibody vehicles, 2, 24, 26
Aluminum, 34–35
  welding of, 85
Anchoring, of vehicle, for frame realignment, 67–68, 68f, 75, 75f
Antenna wire, on/within automotive glass, 42
Antilacerative glass, 42
Antilock brake system (ABS), 236, 237f–238f
Aprons/wheelhouses, damage assessment of, 164–166, 165f–166f
Arc brazing, 36, 82f, 87
Automatic transaxle
  of front wheel drive vehicle, 221, 221f
  of rear wheel drive vehicle, 223, 224f
Automotive Service Association (ASA), 8, 9f
AWS E-70S-6 wire, with MIG welding, for high-strength steels, 36–37

B

Basecoat/clear coat finishes, 100–101, 101f
Batteries, depreciation of, 146
Bed, of pickup truck, 23
Bed side replacement, in pickup truck, 188–190, 190f–191f
Betterment, 145–146, 146f–147f, 267, 267f
"Biscuits," rubber, on vehicle frame, 20, 20f, 21, 52
Blacklisting, of collision repair facilities, as unethical practice, 11
Body filler, application of, 96, 96f
Body-over-frame (BOF) construction, 2, 18, 20f, 20–23, 21f–23f
  anchoring of, 67–68, 68f
  buckles in, 50. *See also* Buckle damage
  characteristics of, 20–21
  components of, 21–23, 22f–23f
  diamond condition of, 51
  impact effects on, 52f, 52–53, 53f–54f
  passenger compartment of, 46
  pulling/alignment of, 76f, 76–77
  sidesway indicated on, 50
  twist damage in, 51. *See also* Twist damage
Body shell, 173, 174f
Body weight (average fleet), past and present, 20
Bonding, of nonstructural panels, 88–90, 89f–90f
Box, of pickup truck, 23
Box-shaped sections, on frame side members, 20, 21f, 60, 60f
Brakes, damage assessment of, 236, 236f–238f
Brazed areas, separation of, 78–79
Brushable seam sealers, 39–41, 41f
Buckle damage, 50f, 50–51, 51f, 55f, 59
  collapsed hinge, 60, 60f
  collapsed rolled, 61, 64, 64f
  simple hinge, 59–60, 60f
  simple rolled, 61, 61f
Buick Riviera, aluminum hood used in, 34
Bumpers
  damage assessment of, 150–155, 150f–155f
  front, 35f
  rear, 74
Business analysis, computerized, 283–284
Business management, for collision repair facilities, 3

C

Cab
  on body-over-frame construction, 23, 23f
  corners, 188, 188f–189f
Camaro (Chevrolet), composite construction of, 2
Camber angle, assessment of, 123–125, 124f
Catalytic converter, 226f, 226–227
Centerline, used in gauging, 118, 119f
Center of gravity, and impact, direction of, 46–47, 47f
Center pillar, damage assessment of, 181, 181f
Changes for profit, unauthorized, 11
Charging system, 214, 216f
Check digit, in VIN, 260
Chemical adhesive bonding, for thermosetting plastics, 37, 102, 103t, 104f
Chevrolet Camaro, composite construction of, 2
Chip-resistant primer, 98
Claims manager, of insurance company, 7
Clamps, attached to vehicle, for frame realignment, 68, 69f
Clear vision position, measurement of, 236
Coil spring suspension, 227–229, 228f–229f
Cold-rolled steels, 34
Cold-working, of metal, 70, 71f
Collapsed hinge buckle, 60, 60f
Collapsed rolled buckle, 61, 64, 64f
Collision
  damage from. *See also* Impact forces
    sequence of, 52, 56–57, 64, 70, 76
    types of, 49–52, 50f–51f
  dissipation area and, 47–48, 48f
  impact strength and, 46f, 46–47, 47f
  and stress concentration, 48f, 48–49, 49f
Collision estimating guides, 6f, 6–7, 7f, 13–15, 14f
  and sectioning, 87, 87f

Collision insurance, 4
Collision repair facilities
  ethical practices of, 10–12
  and insurance companies, professional relationship between, 8–9, 9f
  job tasks of, 1–2
  safety guidelines for visiting, 107
  selection of, 10
Combination crown panels, 62f, 63, 63f
Compaction, during collision, 46
Competency testing, of adjusters/repair facilities, 8
Competition, unethical promotion of, by insurance companies, 12
Component assembly technique, of vehicle assessment, 32, 34f–35f
Composite vehicles (steel/plastic), 2
Comprehensive insurance, 4–5
Computerized estimating systems, 2–3, 5f, 7, 279–280, 280f, 287–288
  advantages of, 288
  components of, 284f–286f, 284–288, 288f–290f
  for damage report writing, 280–282, 281f
  data base for, crash guide information used in, 14
  for job costing, 282
  terminology for, 284–285
Cone concept, of unibody impact, 54, 54f
Continuous welds, 81, 84f
  separation of, 78, 78f
Convoluted support members, in unibody vehicles, 48–49, 49f
Cooling system, damage assessment of, 210–212, 210f–212f
Cornering angle, assessment of, 123
Corporate Average Fuel Economy (CAFE) standard, 19
Corrosion protection
  seam sealers and, 39
  on sheet metals, 34
Corvette, plastic use in, 37
Cost management programs, computerized, 2–3
Cowl/dash assembly, damage assessment of, 173–176, 175f
Crash guides. *See* Collision estimating guides
Crush zones, of unibody vehicles, 54f, 54–55, 55f, 57
Customer
  confidence in collision repair facility, 4, 209
  rights/responsibilities of, 12
  unethical practices of, 11
  vehicle owner as, 9–10

D
Damage reports, 3–4, 5f
  with computerized estimating systems, 280–282, 281f
  fairness of, 253
  by insurance adjuster/appraiser, 6
  logical sequence in preparing, 13, 32, 108–109, 268
  as negotiating tools, 4, 7
  preparation of, 259–267, 262f–267f, 264t-265t, 267f, 267–274, 270f–273f
  tools/equipment needed for, 15, 107–108, 108f
  review of procedure, 277–278
  as selling tools, 4, 6f
Dash panel
  assessment of, 125, 173–176, 175f
  on body-over-frame construction, 22
  repairs to, 72, 73f
Datum, used in gauging, 118, 119f
Deductibles
  in collision insurance policy, 4
  illegal saving of, 11
Defrost glass, 42
Dents
  pulling, 93f, 93–94, 94f
    with studs, 94, 94f
  raising with pick, 93, 93f
  repair of. *See* Metalworking; Spring hammering
Depreciation, of parts. *See* Betterment
Design changes, in vehicles, 2, 17–20, 18f–19f
Detailing, after refinishing, 242–243, 243f
Dewaxing, before refinishing, 97, 97f, 239–240, 240f
Diamond damage, 50f, 51, 77, 77f
Digital imaging, 283–284, 284f
Dimensional measurement methods, of panel positioning, 80, 80f
Direct damage, to sheet metal, 59, 59f, 64
Direct Repair Programs (DRP), 9
Disc brakes, 236, 236f
Disclosure statement, for aftermarket parts, 257–258
Discounts, unethical requests of, by insurance companies, 12
Dissipation area, of impact, 47–48, 48f
Dolly, for spring hammering, 69–70, 71f, 90, 91f, 92, 92f
Doors, 34f, 125f, 125–126
  damage assessment of, 190–193, 191f–192f, 194f
Double crown panels, 63f, 63–64, 64f
Drive axles, of front wheel drive vehicle, 221, 221f
Drive-in claims centers, for appraisal and settlement, 8
Driveshaft, of rear wheel drive, 223, 224f
Drum brakes, 236, 237f
Durability, as repair goal, 9

E
Economy standards, for fuel, 19
Eggshell concept, of unibody construction, 19, 19f
Elastic deformation, of steel, 57, 58f
Electrical system, damage assessment of, 214–218, 216f–218f
Emission control standards, 19
Enamel paints, 99–100
Encapsulated glass, 41
End-to-end stretch pull, 75, 76f
Energy absorbers, within bumpers, 152–155, 152f–155f
Engine
  in front wheel drive vehicle, 218–219, 219f–220f
  in rear wheel drive vehicle, 222, 223f
Engine cradle, of front wheel drive vehicle, 221, 222f
Entire-car painting, 249–250
Environmental Protection Agency (EPA), emission control standard of, 19
Epoxy primer, 98
Equipment
  for damage inspections, 15, 107–108, 108f
  for glass removal, 43
  for new design changes, 2–3, 3f
Estimates
  multiple, requested by insurance company, 10
  verbal/nonwritten, avoidance of, 15
Estimating guides. *See* Collision estimating guides
Estimating systems, computerized, 2–3, 5f, 7
Ethics, in collision repair, 10–12
Exhaust system, 225–227, 226f
Extended-cab, on body-over-frame construction, 23
Exterior trim/finish (Zone 5), assessment of, 126, 126f

F
Falsified claims. *See* Ethics, in collision repair
Fastener method, of panel installation, 79
Featheredging, before refinishing, 97, 97f, 99, 240, 240f
Fenders
  on body-over-frame construction, 22, 22f
  damage assessment of, 162–164, 162f–164f
  indirect buckle damage to, 61f
Finish, as repair goal, 9
Fire damage, 6, 126
Fit, as repair goal, 9
Flare side pickup bed, 23
Flat rate operations, 134–137, 134f–137f. *See also* Labor times/rates
Flood damage, 126
Ford F150, aluminum hood used in, 34
Four wheel drive, 223, 225f
Frame construction design, history of, 2, 17–20, 18f–19f
Frame members
  damage assessment of, 166–168, 167f–171f
  repairs to, 72, 72f, 74
Front body
  assessment of damage to, 149–171
  of FF unitized vehicle, 30, 31f
  of FR unitized vehicle, 26f, 26–27
Front bumper, 35f

Front engine/front drive (FF) unibody vehicles, 29–32, 30f–33f
  characteristics of, 30
  frame realignment of, 73, 74f
Front engine/rear drive (FR) unibody vehicles, 26f, 26–28
  characteristics of, 26
  frame realignment of, 73, 74f
Front fender, on body-over-frame construction, 22, 22f
Front impacts, unibody, 55, 55f–56f, 71f
Front pillar, repairs to, 72, 73f
Front suspension systems, 227–231, 228f–231f
Front wheel drive vehicle. *See also* Front engine/front drive (FF) unibody vehicles
  automatic transaxle of, 221, 221f
  damage assessment of, 218–219, 219f–220f
  drive axles of, 221, 221f
  introduction of, 20
Fuel economy standards, 19
Full-frame vehicles, 2, 18
Fusion welding, 80

G
Galvanized sheet metal, 34
Garnish mouldings, around automotive glass, 42
Gasket technique, of glass installation, 42, 42f
Gas metal arc welding (GMAW) system, 81
Gas torch, metal shrinking with, 95f, 95–96
Gauging, of frame damage, 118–121, 119f–121f
General Motors
  plastics used by, 37
  vans, composite construction of, 2
Glass, 41f, 41–43, 42f–43f, 143f. *See also* Windshield
  antilacerative, 42
  broken, cleanup of, 12, 41
  defrost, 42
  encapsulated, 41
  installation of, 42f, 42–43, 43f
  laminated, 41–42
  modular, 43, 43f
  mouldings around, 42
  parts/prices, listed in collision estimating guides, 13
  rear, 184
  safety, 41–42
  tempered, 42
  tinted, 42
GM
  plastics used by, 37
  vans, composite construction of, 2
Grain structure, of sheet metal, 58, 58f, 69, 70f, 131f
Gravity, center of, and impact direction, 46–47, 47f
Grille, damage assessment of, 155–156, 156f

H
Hail damage, covered by comprehensive insurance, 6
Hammer, body, 69–70, 71f, 90, 91f–92f
Hammer-off-dolly method, of metal repair, 92f, 92–93
Hammer-on-dolly method, of metal repair, 92
Hanging gauges, 119–121, 120f
Hazardous wastes, handling/disposing of, impact on costs, 2
Header panel, damage assessment of, 156f, 156–157
Headlight assemblies, damage assessment of, 157, 157f–158f, 163, 216–217, 217f
Head-on collision, compaction during, 46
Heat
  applied before spring-hammering, 70, 71f
  effect on plastic, 39f
  high-strength steels weakened by, 2, 35–37, 51, 60
Heavy-bodied sealers, 39, 40f
Hidden damage, supplements for, 4, 7
High spots, of pressure, in damaged metal, 61–64, 62f–64f
High-strength, low-alloy (HSLA) steels, 36, 81
High-strength steels, 32, 35–37, 36f, 70
  weakened by heat, 2, 35–37, 51, 60
High-tensile strength (HSS) steels, 36–37
Hinge buckle
  collapsed, 60, 60f
  simple, 59–60, 60f
Hinge/windshield pillars, damage assessment of, 178–179, 179f
Hood panel, damage assessment of, 159–162, 160f–161f
Horizontal force, of impact, 46, 46f
Hot-rolled steels, 34
Hourglass frame shape, 18, 18f
HSLA steels, 36, 81
Hydraulic lift, for damage inspections, 15, 15f

I
Identification
  of paint, 264, 264f–265f
  of plastic type, 37, 40t
Impact forces
  area of, 46f, 46–47, 47f
  effect on body-over-frame vehicles, 52f, 52–53, 53f–54f
  effect on sheet metal, 57–59, 58f–59f
  effect on unibodies, 54f, 54–57, 55f–56f
  strength of, 46f, 46–47, 47f
  transmission of, 49, 49f
Independent collision repair facilities, 2
Independent strut-type front suspension, 30
Indirect damage, to sheet metal, 59f, 59–61
Inertia damage, 57
Information storage/retrieval, computerized, 286
Inner quarter panel, 186
Insurance
  adjusters/appraisers, 6
  companies
    ethical practices of, 10–12
    multiple estimates requested by, 10
    ongoing training in, 8
    procedures followed by, 6–8
    professional relationships with, 8–9, 9f
  information about, on damage report, 260
  types of, required for auto loans, 4–5
Inter-Industry Conference on Auto Collision Repair (I-CAR), 8, 9f, 108
Interior trim/upholstery, damage assessment of, 199–201, 200f, 202f–203f
Inventory control, with computerized system, 282–283
Isolators, 35f

J
Job costing
  with computerized system, 282
  system of, 4
Jounce-rebound test
  for camber measurement, 123–125, 124f
  for wheel alignment, 122–123, 123f
Judgement labor time, 13. *See also* Nonincluded operations

K
Kickup. *See* Sag damage

L
Labor times/rates
  additions to, 138f, 138–139
  adjustments to, 137–139, 138f–139f
  from collision estimating guides, 6–7, 7f, 13, 132t, 132f, 132–133, 133f–134f
  on sealers, 41
  unethical determination/fixing of, 12
Ladder-shaped frames
  on early vehicles, 18, 18f
  modern, 21, 21f
Laminated glass, 41–42
Lateral bending, of unibody vehicle, 55, 55f, 57, 73, 74f
Leaf spring suspension, 232f
Liability insurance, 4
Licensing, of adjusters/repair facilities, 8
Liftbacks/liftgates, rear body section of, 28. *See also* Tailgates
Lighting systems, 214–217, 216f–218f. *See also* Headlight assemblies, damage assessment of
Linkage, steering, 233–234, 234f–235f
Lock pillar, 186, 187f
Longitudinal force, of impact, 46, 46f
Low-carbon (mild) steels, 32, 34

Low spots, of tension, in damaged metal, 61–64, 62f–64f
Luggage lid, damage assessment of, 196, 196f–197f
Lumina van, space frame of, 32, 33f, 37, 39f

M

MacPherson strut suspension, 230, 230f
 introduction of, 20
 realignment of, 73, 74f
 towers used in. *See* Strut towers, damage to
Management, of collision repair facilities, 3
Manual transaxle
 of front wheel drive vehicle, 220–221, 221f
 of rear wheel drive vehicle, 223, 223f
Market value, of vehicle, 254–255, 254f–256f
Martensitic steels, 37
Mash damage, 50f, 51, 51f, 57, 76
Masking, before refinishing, 97, 97f, 241, 241f
Mazda RX7, aluminum hood used in, 34
Measurement
 dimensional, panel positioning via, 80, 80f
 of secondary damage (Zone 2), 112–114, 112f–117f, 115–121, 118f–121f
 of vehicle dimensions, for frame straightening, 73, 73f
Mechanical damage (Zone 3). *See also* Separate assemblies
 analysis of, 121–125, 122f–124f, 209–238
Mechanical parts, depreciation of, 146
Metal adhesive bonding, of nonstructural panels, 88–90, 89f–90f
Metal composition. *See also* Steel
 recent changes in, 20
 used in vehicles, 32–37, 36f
Metal conditioning, before refinishing, 97, 241, 241f
Metal inert gas welding. See MIG welding
Metal shrinking, 95f, 95–96
Metalworking, 90–94, 91f–94f. *See also* Dents; Spring hammering
MIG welding, 80–85, 82f–83f
 equipment needed for, 2
 for high-strength steels, 36–37
 vs. resistance spot welding, 85
 techniques of, 81–84, 84f
 of unibody vehicles, 2
Mild coat primer, 98–99
Mild (low-carbon) steels, 32, 34
Mileage, recorded on damage report, 264, 265f
Mitchell Collision Estimating Guide, 6f, 13, 130f. *See also* Collision estimating guides
Modular glass, 43, 43f
Motor Crash Estimating Guide, 13. *See also* Collision estimating guides
Mouldings, around automotive glass, 42
Mounts, rubber, on vehicle frame, 20, 20f, 21, 52
Multiple estimates, requested by insurance company, 10
Multiple-pull systems, for frame realignment, 68

N

Nash, early unibody design by, 18–19
National Auto Glass Specifications (NAGS), 13, 176
National Institute for Automotive Service Excellence (ASE), 8, 9f
Negotiation
 between insurance company and collision repair facility, 9, 13. *See also* Nonincluded operations
 using damage reports, 4, 7
No-fault insurance, 6
Noise/vibration
 amplified in unibodies, 24, 30
 sealers to reduce, 39
Nonincluded operations
 in bed side replacement, 190
 in bumper repair times, 155
 in collision estimating guides, 133, 133f, 138–139
 in cowl/dash assembly repair, 175–176
 in door skin/shell repair, 192–193
 in frame repair, 168
 in grille repair times, 156
 in header panel repair times, 155
 in hood repair, 161–162
 in luggage lid (rear deck lid) repair, 196
 in quarter panel repair, 188
 in radiator support repair, 159
 in rear panel repair, 195–196
 in refinishing, 243–244
 in remove & replace, 134–135
 in rocker panel repair, 181
 in roof panel repair, 184
Nonstructural parts, replacement of, 87–90, 88f–89f
Nonweldable plastics. *See* Thermosetting plastics (nonweldable)
Nonwritten estimates, avoidance of, 15

O

Operation times. *See* Labor times/rates
Original equipment manufacturer (OEM) parts, 6
 listed in collision estimating guides, 13
 prices for, 141–143, 142f–143f
Outer quarter panel, 184–186, 185f–186f
Overhaul (O/H) times, 136, 136f
Overlap
 of refinishing operations, 245
 of repair operations, 137–138, 138f
Oxyacetylene welding
 for HSS steels, 37
 vs. MIG welding, 81
 not recommended for HSLA steels, 36
 separation of brazed areas with, 78

P

Paint. *See also* Refinishing
 computerized shop operations, 283, 283f
 depreciation of, 146
 identification/coding of, 264, 264f–265f
 removal of, before metal repair, 91, 91f
Panels
 depreciation of, 146
 installation of, 79f, 79–80, 80f
 labor rate estimate for, 140f, 140–141, 141f
 positioning techniques for, 80, 80f
 removal of, 77–79, 78f
Partial boxed sections, 60, 61f
Parts
 prices
  aftermarket, 145, 145f
  listed in collision estimating guides, 13. *See also* Separate assemblies
  original equipment manufacturer (OEM), 141–143, 142f–143f
  salvage, 143f, 143–144, 144f
 selection factors, 259
 types of, 6–7
Passenger compartment (Zone 4)
 assessment of, 125f, 125–126
 impact effects on, 46
 of perimeter frame, 21
 protected by stressed hull structuring, 19, 24, 46
Perimeter frames, 21, 21f–22f, 52f
Photographic record, of collision damage, 274–277, 277f
Pickup trucks, body-over-frame construction of, 2, 18, 22–23, 23f. *See also* Body-over-frame (BOF) construction
Plastic deformation, of steel, 57–58, 58f
Plastic primer, 98, 104
Plastics
 automotive, 2, 37–39, 38f–39f, 40t
 identification of type, 37, 40t
 refinishing of, 104, 104f
 repair/replace decisions regarding, 37–39
 repairs to, 102t, 102–104, 103t, 104f
  estimating labor times for, 141, 141f
 types of, 37, 39f
 welding of, 102–104
Plug weld, 81–82, 84f
Polyurethane enamels, 100
Positioning, of new panels, 80, 80f
Preaccident condition, vehicle repaired to, 9. *See also* Prior damage, assessment of

Preparation
for panel installation, 79, 79f, 80
for refinishing, 97, 97f
before spot welding, 86
Pressure forces, in damaged metal, 61–64, 62f–64f
Pressure welding, 80–81
"Prevailing practice," as reason not to pay, given by insurance company, 12
Primary damage (Zone 1), analysis of, 109f, 109–110, 110f
Primer, 98f, 98–99, 99f, 241
weld-through, 79, 98
Prior damage, assessment of, 266f, 266–267
Professional relationships, between collision repair facilities and insurance companies, 8–9, 9f
Profit margin
for collision repair facilities, 3
and computerized job costing, 282
for insurance companies, 11
Pulling. *See also* Realignment; Straightening
equipment for, 3f
estimating labor time for, 139–140
procedure for, 68–77, 70f–77f

Q

Quarter panel, damage assessment of, 184–190, 184f–191f

R

Rack-and-pinion steering
damage assessment of, 234–235, 235f
introduction of, 20
Radiator support
on body-over-frame construction, 21–22, 23f
damage assessment of, 158f, 158–159, 159f
Realignment
estimating labor time for, 139–140
of frame, 67–68, 68f–69f
of front-end damage, 70–73, 71f–74f
of rear damage, 74–75, 75f
of side damage, 75f, 75–77, 76f–77f
Rear body
damage assessment of, 193–196, 195f
of FF unitized vehicle, 30, 31f
of FR unitized vehicle, 28, 29f
Rear clip/cliptop, repair using, 199, 199f–200f
Rear impacts
straightening of, 74–75, 75f
unibody, 55, 55f–56f
Rear wheel alignment, in front engine/front drive (FF) vehicles, 30
Rear wheel drive vehicles, 222–223, 223f–225f. *See also* Front engine/rear drive (FR) unibody vehicles
front side member impact to, 73, 74f
Reference holes, measurement by, 73, 73f
Refinishing. *See also* Paint
adjustments to labor times, 244f, 244–245, 245f, 245t
application of, 98f, 98–101
cost of materials, 248–249, 248f–250f
included/nonincluded operations in, 243–244
of plastics, 104, 104f
surface preparation for, 97, 97f, 239–241, 240f–241f
Refrigerants, used in air conditioning, 213–214
Registration, of adjusters/repair facilities, 8
Remove and replace (R&R) operations, 134f, 134–135
Remove & install (R&I) operations, 135, 135f
Repair/replace decisions, 37–39, 129–132, 131f, 257
regarding plastic, 37–39
regarding steels, 51, 57
Resistance spot welding, 81, 82f, 85–86, 86f
Restraint systems, assessment of, 125, 125f, 126, 201, 203f–205f
Reveal mouldings, around automotive glass, 42
Rocker panels, damage assessment of, 179f, 179–181, 180f
Rollover, vehicle, 52–53, 54f, 56f
Roof, damage assessment of, 181–184, 182f–183f
Roughing-out stage, before metal repair, 90
Rubber gasket technique, of glass installation, 42, 42f, 176
Rubber mounts ("biscuits"), on vehicle frame, 20, 20f, 21, 52
Rust protection, on sheet metals, 34

S

Safety, as repair goal, 9
Safety glass, 41–42
Sag damage, 50, 50f, 51–52, 57, 77, 77f
Sail panel, 186
Salvage parts, 6–7
estimating prices for, 143f, 143–144, 144f
Sanding, before refinishing, 97, 97f, 240, 240f
Saturn cars
composite construction of, 2
space frame of, 32
Sealers, 39–41, 40f–41f
as primer, 99, 99f, 241
Seam sealers, 39–41, 40f–41f
Seat belts, assessment of, 126, 201, 205f–206f
Secondary damage (Zone 2)
analysis of, 110f, 110–121, 111f
measurement of, 112–114, 112f–117f, 115–121, 118f–121f
as result of inertia, 57
to unibody vehicles, 24, 54, 57
visual signs of, 111, 111f
"Second-generation" unibodies, 23, 23f
Sectioning method, of repair, 2, 45, 86f, 87, 87f, 136–137, 137f
Sedans, rear body section of, 28, 29f
Self-etching primer, 98
Semiunitized construction, 23, 23f
Sequence
of collision damage, 52, 56–57, 64, 70, 76
in preparing damage reports, 13, 32, 108–109, 268
Service identification label, 264, 265f
Sheet metal. *See also* Steel
damage to, 57–64, 58f–64f
used in vehicles, 32–33
Sheet molded compound (SMC), of Lumina van, 32, 33f, 37, 39f
Shock absorber, on body-over-frame vehicle, 20
Short-long arm suspensions, 227–229, 228f–230f
Shrinking, of stretched metal, 95f, 95–96
Side body
of FF unitized vehicle, 32f
of FR unitized vehicle, 27, 27f
panels, damage assessment of, 178–181, 179f–182f
Side impacts
straightening of, 75f, 75–77, 76f–77f
unibody, 56, 56f
Sidesway damage, 47, 49–50, 50f, 52, 57
correcting by pull, 72, 72f, 77, 77f
Simple hinge buckle, 59–60, 60f
Single crown panels, 62, 62f–63f
Skid lockup, as panic reaction, 47, 47f
Skills needed, for new design changes, 3
Slappers, dent repair with, 91
Society of Collision Repair Specialists (SCRS), 8, 9f
Solid seam sealers, 41
Space frame vehicles, 32, 33f, 37, 39f
Spoon, for spring hammering, 69–70, 71f, 90–91, 91f–92f
Sport utility vehicles (SUVs), body-over-frame construction of, 2, 18, 22–23. *See also* Body-over-frame (BOF) construction
Spot welding, 84, 84f
removal of, 77–78, 78f
on unibodies, 24
Springback, of steel. *See* Elastic deformation, of steel
Spring hammering, 69–70, 71f, 90–91
Station wagons, rear body section of, 28, 29f
Steel. *See also* Sheet metal
high-strength, 32, 35–37, 36f
low-carbon (mild), 32, 34
strength of, types of, 36
used in unibody construction, 2, 32
Steering system
damage assessment of, 233–236, 234f–235f
new technology of, for unibody design, 19–20
Steering wheel. *See also* Steering system
assessment of, 125, 233

Stitch weld, 84, 84f–85f
Straightening. *See also* Pulling; Realignment
  equipment for, 3f
  estimating labor time for, 139–140
  hot vs. cold, 51
Stress concentration, during collision, 48f, 48–49, 49f
Stressed hull structure, of unibody vehicles, 19, 19f, 23
Stripping, before refinishing, 97
Structural panels. *See* Panels
Structural parts, replacement of, 86f–88f, 87
Strut suspension
  introduction of, 20
  in unibody vehicles, 30
Strut towers, damage to, 52, 52f
Stub frame, on semiunitized construction, 23, 23f
Styleside pickup bed, 23
Supplements
  on computerized system, 282–283
  for hidden damage, 4, 7
Surfacers, as primer, 99, 99f
Suspension systems, 227–231, 228f–231f
  new technology of, for unibody design, 19–20, 30

T
Tack weld, 81, 84f
Tailgates, 23
  damage assessment of, 197–199, 198f
Tempered glass, 42
Tensile strength, of steel, 36
Tension forces, in damaged metal, 61–64, 62f–64f
Theft, covered by comprehensive insurance, 6
Thermoplastics (weldable), 2, 37
Thermosetting plastics (nonweldable), 2, 37, 39f
Thin-bodied sealers, 39
Three-coat finishes, 101, 102f, 247f, 247–248, 248f
TIG welding, 80, 85
Time control, with computerized system, 282
Tinted glass, 42
Tires, depreciation of, 146
Tools
  for damage inspections, 15, 107–108, 108f
  for glass removal, 43
Topcoats, 99–101, 100f–102f, 242, 242f
Top impacts, unibody, 56, 56f
Topper, of pickup truck, 23
Torque box structure
  of perimeter frame, 21, 21f
  in unibody design, 24, 24f
Torsion bar suspension system, 229, 229f–230f
Total loss ceiling, based on actual cash value, 255–257
Total loss vehicles, 255, 255f–256f
  decision process regarding, 257–258
Trade associations, within collision industry, 8, 9f
Training, ongoing
  in insurance industry, 8
  on new equipment, 3
Tram gauge, for damage inspections, 15, 16f, 51, 118f, 119–120
Transaxle
  of front wheel drive vehicle
    automatic, 221, 221f
    damage assessment of, 219–221, 221f–222f
    manual, 220–221, 221f
  of rear wheel drive vehicle, manual, 223, 223f
Transverse engine
  in front engine/front drive (FF) vehicles, 30, 30f
  introduction of, 20
Tricoat finishes. *See* Three-coat finishes
Tungsten Inert Gas (TIG) welder, 80, 85
Turning ratio, testing of, 123
Twin I-beam suspension, 230–231, 231f
Twist damage, 50f–51f, 51–52, 57, 70, 75–76, 76f
  diagonal measurements to check, 120, 120f
Two-stage finishes, labor times for, 245–246, 246f

U
Unauthorized changes, for profit, 11
Underbody, of FR unitized vehicle, 27–28, 28f
Undercar damage inspection, 15, 15f
Undercoating, 98f, 98–99, 99f, 241, 242f
Unethical behavior, in collision repair, 10–12
Unibody vehicles, 2, 18–20, 19f, 23–25, 24f–25f
  anchoring of, 68, 75, 75f
  buckles in, 50. *See also* Buckle damage
  characteristics of, 24
  convoluted design used in, 48–49, 49f
  front engine/rear drive, 26f, 26–28
  front-hit, 71f
  high-strength steel used on, 35, 36f
  impact effects on, 49, 49f, 54f, 54–57, 55f–56f
  passenger compartment of, 19, 24, 46
  pulling/alignment of, 76f, 76–77
  sidesway indicated on, 50
  suspension system in, 52
  twist condition in, 52. *See also* Twist damage
Uniside quarter panel, 186
Unitized vehicle construction. *See* Unibody vehicles
Upholstery
  damage assessment of, 199–201, 200f, 202f–203f
  depreciation of, 146
U-shaped channel sections, on frame side members, 20, 21f

V
Vandalism damage, 6, 126
Vans
  body-over-frame construction of, 18, 22–23. *See also* Body-over-frame (BOF) construction
  side panels of, 188, 189f
Vehicle Description Section (VDS), 260
Vehicle Identification Number (VIN), 260–263, 263f–264f
Vehicle Identification Section (VIS), 260–261
Verbal estimates, avoidance of, 15
Vertical bending, of unibody vehicle, 55, 55f, 57
Vertical force, of impact, 46, 46f
Vibration/noise
  amplified in unibodies, 24, 30
  sealers to reduce, 39
Visual inspection, panel positioning via, 80

W
Washing, before refinishing, 97, 239, 240f
Wax. *See* Dewaxing, before refinishing
Weight (average fleet), past and present, 20
Weldable plastics. *See* Thermoplastics (weldable)
Welding, 80–90, 82f–90f
  aluminum, 85
  arc, not recommended for HSLA steels, 36
  fusion, 80
  MIG
    equipment needed for, 2
    for high-strength steels, 36–37
    of unibody vehicles, 2
  oxyacetylene
    for HSS steels, 37
    not recommended for HSLA steels, 36
  plastic, 2, 37, 102–104
  pressure, 80–81
  removal of, 77–78, 78f
  spot, on unibodies, 24
  for thermoplastics, 37
  wire used for, 36–37, 81
Wheel alignment, assessment of, 122–123, 123f
Wheelhouse, damage assessment of, 164–166, 165f–166f, 187, 187f
Wheels, damage assessment of, 232–233, 233f
Wind damage, covered by comprehensive insurance, 6
Windshield. *See also* Glass
  damage assessment of, 176f, 176–177, 177f–178f
Windshield pillars, damage assessment of, 178–179, 179f

Work hardening, of sheet metal, 58f, 58–59, 59f, 60
World Manufacturers Identification (WMI), 260

Y
Yield strength, of steel, 36, 57

Z
Zinc chromate primer, 98
Zinc-plated sheet metal, 34
Zinc weld-through primer, 98
Zone 1 (primary damage), analysis of, 109f, 109–110, 110f
Zone 2 (secondary damage), analysis of, 110f, 110–121, 111f
Zone 3 (mechanical damage), analysis of, 121–125, 122f–124f
Zone 4 (passenger compartment), analysis of, 125f, 125–126
Zone 5 (exterior trim/finish), assessment of, 126, 126f
Zone method, of damage analysis, 108–109, 268. *See also separate zones*